Diversity of Inland Wetlands: Important Roles in Mitigation of Human Impacts

Diversity of Inland Wetlands: Important Roles in Mitigation of Human Impacts

Editors

Mateja Germ
Igor Zelnik
Matthew Simpson

Basel • Beijing • Wuhan • Barcelona • Belgrade • Novi Sad • Cluj • Manchester

Editors

Mateja Germ	Igor Zelnik	Matthew Simpson
University of Ljubljana	University of Ljubljana	35percent
Ljubljana, Slovenia	Ljubljana, Slovenia	Stroud, UK

Editorial Office
MDPI
St. Alban-Anlage 66
4052 Basel, Switzerland

This is a reprint of articles from the Special Issue published online in the open access journal *Diversity* (ISSN 1424-2818) (available at: https://www.mdpi.com/journal/diversity/special_issues/Inl_Wet).

For citation purposes, cite each article independently as indicated on the article page online and as indicated below:

Lastname, A.A.; Lastname, B.B. Article Title. *Journal Name* **Year**, *Volume Number*, Page Range.

ISBN 978-3-0365-9164-3 (Hbk)
ISBN 978-3-0365-9165-0 (PDF)
doi.org/10.3390/books978-3-0365-9165-0

Cover image courtesy of Matej Holcar

Contents

About the Editors

Mateja Germ

Mateja Germ (Professor, Ph.D.) obtained her Ph.D. in Biology from the University of Ljubljana in 2000. She is a Full Professor at the Chair for Ecology and Environment conservation at the University of Ljubljana. Her research interests are various aspects of the ecology and physiology of plants, especially macrophytes and agricultural plants. She has published over 100 scientific articles and 10 book chapters and co-edited the book *Macrophytes of the River Danube Basin*. She has co-organized several international symposia on macrophytes, buckwheat, UV radiation and plants, and wetland ecology.

Igor Zelnik

Igor Zelnik (Assistant Professor, Ph.D.) obtained his Ph.D. in Biology from the University of Ljubljana in 2005. He is an Assistant Professor at the Chair for Ecology and Environment conservation at the University of Ljubljana. He gives lectures in General Ecology and Ecology of Aquatic Ecosystems. His research interests include various aspects of the ecology, from wetland and aquatic plants to diatoms and benthic invertebrates. A great part of his research work includes publications about biodiversity of aquatic ecosystems, ecology of different types of wetlands, and vegetation ecology. He has published more than 40 articles and several book chapters.

Matthew Simpson

Matthew Simpson (Ph.D.) obtained his Ph.D. in Geography—A proposed Functional Classification of European Wetlands, at Royal Holloway, University of London, in 2002. He has managed conservation, research, and management projects in Europe, Asia, Africa, South America and North America. He is a Director of 35 Percent Ltd, which delivers ecosystem assessment and management related projects around the world. He has organised international workshops in many parts of the world. He is also an Associate Editor of Marine and Freshwater Research. He is the president of Europe Chapter of the Society of Wetland Scientists (SWS) and Vice Chair of the Ramsar Section.

Preface

It is hard to imagine an ecosystem more controversial and misunderstood than wetlands. Although wetlands provide numerous services to humans, people usually view them as wastelands or even places where disease-carrying mosquitoes thrive and threaten residents. Despite all the knowledge about wetlands, the negative connotation and distrust associated with wetlands still seems to be stronger than the awareness of their benefits. From this, we can conclude that articles, books, and other publications that educate people about wetlands are needed. Perhaps the ongoing problems with floods and droughts that we face as consequences of climate change are also an opportunity to change our mindset and change attitudes toward wetlands and stop their decline. After all, wetlands offer many sustainable and cheap solutions to mitigate these impacts such as floods, droughts, water supply, and biodiversity decline.

Our Special Issue, "Diversity of Inland Wetlands: Important Roles in Mitigation of Human Impacts", with three reviews and fourteen articles, is an attempt to achieve the aforementioned goal. Open access to all articles in this relatively extensive Special Issue for everyone increases the opportunity to share knowledge about wetlands with the public around the world. The collection of fourteen original articles from four continents documenting research in study areas in twelve countries confirms this. We would like to take this opportunity to thank the authors for their contributions, the reviewers for their time and improvement of the manuscripts, and the editors for their generous support.

Mateja Germ, Igor Zelnik, and Matthew Simpson
Editors

diversity

Editorial

Diversity of Inland Wetlands: Important Roles in Mitigation of Human Impacts

Igor Zelnik * and Mateja Germ

Department of Biology, Biotechnical Faculty, University of Ljubljana, Jamnikarjeva 101, 1000 Ljubljana, Slovenia; mateja.germ@bf.uni-lj.si
* Correspondence: igor.zelnik@bf.uni-lj.si; Tel.: +386-1-3203339

Citation: Zelnik, I.; Germ, M. Diversity of Inland Wetlands: Important Roles in Mitigation of Human Impacts. *Diversity* **2023**, *15*, 1050. https://doi.org/10.3390/d15101050

Received: 21 September 2023
Accepted: 27 September 2023
Published: 29 September 2023

1. Introduction

Inland wetlands are one of the most vulnerable ecosystems on Earth and have one of the highest rates of decline in surface and biodiversity. They were often considered as wasted land for agriculture, infrastructure, and a source of disease for people. Several local and national governments have launched huge projects for their extermination, and permanent endeavors have been focused on their conversion to something more neat and obviously useful for humans. There has been a fast rate of wetland loss in the last 120 years, with a loss of 64–71% of wetlands since 1900 AD [1]. Losses have been larger and faster for inland than coastal natural wetlands. Moreover, between 1970 and 2015, inland wetlands declined by approximately 35%, three times the rate of forest loss [2].

However, wetlands are vital for human survival and include some of the world's most productive ecosystems, providing ecosystem services leading to countless benefits [3]. They provide ecosystem services like water purification, runoff and river discharge mitigation, and the production of food and fiber. It was not a coincidence that thousands of years ago, wetlands along large rivers, with their fertile land and high productivity, helped ancient cultures like China along the Yellow river, Egypt along the Nile, or Mesopotamia along the Euphrates and Tigris emerge. These civilizations were constantly supplied with water, food, and building materials. Wetlands served as shelter from enemies as well as transport routes.

Over the centuries, these roles have been forgotten, and wetlands have been often treated as wastelands. It has become apparent that climate changes and their consequences will continue to take their toll, and adaptation to these changes is urgent. Loss of biodiversity, eutrophication of aquatic ecosystems, extinction of rare species, increasing flood risk, and adverse climatic conditions are issues which can be mitigated with proper management and maintenance of various wetlands, from marshes, swamps, bogs, fens, shallow lakes, and ponds to anthropogenic wetlands which need human interventions like wet-meadows, storm-ponds and constructed wetlands.

We use the definition and classification of wetlands of Davidson and Finlayson [4], who defined two classes of inland wetlands according to their origin, which are further divided to types according to the environmental conditions: (a) Inland natural wetlands: rivers and streams, natural lakes and pools, peatlands (bogs, mires, fens), marshes and swamps, and groundwater-dependent wetlands (karst and cave systems, springs); (b) Human-made wetlands: reservoirs, ponds, wet grasslands, and constructed wetlands for wastewater treatment. This classification is very broad, inclusive, and similar to the definition used by the Ramsar Convention.

The Ramsar Convention is the only international legal treaty with a primary focus on wetlands, signed in 1971 in Ramsar and which came into force in 1975 [2]. Over the years, 170 countries have joined as Contracting Parties. According to Davidson et al. [5], the network of 2301 Ramsar sites covers 18.6% of the global wetland area, which represents one of the major successes of the convention.

This Special Issue aims to emphasize the importance of different types of wetlands in our environment to maintain ecological balance and reduce the vulnerability of aquatic systems to various environmental and meteorological conditions. We received many contributions from experts studying a plethora of wetland types and the communities thriving in them. Three review articles were gathered in the first part of the Special Issue, presenting an introduction to the topic with a review of wetlands' most typical species; a review of one of the adaptations to these specific abiotic conditions; and another review showing proof that wetlands host a great number of threatened species, making another argument for preserving these ecosystems. The Special Issue continues with 14 original articles coming from four continents and documenting research in study areas in 12 countries, namely Italy, Slovenia, Croatia, Serbia, Bosnia and Herzegovina, Poland, Lithuania, Kazakhstan, Morocco, China, Thailand, and the United States of America. The list of countries where contributing authors live and work is even longer, as it also includes France, Germany, Sweden, Czechia, Slovakia, Ukraine, Greece, and Israel. This embeds the Special Issue in an international context, and we expect it will reach an audience around the globe.

2. Main Messages of the Special Issue and the Book

The common reed is probably the most characteristic plant species for wetlands. It can be found all around the globe in inland as well as coastal wetlands. Its successful strategy and high productivity also make this species invasive in some vulnerable wetland types. On the other hand, the common reed is also the most common species planted in constructed wetlands in Europe and elsewhere [6]. The review article by Čížková et al. [7] thoroughly presents the important role of the common reed in wetlands throughout Europe. One of the most important adaptations in the common reed and other wetland species to aquatic environments are their ventilation systems, enabling them to thrive in permanently or temporarily flooded soils. Ventilation systems, which offer a solution to hypoxic or even anoxic conditions in the substrata where they are rooted, are thoroughly discussed in Bjorn et al.'s paper [8].

The review by Čížková et al. [7] presents *Phragmites australis* as the most common and dominant species in European wetlands. Reed stands provide habitats for vulnerable bird species to nest, feed, or roost. On the other hand, it is the most frequently used plant species in constructed wetlands. However, its performance may vary in response to different combinations of environmental factors. Conservation measures aim to prevent or halt the decline of *P. australis* populations, known as "reed decline", and increase their microhabitat heterogeneity. Service-oriented actions aiming to create suitable conditions for the use of *Phragmites australis* include its use as roofing, as a renewable energy crop, or the use of reed-dominated habitats for waterfowl hunting and livestock grazing. In situations involving multiple uses, a modelling approach that considers the participation of all affected stakeholders can be a useful tool for resolving conflicts and developing a shared vision of the socioeconomic ecosystem in question.

Permanently or temporarily flooded areas require specific adaptations in plants to survive these special conditions. Even waterlogged soils and the absence of air in soil pores makes the thriving of plant root systems an issue that most plants cannot cope with. Molecular oxygen and carbon dioxide may be limited for wetland plants, but they have several mechanisms to obtain these gases from the atmosphere, soil, or through metabolic processes. Among the most common adaptations of partially or entirely submerged plants are various structures that allow the flow of gasses, which are thoroughly discussed by Björn et al. [8]. In emergent plants, gases can be transferred via molecular diffusion, pressurized gas flow, and Venturi-induced convection. In submerged species, direct exchange of gases occurs between submerged tissues and water, as well as the transfer of gases via aerenchyma. Photosynthetic O_2 flows into the rhizosphere, while CO_2 flows from the soil to the leaves, where it can be used for photosynthesis. Two strategies have emerged for plants with floating leaves anchored in anoxic sediment. In water lilies, for instance, air enters through the stomata of young leaves and flows through channels to the rhizomes and roots and back through the older leaves.

Wetlands are important habitats for hundreds of endangered species. The survival of numerous orchids in Europe largely depends on wetlands and their suitable management. The study by Djordjević et al. [9] provides an overview of the current knowledge on orchids of wetland vegetation in the Central Balkans. The orchids were analyzed from taxonomic, phytogeographical, ecological, and conservation perspectives. The most important taxa include the two Balkan endemics (*Dactylorhiza cordigera* subsp. *bosniaca* and *D. kalopissi* subsp. *macedonica*) and the three subendemics of the Balkans and the Carpathians (*Dactylorhiza cordigera* subsp. *cordigera*, *D. maculata* subsp. *transsilvanica*, and *Gymnadenia frivaldii*), as well as a number of Central European, Eurasian, and boreal species. Several orchid taxa found in the wet meadows and fens of the Central Balkans have a southern limit of their distribution in this part of Europe, suggesting that wetlands are important refuges for them. A total of 33 orchid taxa were recorded in different wetland plant communities. Most orchid taxa grow in the following wetland vegetation types: wet meadows (especially in order Molinietalia caeruleae); fens (in order Caricetalia fuscae); tall-herb vegetation along mountain streams and springs; and marshes and herbaceous vegetation of freshwater or brackish water bodies. In addition, detailed taxonomic, ecological, and chorological studies of wetland orchids need to be conducted in order to develop a successful plan for their conservation.

Diatoms are among the most important primary producers on Earth. It is estimated that they contribute up to a quarter of organic matter production on a global level. However, their role as primary producers in wetlands is rarely of great importance, and they have often been neglected. Cantonati et al. [10] found three new diatom species in two contrasting spring types in the northern Apennines (Italy). They describe three species new to science, belonging to the genera *Eunotia* Ehrenb., *Planothidium* Round and L. Bukht. and *Delicatophycus* M.J. Wynne. These species differ morphologically from the most similar species, but the authors were also able to refine the knowledge of the ecological profiles of these species. *Eunotia crassiminor* sp. nov. appears to live in colder inland waters with a circumneutral pH and a strict oligotrophy, even with respect to nitrogen, compared to *Eunotia minor*. The typical habitat of *Planothidium angustilanceolatum* sp. nov. appears to be oligotrophic flowing springs with low conductivity. *Delicatophycus crassiminutus* sp. nov. is probably restricted to hard water springs and similar habitats where CO_2 degassing results in carbonate precipitation. Springs are a unique but highly threatened wetland type. Thus, in-depth knowledge of the taxonomy and ecology of the characteristic diatom species is important because diatoms are excellent indicators of the quality and integrity of these particular ecosystems in the face of direct and indirect human impacts.

Novak and Zelnik [11] investigated the relations between benthic diatom communities and characteristics of karst ponds in the alpine region of Slovenia. Their objective was to examine the structure of the benthic diatom community and its relationships with selected environmental parameters. Since the predominant substrate in these ponds was clay, the epipelic community was analyzed. Hydromorphological characteristics and physical and chemical conditions were also measured at each site. They found 105 diatom species belonging to 32 genera. The most common taxa were *Gomphonema parvulum* Kützing, *Navicula cryptocephala* Kützing, *Sellaphora pupula* Mereschkowsky, and *Achnanthidium pyrenaicum* Kobayasi. The pond with the lowest diversity was located at the highest elevation, while the pond with the highest species richness was located at the lowest elevation. As for ecological types, the most common were motile diatom species. They calculated a positive correlation between the species number and oxygen saturation of water, while the correlation between species richness and NH4-N was negative. The concentrations of NO3-N and NH4-N explained around 20% of the variability in the epipelic diatom community. Contrary to our expectations, we calculated a negative correlation between macroinvertebrate diversity and diatom diversity, which is probably a result of different responses to environmental conditions.

In wetlands where open water habitats are found, as well as sufficient water retention time and low content of inorganic suspended solids, phytoplanktonic communities can also develop. The study by Barinova et al. [12] presents the spatial distribution of the taxonomic diversity of phytoplankton of the shallow Lake Borovoe in the Burabay National Natural Park (Kazakhstan), which is found in the middle of the steppe and is frequently visited by tourists. The phytoplankton of the protected lake was studied in summer of 2019. They found 72 algal and cyanobacterial species in the phytoplankton. The most species rich group were diatoms, followed by green algae and cyanobacteria. They also assessed the ecological status of the lake using species richness, abundance, biomass, and organic pollution and toxic impact indices. Statistical mapping, community similarity, and correlation analyses identified zones impacted by human activities. These were located in the lake shores and in low-alkaline water with a saprobic index of 1.63–2.00, which indicates the oligotrophic-to-mesotrophic status of the lake. Multivariate analysis enabled the assessment of the ecological status of the lake, which may be the result of the interaction of many external environmental factors, such as climatic conditions, the accumulation of organic matter, anthropogenic pressure, and internal processes in the lake. Their results indicate that the eutrophication of Lake Borovoe tends to increase due to pollution from the human-impacted zones on the shores of the lake.

The plant community structure in wetland ecosystems is often influenced by several environmental factors. In wetlands with temporary aquatic periods, most studies focus on the variation resulting from inundation and desiccation patterns. The responses of plant functional groups can provide insights into patterns of cover and richness. Rios et al. [13] studied these relationships in vernal pools of California, which have an intermittent character. The objective of their study was to evaluate how algae and plant functional groups respond to variations in hydro-regime (stable and unstable), nutrient addition (none and added), and straw addition (none, native plants, alien plants). They measured algal cover, total species richness, and functional group cover over two years. Algal cover increased with unstable hydroperiods and nutrient addition. Algae were also negatively related to aquatic plant cover. Aquatic plant cover increased with a stable hydro-regime and decreased with increased thatch. Terrestrial plant cover increased with an unstable hydro-regime and decreased with the addition of straw. Thatch accumulation and excess nutrients may be associated with human activities that directly and indirectly alter plant community composition. The interactions of these factors with the hydro-regime should be considered when evaluating the response of a plant community to changing environmental conditions. Overall, these results are necessary for the conservation and management of essential wetland functions and services.

Thermally abnormal waters are safe sites for alien invasive plants that require warmer conditions than ambient temperatures in the temperate zone. The mentioned sites are often colonized by tropical and subtropical plants. Šajna et al. [14] studied a case of *Pistia stratiotes*, which occurs in southeastern Slovenia. Based on a literature review, they found that at least 55 alien aquatic plant taxa from 21 families have been found in thermally abnormal waters in Europe. Most of these taxa are submerged or rooted macrophytes. Among them is *Pistia stratiotes*, which occurs in seven European countries, with most evidence of its occurrence being recent. Authors studied the growth of *P. stratiotes* in a thermally abnormal stream, where a persistent population could survive harsh winters. Models showed that the optimal temperature for this species biomass was $28.8 \pm 3.5\,°C$. Here, they show that air temperatures had a greater influence on the photosynthetic efficiency of *P. stratiotes*, estimated using chlorophyll fluorescence measurements, than water temperatures. In general, the growth and, consequently, surface cover of free-floating plants cannot be explained by thermally abnormal water temperatures alone. The authors conclude that although the majority of thermophilic alien plants have occurred through intentional introductions, thermally abnormal waters pose an invasion risk for further deliberate, accidental, or spontaneous dispersal, which is more likely for free-floating macrophytes.

Wetland forests and scrub (WFS) are conditioned by the strong influence of water. They consist of different types of vegetation depending on many factors, such as the type and duration of flooding, the height of the water table and its fluctuations, the velocity of the river current, and the capacity of the substrate to retain water. Koljanin et al. [15] used numerical classification and classified the WFS of the Western Balkans at the level of alliances. Their aim was also the identification of the most important ecological gradients influencing the variation in floristic composition. A large set of all published and available unpublished relevés from Slovenia, Croatia, and Bosnia and Herzegovina was classified using the EuroVegChecklist Expert System. In the second step, each of the four classes (Alno glutinosae-Populetea albae, Salicetea purpureae, Alnetea glutinosae, and Franguletea) were analyzed separately, which revealed eight alliances, namely Salicion albae, Salicion triandrae, Salicion eleagno-daphnoidis, Alno-Quercion, Alnion incanae, Alnion glutinosae, Betulion pubescentis, and Salicion cinereae. Edaphic factors were the most important in determining the species composition and differentiation of the studied WFS.

Vukov et al. [16] studied the influence of environmental factors on the functional diversity of aquatic macrophyte communities. They focused on altered waterbodies and included data from 46 sites, which hosted 59 macrophyte species. They calculated seven functional diversity indices and used a set of multivariate analyses to examine the relations between environmental factors, functional diversity indices, and plant traits. The redundancy analysis showed that the environmental factors explained 47.4% of the variability in the functional diversity. The elevation, hemeroby (a measure of the human intervention influence) of riparian land, and water conductivity were the most important factors. Their research also revealed that floating and emergent plant characteristics are a strategy to increase light absorption efficiency at high nutrient concentrations in lowland waters, while submerged plants dominate in more nutrient-poor waters at higher elevations.

The European Habitats Directive has become an important measure for biodiversity protection in the European Union. Three aquatic macrophyte species protected by the mentioned Directive have been found in Lithuania, such as *Aldrovanda vesiculosa*, *Caldesia parnassifolia*, and *Najas flexilis*. Sinkevičienė et al. [17] studied the past and present distributions of these target species, as well as their conservation status. Surveys were conducted from 2015 to 2021 in 73 natural lakes in Lithuania. They confirmed extant populations of *Aldrovanda vesiculosa* in four lakes, *Caldesia parnassifolia* in two, and *Najas flexilis* in four lakes in the northeastern part of Lithuania. The populations of *A. vesiculosa* were surveyed three times. Population densities of *Aldrovanda vesiculosa* ranged from 193.4 ± 159.7 to 224.0 ± 211.0 individuals/m^2. The number of generative individuals of *Caldesia parnassifolia* varied greatly between years. All populations of *Najas flexilis* were small, although the potential habitats in the studied lakes cover relatively large areas. The authors proposed that all lakes with populations of *Aldrovanda vesiculosa*, *Caldesia parnassifolia*, and *Najas flexilis* should be designated as special protection areas and that action plans for the conservation of these species and their habitats be developed and implemented.

Zelnik et al. [18] surveyed macrophyte communities in natural and human-made waterbodies in the active floodplain of the Drava River in northeastern Slovenia. At the same locations, they also measured selected environmental parameters. The main question of their study was whether the presence of alien invasive species *Elodea canadensis* and *E. nuttallii* affected the diversity of macrophyte communities, which were considered as harbors for a great number of native aquatic species. The number of macrophytes in the surveyed water bodies varied from 1 to 23. In addition, *Elodea nuttallii* and *E. canadensis* were present in 19 out of 32 sample sites, with *E. nuttallii* predominating. The less invasive *E. canadensis* was abundant in side-channels but absent from ponds and oxbows, while *E. nuttallii* was dominant in ponds and present in all types. Correlation analyses showed no negative effect of the invasive alien species *Elodea nuttallii* or *E. canadensis* on the species richness and diversity of native aquatic vegetation. A positive correlation was found between the abundance of *E. nuttallii* and water temperature, which could facilitate its spread in the future.

Li et al. [19] analyzed the effects of different types of sediments on the distribution and diversity of plant communities in the Poyang Lake wetlands. Lake Poyang was designated as a Ramsar Wetland more than 30 years ago. It is the largest freshwater lake in China, but the extent of its water surface varies dramatically due to the changeable hydrological regime of its tributaries. At small scales, sediment types mediate hydrologic changes that influence the wetland vegetation's distribution patterns and species diversity. They divided the soil into three types, namely lacustrine, fluvio-lacustrine, and fluvial sediments, and analyzed the distribution and diversity of plants. They discovered that plant communities with *Carex cinerascens*, *Carex cinerascens–Polygonum criopolitanum*, *Polygonum criopolitanum*, and *Phalaris arundinacea* exist within the elevation range. The multivariate analyses performed showed that soil texture and flood duration in 2017 resulted in a different distribution of wetland plant communities under the conditions of the three sediment types along the littoral zones. They also revealed that plant communities on lacustrine sediments had the highest species diversity among the studied vegetation types.

The Yellow River is the sixth longest river on Earth and is considered the mother river of China. Fertile wetlands along the Yellow River supported the rise of Chinese civilization and culture thousands of years ago. Nowadays, the conservation of biodiversity in the middle and lower reaches of the Yellow River is an urgent concern due to the effects of sediment deposition and, above all, human activities. Yuan et al. [20] analyzed over 800 plots that were established in seven nature reserves. The aim of their study was to survey the diversity of plant communities in wetlands along the mentioned reaches from the perspective of the natural environment and human disturbance. A total of 184 plant species belonging to 135 genera and 52 families were recorded. Variation partitioning analysis showed that the effects of environmental factors, such as elevation, precipitation, evaporation, and temperature on wetland plants' beta diversity were the strongest (15.5% and 17.1%, respectively), followed by the effects of human disturbance factors, expressed as population density, industrial output value, and agricultural output value (15.13% and 16.71%, respectively). The wetland species showed strong associations with the nature reserves protecting the Yellow River wetlands in Henan Province. The results shed light on the conservation of plant diversity in wetlands along the river.

Malinowska and Novak [21] present the results of a study of barium, lithium, and titanium content in six plant species and from the soils sampled in wet mid-field depressions. These occasionally flooded depressions were in fact temporary wetlands surrounded by cropland, permanent grassland, and shrubland. The study area was located in east-central Poland. The following plant species were used in the experiment: *Mentha arvensis* L., *Comarum palustre* L., *Potentilla anserina* L., *Achillea millefolium* L., *Lysimachia vulgaris* L., and *Lycopus europaeus* L. The content of Li, Ba, and Ti in plants, bottom sediment, and soil was determined using the ICP-AES method after dry mineralization. *Mentha arvensis* has the greatest accumulation potential for these chemical elements among the studied species. However, no excessive Ba, Li, or Ti content was found in plants growing at different distances from cropland, permanent grassland, and shrub vegetation. The highest Ba content was found in periodically flooded soil, while the highest contents of Li and Ti were measured in non-flooded soils.

Ruengsawang et al. [22] analyzed sponge-dwelling aquatic insects in the Lower Mekong Basin in eastern Thailand. For the purpose of this study, they established a group of shallow-water sponges attached to a raft in the Pong River, which is a tributary of the Mekong River. They investigated the aquatic insect metacommunity in habitat-forming sponges *Corvospongilla siamensis* (Demospongiae: Spongillidae), which is endemic to Southeast Asia. The model sponges harbored four orders of insects belonging to 10 families and 19 genera/species, capable of living in water-bearing channels, at the body surface, or in the extracellular matrix. Trichoptera and Diptera were the predominant orders. In the studied river, dominated by soft bottoms, sponges play a functional role, since the insects use them as a substrate, nursery, food source, and as shelter from predators and harsh environmental conditions. The feeding behavior of these insects indicates the adaptive trait of recycling sponge siliceous spicules as a source of exogenous material to strengthen larval and pupal cases and the digestive system. These results, together with the global inventory, focus on sponges as key habitat-forming species and ecosystem engineers for river/lake/wetland insect communities and as promising candidates for tropical freshwater ecosystem restoration projects through bioremediation.

Taybi et al. [23] aimed to provide an updated list of alien species, their main pathways of introduction, and their possible threats to native species. Their dataset was based on an extensive literature review, amended by their research work. The main areas where alien animal species occur in Moroccan freshwaters correspond to protected areas (e.g., Ramsar Site). These areas currently host 41 confirmed ASs, belonging to different taxonomic groups. Fishes are the most abundant taxonomic group, with 21 species, followed by 7 species of mollusks, as well as 7 species of arthropods. Almost half of the ASs were intentionally introduced. They correspond to restocking programs and are probably the greatest threat to native species diversity through predation, competition, and hybridization. Commercial activities around aquariums and ornamental fish species appear as the second source encouraging colonization by AS. The introduction and implementation of protective regulations for the import of exotic species in Morocco seems urgently needed to protect native diversity. In addition, detecting and monitoring the expansion of ASs in colonized areas and studies to improve biological and ecological knowledge seem crucial to mitigate their potential impact on native communities.

3. Conclusions

Wetlands, in all of their forms and sizes, could offer a solution to mitigating the consequences of climate change, which seems to be a major threat to our welfare and an issue hard to address. One of the most crucial mitigation measures is to increase the retention time of water in landscapes. Wetlands retain water and prevent excessive runoff, which affects microclimatic conditions and reduces the probability and magnitude of floods. The water which is stored in wetlands can have beneficial effects on various aspects, such as the local climate and water discharge management. In this respect, wetlands help us adapting to climate changes. The evapotranspiration from wetlands could mitigate the increasing temperatures, especially in urban landscapes/areas, where it could have a cooling effect. The water entering the landscape in the form of watercourses or precipitation could be stored in wetlands and provide the water supplies during dry periods and prevent extreme drops in the groundwater table. But what is even more important is that wetlands could serve as buffers to compensate for flood waves threatening populated areas and infrastructure.

Another major role of wetlands in their broadest sense is biodiversity conservation. Different types of wetlands host numerous vulnerable or threatened species, which can survive only if their habitats—wetlands—remain healthy, or in other words, in good ecological status. Our Special Issue and the contributions within it discuss many problems connected to biodiversity conservation, but also give advice on how to manage specific types of wetlands to prevent further loss of biodiversity.

Author Contributions: Writing—original draft preparation, M.G. and I.Z.; writing—review and editing, I.Z. and M.G. All authors have read and agreed to the published version of the manuscript.

Funding: This research was funded by the Slovenian Research Agency, within the core research funding Nr. P1-0212, "Biology of Plants".

Acknowledgments: Thanks to the authors who contributed their articles to this Special Issue and who helped to build a long list of contributions. We would like to thank Editors of the journal *Diversity*. A special thanks goes to the anonymous reviewers, who provided their comments and suggestions and helped to improve the quality of the articles.

Conflicts of Interest: The authors declare no conflict of interest.

References

1. Davidson, N.C. How Much Wetland Has the World Lost? Long-Term and Recent Trends in Global Wetland Area. *Mar. Freshw. Res.* **2014**, *65*, 934. [CrossRef]
2. Gardner, R.C.; Finlayson, C. Global Wetland Outlook: State of the World's Wetlands and Their Services to People. *Ramsar Conv. Secr.* **2018**. Stetson University College of Law Research Paper No. 2020-5. Available online: https://ssrn.com/abstract=3261606 (accessed on 31 August 2023).
3. Russi, D.; Ten Brink, P.; Farmer, A.; Badura, T.; Coates, D. *The Economics of Ecosystems and Biodiversity for Water and Wetlands*; IEEP: London, UK; Brussels, Belgium, 2013.
4. Davidson, N.C.; Finlayson, C.M. Extent, Regional Distribution and Changes in Area of Different Classes of Wetland. *Mar. Freshw. Res.* **2018**, *69*, 1525. [CrossRef]
5. Davidson, N.C.; Fluet-Chouinard, E.; Finlayson, C.M. Global Extent and Distribution of Wetlands: Trends and Issues. *Mar. Freshw. Res.* **2018**, *69*, 620. [CrossRef]
6. Vymazal, J. The Historical Development of Constructed Wetlands for Wastewater Treatment. *Land* **2022**, *11*, 174. [CrossRef]
7. Čížková, H.; Kučera, T.; Poulin, B.; Květ, J. Ecological Basis of Ecosystem Services and Management of Wetlands Dominated by Common Reed (Phragmites Australis): European Perspective. *Diversity* **2023**, *15*, 629. [CrossRef]
8. Björn, L.O.; Middleton, B.A.; Germ, M.; Gaberščik, A. Ventilation Systems in Wetland Plant Species. *Diversity* **2022**, *14*, 517. [CrossRef]
9. Djordjević, V.; Aćić, S.; Kabaš, E.; Lazarević, P.; Tsiftsis, S.; Lakušić, D. The Orchids of Wetland Vegetation in the Central Balkans. *Diversity* **2022**, *15*, 26. [CrossRef]
10. Cantonati, M.; Bilous, O.; Angeli, N.; van Wensen, L.; Lange-Bertalot, H. Three New Diatom Species from Spring Habitats in the Northern Apennines (Emilia-Romagna, Italy). *Diversity* **2021**, *13*, 549. [CrossRef]
11. Novak, K.; Zelnik, I. Relations between Benthic Diatom Community and Characteristics of Karst Ponds in the Alpine Region of Slovenia. *Diversity* **2021**, *13*, 531. [CrossRef]
12. Barinova, S.; Krupa, E.; Khitrova, E. Spatial Distribution of the Taxonomic Diversity of Phytoplankton and Bioindication of the Shallow Protected Lake Borovoe in the Burabay National Natural Park, Northern Kazakhstan. *Diversity* **2022**, *14*, 1071. [CrossRef]
13. Rios, J.; Dibbell, M.; Flores, E.; Kneitel, J.M. Beyond the Hydro-Regime: Differential Regulation of Plant Functional Groups in Seasonal Ponds. *Diversity* **2023**, *15*, 832. [CrossRef]
14. Šajna, N.; Urek, T.; Kušar, P.; Šipek, M. The Importance of Thermally Abnormal Waters for Bioinvasions—A Case Study of Pistia Stratiotes. *Diversity* **2023**, *15*, 421. [CrossRef]
15. Koljanin, D.; Brujić, J.; Čarni, A.; Milanović, Đ.; Škvorc, Ž.; Stupar, V. Classification of Wetland Forests and Scrub in the Western Balkans. *Diversity* **2023**, *15*, 370. [CrossRef]
16. Vukov, D.; Ilić, M.; Ćuk, M.; Igić, R. Environmental Drivers of Functional Structure and Diversity of Vascular Macrophyte Assemblages in Altered Waterbodies in Serbia. *Diversity* **2023**, *15*, 231. [CrossRef]
17. Sinkevičienė, Z.; Kamaitytė-Bukelskienė, L.; Petrulaitis, L.; Gudžinskas, Z. Current Distribution and Conservation Issues of Aquatic Plant Species Protected under Habitats Directive in Lithuania. *Diversity* **2023**, *15*, 185. [CrossRef]
18. Zelnik, I.; Germ, M.; Kuhar, U.; Gaberščik, A. Waterbodies in the Floodplain of the Drava River Host Species-Rich Macrophyte Communities despite Elodea Invasions. *Diversity* **2022**, *14*, 870. [CrossRef]
19. Li, J.; Liu, Y.; Liu, Y.; Guo, H.; Chen, G.; Fu, Z.; Fu, Y.; Ge, G. Effects of Sediment Types on the Distribution and Diversity of Plant Communities in the Poyang Lake Wetlands. *Diversity* **2022**, *14*, 491. [CrossRef]
20. Yuan, Z.; Xiao, M.; Su, X.; Zhao, H.; Li, Y.; Zhang, H.; Zhou, Z.; Qi, R.; Chen, Y.; Wang, W. Effects of Environment and Human Activities on Plant Diversity in Wetlands along the Yellow River in Henan Province, China. *Diversity* **2022**, *14*, 470. [CrossRef]
21. Malinowska, E.; Novak, J. Barium, Lithium and Titanium Content in Herbs of Mid-Field Wet Depressions in East-Central Poland. *Diversity* **2022**, *14*, 189. [CrossRef]
22. Ruengsawang, N.; Sangpradub, N.; Manconi, R. Aquatic Insects in Habitat-Forming Sponges: The Case of the Lower Mekong and Conservation Perspectives in a Global Context. *Diversity* **2022**, *14*, 911. [CrossRef]
23. Taybi, A.F.; Mabrouki, Y.; Piscart, C. Distribution of Freshwater Alien Animal Species in Morocco: Current Knowledge and Management Issues. *Diversity* **2023**, *15*, 169. [CrossRef]

Review

Ecological Basis of Ecosystem Services and Management of Wetlands Dominated by Common Reed (*Phragmites australis*): European Perspective

Hana Čížková [1,*], Tomáš Kučera [2], Brigitte Poulin [3] and Jan Květ [2,4,*,†]

1   Faculty of Agriculture and Technology, University of South Bohemia, Studentská 1668, CZ-370 05 České Budějovice, Czech Republic
2   Faculty of Science, University of South Bohemia, Branišovská 1760, CZ-370 05 České Budějovice, Czech Republic
3   Tour du Valat Research Institute, Le Sambuc, 13200 Arles, France
4   Global Change Research Institute, Academy of Sciences of the Czech Republic, Bělidla 986/4a, CZ-603 00 Brno, Czech Republic
*   Correspondence: hcizkova@fzt.jcu.cz (H.Č.); jan.kvet@seznam.cz (J.K.)
†   Present address: Vrchlického 320, CZ-379 01 Trebon, Czech Republic.

Abstract: The common reed (*Phragmites australis*) is a frequent dominant species in European wetlands. Yet, its performance can vary in response to different combinations of environmental factors. This accounts for *P. australis* decline on deep-water sites, its stable performance in constructed wetlands with subsurface horizontal flow and its expansion in wet meadows. Reed stands provide habitats for nesting, feeding or roosting of vulnerable bird species. Conservation measures aim at preventing or stopping the decline of *P. australis* stands, increasing their micro-habitat heterogeneity and reducing the reed penetration into wet meadows. Service-oriented measures aim at providing suitable conditions for direct use of reed stalks for roof thatching or as a renewable energy crop or the use of the reed-dominated habitats for waterfowl hunting, cattle grazing or fishing. The compatibility between nature conservation and different socioeconomic uses can be promoted by collective agreements, agri-environmental contracts or payments for ecosystem services of the reedbeds. In situations with multiple uses, a modelling approach considering the participation of all the stakeholders concerned can be a useful tool for resolving conflicts and developing a shared vision of the respective socio-ecosystem.

Keywords: biodiversity; conservation measures; Europe; habitats; multiple uses; *Phragmites australis*; socioeconomic uses; wetland

Citation: Čížková, H.; Kučera, T.; Poulin, B.; Květ, J. Ecological Basis of Ecosystem Services and Management of Wetlands Dominated by Common Reed (*Phragmites australis*): European Perspective. *Diversity* **2023**, *15*, 629. https://doi.org/10.3390/d15050629

Academic Editors: Mateja Germ, Igor Zelnik, Matthew Simpson and Mario A. Pagnotta

Received: 7 March 2023
Revised: 7 April 2023
Accepted: 11 April 2023
Published: 5 May 2023

1. Introduction

The common reed (*Phragmites australis* [Cav.] Trin. ex Steud.) is a common wetland plant species with a nearly cosmopolitan distribution, forming monodominant and productive stands under optimal conditions [1]. In its native range, local populations of *P. australis* have formed an integral part of wetland vegetation. Wetlands dominated by *P. australis* have for long provided local human communities with food (waterfowl, venison, fish), fodder and otherwise useful plant materials [2,3].

The current controversy in the perception of *P. australis* on a global scale is linked to its expansion to ecosystems with less competitive dominants and, above all, its invasion outside its native range. The invasion of genotypes of European origin in North American wetlands has stimulated research of the genetic diversity within the species and the whole genus worldwide [4,5], the ecophysiological behaviour of the invasive as compared to native genotypes (e.g., [6,7]), the ecological background of invasiveness of this species (e.g., [8,9]) and the methods of controlling its expansion ([10] and references therein). Such studies partly overshadowed the research progress dealing with its more balanced role and management practices in its native range.

The aim of this paper is, therefore, to give an overview of various uses and ways of managing the *P. australis*-dominated wetlands in Europe, where it is native and its use has a long tradition. In the largely drained European continent, such wetlands still occupy vast areas in northern, southeastern and southwestern parts, and scattered fragments occur in the whole territory. They fulfil regulation, habitat, production, and information functions, as listed by de Groot et al. [11], and provide related ecosystem services [3,12–16]. In response to the continuing wetland drainage, support of biodiversity of wetland biota has increased in priority in the last 50 years. The management goals then reflect human preferences based on the perception of the ecosystem (Figure 1).

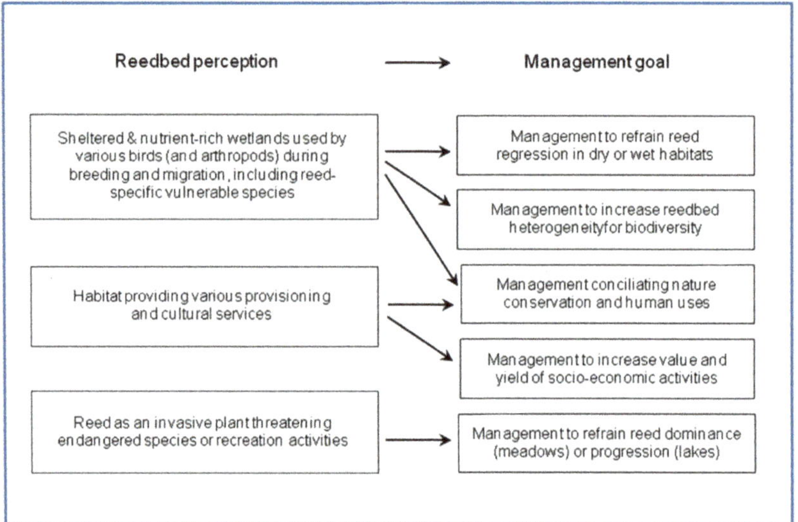

Figure 1. Typical management goals associated with each of the three main perceptions of reedbeds in Europe.

In the following text, we first give a brief overview of the ecological requirements and natural vegetation types with *P. australis* occurring in Europe, as background knowledge needed for their successful management. Then we focus on various uses of the *P. australis*-dominated wetlands and related management measures.

2. The Genetic Delineation and Ecological Niche of *P. australis* in Europe

Recent genetic studies have delimited five species of the genus *Phragmites* [5], of which *Phragmites australis* (Cav.) Trin ex Steudel is the most widely distributed. Within this species, Lambertini et al. [17] identified two genetically distinct groups of populations occurring in Europe: one inhabiting temperate Europe (European *P. australis*) and the other found in the Mediterranean region (Mediterranean *P. australis*). These two genetically delimited groups probably correspond to two respective subspecies: *P. australis* ssp. *australis* (also including the invasive populations of European origin in North America) and *P. australis* ssp. *altissimus*, proposed by Clayton [18]. Because the ecological literature scarcely distinguishes between the lower taxa of *P. australis*, only the species name (*P. australis sensu lato*) is used in this paper, except where the lower taxa are explicitly mentioned in the studies cited.

The response of *P. australis* to its habitats has been treated in detail by at least three monographs [16,19,20], two successive reviews in the series "The Biological Flora of the British Isles" [21,22] and several other review articles on the biology and ecology of *P. australis* worldwide [23,24]. Worthy of special attention is also the conceptual article by Eller et al. [9], focused primarily on the ecological genetics of reed.

Briefly, *P. australis* is a robust perennial grass species with a nearly cosmopolitan distribution and a great capacity to acclimate to a wide range of environmental conditions regarding latitude (up to 70° north), altitude (in Europe up to 1900 m in the Alps), climate (oceanic to continental), water table (more than 2 m depth in European lakes with a great light transparency), substrate (mineral to organic), trophic conditions (oligotrophic to eutrophic), pH (2.5 to 9.8) and salinity (up to 65‰ over short periods). On the other hand, *P. australis* stands do not tolerate a sudden high rise of the water table, avoid strongly reducing organic substrates and are highly sensitive to mechanical damage of any type. Coincidence of marginal values of more factors is even more destructive.

There is also a distinct clinal variation across latitudes. European *P. australis* populations from lower latitudes tend to allocate less aboveground biomass to leaves and more to stems as compared to those from higher latitudes; they also produce fewer shoots. In the Mediterranean region *P. australis* can reach heights of up to 5 m, while in temperate Europe *P. australis* usually reaches maximum stem heights of 2–3.5 m. This relationship, however, is not linear, which is partly due to genetic differences between the temperate and Mediterranean groups of *P. australis* and is further complicated by the existence of several ploidy levels, which are not clearly related to the production and growth characteristics [9,17].

3. Vegetation with *Phragmites australis*

3.1. General Overview

P. australis-dominated communities represent an important long-term stage in successional seres and form important azonal wetland habitats, especially on shores of standing and slowly flowing meso- to eutrophic waters with bottom sediments and/or soils ranging between nutrient-rich and nutrient-poor ones [25]. *P. australis* is a frequent dominant or co-dominant species in communities extending along fresh and brackish running waters from their upper reaches (Figure 2A) downstream and cover large areas in river floodplains (Figure 2B–E). The largest stands of *P. australis* occur in inundated freshwater and brackish reed marshes in deltas of the main European rivers, such as the Rhine, Ebro, Rhone, Danube, Dnipro and Volga [26]. *P. australis* also forms monodominant stands in littoral zones of both natural and artificial shallow lakes (Figure 3). It is a co-dominant or dominant species of marshy fens (Figure 4A–D) and can form patches on temporarily wet hilly slopes (Figure 4E). *P. australis* is a common species in the understory of alder (Figure 2A) and willow carrs or wet pine forests [27]. It also forms an important vegetation component of wetlands significantly altered by humans and novel ecosystems ecosystems in the sense of [28], emerging in response to human activities (Figure 5). They include permanently or temporarily wet landscape elements such as constructed wetlands used for wastewater treatment (Figure 5A), drainage canals and ditches (Figure 5B), abandoned wet meadows, wet parts of spoil heaps and brownfields, and also littoral zones of artificial water bodies serving various purposes (Figure 5C,D).

Based on the phytosociological approach predominantly used in continental Europe, *P. australis*-dominated communities are included in the class *Phragmito-Magnocaricetea* Klika et Novák 1941. This class comprises vegetation types commonly occurring all over Europe and Asian Russia [29]. The class *Phragmito-Magnocaricetea* consists of 11 alliances with a total of 90 associations, out of which *P. australis* is dominant in three, constant in 34 and present in 70 ([30], Table 1). The most common is the alliance *Phragmition australis* Koch 1926, which includes associations dominated by tall helophytes, 11 of them dominated by a single tall helophyte species common in Europe. These plant stands are sometimes referred to as reedbeds *sensu lato* in ecological literature. The association *Phragmitetum australis* is the most widely spread one and often forms a mosaic with associations dominated by other common tall helophytes such as *Typha latifolia*, *T. angustifolia* or *Schoenoplectus lacustris*, and with various sedges (*Carex* spp.) near the reedbeds' landward boundaries.

Figure 2. Habitats with *P. australis* along flowing waters. (**A**)—Alder carr on the upper course of the Rudava River, western Slovakia. (**B**)—vegetation zonation with *P. australis* in a eutrophic riverine habitat: the Danube, southern Slovakia. (**C**)—*P. australis* dominated non-tidal riverine wetlands: Biebrza River, eastern Poland. (**D**)—Fenéki lake, a restored *P. australis*-domianted wetland in the Kis-Balaton water protection system, Hungary. (**E**)—*P. australis* dominated tidal brackish wetlands: the Danube delta, Romania. Photographs by Hana Čížková (**A,B**), Aat Barendregt (**C,D**), Josef Rajchard (**E**).

Figure 3. *P. australis*-dominated littoral wetlands. (**A**)—Lake Ladoga, Russia. (**B**)—declining reed stands of Lake Trasimeno, Italy. (**C**)—*P. australis*-dominated littoral zone of the saline lake Gallocanta, Spain. (**D,E**)—stable and declining reed stands of Lake Fertö/Neusiedlersee, Hungary. (**F**)—regenerating *P. australis* stand of Řeřabinec fishpond, Czech Republic. Photographs by Galina A. Elina (**A**), Aat Barendregt (**B**), Jiří Dušek (**C**), Mária Dinka (**D,E**), Hana Čížková (**F**).

Figure 4. *P. australis*-dominated terrestrial habitats. (**A**)—Upper limit of occurrence of *P. australis* in the Krkonoše mountains, Czech Republic. (**B**)—*P. australis* in a non-tidal acid fen Ilperveld north of Amsterdam, the Netherlands. (**C**)—*P. australis*-dominated fen (nature reserve „U Vomáčků"), Czech Republic. (**D**)—expansion of *P. australis* in a floating fen, Rzeczin. Poland. (**E**)—Expansion of *P. australis* in a littoral sedge marsh: Staňkovský lake, Czech Republic. Photographs by Michaela Čepková (**A**), Aat Barendregt (**B**), Hana Čížková (**C–E**).

Figure 5. *P. australis* dominated habitats created or strongly altered by human activities: (**A**)—constructed wetland used for wastewater tratment of Slavošovice village, Czech Republic. (**B**)—drainage canal below the Gabčíkovo reservoir, Slovakia. (**C**)—*P. australis* domianted littoral of a sand-pit lake, Czech Republic. (**D**)—*P. australis* domianted littoral zone of a small village pond, Czech Republic. Photographs by Hana Čížková.

Table 1. Occurrence of *P. australis* in the alliances of the class *Phragmito-Magnocaricetea* based on the synpotic table published by Chytrý et al. [30].

Sub-Class	Alliance	No. of Associations	Occurrence of *P. australis*			No. of Relevés
			Dominant	Constant	Present	
Phragmitetalia	*Phragmition communis*	19	1	5	18	12,690
Bolboschoenetalia	*Scirpion maritimi*	7	1	7	7	1682
	Bolboschoeno maritimi-Schoenoplection tabernaemontani	6	1	6	3	1796
Magnocaricetalia	*Magnocaricion elatae*	17	0	9	17	4452
	Magnocaricion gracilis	6	0	3	6	5181
	Carici-Rumicion hydrolapathi	3	0	3	2	983
Nasturtio-Glycerietalia	*Glycerio-Sparganion*	9	0	0	0	3177
	Caricion broterianae	3	0	0	0	367
Oenanthetalia and *Arctophiletalia*	*Eleocharito palustris-Sagittarion sagittifoliae*	18	0	1	17	4956
	Alopecuro-Glycerion spicatae	1	0	0	0	30
	Arctophilion fulvae	1	0	0	0	19
Total	11	90	3	34	70	35,333

3.2. Regional Survey

In northwestern Europe, characterized by the Atlantic climate, *P. australis* grows mostly in shallowly inundated or permanently waterlogged habitats, especially in the communities of the alliance *Phragmition communis*. An overview of the vegetation with *P. australis* on the British Isles has recently been published by Rodwell [31] and Packer et al. [22]. Briefly, it is dominant in four types of wet habitats (Table 2):

1. Freshwater reedbeds usually hosting species-poor plant communities including *P. australis*, other marsh dominants such as *Typha latifolia*, *T. angustifolia*, *Schoenoplectus lacustris*, *Bolboschoenus maritimus* and tall sedges.
2. Tall-herb species-rich fens with *Cladium mariscus* and *Calamagrostis canescens* or some other species (*Juncus subnodulosus*, *Carex elata*, *C. acutifomis*, *C. appropinquata*, *C. lasiocarpa*, *C. diandra*) as co-dominants.
3. Saline brackish marshes in which more halophlous species such as *Atriplex prostrata*, *Juncus gerardii*, and *Aster tripolium* co-occur with *P. australis*.
4. A tall-herb vegetation of abandoned moist-to-wet meadows, including tall herbaceous dicotyledons such as *Eupatorium cannabinum*, *Angelica sylvestris*, *Lythrum salicaria*, *Cirsium palustre*, *Filipendula ulmaria*, and *Epilobium hirsutum*.

Additionally, *P. australis* grows sparsely in some other habitats such as salt marshes and dune slack communities on peaty mineral soils with *Salix repens*. It also frequently outcompetes sedges in fen and wet meadow vegetation in lowland regions. It occurs also in the Atlantic wet heath vegetation in the underlayer of *Hippophaee rhamnoides* scrubs on moving coastal dunes, in the understory of willow carrs, alder and willow woodlands, and birch and pine open-bog woodlands. In vegetation affected by human activities, *P. australis* occurs in tall-herb "nitrophilous" stands with *Urtica dioica*, *Cirsium arvense* and *Epilobium hirsutum* [31].

Table 2. Synopsis of vegetation with dominant *P. australis* in Europe: survey of habitats based on the regional vegetation monographs.

Region/Country	Freshwater Reed Beds	Brackish Swamps	Tall-Herb Fens and Moist Meadows
N and NW Europe			
Scandinavia [32]	*Schoenoplecto-Phragmitetum*	*Bolboschoenetum maritimi*	*Magnocaricion*
Great Britain [22,31]	*Phragmites australis* comm.	*Halo-Scirpion* *Elymion pycnanthi* *Ammophilion arenariae*	*Phragmites australis-Peucedanum palustre* comm. *Phragmites australis-Eupatorium cannabinum* comm.
Netherlands [33]	*Typho-Phragmitetum*	*Phragmition*	In more communities
Central Europe			
Germany [34]	*Scirpo-Phragmitetum* *Phragmiti-Euphorbietum palustris*	In more communities	*Thelypterido-Phragmitetum* *Phragmiti-Caricetum lasiocarpae*
Poland [35]	*Phragmitetum australis*	*Phragmition*	*Thelypteridi-Phragmitetum*
Czech Republic [36]	*Phragmitetum australis* *Phragmition australis*	*Astero pannonici-Bolboschoenetum compacti* *Schoenoplectetum tabernaemontani*	*Thelypterido palustris-Phragmitetum australis* *Magno-Caricion elatae* *Cladietum marisci*
Austria [37]	*Phragmitetum vulgaris* *Phragmiti-Euphorbietum palustris*	*Bolboschoeno-Phragmitetum communis* (inland salt marshes)	*Caricion lasiocarpae*
SE Europe			
Hungary [38,39]	*Phragmitetum communis* *Scirpo-Phragmitetum*	–	–
Romania [20,40]	*Scirpo-Phragmitetum*	*Phragmition*	–
Croatia [41]	*Phragmition*		*Caricetum vesicariae* *Phalaridetum arundinaceae*

Table 2. *Cont.*

Region/Country	Freshwater Reed Beds	Brackish Swamps	Tall-Herb Fens and Moist Meadows
		E Europe	
Ukraine [42]	*Phragmitetum communis*	*Phragmiti-Juncetum maritimi*	*Phragmiteto-Schoenetum ferrugunei* [43]
Russia [44–46] (Volga [29])	*Phragmition communis* *Calystegio-Phragmitetum*	*Puccinellio Phragmition* *Argusio-Phragmitetum*	*Phragmiti-Magnocaricion* –
		S and SW Europe	
France [47]	*Phragmition* *(Scirpo-Phragmitetum)*	*Phragmites communis-Juncus maritimus-Scirpus maritimus* comm.	–
Italy [48]	*Phragmitetum australis*	*Bolboschoenus maritimus* agg. community *Schoenoplectetum tabernaemontani*	*Magno-Caricion elatae*
Spain [47]	*Typho angustifoliae-Phragmitetum australis* *Scirpo lacustris-Phragmitetum*	*Scirpo compacti-Phragmitetum australis*	–

Notes: *Phragmites communis* and *P. australis* are synonyms; we use the community name in the original form as used in the regional vegetation survey without any correction according to the Code of phytosociological nomenclature. Other synonyms: *Scirpus lacustris* = *Schoenoplectus lacustris*, *Scirpus maritimus* and *S. compactus* = *Bolboschoenus maritimus* (syn. *B. compactus*). The more detailed a regional vegetation survey is, the greater number of associations is distinguished. The negative information (–) means that either the community is not present in the region, or if present, has not been recognized and classified.

In the Netherlands, phytosociologists report *P. australis* from nearly all the habitat types described for the British Isles. In addition, they mention its occurrence in pioneer vegetation on strandlines of sand beaches and ephemeral vegetation on salt mud and sand flats [32].

In northeastern and central Europe, characterized by sub-Atlantic climate, *P. australis* grows in much the same habitats as described for northwestern Europe.Dense monodominant stands in mesotrophic to eutrophic shallow or standing water bodies are typical, alternating with other communities of tall helophytes, such as *T. angustifolia, T. latifolia* and *Schoenoplectus lacustris* (Table 2). Such stands occupy the transition (ecotone) between the terrestrial and aquatic zone (eulittoral to infralittoral in the sense of Hutchinson [49]). Towards open water, *P. australis* is successively replaced by diverse floating-leaved and submerged species such as *Nuphar lutea, Potamogeton* sp. div., *Hydrocharis morsus-ranae, Ceratophyllum demersum* and duckweeds (*Lemna* spp.) ([45], Figure 2B). Towards the terrestrial end of the zonation, the *P. australis*-dominated communities typically change to vegetation dominated by sedges species such as *Carex elata, C. acuta* or *C. riparia* ([45,50], Figure 4E). It occurs also in swamps dominated by alder (*Alnus*) and ash (*Fraxinus*), as well as in willow (*Salix* spp.) and alder carrs (stands; Figure 2A). It is a co-dominant of a variety of minerotrophic peat habitats, together with sedges such as *Carex nigra* or *C. rostrata* ([36,43,45], Figure 2C). Throughout central Europe, *P. australis* forms successional stages in abandoned meadows ([25], Figure 3E) and invades *Carex*-dominated marshes and fens in response to eutrophication ([51], Figure 4E).

In European regions with continental or Mediterranean climate (much of southern, southeastern, and eastern Europe), extensive reedbeds are associated with standing or slowly flowing fresh waters and brackish estuaries (Figure 2E), where *P. australis* can be as tall as 9 m. Floating islands dominated by *P. australis*, first described from the Danube delta [20,52], are a characteristic phenomenon in the lower reaches of large eastern European rivers [52,53]. *P. australis* is also present in inland salt marshes dominated by tall herb vegetation and on stabilized sand dunes along the sea coast (Figures 3C and 6) [54]. *P. australis* is present in all communities of the *Phragmition* alliance of the Volga River floodplain as well as in alluvial salt meadows, where it occurs together with *Argusia sibirica, Suaeda confusa, Atriplex calotheca, Lepidium latifolium, Crypsis schoenoides, C. aculeata, Bolboschoenus maritimus agg., Althaea officinalis,* and *Aeluropus prudens* [44].

Figure 6. Local variability of *Phragmites australis* salt wetlands near Ravenna, Italy: (**A,B**)—salt marshes in back dunes with *Tamarix, Juncus maritimus, Bolboschoenus maritimus* agg. and *Limonia* [55]. (**C,D**)—Valle di Comacchio, brackish marshes in a coastal basin separated from the sea by a sandbank, the colony of flamingo can be seen in the middle of (**C**). (**D**)—the zonation of bank salt marsh vegetation (*Salicornia, Spartina* and *Sarcocornia* zones below the *Phragmites* zone. (**E**)—Lido di Dante, resprouting of *P. australis* stems five weeks after a forest fire, occurring in small local depressions with *Sallix* in *Pinus pinaster* stands in Pineta Ramazzoti (August 2012) (see also [56]). Photographs by Tomáš Kučera.

This overview indicates that *P. australis* dominates mesotrophic to eutrophic habitats subjected to long-term waterlogging or flooding, where it has its ecological optimum. From such habitats it spreads to marginal ones, suboptimal with respect to water or nutrient supply. According to the information available, its occurrence in marginal habitats is common in areas with an oceanic climate, where it is found in almost all types of wetlands, also including vegetation affected by former or current human activities (fens, abandoned wet meadows). In contrast, in areas of Europe with a continental climate, *P. australis* seems to be largely confined to habitats with sufficient water and nutrient supply.

4. Use and Management of *P. australis* Habitats for Biodiversity

4.1. P. australis Stands as Habitats of Birds and Invertebrates

Due to its vigorous growth and effective vegetative spreading, *P. australis* forms dense stands providing sheltered and nutrient-rich habitats suiting various birds and invertebrates [2,57–76]. They serve as breeding or overwintering habitats or migration stopover areas for numerous bird species including rare and endangered ones [77]. Some bird species almost exclusively use reedbeds for these purposes. They include several species of *Acrocephalus* warblers (*A. melanopogon, A. arundinaceus, A. scirpaceus, A. schobaneus*) and, notably, the aquatic warbler (*A. paludicola*) which is vulnerable at the global level, as well as various heron species of which many populations are depleted or still declining [77], such as the little bittern (*Ixobrychus minutus*), the Eurasian bittern (*Botaurus stellaris*), and the purple heron (*Ardea purpurea*). Reedbeds are also extensively used as night roosts by passerines [58,78,79] and provide foraging and nesting sites to ducks and coots, the abundance of which is correlated with the reedbed area [80].

The value of *P. australis* stands as biotopes of waterfowl and other animals has been increasingly appreciated in Europe during recent decades. The restoration, or even creation, of *P. australis*-dominated wetlands has taken place mainly in western Europe, where large reedbeds have disappeared [81]. Most of the European large reed stands are now included in the inventory of Ramsar wetlands of international importance or Special Protection Areas under the European Union Bird Directive. Examples of highly valuable Ramsar sites are the Broadlands in eastern England [82], Lake Constance in Germany, Austria and Switzerland [83], Lake Neusiedlersee/Fertö in Austria and Hungary [84–86], the Lednice and Třeboň fishponds in the Czech Republic [87,88], the Rhone delta (Camargue) in France [89], and the Danube delta in Romania and the Ukraine [90]. Also, the largescale semi-natural treatment wetland, the Kis-Balaton Water Protection System in Hungary, is protected as a Ramsar wetland [91] because of its well-developed zonation of local wetland vegetation [92], supporting rich wildlife. In addition, there are numerous smaller sites protected by the legislation of individual countries.

4.2. Management to Stop P. australis Regression in Dry Habitats

If left unmanaged, moist areas overgrown with *P. australis* tend to change into terrestrial habitats (woodlands or grasslands depending on the regional climate) in a natural hydroseral succession process of wetland terrestrialization (landfilling). The terrestrialization of reed-dominated wetlands is primarily caused by their high net primary production. The annual production of both above- and belowground biomass of *P. australis* is usually greater than its decomposition and export [93–96]. As a result, dead biomass at different stages of decomposition accumulates on the site, and a substantial part of it is transformed into the reed peat [16].

Habitat maintenance at a reed-dominated successional stage is the basic approach to reedbed management [58]. The most common management practices are preventive and consist of reducing the biomass accumulation by removing the reed biomass by its mowing, burning or by litter removal in winter [59,94,97,98], ideally according to a short-term rotational scheme to reduce unfavourable impacts of such operations on birds and invertebrates [15]. At a more advanced successional stage, cutting or burning have only a small impact [99] and restoration through scrub grubbing and bed lowering may become necessary [100]. Stripping the topsoil followed by reed establishment through rhizome transfer, planting seedlings, and natural regeneration by raising water levels has been tested experimentally at several sites in the United Kingdom [101,102]. The management works have returned the reedbeds to an early successional stage to which Eurasian bitterns have responded rapidly [102].

4.3. Management to Revert the Regression of Reed in Wet Habitats

The causes of *P. australis* decline in aquatic habitats can be separated into three groups: (1) eutrophication, (2) high water levels, and (3) mechanical damage by various agents (see [103] for a review of case studies). In many instances, they operate simultaneously, and all have a joint hidden effect (Figure 7): insufficient aeration of belowground parts (roots and rhizomes), which ultimately leads to their death. Due to a lack of oxygen in the rooting substrate, an increasing amount of organic matter is decomposed by anaerobic bacteria, which is associated with the production of toxic metabolites such as organic acids, reduced forms of iron and manganese and hydrogen sulphide. These processes can form a self-perpetuating cycle, which can proceed long after the primary causes faded away.

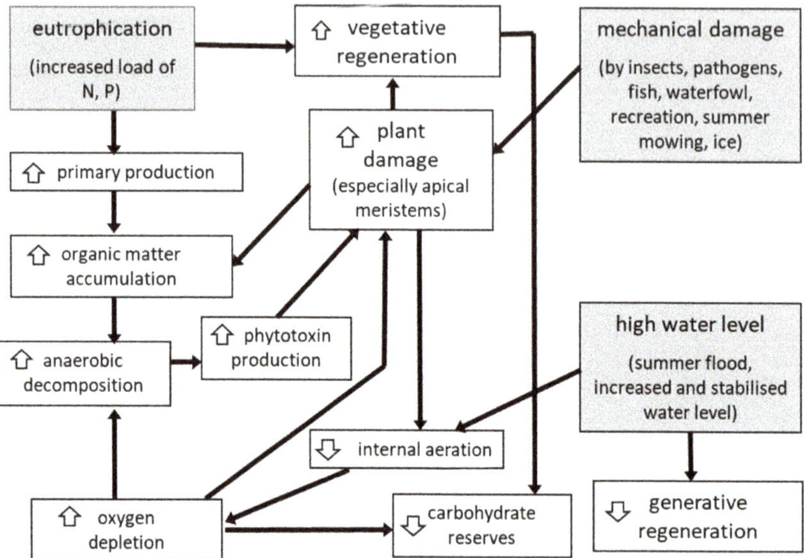

Figure 7. Conceptual model of factors affecting *P. australis* decline. Grey rectangles denote key environmental factors. Black arrows indicate the links between a cause and its effect. Open rectangles and open arrows indicate the processes involved and their direction, respectively.

A variety of measures have been used to reduce the nutrient load to aquatic habitats:

1. More efficient purification of wastewater discharged into the lake [104];
2. Reduction of nutrient input from neighbouring agricultural areas [105];
3. Increased nutrient stripping in the inflowing water by enhancing the mineral nutrient uptake by a dense water and bank vegetation upstream; its thereby enhanced cumulative nutrient uptake deprives the reeds growing downstream of a part of their mineral nutrient supply [106];
4. Removal of accumulated nutrient-rich mud by suction dredging [102].

The last measure can have a most rapid effect, visible within the same vegetation season, but needs to be combined with reduction of nutrient input to make the effect long-lasting.

The detrimental effects of eutrophication or of high water levels can be alleviated by winter or summer drawdown [107]. A severe summer drawdown with the water table reaching 0.5 m below ground surface during at least one month appears as the most sustainable and efficient way to reverse anaerobic conditions, especially strong in nutrient rich organic sediments. Temporary drawdown brings oxygen into the soil and thus reverses the toxicity of reduced compounds [108,109], which in turn supports the stability of the reed stands [110–113]. Experiments have shown that such a deep drying of the sediment rapidly stimulates recolonization of reeds [107]. This management is recommended at least every 5–10 years in southern France to prevent reed regression in marshes flooded permanently in order to attract waterfowl and, especially, stimulate the formation of colonies of nesting purple herons.

P. australis stands can also be destroyed by mechanical damage caused by human recreational activities, boat transport or mechanical effects of waves. Mechanical damage is also caused by insect infestation [22,114,115]), grazing by geese or swans, extremely dense fish stocks in fishponds, or proliferation of exotic mammals such as the muskrat (*Ondatra zibethicus*) and coypu (*Myocastor coypus*). These mammals can destroy significant amounts of reed and will severely limit its vegetative regrowth [116]. For instance, a breeding pair of muskrats can destroy nearly 1000 kg of reed per hectare to satisfy their food and shelter demands [117]. Control programmes to limit their proliferation are often part of reedbed management [102,118].

4.4. Management to Increase Reedbed Heterogeneity

Ecological requirements in terms of hydrology and vegetation structure differ among reed bird species, especially during the breeding season [58,66,76,119–123]. Habitat heterogeneity is hence the most important factor influencing reed birds, next to reedbed size [59,64,124,125]. Management practices aimed at increasing the habitat heterogeneity for wildlife commonly involve:

1. Water control to provide diverse hydrological conditions over the seasons, including spring/summer flooding for nesting birds [126,127].
2. Winter reed cutting or burning according to a rotational scheme to provide reed patches of different 'ages', offering a vegetation structure that complies with the needs of all species in the long term [15,67,128,129].
3. Creation and reprofiling of gently sloping ditches and pools to provide bird foraging habitat [69,101].
4. Hydraulic works to increase habitat connectivity for migrating fish species that use reedbeds as spawning areas [130,131].

Small reed areas offer limited possibilities for spatial heterogeneity. In such situations, priorities must be set regarding which species could be favoured based on the initial state of a site and the management options available.

4.5. Management to Stop the Spread of Reed in Wet Grasslands

While substantial effort has been spent on protecting or restoring *P. australis* vegetation in some deep-water littoral habitats, *P. australis* is considered a nuisance because of its expansive behaviour in some originally nutrient-poor wet grasslands [13], protected because of their floristic diversity or as habitats of vulnerable birds [132]. This happens on sites subjected recently to human-induced eutrophication. The competitive success of *P. australis* under such conditions is ascribed to its ability to make better use of surplus nutrients than the sedge species can.

Common practices to reduce reed dominance in these habitats are cattle grazing at different times of year, as well as summer, autumn or winter mowing [133,134]. A 6-year field experiment carried out in Swiss fen meadows showed that *P. australis* plants retreated from the community as a result of mowing twice a year, namely in June and September [51].

Reed progression in freshwater ecosystems is best controlled by maintaining deep vertical slopes that prevent reed colonization or by mechanically damaging the reed rhizomes. The use of a cage-wheel tractor is a common practice in the Camargue, which has been successful for 10 years. Cutting of reeds several times during the growing season exhausts the rhizome reserves. Even more effective is the multiple cutting of reeds below the water surface during the growing season, which deprives the rhizomes of oxygen [135].

5. Use and Management for Direct Economic Benefits

5.1. Overview of Economic Benefits

In the past, *P. australis* was used as a resource of material for various crafts and as a technological resource. *P. australis*-dominated wetlands also served as environments providing food such as birds and fish [3,16,136]. Some of the historical uses have lasted till now, some others have been modified or abandoned. A new impetus for *P. australis* use has been given by paludiculture, i.e., the agricultural management of peaty soils, aimed at preventing carbon loss resulting from their drainage [137–139].

The use of dry reed has a long-lasting tradition for roof thatching, fabrication of mats and production of building materials [20,122,136]. Although roof thatching declined at the end of the 19th century, it has gained in popularity over the last few decades, especially in the U.K., Ireland, Denmark, Belgium, Germany, and the Netherlands. In these regions, local reed is predominantly processed by small local producers. Much of the thatching material for western Europe, however, comes nowadays from southeastern Europe because of its higher quality and cheaper labour [140,141].

Common reed was also an important forage crop for cattle before the agricultural revolution [122] and, locally, still is. Summer harvest has become rare, but extensive grazing remains a common practice, especially in the Mediterranean area [142].

After World War II, *P. australis* was used in pulp manufacturing in some countries of the former Soviet bloc, namely the former USSR (Krotkevich 1970 in [136]), Romania [20], Bulgaria (G. Georgiev, pers. comm.), and the former German Democratic Republic (J. Köbbing, pers. comm). However, this industry was closed after the shift of these countries to the free-market economy in the 1990s, mainly owing to high harvesting costs [3]. Sustainable harvesting is also limited by a low regeneration ability of reed stands, whose terminal buds get easily damaged by the harvesting machines unless special precautions are taken [20]. Reed harvesting is now limited also by warmer winters preventing the formation of sufficiently thick ice that would support the cutting machines, or persons carrying out the reed harvest manually.

The interest in alternative energy sources has promoted the study of *P. australis* biomass yield [143–146]) and its applicability for combustion [147–151] or biomethane production [152]. An economic evaluation revealed that profitable use of harvested reed is confined to areas with relatively cheap labour and lacking long-distance energy supply or where reed is harvested as part of habitat management [153,154]).

P. australis cultivation also constituted the basis of the so-called biological drainage of wet areas. It was widely employed in the conversion of drained Dutch polders to agricultural land. After the polder drawdown, reed caryopses were sown (from the air) on the bare wet sediments. Within 2 to 3 years, it became completely overgrown with dense *P. australis* stands which were then left intact for several years until they were burned (as dead shoots in winter) and afterwards ploughed into the new organic-rich soil. Afterwards, rape (*Raphanus sativus*) was cultivated there, usually for two successive seasons, gradually suppressing the remaining viable reed shoots. Agricultural use of this newly gained land could start only after this stage, and the subsequent crop rotation was adjusted to eliminate almost all remaining sparsely occurring viable reed plants (e.g., [155,156].

In many European countries, (Czech Republic, France, Lithuania, Poland, Ukraine, Hungary, Serbia, etc.) fishponds were constructed for fish farming several centuries ago. Many of them, especially large ones with extensive littoral reed stands, provide habitats of great importance for the conservation of waterfowl [157–161]. Nevertheless, those without a legal conservation status are increasingly used for waterfowl hunting [157,160,162].

Sport fishing and ecotourism are also associated with the reedbeds and littoral belts as important structural elements of the landscape. This role of the reedbeds is additional to their importance as spawning areas for fish and as sites suitable for birdwatching. These provisioning services of the reedbeds have facilitated the conservation of several large reedbed areas.

5.2. Management for Reed Harvesting

Reed harvesting is a specific, sustainable and socially valued economic use of reedbeds. However, cutting all dry stalks in winter deprives wintering animal species of their habitat, as well as many migratory bird species of a sufficient reed cover for breeding after their return in spring, especially in continental and northern areas [63,67]. Several management options have been proposed to counteract the negative effects of reed harvesting on wildlife. A predominance of reed harvested every other year, coupled with the retention of patches harvested on a longer rotation, is considered as an effective compromise between conservation and commercial interests in the U.K. [163]. Because dry one-year stalks protect emerging next season's green shoots from late frost, biennial cutting has been shown to produce 50–75% more reed than annual cutting in the U.K. [60]. The situation, however, is different in countries such as France, where harvesting has locally remained an important commercial activity.

As two-year stalks are considered as waste material, biennial cutting requires sorting out first- from second-year stalks, therefore being no longer economically profitable. Likewise, maintaining a mosaic of reed patches of different ages with unmanaged fragments is not commercially feasible, although it is optimal for biodiversity [15,98,125]. A 5-year experiment conducted in southern France has demonstrated that optimal dry-reed density for Eurasian bitterns is obtained one year after reed cutting, especially in marshes with a homogeneous reed cover (Figure 3). Based on these results, management recommendations to reed harvesters consist of leaving 10% of uncut reed on a rotational basis (Mediterranean region) or 20% on fixed areas (northern region). It is also recommended to maintain a dry reed fringe around water bodies to preserve important bird foraging areas and reduce local damage to the rhizomes by reed-harvesting machines. Implementation of these measures has been encouraged through Natura 2000 contracts and agro-environmental schemes but could also be promoted through ecological marketing (eco-labels).

Reed harvesters need dense homogeneous stands of current-year shoots. Water management resulting in favourable conditions for reed harvest generally consists of (1) freshwater input in spring to favour reed growth, (2) summer drawdown to improve reedbed health and ground hardness (in the Mediterranean region) and prevent rhizome buds from their growing close to or above the ground surface [20], (3) low water levels in winter to increase the length of harvested stalks and facilitate access of cutting engines. Dry and leafless reed is cut before emergence of new shoots in spring and above the water (or ice) to allow dry stalks to pursue their role of rhizome oxygenation (Venturi effect).

5.3. Management for Waterfowl Hunting

Presence of water is essential to ducks, but permanent flooding of ponds with little water renewal often results in eutrophication and subsequent degradation of emergent and submerged macrophytes over time [163]. Periodically exposed soil is recommended to maintain appropriate conditions for sustainable management of duck populations in standing waters (J.-B. Mouronval, pers. comm.). For instance, drying of reed beds from March to September every 2–3 years will favour the dominance of annual hydrophytes and development of graminoid and amphibious plants at the marsh edge, ensuring a good seed bank for granivorous species. A short drawdown in February–March every year or at least every 3–4 years will favour the maintenance of perennial hydrophytes that are an important food source to herbivorous birds during the winter months while reducing the eutrophication rate.

Water management associated with waterfowl hunting obviously requires flooding during (and shortly before) the hunting season. However, the most common management practices involve permanent flooding or semipermanent flooding with drawdown after the hunting season (February-March). Another important aspect of the management is the creation and maintenance of large open-water areas in the vegetation to attract ducks.

5.4. Sustainable Grazing

Shoots of *P. australis*, especially their youngest parts, represent a favourable source of food for both domestic and wild herbivores. Wetlands provide a valuable forage crop especially in hot and dry areas such as the Mediterranean region, where the growth of terrestrial vegetation is reduced by lack of soil water from early summer.

Grazing of reedbeds by cattle is only possible when water levels are well below the soil surface. Even so, stocking rates should be less than one animal unit per hectare to be sustainable. Flooding after grazing should be avoided in order to ensure soil oxygenation necessary for rhizome recovery [164]. With one animal per hectare from June to September, the consumption of aboveground reed biomass can reach 42%, with up to 98% of biomass loss due to trampling and additional damage [165].

In view of its deleterious effect on reeds, the compatibility of grazing and nature conservation mostly consists in reed control with respect to the reed dominance and progression. Low grazing pressures on reedbeds or adjacent habitats can contribute to their floristic diversity and provision of habitats suitable for the aquatic warbler [166] or waders [165]. The duration and periodicity of grazing (or mowing) depend on the trade-offs between the aims of vegetation control and the resulting degree of disturbance on breeding birds, sometimes translating into a rotational scheme, insuring the provision of adequate bird habitats on a long-term basis [133].

5.5. Compatibility with Fish Farming

The traditional fish farming in ponds takes place in 2–3-year long cycles supporting the growth of fish from fingerlings to the market size. After the end of each cycle, the fishponds are emptied for the fish harvest. Before the agricultural revolution, they were typically dried and sowed with a summer crop every 3 to 7 years to aerate the bottom sediments and thus mineralize a large proportion of organic components of the pond mud, which becomes a sink of oxygen when it is saturated with water. Relatively high amounts of mineral nutrients are exported from each fishpond during its drawdown preceding the fish harvest. Nowadays, intensification of practices aimed at increasing the fish yields include scraping of shallow littoral areas to augment the water volume for fish (at the expense of littoral vegetation), fertilization, supply of fish feed and also water oxygenation [167–170].

Compatibility between fish farming and nature conservation involves mostly the maintenance of gently sloping shores to permit the development of the littoral belt of common reed and other helophytes so that they represent at least 15% of pond area [169,170]. Intensive management practices involving the use of fertilizers, predominance of carps with less than 10% of carnivorous species and yields above 200 kg per hectare have also been shown to decrease the conservation value of the fishponds [169,171].

6. Restoration and Construction of *P. australis*-Dominated Wetlands

6.1. Rewetting of Agricultural Peat Soils

In many lowland areas in Europe, peatlands in river floodplains, as well as along the shores of lakes and seas, were drained and converted to arable land. After the soil profiles were aerated, the organic matter accumulated during previous flooding began to decompose and the soil surface began to sink. Keeping the water table low required damming of the area and continuous pumping, which is expensive and economically unfeasible in less fertile areas. Some such areas in northern Europe were therefore rewetted and then left unmaintained. The aim of these measures was first to halt peat loss [172] and then to restore spontaneously developing ecosystems accumulating peat. As the sites were usually heavily eutrophicated as a result of mineral fertiliser application during previous agricultural use, it was usually not possible to restore the original species composition, adapted to oligo- to mesotrophic conditions. The intention was therefore to create a mosaic of helophyte communities (i.e., tall sedges and reeds) and open habitats for waterfowl [173,174]. This approach has been used, for example, in northeastern Germany in the Peene River floodplain and in northwestern Hungary in the Hanság area (which was part of Lake Neusiedl until the 18th century) [175].

6.2. Constructed Wetlands for Wastewater Treatment

Use of constructed wetlands for wastewater treatment has gained in popularity over the last few decades [176–178]. Most constructed wetlands in Europe are planted with *P. australis*. In the 1980s and 1990s, *P. australis* was the most frequent species planted in constructed wetlands designed with continuous subsurface horizontal flow that were used to treat wastewater in small settlements and communities [179]. Much attention was also devoted to the assessment of various functions of the plants in the treatment process. To date there is much agreement that *P. australis* affects the wetland functions positively by thermally insulating the bed surface in winter, protecting it against water erosion as well as preventing clogging, and creating microhabitats for microorganisms present in the treatment bed [180,181]. In addition, *P. australis* provides a source of organic carbon for microbial processes [182,183].

Following the finding that oxygen supply to the bed by internal ventilation systems of plants is too low as to fully meet the oxygen demand for the treatment process [184–186], attention has been devoted to systems with vertical flow in which oxygen transfer to the bed is promoted by vertical percolation of the wastewater [181,187,188]. The next technological stage, i.e., hybrid systems combining the positives of both the horizontal and vertical flow [189], have retained *P. australis* as a suitable plant species.

P. australis is also a common plant species of surface-flow constructed wetlands, aimed mainly at nutrient removal from nonpoint sources (e.g., [190,191]). The problem of nitrogen abatement is vital especially in marine coastal areas, where nitrogen appears to be the limiting nutrient in many situations [192,193]. The main management practice associated with this use consists of cutting and removal of aboveground reed biomass. The amounts of nutrients trapped change during the growing season, with maxima attained at the peak of the aboveground biomass in the summer months [87,194].

Besides small- to medium-scale constructed wetlands, *P. australis*-dominated vegetation covers an area of about 10 km^2 of Fenéki Lake, forming part of one of the largest constructed wetlands of the world, the Kis-Balaton Water Protection System, Hungary. This system of a total area of about 70 km^2 has been constructed on the place of former natural wetlands in the mouth of the Zala River in order to trap nutrients and suspended solids carried by its waters before they are discharged to Lake Balaton [195].

7. Multiple Uses

Preference for particular uses of *P. australis* stands leads to conflicts of interest among groups of various stakeholders. Problems occur mainly in harmonising the management of reed for biodiversity on the one hand and its uses for direct economic benefits on the other hand. The timing and amplitude of water-level fluctuations represent the most important complex abiotic factor. Water requirements of many breeding birds are compatible with hydrological conditions that favour reed growth in spring and support the overall stability of the plant stands. On the other hand, *P. australis* stands can retreat as a result of permanent flooding required to attract ducks for hunting or stabilized high water tables in fishponds aimed at maximising fish production. High stocks of cyprinid fish also compete with ducks for food such as zooplankton or benthos.

Many of the conflicts can, however, be prevented or overcome with management actions considering multiple benefits. Implementation of a collective agreement regarding water management rules can be necessary to favour diversity of uses and avoid ecosystem degradation. The plastic morphology of reeds, as well as the rapid yet reversible responses of reedbed structure to environmental conditions, makes it an ideal system for implementing evidence-based, adaptive co-management approaches by their users.

In situations of multiple uses with potential negative impact on ecosystem health, a companion modelling approach involving scientists and stakeholders can be useful to solve conflicts and build a shared vision of the socio-ecosystem [196,197]. The simple ecological functioning of reedbeds makes this ecosystem particularly suited for modelling [123]. An agent-based model called REEDSIM was developed in the Camargue [198] for testing long-term effects of various management schemes, climatic scenarios and market contexts on the health, biodiversity and economic yield of reedbeds (Figure 8). It comprises three sub-modules: (a) a topographical and hydrological module that defines the structural properties of a virtual wetland flooded by seasonal water levels, (b) an ecological module that sets reedbed and bird population dynamics, and (c) a decision module specific to each kind of activity, defined through semidirective interviews with each type of users (farmers, reed harvesters, hunters, and naturalists). A simplified version of the model has further been developed into a role-playing game (RPG), called BUTORSTAR, which simulates the impacts of reedbed management resulting from decisions made by the farmers, reed harvesters, hunters, and naturalists [199]. This RPG is based on an archetypal wetland made of a virtual landscape. Four different water regimes are proposed, each one adapted to a particular wetland use. Land-use and water management decisions are made by the players at both estate and management-unit levels. These decisions are entered into the model each year as the results of the negotiation process between the players. This RPG creates a continuum of learning that crosses the traditional boundaries between disciplines and allows the players to conduct multipurpose experiments that contribute to their comprehensive understanding of the socio-ecosystem. Typically, a hunter is asked to play the role of a reed harvester and so forth, facilitating dialogue among users in situations of conflict and providing a transdisciplinary knowledge-based tool to support collective thinking and decision processes.

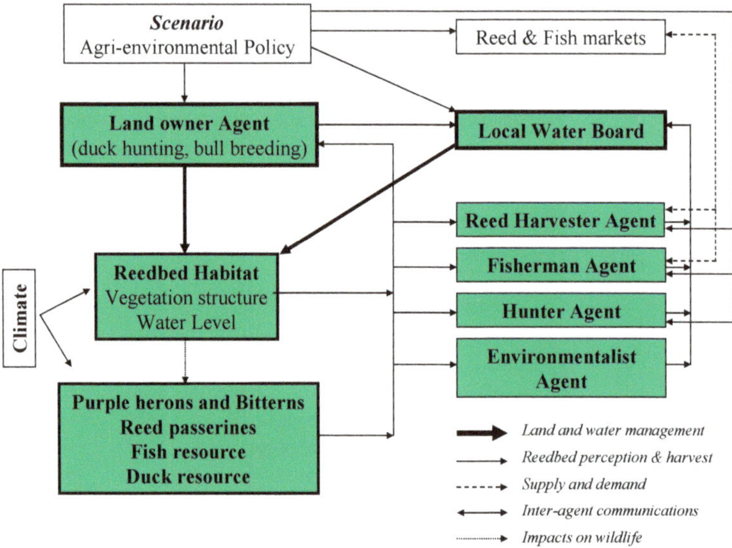

Figure 8. Conceptual frame of the REEDSIM agent-based simulation, (adapted with permission from Mathevet et al. [124]). This model comprises three sub-modules: the physical environment, bird population dynamics and socioeconomic decisions of stakeholders.

8. Future Prospects of *P. australis* in Europe

The present intensive land-use and search for adaptation measures to climate change represent new drivers of ecological development of European landscapes. If incautiously applied, they may inflict negative effects on all types of wetlands [200]. A holistic approach needs to be developed in order to counteract or, at least, minimise them.

The information reviewed in this paper clearly documents the diversity of *P. australis* habitats and human uses. This knowledge may help us predict possible changes in its status in Europe in connection with the ongoing climate change. Čížková et al. [200] have considered the likely changes to wetland biotopes. The following impacts may specifically concern *P. australis* biotopes: (1) In coastal areas, sea level rise might result in a reduction of the area of *P. australis*-dominated wetlands in estuaries of large rivers. (2) In continental areas of southeastern Europe, littoral wetlands dominated by *P. australis* may be negatively affected by anticipated water shortages. (3) In central and western Europe, the anticipated increase in the frequency and duration of flooding are likely to become a continuous threat to *P. australis* stands in lakes. (4) In northern Europe, the predicted increase in temperature might favour the expansion of *P. australis* in two ways: directly by stimulating *P. australis* growth and indirectly by increasing nutrient availability as a result of accelerated decomposition of soil organic matter. These mechanisms may be important especially in littorals of oligotrophic lakes and in wet grasslands.

As an opportunistic species of a highly competitive potential, *P. australis* will continue to occupy wet unmanaged biotopes in agricultural landscapes and occur in wet succession seres on abandoned land such as spoil heaps.

Lefebvre et al. [201] simulated the future evolution of water balance, wetland condition and water volumes necessary to maintain *P. australis* habitats at mid- and late- 21st century at 135 localities in Mediterranean Europe under two scenarios assuming a stabilization (RCP 4.5) or increase (RCP 8.5) of greenhouse gases emissions. The simulations performed under current conditions show that wetland habitats would remain in good condition at 97% of localities. However, by 2050 this proportion would decrease to 87% and 66% under the RCP 4.5 and RCP 8.5 scenarios, and even further to 78% and 36% by 2100. The simulations suggest that wetlands could persist with up to a 400 mm decrease of annual precipitation. Such resilience to climate change was attributed to the semipermanent character of wetlands (lower evaporation on dry ground) and their capacity to act as water reservoirs (higher precipitation expected in some countries during winter). The countries at highest risk of wetland degradation and loss were Portugal and Spain. Degradation of *P. australis* stands due to climate change will negatively affect their biodiversity and the services they provide as animal refuges and primary resources for industry and tourism. Preservation of their catchment areas and proactive management to reduce nonclimate stressors is urgently needed to preserve these wetlands.

As follows from previous sections, human preferences in landscape management may be equally important as environmental determinants for the further fate of *P. australis*-dominated wetlands. As pointed out by Čížková et al. [200], this holds for the future condition of European wetlands in general. Focusing on *P. australis*-dominated wetlands, the role of the species as a habitat former is particularly important in wetlands of international importance [202] and in constructed wetlands. The knowledge of ecophysiological mechanisms underlying *P. australis* performance forms a useful theoretical background for effective management of such *P. australis* wetlands. The use of *P. australis* as potential raw material and alternative energy resource appears to benefit from association of the uses with biotope care (e.g., [203]).

9. Conclusions

1. This review of knowledge on European *P. australis* populations indicates that it is a plastic and versatile species, forming part of varied plant communities all over Europe.
2. The analysis of the ecophysiological response to multiple stressors is used as a tool for understanding the population dynamics of *P. australis* in the main habitat types in Europe. Its decline at deep-water sites, stable performance in constructed wetlands with subsurface horizontal flow and expansion in wet grasslands are given as examples.
3. Of various human uses, the role of *P. australis* as a habitat former has gained an increasing value. Vulnerable birds are major drivers of reedbed management, especially in northwestern Europe, where large reedbeds have deteriorated or disappeared, which was followed by intensive habitat management ('gardening'), restoration and creation. Traditional socioeconomic uses are being abandoned, intensified or replaced by more lucrative activities (e.g., waterfowl hunting). Uses of common reed as energy crop and renewable eco-material for green buildings are limited but promising.
4. Each of the uses should be based on management practices that include both natural and human-driven processes. Nevertheless, the long-term maintenance or intensification of the economic uses often leads to practices that are not sustainable and get into conflict with nature conservation. Harmonisation of multiple uses with the help of innovative approaches (modelling) can assure a more sustainable future of *P. australis* wetlands.

Generally, *P. australis* will continue to be an important wetland species both in the ecological and social contexts in Europe, owing to its importance in both natural and human-altered vegetation, as well as its other ecosystem and economic values.

Author Contributions: Conceptualization, H.Č. and J.K.; vegetation, T.K.; conservation, B.P.; ecosystem services, H.Č. and B.P.; review and editing, J.K. All authors have read and agreed to the published version of the manuscript.

Funding: This research received no external funding.

Institutional Review Board Statement: Not applicable.

Informed Consent Statement: Not applicable.

Data Availability Statement: Data sharing not applicable.

Acknowledgments: We thank Dennis Whigham for his comments on the manuscript and Aat Barendregt, Jiří Dušek, and Josef Rajchard for providing photographs for Figures 2–5.

Conflicts of Interest: The authors declare no conflict of interest.

References

1. den Hartog, C.; Květ, J.; Sukopp, H. Reed. A common species in decline. *Aquat. Bot.* **1989**, *35*, 1–4. [CrossRef]
2. Kiviat, E. Organisms using *Phragmites australis* are diverse and similar on three continents. *J. Nat. Hist.* **2019**, *53*, 1975–2010. [CrossRef]
3. Köbbing, J.F.; Thevs, N.; Zerbe, S.; Wichtmann, W.; Couwenberg, J. The utilisation of reed (*Phragmites australis*): A review. *Mires Peat* **2013**, *13*, 1–14.
4. Koppitz, H. Analysis of genetic diversity among selected populations of *Phragmites australis* world-wide. *Aquat. Bot.* **1999**, *64*, 209–221. [CrossRef]
5. Lambertini, C.; Gustafsson, M.H.G.; Frydenberg, J.; Lissner, J.; Speranza, M.; Brix, H. A phylogeographic study of the cosmopolitan genus *Phragmites* (*Poaceae*) based on AFLPs. *Plant Syst. Evol.* **2006**, *258*, 161–182. [CrossRef]
6. Tho, B.T.; Sorrell, B.K.; Lambertini, C.; Eller, F.; Brix, H. *Phragmites australis*: How do genotypes of different phylogeographic origins differ from their invasive genotypes in growth, nitrogen allocation and gas exchange? *Biol. Invasions* **2016**, *18*, 2563–2576. [CrossRef]
7. Pyšek, P.; Skálová, H.; Čuda, J.; Guo, W.; Doležal, J.; Kauzál, O.; Lambertini, C.; Pyšková, K.; Brix, H.; Meyerson, L.A. Physiology of a plant invasion: Biomass production, growth and tissue chemistry of invasive and native *Phragmites australis* populations. *Preslia* **2019**, *91*, 51–75. [CrossRef]
8. Kettenring, K.M.; McCormick, M.K.; Baron, H.M.; Whigham, D.F. Mechanisms of *Phragmites australis* invasion: Feedbacks among genetic diversity, nutrients, and sexual reproduction. *J. Appl. Ecol.* **2011**, *48*, 1305–1313. [CrossRef]
9. Eller, F.; Skálová, H.; Caplan, J.S.; Bhattarai, G.P.; Burger, M.K.; Cronin, J.T.; Guo, W.Y.; Guo, X.; Hazelton, E.L.; Kettenring, K.M.; et al. Cosmopolitan species as models for ecophysiological responses to global change: The common reed *Phragmites australis*. *Front. Plant Sci.* **2017**, *8*, 1833. [CrossRef]
10. Hazelton, E.L.; Mozdzer, T.J.; Burdick, D.M.; Kettenring, K.M.; Whigham, D.F. *Phragmites australis* management in the United States: 40 years of methods and outcomes. *AoB Plants* **2014**, *6*, plu001. [CrossRef] [PubMed]
11. De Groot, R.S.; Wilson, M.A.; Boumans, R.M. A typology for the classification, description and valuation of ecosystem functions, goods and services. *Ecol. Econ.* **2002**, *41*, 393–408. [CrossRef]
12. Marks, M.; Lapin, B.; Randall, J. Phragmites australis (p communis): Threats, management, and monitoring. *Nat. Areas J.* **1994**, *14*, 285–294.
13. Güsewell, S.; Klötzli, F. Assessment of aquatic and terrestrial reed (*Phragmites australis*) stands. *Wetl. Ecol. Manag.* **2000**, *8*, 367–373. [CrossRef]
14. Ludwig, D.E.; Iannuzzi, T.J.; Esposito, A.N. *Phragmites* and environmental management: A question of values. *Estuaries* **2003**, *26*, 624–630. [CrossRef]
15. Valkamaa, E.; Lyytinena, S.; Koricheva, J. The impact of reed management on wildlife: A meta-analytical review of European studies. *Biol. Conserv.* **2008**, *141*, 364–374. [CrossRef]
16. Haslam, S.M. *A Book of Reed:* (Phragmites australis *(Cav.) Trin. ex Steudel, Formerly* Phragmites Communis *Trin.)*; Forrest Press: Cardigan, UK, 2010.
17. Lambertini, C.; Sorrell, B.K.; Riis, T.; Olesen, B.; Brix, H. Exploring the borders of European *Phragmites* within a cosmopolitan genus. *AoB Plants* **2012**, pls020. [CrossRef]
18. Clayton, W.D. The correct name of the common reed. *Taxon* **1968**, *17*, 168–169. [CrossRef]
19. Björk, S. Ecological investigations of *Phragmites communis*. *Folia Limnol. Scand.* **1967**, *14*, 1–248.
20. Rodewald-Rudescu, L. Das Schilfrohr, *Phragmites communis* Trin. In *Binnengewässer Bd. 27*; Elster, H.D., Ohle, J.W., Eds.; Schweizerbart'scher Verlag: Stuttgart, Germany, 1974; 307p.
21. Haslam, S.M. Biological flora of the British Isles. *Phragmites communis* Trin. *J. Ecol.* **1972**, *60*, 585–610.
22. Packer, J.G.; Meyerson, L.A.; Skálová, H.; Pyšek, P.; Kueffer, C. Biological flora of the British Isles: *Phragmites australis*. *J. Ecol.* **2017**, *105*, 1123–1162. [CrossRef]

23. Mal, T.K.; Narine, L. The biology of Canadian weeds. 129. *Phragmites australis* (Cav.) Trin. ex Steud. *Can. J. Plant Sci.* **2004**, *84*, 365–396. [CrossRef]

24. Engloner, A.I. Structure, growth dynamics and biomass of reed *(Phragmites australis)*—A review. *Flora* **2009**, *204*, 331–346. [CrossRef]

25. Ellenberg, H. *Vegetation Ecology of Central Europe*; Cambridge University Press: Cambridge, UK, 1988.

26. Hejný, S.; Remy, D.; Pott, R. Tall reed vegetation and tall sedge swamps, aquatic vegetation. In *Karte der Natürlichen Vegetation Europas/Map of the Natural Vegetation of Europe. Maßstab/Scale 1:2,500,000*; Bohn, U., Gollub, G., Hettwer, C., Neuhäuslová, Z., Raus, T., Schlüter, H., Eds.; Bundesamt für Naturschutz: Bonn, Germany, 2004; pp. 437–449.

27. Bohn, U.; Gollub, G.; Hettwer, C.; Neuhäuslová, Z.; Raus, T.; Schlüter, H. (Eds.) *Karte der Natürlichen Vegetation Europas/Map of the Natural Vegetation of Europe. Maßstab/Scale 1:2,500,000*; Bundesamt für Naturschutz: Bonn, Germany, 2004.

28. Hobbs, R.J.; Arico, S.; Aronson, J.; Baron, J.S.; Bridgewater, P.; Cramer, V.A.; Epstein, P.R.; Ewel, J.J.; Klink, C.A.; Lugo, A.E.; et al. Novel ecosystems: Theoretical and management aspects of the new ecological world order. *Glob. Ecol. Biogeogr.* **2006**, *15*, 1–7. [CrossRef]

29. Solomeshch, A.I.; Mirkin, B.M.; Ermakov, N.; Ishbirdin, A.; Golub, V.; Saitov, M.; Zhuravliova, S.; Rodwell, J. *Red Data Book of Plant Communities in the Former USSR*; Unit of Vegetation Science: Lancaster, UK, 1997.

30. Chytrý, M.; Hennekens, S.M.; Jiménez-Alfaro, B.; Knollová, I.; Dengler, J.; Jansen, F.; Landucci, F.; Schaminée, J.H.; Aćić, S.; Agrillo, E.; et al. European Vegetation Archive (EVA): An integrated database of European vegetation plots. *Appl. Veg. Sci.* **2016**, *19*, 173–180. [CrossRef]

31. Rodwell, J. (Ed.) *British Plant Communities*; Cambridge University Press: Cambridge, UK, 2000; Volume 1–5.

32. Schaminée, J.H.J.; Stortelder, A.H.F.; Westhoff, V.; Weeda, E.J.; Hommel, P.W.F.M. (Eds.) *De Vegetatie van Nederland. Deel 1–5*; Opulus Press: Uppsala, Sweden; Leiden, The Netherlands, 1995–1999.

33. Dierssen, K. *Vegetation Nordeuropas*; Verlag Eugen Ulmer: Stuttgart, Germany, 1996.

34. Pott, R. *Die Pflanzengesellschften Deutschlands*; Verlag Eugen Ulmer: Stuttgart, Germany, 1995.

35. Matuszkiewicz, W. *Przewodnik do Oznaczania Zbiorowisk Roslinnych Polski*; Państwowe Wydawnictwo Naukowe: Warszawa, Poland, 2002.

36. Chytrý, M. (Ed.) *Vegetation of the Czech Republic*; Academia: Praha, Czech Republic, 2007–2011; Volume 1–3. (In Czech)

37. Mucina, L.; Grabherr, G.; Ellmauer, T.; Wallnöfer, S. (Eds.) *Die Pflanzengesellschaften Österreichs 1–3*; Gustav Fischer: Jena, Germany, 1993.

38. Borhidi, A. An annotated checklist of the Hungarian plant communities 1. The non-forest vegetation. In *Critical Revision of the Hungarian Plant Communities*; Borhidi, A., Ed.; Janus Pannonius University: Pécs, Hungary, 1996; pp. 43–94.

39. Pomogyi, P. *Changes in the Reed Communities of Lake Fenéki in the Little Balaton Protection System Based on the Results of the Vegetation Mapping*; Research Report; West Danube Water Directorate: Készthély, Hungary, 1996. (In Hungarian)

40. Horvat, I.; Glavač, V.; Ellenberg, H. *Vegetation Südosteuropas*; Gustav Fischer Verlag: Stuttgart, Germany, 1974.

41. Stančić, Z. Marshland vegetation of the class *Phagmito-Magnocaricetea* in Croatia. *Biologia* **2007**, *62*, 297–314. [CrossRef]

42. Solomakha, V.A. *The Syntaxonomy of Vegetation of the Ukraine*; Ukrainian Phytosociological Centre: Kyiv, Ukraine, 1996.

43. Shelyag-Sosonko, J.R. *Zelenaya kniga Ukrajinskoi SSR*; Naukova Dumka: Kiev, Ukraine, 1987. (In Russian)

44. Golub, V.B.; Mirkin, B.M. Grasslands of the lower Volga valley. *Folia Geobot. Et Phytotaxon.* **1986**, *21*, 337–395. [CrossRef]

45. Raspopov, I.M. *Higher Aquatic Plant Communities of the Great Lakes in the Northwestern USSR*; Izdatelstvo Nauka: Leningrad, Russia, 1985. (In Russian)

46. Gejny, S.; Sytnik, K.M. *Macrophytes as Environmental Indicators*; Naukova Dumka: Kiev, Ukraine, 1993. (In Russian)

47. Rivas-Martínez, S.; Díaz, T.E.; Fernández-González, F.; Izco, J.; Loidi, J.; Lousã, M.; Penas, Á. Vascular plant communities of Spain and Portugal. Addenda to the Syntaxonomical checklist of 2001. *Itinera Geobot.* **2002**, *15*, 5–922.

48. Landucci, F.; Gigante, D.; Venanzoni, R.; Chytrý, M. Wetland vegetation of the class Phragmito-Magno-Caricetea in central Italy. *Phytocoenologia* **2013**, *43*, 67–100. [CrossRef]

49. Hutchinson, G.E. A treatise on limnology. In *Limnological Botany*; John Wiley & Sons: New York, NY, USA; London, UK; Sydney, Australia; Toronto, ON, Canada, 1977; Volume 3, 660p.

50. Hroudová, Z.; Zákravský, P. Vegetation dynamics in a fishpond littoral related to human impact. *Hydrobiologia* **1999**, *415*, 139–145. [CrossRef]

51. Güsewell, S. Management of *Phragmites australis* in Swiss fen meadows by mowing in early summer. *Wetl. Ecol. Manag.* **2003**, *11*, 433–445. [CrossRef]

52. Rudescu, L. Neue biologische Probleme bei den Phragmiteskulturarbeiten im Donaudelta. *Arch. Für Hydrobiol. Suppl.* **1965**, *30*, 80–111. [CrossRef]

53. Kokin, K.A. *Ecology of Higher Aquatic Plants*; Izdatelstvo Moskovskogo Universiteta: Moskva, Russia, 1982. (In Russian)

54. Dubyna, D.V.; Neuhäuslová, Z.; Šeljag-Sosonko, J.R. Coastal vegetation of the Birjučij Island Spit in the Azov Sea, Ukraine. *Preslia* **1994**, *66*, 193–216.

55. Antonellini, M.; Mollema, P.N. Impact of groundwater salinity on vegetation species richness in the coastal pine forests and wetlands of Ravenna, Italy. *Ecol. Eng.* **2010**, *36*, 1201–1211. [CrossRef]

56. Giambastiani, B.M.; Greggio, N.; Nobili, G.; Dinelli, E.; Antonellini, M. Forest fire effects on groundwater in a coastal aquifer (Ravenna, Italy). *Hydrol. Process.* **2018**, *32*, 2377–2389. [CrossRef]

57. Dobrowolski, K.A. Role of birds in Polish wetland ecosystems. *Pol. Arch. Hydrobiol.* **1973**, *20*, 217–221.

58. Bibby, C.J.; Lunn, J. Conservation of reed beds and their avifauna in England and Wales. *Biol. Conserv.* **1982**, *23*, 167–186. [CrossRef]

59. Burgess, N.D.; Evans, C.E. *The Management of Reedbeds for Birds*; The Royal Society for the Protection of Birds: Sandy, UK, 1989.

60. Andrews, J.; Ward, D. The management and creation of Reedbeds—Especially for rare birds. *Br. Wildl.* **1991**, *3*, 81–91.

61. Fojt, W.; Foster, A. Reedbeds, their wildlife and requirements. A: Botanical and invertebrate aspects of reedbeds, their ecological requirements and conservation significance. In *Reedbeds for Wildlife*; Ward, D., Ed.; The Royal Society of Protection for Birds: Oxford, UK; University of Bristol Information Press: Oxford, UK, 1992; pp. 49–56.

62. Tyler, G. Reedbeds their wildlife requirements, B. Requirements of birds in reedbeds. In *Reedbeds for Wildlife*; Ward, D., Ed.; The Royal Society for the Protection of Birds: Oxford, UK; University of Bristol Information Press: Oxford, UK, 1992; pp. 57–64.

63. Graveland, J. Effects of reed cutting on density and breeding success of Reed Warbler *Acrocephalus scirpaceus* and Sedge Warbler *Acrocephalus schoenobaenus*. *J. Avian Biol.* **1999**, *30*, 469–482. [CrossRef]

64. Báldi, A.; Kisbenedek, T. Bird species numbers in an archipelago of reeds at Lake Velence, Hungary. *Glob. Ecol. Biogeogr.* **2000**, *9*, 451–461. [CrossRef]

65. Barbraud, C.; Lepley, M.; Mathevet, R.; Mauchamp, A. Reedbed selection and colony size of breeding purple herons *Ardea purpurea* in southern France. *Ibis* **2002**, *144*, 227–235. [CrossRef]

66. Martinez-Vilalta, J.; Bertolero, A.; Bigas, D.; Paquet, J.Y.; Martinez-Vitalta, A. Habitat selection of passerine birds nesting in the Ebro Delta reedbeds (NE Spain): Management implications. *Wetlands* **2002**, *22*, 318–325. [CrossRef]

67. Poulin, B.; Lefebvre, G. Effect of winter cutting on the passerine breeding assemblage in French Mediterranean reedbeds. *Biodivers. Conserv.* **2002**, *11*, 1567–1581. [CrossRef]

68. Adamo, M.C.; Puglisi, L.; Baldaccini, N.E. Factors affecting Bittern Botaurus stellaris distribution in a Mediterranean wetland. *Bird Conserv. Int.* **2004**, *14*, 153–164. [CrossRef]

69. Gilbert, G.; Tyler, G.A.; Dunn, C.J.; Smith, K.W. Nesting habitat selection by bitterns *Botaurus stellaris* in Britain and the implications for wetland management. *Biol. Conserv.* **2005**, *124*, 547–553. [CrossRef]

70. Self, M. A review of management for fish and bitterns, *Botaurus stellaris*, in wetland reserves. *Fish. Manag. Ecol.* **2005**, *12*, 387–394. [CrossRef]

71. Pasinelli, G.; Schiegg, K. Fragmentation within and between wetland reserves: The importance of spatial scales for nest predation in reed buntings. *Ecography* **2006**, *29*, 731–732. [CrossRef]

72. Poulin, B.; Lefebvre, G.; Allard, S.; Mathevet, R. Reed harvest and summer drawdown enhance bittern habitat in the Camargue. *Biol. Conserv.* **2009**, *142*, 689–695. [CrossRef]

73. Provost, P.; Kerbiriou, C.; Jiguet, F. Foraging range and habitat use by Aquatic Warblers *Acrocephalus paludicola* during a fall migration stopover. *Acta Ornithol.* **2010**, *45*, 173–180. [CrossRef]

74. Andersen, L.H.; Nummi, P.; Rafn, J.; Frederiksen, C.M.S.; Kristjansen, M.P.; Lauridsen, T.L.; Trojelsgaard, K.; Pertoldi, C.; Bruhn, D.; Bahrndorff, S. Can reed harvest be used as a management strategy for improving invertebrate biomass and diversity? *J. Environ. Manag.* **2021**, *300*, 113637. [CrossRef]

75. Musseau, R.; Crépin, M.; Brugulat, C.; Kerbiriou, C. Conservation and Restoration of Coastal Reed Beds in the Context of Global Change: Potential Effects of Habitat Fragmentation for Specialist Marshland Passerines. *Wetlands* **2021**, *41*, 70. [CrossRef]

76. Nemeth, E.; Dvorak, M. Reed die-back and conservation of small reed birds at Lake Neusiedl, Austria. *J. Ornithol.* **2022**, *163*, 683–693. [CrossRef]

77. BirdLife International. *European Birds of Conservation Concern: Populations, Trends and National Responsibilities*; BirdLife International: Cambridge, UK, 2017.

78. Smiddy, P.; Cullen, C.; O'Halloran, J. Time of roosting of Barn Swallows *Hirundo rustica* at an Irish reedbed during autumn migration. *Ringing Migr.* **2007**, *23*, 228–230. [CrossRef]

79. Buckland, S.T.; Hereward, A.C. Trap shyness of Yellow Wagtails *Motacilla flava flavissima* at a premigratory roost. *Ringing Migr.* **1982**, *4*, 15–23. [CrossRef]

80. Broyer, J.; Calenge, C. Influence of fish-farming management on duck breeding in French fish pond systems. *Hydrobiologia* **2010**, *637*, 173–185. [CrossRef]

81. Wotton, S.; Brown, A.; Burn, A.; Dodd, A.; Droy, N.; Gilbert, G.; Hardiman, N.; Rees, S.; White, G.; Gregory, R. Boom or bust—A sustainable future for reedbeds and Bitterns? *British Wildlife* **2005**, *20*, 305–315.

82. Information Sheet on Ramsar Wetlands. Broadland. Available online: http://jncc.defra.gov.uk/pdf/RIS/UK11010.pdf (accessed on 10 April 2023).

83. Jacoby, H. *Die Vögel des Bodenseegebietes*; ALA/Schweizerische Gesellschaft für Vogelkunde und Vogelschutz: Sempach, Switzerland, 1970; 260p.

84. Löffler, H. Neusiedlersee. The Limnology of a Shallow Lake in Central Europe. In *Monographiae Biologicae*; Dr. W. Junk bv: The Hague, The Netherlands, 1979; 543p.

85. Dorogman, C. Information Sheet on Ramsar Wetlands: Lake Fertö. 2007. Available online: https://rsis.ramsar.org/RISapp/files/RISrep/HU420RISformer_150330.pdf (accessed on 10 April 2023).

86. Metz, H. Information Sheet on Ramsar Wetlands: Neusiedler See–Seewinkel. 2005. Available online: https://rsis.ramsar.org/RISapp/files/RISrep/AT271RIS.pdf (accessed on 6 April 2012).

87. Květ, J. Mineral nutrients in shoots of reed (*Phragmites communis* Trin.). *Pol. Arch. Hydrobiol.* **1973**, *20*, 137–147.

88. Dykyjová, D.; Květ, J. (Eds.) *Pond Littoral Ecosystems. Structure and Functioning*; Ecological Studies 28; Springer: Berlin/Heidelberg, Germany; New York, NY, USA, 1978; 464p.

89. Information Sheet on Ramsar Wetlands: The Camargue. 1993. Available online: https://rsis.ramsar.org/RISapp/files/RISrep/FR786RISformer1993_EN.pdf (accessed on 10 April 2023).

90. The Annotated Ramsar List: Romania. Available online: https://rsis.ramsar.org/sites/default/files/rsiswp_search/exports/Ramsar-Sites-annotated-summary-Romania.pdf?1680615226 (accessed on 10 April 2023).

91. The Annotated Ramsar List: Hungary. Available online: https://rsis.ramsar.org/sites/default/files/rsiswp_search/exports/Ramsar-Sites-annotated-summary-Hungary.pdf?1622485211 (accessed on 10 April 2023).

92. Dömötörfy, Z. Changes in macro-vegetation of the Kis-Balaton wetlands over the last two centuries: A GIS perspective. *Hydrobiologia* **2003**, *506–509*, 671–679. [CrossRef]

93. Lee, S.Y. Net aerial primary productivity, litter production and decomposition of the reed *Phragmites communis* in a nature reserve in Hong Kong: Management implications. *Mar. Ecol. Prog. Ser.* **1990**, *66*, 161–174. [CrossRef]

94. Cowie, N.R.; Sutherland, W.J.; Ditlhogo, M.K.M.; James, R. The effects of conservation management of reed beds II. The flora and litter disappearance. *J. Appl. Ecol.* **1992**, *29*, 277–284. [CrossRef]

95. Day, J.W., Jr.; Pieczyńska, E.; Úlehlová, B. Further fate of organic matter in wetlands. In *The Production Ecology of Wetlands*; Westlake, D.F., Květ, J., Szczepański, A., Eds.; Cambridge University Press: Cambridge, UK, 1978; pp. 169–191.

96. Úlehlová, B. The role and fate of decomposers in wetlands. In *The Production Ecology of Wetlands*; Westlake, D.F., Květ, J., Szczepański, A., Eds.; Cambridge University Press: Cambridge, UK, 1998; pp. 192–210.

97. van der Toorn, J.; Mook, J.H. The influence of environmental factors and management on stands of *Phragmites australis*. I. Effects of burning, frost, and insect damage on shoot density and shoot size. *J. Appl. Ecol.* **1982**, *19*, 477–499. [CrossRef]

98. Ditlhogo, M.K.; James, M.R.; Laurence, B.R.; Sutherland, W.J. The effects of conservation management of reed beds I. The invertebrates. *J. Appl. Ecol.* **1992**, *29*, 265–276. [CrossRef]

99. Ostendorp, W. Impact of winter reed harvesting and burning on the nutrient economy of reed beds. *Wetl. Ecol. Manag.* **1995**, *3*, 233–248. [CrossRef]

100. Schulze Hagen, K.; Pleines, S.; Sennert, G. Rapid increase of Cuckoo parasitism in a local Reed Warbler population. *Vogelwelt* **1996**, *117*, 83–86.

101. Hawke, C.J.; Jose, P.V. *Reedbed Management for Commercial and Wildlife Interests*; The Royal Society for the Protection of Birds: Sandy, UK, 1996.

102. White, G.; Purps, J.; Alsbury, S. *The Bittern in Europe: A Guide to Species and Habitat Management*; The Royal Society for the Protection of Birds: Sandy, UK, 2006.

103. Ostendorp, W. 'Die-back' of reeds—A critical review of literature. *Aquat. Bot.* **1989**, *35*, 5–26. [CrossRef]

104. Kramer, I.; Kapfer, A. Naturnahe Uferbereiche und Flachwasserzonen des Bodensees. *Biotope Baden-Württemberg* **2001**, *13*, 1–47.

105. HELCOM. The review of more specific targets to reach the goal set up in the 1988/1998 Ministerial Declarations regarding nutrients. *Balt. Sea Environ. Proc.* **2003**, *89*.

106. Stamati, F.E.; Chalkias, N.; Moraetis, D.; Nikolaidis, N.P.F. Natural attenuation of nutrients in a Mediterranean drainage canal. *J. Environ. Monit.* **2010**, *12*, 164–171. [CrossRef]

107. Alvarez, M.G.; Tron, F.; Mauchamp, A. Sexual versus asexual colonization by *Phragmites australis*: 25-year reed dynamics in a Mediterranean marsh, Southern France. *Wetlands* **2005**, *25*, 639–647. [CrossRef]

108. Linthurst, R.A. The effect of aeration on the growth of *Spartina alterniflora* Loisel. *Am. J. Bot.* **1979**, *66*, 685–691. [CrossRef]

109. van Wijck, C.; De Groot, C.J. The impact of desiccation of a freshwater marsh (Garcines Nord, Camargue, France) on sediment–water–vegetation interactions. *Hydrobiologia* **1993**, *252*, 95–103. [CrossRef]

110. Morris, J.T.; Dacey, J.W.H. Effect of oxygen on ammonium uptake and root respiration by *Spartina alternifolia*. *Am. J. Bot.* **1984**, *71*, 979–985. [CrossRef]

111. Weisner, S.E.; Granéli, W. Influence of substrate conditions on the growth of *Phragmites australis* after a reduction in oxygen transport to below-ground parts. *Aquat. Bot.* **1989**, *35*, 71–80. [CrossRef]

112. Armstrong, J.; Armstrong, W.; Van der Putten, W.H. *Phragmites* die-back: Bud and root death, blockages within the aeration and vascular systems and the possible role of phytotoxins. *New Phytol.* **1996**, *133*, 399–414. [CrossRef]

113. Brix, H.; Sorrel, B.K. Oxygen stress in wetland plants: Comparison of de-oxygenated and reducing root environments. *Funct. Ecol.* **1996**, *10*, 521–526. [CrossRef]

114. Fuchs, C. The beetle *Donacia clavipes* as possible cause for the reed decline at lake Constance (Untersee). *Limnol. Aktuell* **1993**, *5*, 41–48.

115. Tscharntke, T. Insects on common reed (*Phragmites australis*): Community structure and the impact of herbivory on shoot growth. *Aquat. Bot.* **1999**, *64*, 399–410. [CrossRef]

116. Boorman, L.A.; Fuller, R.M. The changing status of reedswamp in the Norfolk Broads. *J. Appl. Ecol.* **1981**, *18*, 241–269. [CrossRef]
117. Skuhravý, V.; Pokorný, V.; Pelikán, J.; Skuhravá, M.; Hudec, K.; Rychnovský, B. *Invertebrates and Vertebrates Attacking Common Reed Stands (Phragmites communis) in Czechoslovakia*; Studie ČSAV; Academia: Praha, Czech Republic, 1981; Volume 1, pp. 1–112.
118. Bertolino, S.; Perrone, A.; Gola, L. Effectiveness of coypu control in small Italian wetland areas. *Wildl. Soc. Bull.* **2005**, *33*, 714–720. [CrossRef]
119. van der Hut, R.M.G. Habitat choice and temporal differentiation in reed passerines of a Dutch marsh. *Ardea* **1986**, *74*, 159–176.
120. Leisler, B.; Ley, H.W.; Winkler, H. Habitat, behavior and morphology of *Acrocephalus* warblers: An integrated analysis. *Ornis Scand.* **1989**, *20*, 181–186. [CrossRef]
121. Jedraszko-Dabrowska, D. Reeds as construction supporting great reed warbler (*Acrocephalus arundinaceus* L.) and reed warbler (*Acrocephalus scirpaceus* Herm.) nests. *Ekol. Pol.* **1992**, *39*, 229–242.
122. Ward, D. *Reedbeds for Wildlife*; The Royal Society for the Protection of Birds: Oxford, UK; University of Bristol Information Press: Oxford, UK, 1992.
123. Poulin, B.; Davranche, A.; Lefebvre, G. Ecological assessment of *Phragmites australis* wetlands using multi-season SPOT-5 scenes. *Remote Sens. Environ.* **2010**, *114*, 1602–1609. [CrossRef]
124. Schiess, H. Schilfbestände als Habitatinseln von Vögeln. *Vierteljahrsschr. Der Nat. Ges. Zuerich* **1990**, *135*, 259–265.
125. Tscharntke, T. Fragmentation of *Phragmites habitats*, minimum viable population size, habitat suitability, and local extinction of moths, midges, flies, aphids, and birds. *Conserv. Biol.* **1992**, *6*, 530–536. [CrossRef]
126. Poulin, B.; Lefebvre, G.; Mathevet, R. Habitat selection by booming bitterns *Botaurus stellaris* in French Mediterranean reed-beds. *Oryx* **2005**, *39*, 265–274. [CrossRef]
127. Poulin, B.; Lefebvre, G.; Mauchamp, A. Habitat requirements of passerines and reedbed management in southern France. *Biol. Conserv.* **2002**, *107*, 315–325. [CrossRef]
128. Puglisi, L.; Adamo, M.C.; Baldaccini, N.E. Man-induced habitat changes and sensitive species: A GIS approach to the Eurasian Bittern *(Botaurus stellaris)* distribution in a Mediterranean wetland. *Biodivers. Conserv.* **2005**, *14*, 1909–1922. [CrossRef]
129. Schmidt, M.H.; Lefebvre, G.; Poulin, B.; Tscharntke, T. Reed cutting affects arthropod communities, potentially reducing food for passerine birds. *Biol. Conserv.* **2005**, *121*, 157–166. [CrossRef]
130. Knights, B. Enhancing stocks of European eel, *Anguilla anguilla*, to benefit bittern, *Botaurus stellaris*. In *Interactions between Fish and Birds: Implications for Management*; Cowx, I.G., Ed.; Blackwell Publishing Ltd.: Oxford, UK, 2003; Chapter 22; pp. 298–313.
131. Noble, R.A.A.; Harvey, J.P.; Cowx, I.G.G. Can management of freshwater fish populations be used to protect and enhance the conservation status of a rare, fish-eating bird, the bittern, *Botaurus stellaris*, in the UK? *Fish. Manag. Ecol.* **2004**, *11*, 291–302. [CrossRef]
132. Kloskowski, J.; Krogulec, J. Habitat selection of Aquatic Warbler *Acrocephalus paludicola* in Poland: Consequences for conservation of the breeding areas. *Vogelvelt* **1999**, *120*, 113–120.
133. Tanneberger, F.; Flade, M.; Preiksa, Z.; Schröder, B. Habitat selection of the globally threatened Aquatic Warbler *Acrocephalus paludicola* at the western margin of its breeding range and implications for management. *Ibis* **2010**, *152*, 347–358. [CrossRef]
134. Dijkema, K.S. Salt and brackish marshes around the Baltic Sea and adjacent parts of the North Sea: Their vegetation and management. *Biol. Conserv.* **1990**, *51*, 191–210. [CrossRef]
135. Hellings, S.E.; Gallagher, J.L. The effects of salinity and flooding on *Phragmites australis*. *J. Appl. Ecol.* **1992**, *29*, 41–49. [CrossRef]
136. Haslam, S.M.; Klötzli, F.; Sukopp, H.; Szczepański, A. The management of wetlands. In *The Production Ecology of Wetlands*; Westlake, D.F., Květ, J., Szczepański, A., Eds.; Cambridge University Press: Cambridge, UK, 1998; pp. 405–464.
137. Wichtmann, W.; Schröder, C.; Joosten, H. *Paludiculture—Productive Use of Wet Peatlands. Climate Protection—Biodiversity—Regional Economic Benefits*; Schweizerbart Science Publishers: Stuttgart, Germany, 2016.
138. Ziegler, R.; Wichtmann, W.; Abel, S.; Kemp, R.; Simard, M.; Joosten, H. Wet peatland utilisation for climate protection–An international survey of paludiculture innovation. *Clean. Eng. Technol.* **2021**, *5*, 100305. [CrossRef]
139. Lahtinen, L.; Mattila, T.; Myllyviita, T.; Seppälä, J.; Vasander, H. Effects of paludiculture products on reducing greenhouse gas emissions from agricultural peatlands. *Ecol. Eng.* **2022**, *175*, 106502. [CrossRef]
140. Wichmann, S.; Köbbing, J.F. Common reed for thatching—A first review of the European market. *Ind. Crops Prod.* **2015**, *77*, 1063–1073. [CrossRef]
141. Becker, L.; Wichmann, S.; Beckmann, V. Common Reed for Thatching in Northern Germany: Estimating the Market Potential of Reed of Regional Origin. *Resources* **2020**, *9*, 146. [CrossRef]
142. Mesléard, F.; Perennou, C. *Aquatic Emergent Vegetation, Ecology and Management*; Conservation of Mediterranean Wetlands, No. 6; Tour du Valat: Arles, France, 1996.
143. Wichtmann, W.; Wichmann, S. Environmental, social and economic aspects of a sustainable biomass production. *J. Sustain. Energy Environ. Spec. Issue* **2011**, *1*, 77–81.
144. Dragoni, F.; Giannini, V.; Ragaglini, G.; Bonari, E.; Silvestri, N. Effect of harvest time and frequency on biomass quality and biomethane potential of common reed (*Phragmites australis*) under paludiculture conditions. *Bioenergy Res.* **2017**, *10*, 1066–1078. [CrossRef]

145. Eller, F.; Ehde, P.M.; Oehmke, C.; Ren, L.; Brix, H.; Sorrell, B.K.; Weisner, S.E. Biomethane Yield from different European Phragmites australis genotypes, compared with other herbaceous wetland species grown at different fertilization regimes. *Resources* **2020**, *9*, 57. [CrossRef]

146. Granéli, W. Reed (*Phragmites australis* (Cav.) Trin. ex Steudel) as an energy source in Sweden. *Biomass* **1984**, *4*, 183–208. [CrossRef]

147. Kitzler, H.; Pfeifer, C.; Hofbauer, H. Combustion of reeds in a 3 MW district heating plant. *Int. J. Environ. Sci. Dev.* **2012**, *3*, 407–411. [CrossRef]

148. Lica, D.; Coşereanu, C.; Budău, G.; Lunguleasa, A. Characteristics of reed briquettes–biomass renewable resource of the Danube delta. *Pro Ligno* **2012**, *8*, 44–51.

149. Patuzzi, F.; Mimmo, T.; Cesco, S.; Gasparella, A.; Baratieri, M. Common reeds (*Phragmites australis*) as sustainable energy source: Experimental and modelling analysis of torrefaction and pyrolysis processes. *GCB Bioenergy* **2012**, *5*, 367–374. [CrossRef]

150. Ren, L.; Eller, F.; Lambertini, C.; Guo, W.Y.; Brix, H.; Sorrell, B.K. Assessing nutrient responses and biomass quality for selection of appropriate paludiculture crops. *Sci. Total Environ.* **2019**, *664*, 1150–1161. [CrossRef]

151. Kuptz, D.; Kuchler, C.; Rist, E.; Eickenscheidt, T.; Mack, R.; Schön, C.; Hartmann, H. Combustion behaviour slagging tendencies of pure blended kaolin additivated biomass pellets from fen paludicultures in two small-scale boilers < 30 kW. *Biomass Bioenergy* **2022**, *164*, 106532.

152. Hartung, C.; Andrade, D.; Dandikas, V.; Eickenscheidt, T.; Drösler, M.; Zollfrank, C.; Heuwinkel, H. Suitability of paludiculture biomass as biogas substrate— biogas yield and long-term effects on anaerobic digestion. *Renew. Energy* **2020**, *159*, 64–71. [CrossRef]

153. Wichtmann, W.; Tanneberg, F. Land use options for rewetted peatlands. In *Carbon Credits from Peastland Rewetting. Climate–Biodiversity–Land Use*; Tanneberg, F., Wichtmann, W., Eds.; Schweitzerbart Science Publishers: Stuttgart, Austria, 2011; pp. 107–132.

154. Wichmann, S. Commercial viability of paludiculture: A comparison of harvesting reeds for biogas production, direct combustion, and thatching. *Ecol. Eng.* **2017**, *103*, 497–505. [CrossRef]

155. Toorn, J. van der Variability of *Phragmites australis* (Cav.) Trin. Ex Steudel in relation to the environment. *Van Zee Tot Land* **1972**, *48*, 1–122.

156. Mook, J.H.; van der Toorn, J.W. Experiment on the development of reed vegetation in the Zuidflevoland polder. In *Progress Report, 1974*; Institute of Ecological Research: Heteren, The Netherlands, 1974; pp. 11–15.

157. Lutz, M. Les étangs de pisciculture en Europe centrale. Typologie des systèmes d'exploitation et impacts des modalités de gestion sur l'avifaune. Ph.D. Thesis, Université Louis Pasteur, Strasbourg, France, 2001.

158. Dyrcz, A. Breeding ecology of the great Reed Warlber (*Acrocephalus arundinaceus*) and Reed Warbler (*A. scirpaceus*) at fish ponds in SW-Poland and lakes in NW-Switzerland. *Acta Ornithol. Warszawa* **1981**, *18*, 307–334.

159. Broyer, J.; Varagnat, P.; Constant, G.; Caron, P. Habitat du Héron pourpré *Ardea purpurea* sur les étangs de pisciculture en France. *Alauda* **1998**, *66*, 221–228.

160. Brochet, A.L.; Gauthier-Clerc, M.; Mathevet, R.; Béchet, A.; Mondain-Monval, J.Y.; Tamisier, A. Marsh management, reserve creation, hunting periods and carrying capacity for wintering ducks and coots. *Biodivers. Conserv.* **2009**, *18*, 1879–1894. [CrossRef]

161. Prokešová, J.; Kocian, L. Habitat selection of two *Acrocephalus* warblers breeding in reed beds near Malacky (Western Slovakia). *Biologia* **2004**, *59*, 637–644.

162. Grim, T.; Honza, M. Effect of habitat on the diet of reed warbler (*Acrocephalus scirpaceus*) nestlings. *Folia Zool.* **1996**, *45*, 31–34.

163. Tamisier, A.; Grillas, P. A review of habitat changes in the Camargue: An assessment of the effects of the loss of biological diversity on the wintering waterfowl community. *Biol. Conserv.* **1994**, *70*, 39–47. [CrossRef]

164. Wheeler, B.D. Integrating wildlife with commercial uses. In *Reedbeds for Wildlife*; Ward, D., Ed.; The Royal Society for the Protection of Birds: Oxford, UK; University of Bristol Information Press: Oxford, UK, 1992; pp. 79–89.

165. van Deursen, E.J.M.; Drost, H.J. Defoliation and treading by cattle of reed *Phragmites australis*. *J. Appl. Ecol.* **1990**, *27*, 284–297. [CrossRef]

166. Poulin, B.; Duborper, E.; Lefebvre, G. Spring stopover of the globally threatened aquatic warbler *Acrocephalus paludicola* in Mediterranean France. *Ardeola* **2010**, *57*, 167–173.

167. Hejný, S.; Husák, Š. Effect of fishpond management on the littoral communities. Exploitation of reed. In *Pond Littoral Ecosystems. Structure and Functioning*; Dykyjová, D., Květ, J., Eds.; Springer: Berlin/Heidelberg, Germany; New York, NY, USA, 1978; pp. 397–415.

168. Kubů, F.; Květ, J.; Hejný, S. Fishpond management (Czechoslovakia). In *Wetlands and Shallow Continental Water Bodies*; Patten, B.C., Jorgeusen, S.E., Dumont, H.J., Gopal, B., Koryavov, P., Kvet, J., Loftier, H., Sverizhev, Y., Tundisi, J.G., Eds.; SPB Academic Publishing: The Hague, The Netherlands, 1994; Volume 2, pp. 391–404.

169. Bernard, C. L'étang, l'homme et l'oiseau Incidences des modes de gestion des étangs piscicoles sur les ceintures de végétation et l'avifaune nicheuse en Sologne, Brenne, Bresse, Territoire de Belfort et Champagne humide. Ph.D. Thesis, Ecole Normale Supérieure des Lettres et Sciences Humaines de Lyon, Lyon, France, 2018.

170. Broyer, J.; Curtet, L. The influence of fish farming intensification on taxonomic richness and biomass density of macrophyte-dwelling invertebrates in French fishponds. *Knowl. Manag. Aquat. Ecosyst.* **2011**, *400*, 10. [CrossRef]

171. Janda, J.; Květ, J. Measuring ecological change in Czechoslovak wetlands. In *Waterfowl and Wetland Conservation in the 1990s: A Global Perspective, Proceedings of an IWRB Symposium, St Petersburg Beach, FL, USA, 12–19 November 1992*; Moser, M., Prentice, R.C., van Vessen, J., Eds.; IWRB: Slimbridge, UK, 1993; pp. 77–82.

172. Joosten, H.; Clarke, D. *Wise Use of Mires and Peatlands*; International Mire Conservation Group: Kiel, Germany; International Peat Society: Jyväskylä, Finland, 2002; Volume 304.

173. Pfadenhauer, J.; Sliva, J.; Marzelli, M. *Renaturierungvon Landwirtschaftlich Genutzten Niedermooren Undabgetorften Hochmooren*; Schriftenreihe Bayerisches Landes-amt für Umweltschutz 148: Augsburg, Germany, 2000.

174. Hennicke, F. Das Naturschutzgroßprojekt Peenetal-Landschaft. In *Landschaftsökologische Moorkunde*, 2nd ed.; Succow, M., Joosten, H., Eds.; Schweizerbart: Stuttgart, Germany, 2001; pp. 487–492.

175. Timmermann, T.; Margóczi, K.; Takács, G.; Vegelin, K. Restoration of peat-forming vegetation by rewetting species-poor fen grasslands. *Appl. Veg. Sci.* **2006**, *9*, 241–250. [CrossRef]

176. Toet, S.; Van Logtestijn, R.S.P.; Schreijer, M.; Kampf, R.; Verhoeven, J.T.A. The functioning of a wetland system used for polishing effluent from a sewage treatment plant. *Ecol. Eng.* **2005**, *25*, 101–124. [CrossRef]

177. Meerburg, B.G.; Vereijken, P.H.; de Visser, W.; Verhagen, J.; Korevaar, H.; Querner, E.P.; de Blaeij, A.T.; van der Werf, A. Surface water sanitation and biomass production in a large constructed wetland in the Netherlands. *Wetl. Ecol. Manag.* **2010**, *18*, 463–470. [CrossRef]

178. Moreno-Mateos, D.; Pedrocchi, C.; Comín, F.A. Effects of wetland construction on water quality in a semi-arid catchment degraded by intensive agricultural use. *Ecol. Eng.* **2010**, *36*, 631–639. [CrossRef]

179. Vymazal, J.; Brix, H.; Cooper, P.F.; Green, M.B.; Haberl, R. (Eds.) *Constructed Wetlands for Wastewater Treatment in Europe*; Backhuys Publishers: Leiden, The Netherlands, 1998.

180. Brix, H. Do macrophytes play a role in constructed treatment wetlands? *Water Sci. Technol.* **1997**, *35*, 11–17. [CrossRef]

181. Cooper, P. What can we learn from old wetlands? Lessons that have been learned and some that may have been forgotten over the past 20 years. *Desalination* **2009**, *24*, 1–26. [CrossRef]

182. Picek, T.; Čížková, H.; Dušek, J. Greenhouse gas emissions from a constructed wetland—Plants as important sources of carbon. *Ecol. Eng.* **2007**, *31*, 98–106. [CrossRef]

183. Zhai, X.; Piwpuan, N.; Arias, C.A.; Headley, T.; Brix, H. Can root exudates from emergent wetland plants fuel denitrification in subsurface flow constructed wetland systems? *Ecol. Eng.* **2013**, *61*, 555–563. [CrossRef]

184. Brix, H.; Schierup, H.-H. Soil oxygenation in constructed reed beds: The role of macrophyte and soil-atmospfere interface oxygen transport. In *Constructed Wetlands in Water Pollution Control*; Cooper, P.F., Findlater, B.C., Eds.; Pergamon Press: Oxford, UK, 1990; pp. 53–66.

185. Armstrong, W.; Armstrong, J.; Beckett, P.M. Measurement and modelling of oxygen release from roots of *Phragmites australis*. In *Constructed Wetlands in Water Pollution Control*; Cooper, P.F., Findlater, B.C., Eds.; Pergamon Press: Oxford, UK, 1990; pp. 41–51.

186. Tyroller, L.; Rousseau, D.P.L.; Santa, S.; García, J. Application of the gas tracer method for measuring oxygen transfer rates in subsurface flow constructed wetlands. *Water Res.* **2010**, *44*, 4217–4225. [CrossRef]

187. Cooper, P. A review of the design and performance of vertical-flow and hybrid reed bed treatment systems. *Water Sci. Technol.* **1999**, *40*, 1–9. [CrossRef]

188. Cooper, P. The performance of vertical flow constructed wetland systems with special reference to the significance of oxygen transfer and hydraulic loading rates. *Water Sci. Technol.* **2005**, *51*, 81–90. [CrossRef]

189. Vymazal, J.; Kröpfelová, L. A three-stage experimental constructed wetland for treatment of domestic sewage: First two years of operation. *Ecol. Eng.* **2011**, *37*, 90–98. [CrossRef]

190. Borin, M.; Tocchetto, D. Five year water and nitrogen balance for a constructed surface flow wetland treating agricultural drainage waters. *Sci. Total Environ.* **2007**, *380*, 38–47. [CrossRef]

191. García-Lledó, A.; Ruiz-Rueda, O.; Vilar-Sanz, A.; Sala, L.; Bañeras, L. Nitrogen removal efficiencies in a free water surface constructed wetland in relation to plant coverage. *Ecol. Eng.* **2011**, *37*, 678–684. [CrossRef]

192. Cloern, J.E. Our evolving conceptual model of the coastal eutrophication problem. *Mar. Ecol. Prog. Ser.* **2001**, *210*, 223–253. [CrossRef]

193. Conley, D.J.; Paerl, H.W.; Howarth, R.W.; Boesch, D.F.; Seitzinger, S.P.; Havens, K.E.; Lancelot, C.; Likens, G.E. Controlling eutrophication: Nitrogen and phosphorus. *Science* **2009**, *323*, 1014–1015. [CrossRef] [PubMed]

194. Dykyjová, D. Nutrient uptake by littoral communities of helophytes. In *Pond Littoral Ecosystems. Structure and Functioning*; Dykyjová, D., Květ, J., Eds.; Ecological Studies 28; Springer: Berlin/Heidelberg, Germany; New York, NY, USA, 1978; pp. 257–291.

195. Tátrai, I.; Mátyás, K.; Korponai, J.; Paulovits, G.; Pomogyi, P. The role of the Kis-Balaton Water Protection System in the control of water quality of Lake Balaton. *Ecol. Eng.* **2000**, *16*, 73–78. [CrossRef]

196. Bousquet, F.; Le Page, C. Multi-agent simulations and ecosystem management: A review. *Ecol. Model.* **2004**, *176*, 313–332. [CrossRef]

197. Mathevet, R.; Bousquet, F. *Résilience & Environnement, Penser les Changements Socio-Écologiques*; Buchet-Chastel: Paris, France, 2014.

198. Mathevet, R.; Mauchamp, A.; Lifran, R.; Poulin, B.; Lefebvre, G. ReedSim: Simulating ecological and economical dynamics of Mediterranean reedbeds. In *Integrative Modelling of Biophysical, Social and Economic Systems for Resource Management Solution*; Post, D., Ed.; Modelling and Simulation Society of Australia and New Zealand: Townsville, Australia, 2003; pp. 1007–1012.

199. Poulin, B.; Mathevet, R.; Le Page, C.; Etienne, M.; Lefebvre, G.; Poulin, B.; Gigot, G.; Proréol, S.; Mauchamp, A. BUTORSTAR: A role-playing game for collective awareness of wise reedbed use. *Simul. Gaming* **2007**, *38*, 233–262.

200. Čížková, H.; Květ, J.; Comín, F.A.; Laiho, R.; Pokorný, J.; Pithart, D. Actual state of European wetlands and their possible future in the context of global climate change. *Aquat. Sci.* **2013**, *75*, 3–26. [CrossRef]

201. Lefebvre, G.; Redmond, L.; Germain, C.; Palazzi, E.; Terzago, S.; Willm, L.; Poulin, B. Predicting the vulnerability of seasonally-flooded wetlands to climate change across the Mediterranean Basin. *Sci. Total Environ.* **2019**, *692*, 546–555. [CrossRef]
202. Convention on Wetlands. Available online: www.ramsar.org (accessed on 10 April 2023).
203. Tanneberg, F.; Wichtmann, W. (Eds.) *Carbon Credits from Peastland Rewetting. Climate–Biodiversity–Land Use*; Schweitzerbart Science Publishers: Stuttgart, Austria, 2011.

Review

Ventilation Systems in Wetland Plant Species

Lars Olof Björn [1], Beth A. Middleton [2], Mateja Germ [3,*] and Alenka Gaberščik [3]

[1] Molecular Cell Biology, University of Lund, SE-223 62 Lund, Sweden; lars_olof.bjorn@biol.lu.se
[2] U.S. Geological Survey, Wetland and Aquatic Research Center, 700 Cajundome Boulevard, Lafayette, LA 70506, USA; middletonb@usgs.gov
[3] Biotechnical Faculty, University of Ljubljana, Jamnikarjeva 101, SI-1000 Ljubljana, Slovenia; alenka.gaberscik@bf.uni-lj.si
*   Correspondence: mateja.germ@bf.uni-lj.si

Abstract: Molecular oxygen and carbon dioxide may be limited for aquatic plants, but they have various mechanisms for acquiring these gases from the atmosphere, soil, or metabolic processes. The most common adaptations of aquatic plants involve various aerenchymatic structures, which occur in various organs, and enable the throughflow of gases. These gases can be transferred in emergent plants by molecular diffusion, pressurized gas flow, and Venturi-induced convection. In submerged species, the direct exchange of gases between submerged above-ground tissues and water occurs, as well as the transfer of gases via aerenchyma. Photosynthetic O_2 streams to the rhizosphere, while soil CO_2 streams towards leaves where it may be used for photosynthesis. In floating-leaved plants anchored in the anoxic sediment, two strategies have developed. In water lilies, air enters through the stomata of young leaves, and streams through channels towards rhizomes and roots, and back through older leaves, while in lotus, two-way flow in separate air canals in the petioles occurs. In *Nypa* Steck palm, aeration takes place via leaf bases with lenticels. Mangroves solve the problem of oxygen shortage with root structures such as pneumatophores, knee roots, and stilt roots. Some grasses have layers of air on hydrophobic leaf surfaces, which can improve the exchange of gases during submergence. Air spaces in wetland species also facilitate the release of greenhouse gases, with CH_4 and N_2O released from anoxic soil, which has important implications for global warming.

Keywords: metabolic gases; greenhouse gases; aerenchyma; anoxic soil

Citation: Björn, L.O.; Middleton, B.A.; Germ, M.; Gaberščik, A. Ventilation Systems in Wetland Plant Species. *Diversity* **2022**, *14*, 517. https://doi.org/10.3390/d14070517

Academic Editor: Corrado Battisti

Received: 28 April 2022
Accepted: 14 June 2022
Published: 27 June 2022

1. Introduction

The aquatic environment holds special challenges for plant survival. The diffusion of gases in water is about 10^4-fold slower than in air, so that aquatic plants must perform photosynthesis in water, and maintain aerobic respiration in flooded conditions [1,2]. Herbaceous wetland plants differ significantly, according to the accessibility of gases for their metabolism regarding their position in the water column. Researchers define various functional groups, namely, (1) emergent macrophytes or helophytes that are rooted in water-saturated soil, with foliage extending into the air (e.g., *Typha latifolia, Phragmites australis*); (2) floating-leaved macrophytes that are living in water rooted in hypoxic or anoxic sediment, with leaves floating on the water surface (e.g., *Nuphar luteum, Nymphaea alba*); (3) submerged macrophytes that grow completely submerged under the water, with roots or rhizoids attached to the substrate (e.g., *Myriophyllum spicatum, Potamogeton crispus*); and (4) free-floating macrophytes that float on or under the water surface, and are usually not rooted in the sediment (e.g., *Ceratophyllum demersum*) [3]. In addition, wetlands also host many different woody plants that are permanently or occasionally rooted in water-saturated sediment [4]. These species, belonging to different groups, often possess adaptations to overcome oxygen and carbon dioxide deficiencies, in order to maintain optimal conditions for photosynthesis and respiration. Emergent and floating-leaved species have an advantage over submerged species because their above-ground parts are fully or partly exposed to air. Aerial leaves have stomata in their epidermis, which can be adjusted to optimize exposure of internal tissues to the atmosphere and the exchange of gases. Thus, aerial plant parts are well supplied with oxygen, but for roots and rhizomes anchored in water-saturated soils, oxygen for respiration can be limited. Therefore, efficient ventilation systems are crucial for their survival. Ventilation systems rely on a passive molecular diffusion process, on pressurized gas flow, or Venturi-induced convection [5]; however, in submerged plant tissue, the direct exchange of gases between these tissues and water also occurs [6]. In most aquatic species, ventilation is enabled by an extended system of air canals and intercellular spaces called aerenchyma, which develop in different plant organs from roots, to stems and leaves. [7,8]. Gases in aerenchyma can originate from the atmosphere, rhizosphere, or plant metabolism [9]. Laing [10] shows a strong relationship between the leaf area and the extent of changes of oxygen and carbon dioxide concentrations in aerenchyma during periods of illumination; thus, the contribution of metabolic gases may vary significantly among species.

Aquatic plants mainly form aerenchyma constitutively in different organs, namely, roots, leaves, and stems, while some amphibious and terrestrial plants produce aerenchyma in response to an oxygen shortage [7]. The presence of aerenchyma may differ among species. Independent of habitat, aerenchyma patterns are stable at the genus level, and the consistency of pattern is stronger in the roots than in the shoots [11]. In addition to the atmosphere, gases in aerenchyma can originate from the rhizosphere or plant metabolism [9]. The formation of aerenchyma may not depend on environmental conditions, or be induced by flooding [1]. Aerenchyma cells are formed lysigenously by programmed cell death, as is the case of rice roots; schizogenously by the expansion of intercellular spaces [11]; and expansigenously (secondary aerenchyma) by cell division or enlargement, without cell separation or death [12]. These enlarged spaces may develop either in primary tissues (primary aerenchyma), or in secondary tissues (secondary aerenchyma) [13]. According to Doležal et al. [14], lysigenous aerenchyma are mostly produced by submerged plants, schizogenous aerenchyma by terrestrial and perennial wetland plants, and expansigenous honeycomb aerenchyma by aquatic floating-leaved plants. The amount of intercellular spaces varies significantly among species. In aquatic species, these intercellular spaces contribute up to 60% of the leaf volume [15], while in mesophytes, their volumes range from 2–7% [16]. Thus, in non-tolerant species, flooding may result in the demise of the plant.

Beyond ventilation, aerenchyma cells have other important ecological functions, including acting to store gases and increasing their internal conductance to roots and shoots [7]. The transfer of oxygen to underground organs, via aerenchyma during soil flooding, may prevent the suffocation of plants. Oxygen can also be transferred from roots to the rhizosphere, via aerenchyma. This critically important oxygen to oxidize and detoxify toxic chemicals formed in sediments in environments with low redox potential [17,18], noting that a lack of oxygen is associated with reduced forms of sulfur, manganese, and iron that may reach toxic levels in the soil [6].

In wetland soils, gas concentrations of several gases, such as carbon dioxide and methane, exceed atmospheric concentrations. Thus, aerenchyma can also be a path for greenhouse gas emissions from the plant, as methane and nitrous dioxide are released via plants from waterlogged sediments to the atmosphere [19,20].

Some photosynthetic O_2 produced by submerged plants oxygenates the water column, while natant plants can prevent oxygen diffusion from the atmosphere to water [21–24]. Aerenchyma cells lend buoyancy and mechanical resistance to breakage, with a relatively small investment in biomass by aquatic plants [25].

The ventilation in wetland plants takes place via various plant structures, and is enabled by the presence of aerenchyma in these structures. The source of gases and influx and efflux locations may differ significantly among different species and plant groups. In this review, we summarize the outcomes of research, and point out the similarity, diversity, and functional features of ventilation systems in various functional groups of wetland plants, and point out their potential effects on the wider environment.

2. Ventilation Mechanisms in Various Wetland Plant Groups

2.1. Submerged Species

In submerged aquatic species, the ventilation system is especially important since these have no direct connection with the atmosphere [26]. Extensive aerenchyma in the stems of all submerged plants that may be lysigenous and schizogenous (Figure 1) enable buoyancy for their flexible stems in water; however, this may limit the distribution of aquatic vascular plants to the depth where hydrostatic pressure does not compress stems [27]. Gases in submerged species come from plant metabolism, water, and sediment. The aerenchyma function as a reservoir for metabolic gases. Hartman and Brown [28], studying gas dynamics in *Elodea canadensis* Michx. and *Ceratophyllum demersum* L., detect a lag between peak values for dissolved oxygen in the surrounding water, and oxygen in the internal atmosphere. This situation is also true of the representatives of the genus *Lobelia* L., *Lilaeopsis* Greene, and *Vallisneria* L., which obtain more than 75% of their CO_2 needs from the sediment [29,30]. However, photosynthesis is important for internal O_2 status and aeration of anoxic sediments [31–33]. Sand-Jensen et al. [34] show that submerged macrophytes release oxygen from their roots during illumination, and at lower rates during darkness. The amounts of oxygen released varies among species from 0.04 to 3.12 µg O_2/mgDM/h; the species *Lobelia dortmanna* is the most effective. This involvement of photosynthesis enables aerobic metabolism of roots, sediment oxygenation due to radial oxygen loss, and improves nutrient uptake from the sediment [26].

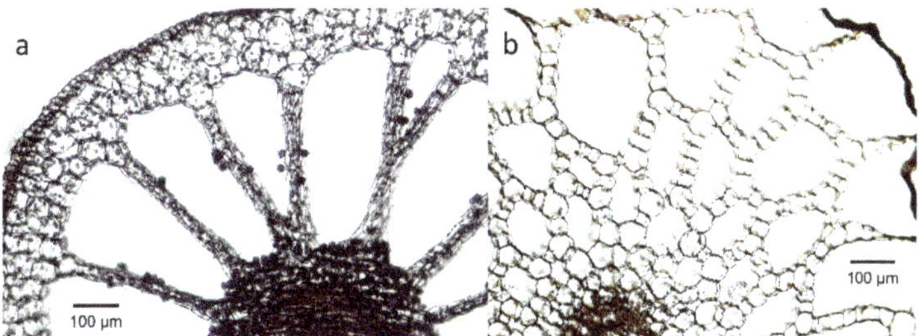

Figure 1. Transection of stems of two submerged species showing different types of aerenchyma; (**a**) lysigenous in *Myriophyllum spicatum* and (**b**) schizogenous in *Potamogeton crispus*. Photo: Matej Holcar.

A few outstanding representatives of submergent species include quillworts, *Isoëtes* L., which are the only surviving genus of a large plant group, distantly related to the clubmosses. This group emerged about 300 million years ago in ephemeral pools and oligotrophic lakes, and later radiated into terrestrial habitats [35]. Species in this group are well-adapted to cold environments, and limited light and carbon dioxide supply [36]. This genus represents the oldest lineage of plants with Crassulacean acid metabolism (CAM) [33].

At least some of species of the genus *Isoëtes* have air canals from the leaves through the periderm to the cortex aerenchyma in the stem [37]. Green [38] compares the air canals and aerenchyma in *Isoëtes* with the corresponding structures in Carboniferous–Permian arborescent lycopsids, such as *Lepidodendron* Stern. spp. The extinct lycopsid swamp trees called *Stigmaria* Stern. have similar air channels in their roots [39].

Most modern species of *Isoëtes* grow in wet environments, and use CAM to fix carbon. Due to the mineralization of organic matter, the CO_2 concentration in sediment is 10–250 times higher than atmospheric equilibrium [40]. Thus, these species uptake sedimentary CO_2 that is transferred via the large gas channels [38], which are present in various *Isoëtes* organs [41] (Figure 2). CO_2 leakage from gas spaces is prevented by the extremely high diffusional resistance of water. However, some *Isoëtes* species have an amphibious character, and may develop leaves with stomata and switch to C_3 photosynthesis if plant parts emerge into the air from the water [35]. In addition, Pedersen et al. [42,43] show that various isoetids (small, aquatic plants with thick, stiff leaves or stems that form basal rosettes [42]), seagrasses, and rosette-bearing wetland species may have leaves buried in sediments, and their achlorophyllous bases may function as an exit for O_2 to the sediment, and entrance for CO_2 from the sediment [42,43]. This system may also affect the concentrations of gases in aerenchyma and photosynthesis [44].

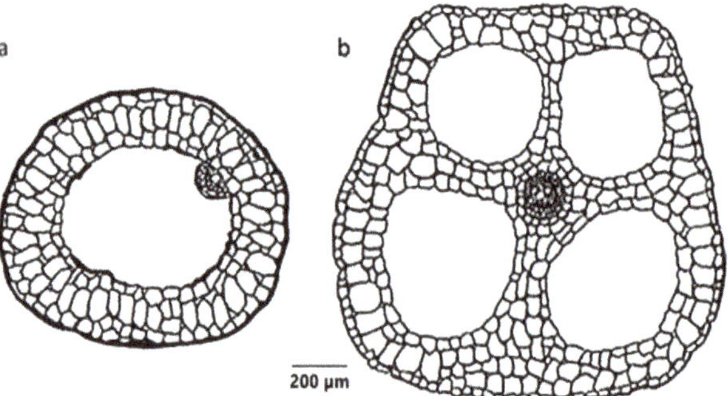

Figure 2. The root (**a**) and leaf (**b**) cross-section of *Isoëtes* L. sp. with large central air spaces. Drawings: Alenka Gaberščik.

2.2. Species with Natant Leaves

The advantage for these species is a direct connection with the atmosphere via natant leaves; however, since some of them grow in water as deep as 3 m [27,45], they need also effective aeration of the anoxic sediment.

Water lilies of the genus *Nymphaea* L. and *Nuphar* Sm. possess well-developed ventilation systems (Figure 3). Air enters through the stomata of young leaves that have just reached the water surface, streams through the channels of long petioles, through rhizomes and roots, and back to the external air through older leaves (Figure 4). This system accelerates O_2 flow from the atmosphere to the roots, and CO_2 and CH_4 flow from the roots to the atmosphere. This remarkable ventilation system has significant physiological and ecological consequences [46]. Laing [10] measures the daily dynamics of internal gas composition in various tissues of *Nuphar advena* (Aiton) W.T. Aiton, and finds a decrease in carbon dioxide and an increase in oxygen during the day, while the opposite is true at night. These findings show the important role of the metabolic process in influencing gas concentrations in various tissues.

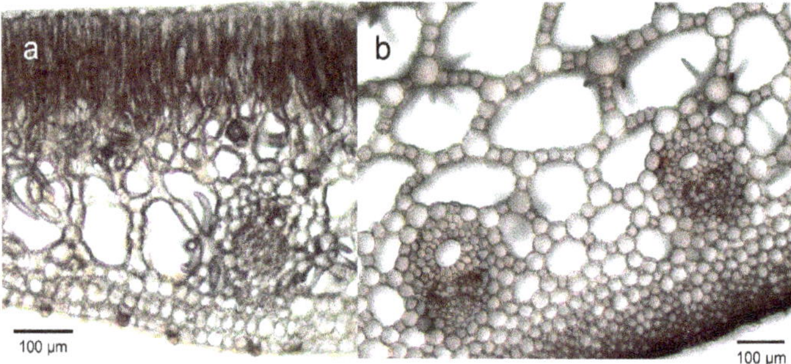

Figure 3. Transection of (**a**) natant leaf and (**b**) petiole of *Nuphar luteum*. Photo: Matej Holcar.

The ventilation is driven by the pressurization of the gas in air spaces of young natant leaves, due to thermo-osmosis of gases during cooling by transpiration [45,46]. Cooling by transpiration of the upper leaf surface enables a steeper temperature gradient within the leaf, intensifying internal aeration [47]. The leaves are warmed by the sun during the day, and by water in the surrounding environment at night, which enables ventilation at night [48]. This process is also supported by specific morphological features that may differ among species [49]. One important feature is that gas transport through a small opening varies with the dimensions of the opening, depending on whether the gas transport takes place by diffusion or by mass flow. The rate of gas transport by diffusion is proportional to the diameter (or some other linear dimension if the opening or aperture is not circular), and to the difference in concentration of the gas on the two sides of the opening, with the slower movement of heavy vs. light gas molecules.

The rate of mass flow follows Poiseuille's law. The flow rate is proportional to the total pressure gradient between younger and older leaves [5], and the diameter (or corresponding measure) is raised to the fourth power. Therefore, diffusion often dominates in plant parts with small openings, and mass flow when the openings are large [46]. The openings in young leaves are narrow. According to Grosse and Schröder [50], the important openings, in this case, are not the stomata ($\approx$5.6 µm $\times$ 2.4 µm), but narrow passages (intercellulars) ($\approx$1 µm) between cells inside the leaves. The oxygen concentration outside the leaves is higher than in the airspaces inside the leaves. This situation is partly because the air inside is diluted by water vapor, and is almost saturated in the intercellular spaces. Therefore, oxygen diffuses into the young leaves, and the total gas pressure there exceeds that of the external air. As the passages to the higher-oxygen external air are so narrow, little air escapes from the young leaves to the external atmosphere. It is easier for air to flow through the wider "pipes" in the petioles down to the rhizomes and roots, where oxygen is consumed by respiration. The rest of the air streams out to the atmosphere through the older leaves, where the passages are wider than in the younger leaves.

The streaming of air in the yellow water lily, *Nuphar luteum* (L.) Sm. is described not only by Schröder et al. [51], but also in a series of articles by John Dacey and coworkers, and the results are summarized by Dacey [46]. Richards et al. [45] perform similar studies on *Nymphaea odorata* Aiton. In the case of *N. luteum*, the network of internal gas spaces enables a pressurized flow-through system, that directs air rich with oxygen from young leaves to the underground organs, and with air rich in carbon dioxide back via the older emergent leaves to the atmosphere [46]. Konnerup et al. [52] compare pressure differences and convective gas flow in twenty species of tropical angiosperms (Table 1). The highest flow rates are found for *Nymphaea rubra* Roxb. and *Nelumbo nucifera* Gaertn.. Some species achieving large pressure differences have low flow rates. The two *Eleocharis* R. Br. species differ remarkably with respect to the pressure differences produced.

Table 1. Comparison of pressure differences and convective gas flow in twenty species of tropical angiosperms (adapted from Konnerup et al. [52]).

Plant	ΔP(Pa)	Air Flow (mL/min)	Plant	ΔP(Pa)	Air Flow (mL/min)
Monocotyledons			**Dicotyledons**		
Cyperaceae			*Nymphaeaceae*		
Cyperus compactus Retz.	20	0.60	*Nymphaea rubra* Roxb.	236	140
Cyperus digitatus Roxb.	14	1.29	*Nymphaea nouchali* Burm. f.	116	15.2
Eleocharis dulcis (Burm. f.) Trin. ex Hensch	628	11.9	*Nelumbonaceae*		
Eleocharis acutangula Schultes	15	0.10	*Nelumbo nucifera* Gaertn.	295	288
Scirpus grossus L. f.	3	0.22	*Menyanthaceae*		
Scirpus littoralis Shrad.	83	0.39	*Nymphoides indica* (L.) Kuntze	485	36
Scleria poaeformis Retz.	22	1.11	*Convolvulaceae*		
Poaceae			*Ipomoea aquatica* Forssk.	3	0.18
Phragmites vallatoria (L.) Veldkamp	482	1.59			
Urochloa mutica (Forsk T.Q. Nguyen	11	0.09			
Hymenachne acutigluma (Steud.) Gilliland	141	0.55			

Table 1. *Cont.*

Plant	ΔP(Pa)	Air Flow (mL/min)	Plant	ΔP(Pa)	Air Flow (mL/min)
Monocotyledons			Dicotyledons		
Oryza rufipogon Griff.	23	0.32			
Leersia hexandra Swartz.	62	0.15			
Pontederiaceae					
Eichhornia crassipes (Mart.) Solms	8	0.12			
Araceae					
Colocasia esculenta (L.) Schott	3	0.10			i
Limnocharitaceae					
Limnocharis flava (L.) Buchenau	6	0.81			

Lotus (*N. nucifera*) is not closely related to water lilies, but has a similar growth form, and ventilates its organs in oxygen-deficient sediment by streaming air. The thermo-osmotic gas transport depends on a temperature differential between the air in the aerenchyma of the leaves and the surrounding atmosphere [53]. Stomata located in the central parts of leaves have an important function in the active regulation of airflow [54]. These central stomata are three times larger than stomata of the other part of the leaf lamina, and control the gas release by opening and closing [55]. Lotus differs from other groups because each petiole has separate channels for descending and ascending air [53,56]. Air is, thus, transferred from leaves, down through petioles, rhizomes, and then back to the atmosphere through large stomata [54] (Figure 4).

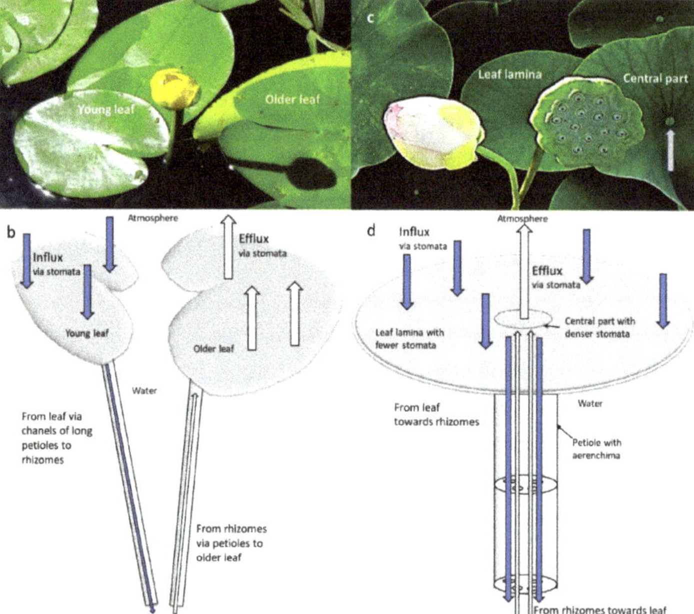

Figure 4. Ventilation system in *Nuphar luteum* (**a,b**) *Nelumbo nucifera* (**b,c**). (**a**) *N. luteum* young and older leaves; (**b**) the scheme of ventilation system where arrows shows that air from atmosphere enters via young leaves through stems with aerenchyma to rhizomes, and via older leaves back to the atmosphere; (**c**) *N. nucifera* leaves with visible central part with denser stomata; (**d**) the scheme of ventilation system with arrows indicating the direction of flow (blue—influx from atmosphere via stomata in leaf lamina through aerenchyma in leaf and petiole to aerenchyma in rhizomes; light grey—efflux from rhizomes through aerenchyma in petiole to the central part of the lamina, and finally to atmosphere). Photos and drawings: Alenka Gaberščik.

Polygonum amphibium L. is an amphibious plant with natant leaves. and thrives in water up to 2 m, with its hollow stems that enable the aeration of underground organs (Figure 5). However, gases trapped in the aerenchyma may also benefit the photosynthesis and lower photorespiration rate [57], due to changes in the ratio of internal CO_2/O_2 concentrations.

Figure 5. *Polygonum amphibium* L. f. natans; (**a**) natant leaves and (**b**) section of a hollow stem. Photo: Matej Holcar.

2.3. Helophytes

Helophytes live anchored in water-saturated sediment, while their above-ground organs are usually in air. Most of them have well-developed aerenchyma [1]. When leaves of different helophytes are submerged, they produce gas films at the leaf surface to improve photosynthesis. Colmer and Pedersen [58] show that *Phalaris arundinacea, Phragmites australis,* and *Typha latifolia* form gas films on both leaf sides, *Glyceria maxima* form gas films on the adaxial only, and *Acorus calamus* and *S. emersum* do not form gas films.

Rhizomes and stems of wetland representatives of horsetail (*Equisetum*) have large canals [59]. Internal ventilation systems differ among *Equisetum* species, and are best developed in the great horsetail, *E. telmateia* Ehrh. In this species, an airflow of 120 mL/min and a wind speed of 10 cm/s were measured [60], and this flow may be evaporation-driven. Air enters via stomata on branches, passes to substomatal cavities, and then via intercellulars to aerenchyma channels in branches, to the main stem, rhizome, and cortical intercellular spaces of roots [61].

Methane release from *E. fluviatile* L. stands suggests that this species lacks a pressurized ventilation system because no diurnal changes are detected [59]. In addition, no or very low airflow is measured in this species, even though it thrives in deep water [62]. The power for the airflow comes from the evaporation of water inside the plant. If the air surrounding the plant is saturated with water vapor, the flow stops, and it is sped up by wind that evaporates water vapor from the plant surface [61]. For example, *E. palustre* L. maintains an inner mass flow of up to 13 mL/min (Figure 6). No flow is found in *E. sylvaticum* L. or *E. arvense* L., perhaps because these species lack air channels in the branches [60]. These species only rely on diffusion for oxygen provision to their below-ground parts, which is facilitated in environments of well-aerated soil.

Figure 6. The cross-section of marsh horsetail (*Equisetum palustre* L.) stem. Photo Matej Holcar.

Common reed and other *Phragmites* species frequently grow in water-saturated anoxic soil. *Phragmites* are ventilated by "compressed air" in a similar way to water lilies and lotus [63,64], but additional aeration may be obtained by suction from old broken reeds when the wind is blowing over them, a Venturi-effect [65], and also by oxygen from photosynthesis. The oxygen pressure in the rhizomes rises rapidly in the morning, peaks near midday, and declines slowly to a minimum at 6 am [66]. Three factors probably contribute to the midday peak including light (photosynthesis), temperature, and the difference in water vapor pressure between the external air and the intercellulars. External humidity is at its lowest in the early afternoon, when humidity in leaf intercellulars (due to insolation) is highest. Oxygen from roots is released in the nearby rhizosphere, which is known as radial oxygen loss [67], and oxidizes different toxic substances produced in anoxic soils that can harm soil biocenosis. This situation is also the case in *Phragmites australis* (Cav.) Trin. ex Steud, where the aeration system increases rhizosphere oxygenation and lowers toxic sulfide concentrations [68]. Some species prevent excessive oxygen loss from roots by forming a barrier in their epidermis, exodermis, or subepidermal layers [69].

Konnerup et al. [52] also find convective ventilation in *P. vallatoria* (L.) Veldkamp, and eight other wetland plant species. The only other monocotyledon species for which a similar pressure difference is found is *Eleocharis dulcis* (Burm. f.) Trin. ex Hensch (*Cyperaceae*), but the flow rate is lower.

In *Scirpus lacustris* (L.), sediment-derived CO_2 in aerenchyma (Figure 7) presents an important source of inorganic carbon used for photosynthesis in submerged green stems, especially before they reach the water surface [70]. In this species, leaves are usually reduced, however, leaf blades up to 100 cm long can develop under water [71].

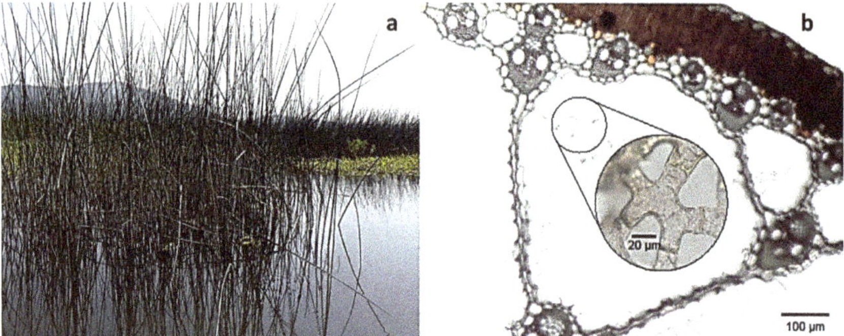

Figure 7. (**a**) Leafless *Schoenus lacustris* stems; (**b**) transection of stem with lacunae filled with thin-wall aerenchymal parenchyma. The magnification of parenchymatic cells is shown in the magnified circular insert. Photo: Matej Holcar.

In plants with reduced pressurized ventilation, the growth rate is diminished, possibly due to reduced mineral uptake and availability, a poorly developed root system, or impaired root function [63]. Studies in the USA reveal differences in function between native and non-native *P. australis* types. Ventilation efficiency is 300% higher per unit area for non-native types in comparison to native types, due to the increased oxidation of the rhizosphere [72].

The cattails (*Typha* L., Figure 8) are phylogenetically related to reeds. *Typhaceae* and reeds ventilate their below-ground parts in similar ways. An air stream of 8 mL/min per leaf is observed for *T. latifolia*, and 3.5 mL/min for *T. angustifolia* L. [73]; however, some leaves can function with influx or efflux in changing environments. A flux of 11 mL/min is observed, compared to a maximum of 60 mL/min through a water lily petiole [46]. Contrary to the situation in lotus, for *Typha*, the stomata constitute the pressure-generating pores [74]. The stomata are most effective for creating an inner pressure when they are partially closed, and provide almost no pressure when they are open to 0.3 µm wide. Air enters through middle-aged leaves, and exits through the oldest ones [73].

Figure 8. Cattail, *Typha* L. sp. Top left, shoots seen from below. Top right, a cross-section of a leaf with large air spaces, which are connected to air spaces throughout the plant. Bottom, rhizomes. Reproduced with permission from Bansal et al. [75].

Duarte et al. [76] claim that oxygen in the root area is not obtained exclusively from the atmosphere or photosynthesis in the leaves, but, in addition, oxygen in the root is supplemented by the decomposition of hydrogen peroxide by catalase. The study of Laing [10] reveals that the shading of *T. latifolia* has little effect on gas composition in air spaces of the submerged stolon; however, a slight decrease in oxygen concentration is observed in leaves. In hypoxic growth conditions, various *Typha* species behave in diverse ways; *T. angustifolia* increases its root porosity and root mass ratio, while *T. latifolia* increases its root diameter [77].

Rice (*Oryza sativa* L.) can grow in deep water ([78]). Under complete submergence, rice leaves and stems elongate significantly to reach the air–water interface; however, this may exhaust its energy reserves, and cause death if deep water persists for a long time [79]. Rice mitigates waterlogging stress by forming lysigenous aerenchyma and a barrier to prevent radial O_2 loss from roots, in order to supply O_2 to the root tip [80]. However, it differs from the species described above in the transport of air. Air is partly transported through a layer of air on the surface of the leaves. At least in the majority of rice cultivars, the leaves are water-repellent on both sides. A layer of air is formed on these leaves, up to 25 μm thick [81], and with a volume that is 44% of that of the leaf (Figure 9). Gas molecules in the thin layer are transported forward by diffusion at a greater pace than if they could move in all directions, but there may also be some mass flow. A layer of air at the leaf surface also contributes to the tolerance against submergence in the case of *Spartina anglica* C.E. Hubb. [44]. The floating leaves of the grass *Glyceria fluitans* (L.) R. Br. have a lower cuticle resistance for dark O_2 uptake on the wettable abaxial side, compared to the hydrophobic adaxial side [82].

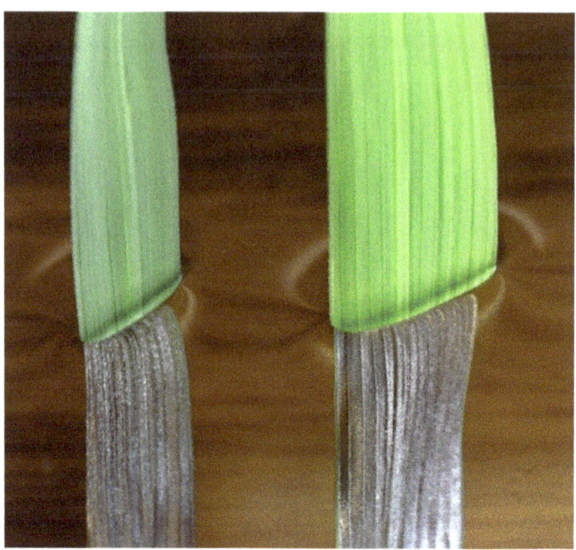

Figure 9. Rice (*Oryza longistaminata* A. Chev. & Roehr.) leaves protrude through the water surface, covered below the water surface by a layer of air. Photo by Ole Pedersen, University of Copenhagen. Cf. the cover of Science Vol. 228, No. 4697, 19 April 1985.

Rice also has large air spaces inside both leaves and roots, while the transition between shoot and root provides greater resistance to flow [83]. In rice, aerenchyma forms constitutively under aerobic conditions, while their formation is further induced under oxygen-deficient conditions [84]. Aerenchyma formation is promoted by ethylene production in adventitious roots [8] and in internodes, and this aerenchyma formation is mediated by reactive oxygen species (ROS) [85]. Little is known about the signaling and molecular regulatory mechanisms of ROS during plant-programmed cell death [86]. In the case of the fast-elongating 'Arborio Precoce' variety, ethylene controls aerenchyma formation, while in the variety 'FR13A', ROS accumulation plays an important role [87]. The study shows that the SNORKEL1 and SNORKEL2 genes that promote internode elongation are exclusively present in the genomes of deep-water rice [88].

Under O_2 deficiency, the porosity of adventitious roots enhances O_2 diffusion towards the root tip [89]. Flow takes place because the respiring plant replaces oxygen with carbon dioxide, which at pH = 7 and 25 °C has a solubility in water that is 140 times that of oxygen [78]. The study of Colmer and Pedersen [90] reveals that the dynamics in pO_2 within shoots and roots of submerged plants are related to periods of darkness and light, which supports the connection of underwater photosynthesis and internal aeration.

In *Pontederia cordata* L., the innermost layer of ground meristem in adventitious roots forms the endodermis and aerenchymatous cortex [91]. The comparison of pO_2 in submerged roots of *Rumex palustris* Sm. (flood-tolerant) and *R. acetosa* L. (flood-intolerant) reveals that roots of the former remain oxic, while in the latter, root pO_2 drops significantly (to 4.6 and 0.8 kPa, respectively) [92].

2.4. Mangrove Forest

Ferns of the genus *Acrostichum* L. (closely related to *Ceratopteris*) live rooted in mangrove vegetation and other settings [93]. All parts of these plants have large air spaces, including the roots. Aerenchyma are found in the roots of *Acrostichum* fossils, which suggests a coastal paleoenvironment [94]. Fonini et al. [95] also report large aerenchyma in the leaves of species *A. danaeifolium* Langsd. & Fisch., which is a feature endemic to mangrove.

A species of palm, *Nypa fruticans* Wurmb. or "snorkeling palm" adapted to life in mangrove forests; usually only the leaves appear above the water surface. These leaves eventually abscise when aging, but the leaf bases remain, and function as air inlets by developing a network of lenticels covering the leaf base [96]. These lenticels are connected with expansigenous aerenchyma [96]. The lenticels allow gases through, but prevent the entry of fluid water. The air spaces of the lenticels are connected to air spaces via the underwater stem to the roots. The first- and second-order roots have a small central stele surrounded by a wide zone of cortex with very spacious air channels, which develop after the separation and dissolution of cells. The mature crown may contain 6 to 8 living leaves, and 12 to 15 bulbous leaf bases at a time [97]. These leaf bases can function for up to 4 years after leaf abscission [96].

The main constituent of mangrove forests includes various types of mangrove species, which are not all closely related. Various mangrove species developed a variety of unique adaptations [98]. Most species possess structures, such as pneumatophores, knee roots, or stilt roots, which provide ventilation during low tides [99]. Studies by Scholander et al. [100] reveal that high oxygen pressure in the roots of *Rhizophora* L. is maintained via ventilation through the lenticels of the stilt roots. The cortex of *Rhizophora* roots have interconnected schizogenous intercellular spaces, extending to within 150 μm of the tip [101].

Pneumatophores act as "snorkels" from the roots of *Sonneratia apetala* Banks (Figure 10). Pneumatophores have numerous lenticels connected with aerenchyma; however, the pneumatophore development phase significantly affects the volume of gas spaces, since young pneumatophores have a less-developed lacunose cortex, in comparison to mature pneumatophores [98]. In *S. alba* Sm., root porosity in different root types ranges from 0 to 60%. Cable roots and pneumatophores have the highest ratio of aerenchyma per area (50–60%) [101]. The changes in cortex cells with developing gas spaces suggests schizogenous and lyzogenous formation of aerenchyma [102].

Figure 10. Pneumatophores of the mangrove tree *Sonneratia apetala* Buch. Ham. From Zhang et al. [103]. Creative Attribution license.

Comparing mangrove species belonging to various orders, Cheng et al. [104,105] find that the more air spaces the roots possesses, the greater the flooding tolerance. Curran et al. [106] conclude that *Avicennia marina* (Forsk.) Vierh. does not need airflow because the diffusion of oxygen satisfies the oxygen requirement. Oxygen enters the pneumatophores of this species through lenticels and "horizontal structures" on the pneumatophores [107]. The volume of gas spaces determined in the roots of this species is 40–50% [108].

2.5. Other Wetland Species

Baldcypress (*Taxodium distichum* (L.) Rich) is a conifer in the south–east of the USA, which can grow under both very wet and dry conditions. Growing in water, *T. distichum* develops "knees" emerging from the roots, which may be involved in oxygen transfer (Figure 11) and carbohydrate storage [109]. Under root and knee submergence, internal O_2 concentrations are significantly higher than during drawdown [110]. The knees do not possess gas conduits, so Rogers [109] concludes that knees serve to aerate the phloem in the inner bark for the oxygen requirements of the phloem, and downward conduction of oxygen dissolved in the phloem sap. Wang and Cao [111] report that flooding in *T. distichum* increases the porosity of the roots, stems, and leaves and, consequently, enhances O_2 diffusion to roots.

Figure 11. "Knees" emerging from the roots of baldcypress, *Taxodium distichum* (L.) Rich. in Arkansas, USA. From Middleton [112]. CC BY 4.0 license.

Some tree species in wetlands develop adventitious roots from the trunk during the rainy season, as an adaptation to the flooded environment. In *Syzygium kunstleri* (King) Bahadur and R.C. Gaur that is native to Borneo and Malaya, oxygen transportation occurs through aerenchyma in the root tips, periderm near the root base, and secondary aerenchyma between layers of phellem [113].

The roots of species colonizing flood forests and riparian zones are often subjected to water-saturated soil. Schröder [51] and Grosse and Schröder [4,50] find a thermo-osmotically driven gas flow in *Alnus glutinosa* (L.) Gaertn. from the external air, through the stems to the roots. A temperature difference of up to 3.6 degrees exists in other *Alnus* Mill. species between the intercellular spaces and the external air [114]. As the temperature difference increases from 0.6 to 3.6 °C, due to increasing light (from 50 to 200 μmol m^{-2} s^{-1}), the pressure difference between the external air and the intercellular spaces in the stem increases from 5 to 17 Pa. An efficient thermo-osmosis process needs a narrow passage, comparable in width to the mean free path lengths of the gas molecules (ca. 0.1 μm), between two compartments of different temperatures. However, some pressurization is possible with larger pores [115]. Details of thermo-osmosis of gases are provided by Denbigh [116], and Denbigh and Raumann [117]. In short, the pressures p1 and p2, which can develop in two compartments (designated 1 and 2), are determined by the equation:

$$\ln\left(\frac{p1}{p2}\right) = K*\left(\frac{1}{T1} - \frac{1}{T2}\right)$$

where T1 and T2 are the absolute temperatures of the two compartments, and K a constant.

The constrictions in the pathway that are needed for the development of thermo-osmotic gas flow in alder are related to the lenticels in the bark of the stem [118,119]. Gas flows from the colder external atmosphere to the warmer intercellular spaces.

In contrast, Armstrong and Armstrong [120] use *A. glutinosa* as an experimental species, and find that the thermo-osmotic gas flow is incompatible with the species plant anatomy, and so claim that roots are supplied with oxygen produced by chlorophyll-containing stems. The apparent discrepancy in the results of Armstrong and Armstrong vs. Grosse's group may be due to a difference in the age of the plants used by the two research groups, or to differences in submergence times. But the experiments by Dittert et al. [119] offer another explanation. They find that under experimental conditions, when the temperature of the trunk is kept above the temperature of the air, thermo-osmotic gas transport takes place, while under natural conditions (in Germany), there is only transport by diffusion, at all times of the year. Batzli and Dawson [121] find that in *Alnus rubra* Bong., lenticels develop (in this case on roots and root nodules) after 50 days of flooding.

3. Diversity of Ventilation Systems

Similar physical processes of ventilation occur in different taxonomical and functional plant groups that thrive in oxygen-deficient sediment; however, their morphological adaptations differ significantly (Table 2). Functional traits of these plants are not only the species' response to specific environmental conditions, but they also depend on their phylogeny. Jung et al. [11] find specific trends of aerenchyma patterns in several taxa of higher plants, and show that these patterns partially coincide with their phylogeny. The study of Cape reeds reveals that the presence of aerenchyma correlates with the eco-hydrological niche at the population and species level, indicating that waterlogging presents an environmental filter that excludes species without aerenchyma [122]. Bedoya and Madrinán [123], studying the evolution of the aquatic habit in genus *Ludwigia* L., find a convergence towards the absence of secondary growth in roots, smaller proportion of lignified tissue area in underground organs, and the presence of primary aerenchyma. However, there are also studies that are not consistent with these results. For example, the study of different *Carex* L. species in a phylogenetic context, with an even sampling across the different clades, shows that the size of the aerenchyma has only a weak relation to soil moisture [124]. *Carex* species with poorly developed aerenchyma have low performance in flooded soil, while partial submergence may even affect species with a larger amount of root aerenchyma [125].

Table 2. Ventilation mechanisms in different taxonomical and functional plant groups that thrive in oxygen-deficient sediment.

Plant Group	Taxonomic Group	Source of Gasses	Ventilation Principle	Special Features	Reference
Submerged	Isoetids	Water, metabolism, Sediment	Diffusion, aeration of rhizosphere via buried leaves	Aerenchyma, CAM	[38,44]
	Angiosperms	Water, metabolism, Sediment	Diffusion	Metabolic gasses trapped in aerenchyma	[29,31]

Table 2. *Cont.*

Plant Group	Taxonomic Group	Source of Gasses	Ventilation Principle	Special Features	Reference
Floating	*Nuphar* spp., *Nymphaea* spp.	Air, metabolism, Sediment	Pressurized ventilation, thermo-osmotic gas transport,	'Heat pump' drives gasses from the atmosphere via young natant leaves, petioles to roots and back, via older leaves to the atmosphere	e.g., [45,48,51]
	Nelumbo nucifera	Air, metabolism, Sediment	Pressurized ventilation, influx via laminal stomata of natant leaves through aerenchyma to rhizomes; back from rhizomes through aerenchyma in petiole through stomata in leaf central part	Leaf lamina with fewer and smaller stomata, leaf central part with larger and denser stomata, which actively regulate the airflow by opening and closing	[53–55]
Helophytes	*Equisetum* spp.—4 out of 9 have through-flow convection	Air, possibly also metabolism, sediment	Pressurized ventilation, humidity-induced diffusion,	Air moves through stomata through branches, via interconnecting aerenchyma channels in stem and rhizomes, with venting through the previous year's stubble or damaged shoot.	[60,61]
	Phragmites spp.	Air, possibly also metabolism, sediment	Pressurized ventilation, suction via old broken stems (Venturi-effect), air films on leaves when submerged	Via leaves, stems to root system, partly to sediment (ROL), and back to stems, leaves, and atmosphere	[58,63–65,126]
	Typha spp.	Air, sediment, possibly metabolism, oxygen in the rhizosphere may be obtained from the decomposition of hydrogen peroxide by catalase	Pressurized ventilation, leaf stomata create inner pressure, air films on leaves when submerged	Air enters through middle-aged leaves, and exits through the oldest ones	[58,73,74,76]
	Oryza sativa	Air films on leaves when submerged	Flow from above-ground parts via roots by diffusion, and possibly also by mass flow	Water-repellent leaf surface; air layer up to 25 μm, large air spaces inside leaves and roots, the porosity of adventitious roots, a barrier in roots to prevent radial O_2 loss from roots	[81,83,89]
Species of mangrove forest	*Acrostichum* spp.			All plant parts have large air spaces	[94,95]
	Nypa fruticans		Bases of abscised leaves function as air inlets, by developing a network of lenticels covering the leaf base connected to aerenchyma	"snorkeling palm" leaf bases function up to 4 years after leaf abscission	[96]
	Mangroves		High oxygen pressure in the roots is maintained via ventilation through the lenticels on different root structures connected with aerenchyma	Special structures, i.e., pneumatophores, knee roots, stilt roots, or plant roots, provide ventilation during low tides	[99]

Table 2. *Cont.*

Plant Group	Taxonomic Group	Source of Gasses	Ventilation Principle	Special Features	Reference
Other wetland species	*Alnus* spp.	Thermo-osmotically driven gas flow	In *Alnus glutinosa*, the flow is from the external atmosphere through the stems to the roots	Thermo-osmotic flow in alder is related to the lenticels in the bark of the stem, stem photosynthesis	[4,51,120,127]
	Taxodium distichum		"knees" emerging from the roots to the surface of the water, flooding increases the porosity of roots, stems, and leaves, and enhanced O_2 diffusion to roots.	Snorkeling	[109]
	Syzygium kunstleri		Oxygen transportation occurs through aerenchyma in the root tips, periderm near the root base, and secondary aerenchyma between layers of phellem		[113]

Most species use aerenchyma as a ventilation path, and as a reservoir for gases originating from the atmosphere, soil, and metabolic processes (e.g., respiration, photorespiration, and photosynthesis). The differences among groups are related either to specific environmental conditions (e.g., emergence, submergence), to specific species anatomy, as is the case for the lotus, or to metabolic processes and properties related to their growth form (e.g., emergent, submerged, floating-leaved, rosette, woody plants).

4. Plant Role in Emission of Greenhouse Gases

Ventilation systems are also important in the emission of greenhouse gases. We concentrate on methane (CH_4), although nitrous oxide (N_2O) is also important. Although the global warming potential (GWP) of methane is much greater than that of nitrous oxide over brief time intervals, nitrous oxide has about twice the warming potential of methane at a time horizon of 100 years (IPPC 2013), because it persists longer in the atmosphere. This section focuses on the role of plants rooted in water-saturated soils in the emission of greenhouse gases. The production of CH_4 in wetlands depends on carbon availability, soil redox potential, availability of O_2, pH, and temperature, and it is released from wetland soils by ebullition, convection, diffusion, and ventilation [128,129]. Thus, CH_4 emissions depend on the plant species present, and their abundance [130–132].

A key role in greenhouse gas emission is also played by different growth forms of plants. On one hand, Kosten et al. [133] find that free-floating plants may reduce methane emissions. On the other hand, some floating-leaved species contribute significantly to methane emission, including *N. odorata* [20], *N. nucifera* [19], and *N. luteum* L. [20], and helophytes such as *Juncus effusus* L. [134], *Spartina anglica* and *P. australis* [135,136], *P. mauritianus* Kunth [137], *T. latifolia* [137], *Typha* spp. [20,130], and *Cyperus papyrus* L. [137,138]. In areas colonized by *P. australis* and *S. lacustris*, methane is emitted exclusively by plant-mediated transport, with the highest emission rates at daytime, and emission peaks following sunrise [139].

The effects of aquatic plants on methane emission are complex [140]. Their dead and decaying remains provide the raw material for methane production, and their aerenchyma and gas canals facilitate the transport of methane to the atmosphere. At the same time, the aeration of their rhizospheres, and their methane-oxidizing microorganisms, decrease the amount of methane that escapes into the atmosphere [141]. In rice, two mutants with decreased root aerenchyma cause 70% less methane emission to the atmosphere than the wild type, while the oxygen transport is only reduced by 50% [142]. However, it is unlikely that such mutants will be a practical solution to the very high contribution of rice paddies to the greenhouse gas budget.

Wetland trees contribute to methane emission, while rainforest trees may function as sinks [143], except when flooded [144,145]. In *Phragmites* spp. [136], pressurized gas flow is important in emission; in the case of these species, the population of methane-producing microbes in the rhizosphere may be the most crucial factor. The relative importance of these factors is discussed by Laanbroek [146].

5. Conclusions

Ventilation in different groups of wetland plants that thrive in oxygen-deficient soils is related to aerenchyma as a ventilation path; however, the mechanisms may differ significantly. This diversity is a consequence of (1) phylogeny, as waterlogging presents an environmental filter that excludes certain species; (2) specific environmental conditions in plant/species habitat; (3) specific morphological adaptations at plant/species level (as is the case for water lily and lotus); (4) growth form (e.g., emergent, submerged, floating-leaved, rosette, woody plants) is related to plant medium (air, water, or both); and (5) metabolic processes (photosynthesis, respiration, photorespiration) that may act as a source or a sink of gases. These diverse structures and processes are related to ventilation support, photosynthesis, and respiration processes. Along with plant ventilation, they contribute to the oxygenation of otherwise anoxic wetland soils by radial oxygen loss from roots to the rhizosphere, and, thus, its detoxification, and also facilitate the release of greenhouse gases from these soils, which sets off a cascade of environmental consequences.

Author Contributions: Conceptualization, L.O.B. and A.G. writing—original draft preparation, L.O.B., B.A.M., M.G. and A.G., writing—review and editing, L.O.B., B.A.M., M.G. and A.G.; visualization, L.O.B., A.G. and B.A.M.; supervision, L.O.B., B.A.M. and A.G.; funding acquisition, A.G. and B.A.M. All authors have read and agreed to the published version of the manuscript.

Funding: This study was supported by the Slovenian Research Agency, through the Plant Biology program (P1-0212), and the U.S. Geological Survey Ecosystems Mission Area.

Institutional Review Board Statement: Not applicable.

Informed Consent Statement: Not applicable.

Data Availability Statement: Not applicable.

Conflicts of Interest: The authors declare no conflict of interest.

References

1. Jackson, M.B. Ethylene and Responses of Plants to Soil Waterlogging and Submergence. *Annu. Rev. Plant Physiol.* **1985**, *36*, 145–174. [CrossRef]
2. Voesenek, L.A.C.J.; Colmer, T.D.; Pierik, R.; Millenaar, F.F.; Peeters, A.J.M. How Plants Cope with Complete Submergence. *New Phytol.* **2006**, *170*, 213–226. [CrossRef] [PubMed]
3. Janauer, G.; Germ, M.; Kuhar, U.; Gaberščik, A. Ecology of Aquatic Macrophytes and Their Role in Running Waters. In *Macrophytes of the River Danube Basin*; Janauer, G., Gaberščik, A., Květ, J., Germ, M., Exler, N., Eds.; Academia, Živá Příroda: Prague, Czech Republic, 2018; pp. 13–33. ISBN 9788020027436.
4. Große, W.; Schröder, P. Oxygen Supply of Roots by Gas Transport in Alder-Trees. *Z. Für Nat. C* **1984**, *39*, 1186–1188. [CrossRef]
5. Colmer, T.D. Long-Distance Transport of Gases in Plants: A Perspective on Internal Aeration and Radial Oxygen Loss from Roots. *Plant Cell Environ.* **2003**, *26*, 17–36. [CrossRef]
6. Mitsch, W.J.; Gosselink, J.G. *Wetlands: Human History, Use, and Science*, 4th ed.; Wiley: Hoboken, NJ, USA, 2007; ISBN 9780471699675.
7. Jackson, M.B.; Armstrong, W. Formation of Aerenchyma and the Processes of Plant Ventilation in Relation to Soil Flooding and Submergence. *Plant Biol.* **1999**, *1*, 274–287. [CrossRef]
8. Justin, S.H.F.W.; Armstrong, W. The Anatomical Characteristics of Roots and Plant Response to Soil Flooding. *New Phytol.* **1987**, *106*, 465–495. [CrossRef]
9. Evans, D.E. Aerenchyma Formation. *New Phytol.* **2004**, *161*, 35–49. [CrossRef]
10. Laing, H.E. The Composition of the Internal Atmosphere of Nuphar Advenum and Other Water Plants. *Am. J. Bot.* **1940**, *27*, 861–868. [CrossRef]
11. Jung, J.; Lee, S.C.; Choi, H.-K. Anatomical Patterns of Aerenchyma in Aquatic and Wetland Plants. *J. Plant Biol.* **2008**, *51*, 428–439. [CrossRef]

12. Bailey-Serres, J.; Voesenek, L.A. Life in the Balance: A Signaling Network Controlling Survival of Flooding. *Curr. Opin. Plant Biol.* **2010**, *13*, 489–494. [CrossRef]
13. Takahashi, H.; Yamauchi, T.; Colmer, T.D.; Nakazono, M. Aerenchyma Formation in Plants. In *Plant Cell Monographs*; Springer: Berlin/Heidelberg, Germany, 2014; Volume 21, pp. 247–265.
14. Doležal, J.; Kučerová, A.; Jandová, V.; Klimeš, A.; Říha, P.; Adamec, L.; Schweingruber, F.H. Anatomical Adaptations in Aquatic and Wetland Dicot Plants: Disentangling the Environmental, Morphological and Evolutionary Signals. *Environ. Exp. Bot.* **2021**, *187*, 104495. [CrossRef]
15. Laan, P.; Berrevoets, M.J.; Lythe, S.; Armstrong, W.; Blom, C.W.P.M. Root Morphology and Aerenchyma Formation as Indicators of the Flood-Tolerance of Rumex Species. *J. Ecol.* **1989**, *77*, 693–703. [CrossRef]
16. Larcher, W. *Physiological Plant Ecology: Ecophysiology and Stress Physiology of Functional Groups*, 4th ed.; Springer: Berlin, Germany, 2003.
17. Jia, Y.; Huang, H.; Chen, Z.; Zhu, Y.-G. Arsenic Uptake by Rice Is Influenced by Microbe-Mediated Arsenic Redox Changes in the Rhizosphere. *Environ. Sci. Technol.* **2014**, *48*, 1001–1007. [CrossRef]
18. Smith, K.E.; Luna, T.O. Radial Oxygen Loss in Wetland Plants: Potential Impacts on Remediation of Contaminated Sediments. *J. Environ. Eng.* **2013**, *139*, 496–501. [CrossRef]
19. Riya, S.; Sun, H.; Furuhata, H.; Shimamura, M.; Nasu, H.; Imano, R.; Zhou, S.; Hosomi, M. Role of Leaves in Methane Emission from Sacred Lotus (*Nelumbo nucifera*). *Aquat. Bot.* **2020**, *163*, 103203. [CrossRef]
20. Villa, J.A.; Ju, Y.; Stephen, T.; Rey-Sanchez, C.; Wrighton, K.C.; Bohrer, G. Plant-mediated Methane Transport in Emergent and Floating-leaved Species of a Temperate Freshwater Mineral-soil Wetland. *Limnol. Oceanogr.* **2020**, *65*, 1635–1650. [CrossRef]
21. Caraco, N.F.; Cole, J.J. Contrasting Impacts of a Native and Alien Macrophyte on Dissolved Oxygen in a Large River. *Ecol. Appl.* **2002**, *12*, 1496. [CrossRef]
22. Carpenter, S.R.; Lodge, D.M. Effects of Submersed Macrophytes on Ecosystem Processes. *Aquat. Bot.* **1986**, *26*, 341–370. [CrossRef]
23. Goodwin, K.; Caraco, N.; Cole, J. Temporal Dynamics of Dissolved Oxygen in a Floatingleaved Macrophyte Bed. *Freshw. Biol.* **2008**, *53*, 1632–1641. [CrossRef]
24. Horvat, B.; Urbanc-Berčič, O.; Gaberščik, A. Water Quality and Macrophyte Community Changes in the Komarnik Accumulation Lake (Slovenia). In *Wastewater Treatment, Plant Dynamics and Management in Constructed and Natural Wetlands*; Springer: Dordrecht, The Netherlands, 2008; pp. 13–22.
25. Raven, J.A. Into the Voids: The Distribution, Function, Development and Maintenance of Gas Spaces in Plants. *Ann. Bot. Co.* **1996**, *78*, 137–142. [CrossRef]
26. Rascio, N. The Underwater Life of Secondarily Aquatic Plants: Some Problems and Solutions. *Crit. Rev. Plant Sci.* **2002**, *21*, 401–427. [CrossRef]
27. Sculthorpe, C.D. *The Biology of Aquatic Vascular Plants*; American Association for the Advancement of Science: London, UK, 1967.
28. Hartman, R.T.; Brown, D.L. Changes in Internal Atmosphere of Submersed Vascular Hydrophytes in Relation to Photosynthesis. *Ecology* **1967**, *48*, 252–258. [CrossRef]
29. Winkel, A.; Borum, J. Use of Sediment CO_2 by Submersed Rooted Plants. *Ann. Bot.* **2009**, *103*, 1015–1023. [CrossRef] [PubMed]
30. Madsen, T.v.; Olesen, B.; Bagger, J. Carbon Acquisition and Carbon Dynamics by Aquatic Isoetids. *Aquat. Bot.* **2002**, *73*, 351–371. [CrossRef]
31. Borum, J.; Pedersen, O.; Greve, T.M.; Frankovich, T.A.; Zieman, J.C.; Fourqurean, J.W.; Madden, C.J. The Potential Role of Plant Oxygen and Sulphide Dynamics in Die-off Events of the Tropical Seagrass, Thalassia Testudinum. *J. Ecol.* **2005**, *93*, 148–158. [CrossRef]
32. Sand-Jensen, K.; Pedersen, O.; Binzer, T.; Borum, J. Contrasting Oxygen Dynamics in the Freshwater Isoetid Lobelia Dortmanna and the Marine Seagrass Zostera Marina. *Ann. Bot.* **2005**, *96*, 613–623. [CrossRef]
33. Pedersen, O.; Rich, S.M.; Pulido, C.; Cawthray, G.R.; Colmer, T.D. Crassulacean Acid Metabolism Enhances Underwater Photosynthesis and Diminishes Photorespiration in the Aquatic Plant Isoetes Australis. *New Phytol.* **2011**, *190*, 332–339. [CrossRef]
34. Sand-Jensen, K.; Prahl, C.; Stokholm, H. Oxygen Release from Roots of Submerged Aquatic Macrophytes. *Oikos* **1982**, *38*, 349–354. [CrossRef]
35. Keeley, J.E. CAM Photosynthesis in Submerged Aquatic Plants. *Bot. Rev.* **1998**, *64*, 121–175. [CrossRef]
36. Břízová, E. Quillwort (Isoëtes), a Mysterious Plant from the Czech Republic. *Acta Mus. Nat. Pragae Ser. B Hist. Nat.* **2011**, *67*, 25–34.
37. Green, W.A. The Parichnos Problem and the Function of Aerenchyma in the Lycopsida. *Bull. Peabody Mus. Nat. Hist.* **2014**, *55*, 191–200. [CrossRef]
38. Green, W.A. The Function of the Aerenchyma in Arborescent Lycopsids: Evidence of an Unfamiliar Metabolic Strategy. *Proc. R. Soc. B Biol. Sci.* **2010**, *277*, 2257–2267. [CrossRef] [PubMed]
39. Breecker, D.O.; Royer, D.L. The Pedogenic Formation of Coal Balls by CO_2 Degassing through the Rootlets of Arborescent Lycopsids. *Am. J. Sci.* **2019**, *319*, 529–548. [CrossRef]
40. Pedersen, O.; Sand-Jensen, K.; Revsbech, N.P. Diel Pulses of O_2 and CO_2 in Sandy Lake Sediments Inhabited by Lobelia Dortmanna. *Ecology* **1995**, *76*, 1536–1545. [CrossRef]

41. Smits, A.J.M.; Laan, P.; Thier, R.H.; van der Velde, G. Root Aerenchyma, Oxygen Leakage Patterns and Alcoholic Fermentation Ability of the Roots of Some Nymphaeid and Isoetid Macrophytes in Relation to the Sediment Type of Their Habitat. *Aquat. Bot.* **1990**, *38*, 3–17. [CrossRef]

42. Smolders, A.J.P.; Lucassen, E.C.H.E.T.; Roelofs, J.G.M. The Isoetid Environment: Biogeochemistry and Threats. *Aquat. Bot.* **2002**, *73*, 325–350. [CrossRef]

43. Pedersen, O.; Pulido, C.; Rich, S.M.; Colmer, T.D. In Situ O_2 Dynamics in Submerged Isoetes Australis: Varied Leaf Gas Permeability Influences Underwater Photosynthesis and Internal O_2. *J. Exp. Bot.* **2011**, *62*, 4691–4700. [CrossRef] [PubMed]

44. Winkel, A.; Colmer, T.D.; Pedersen, O. Leaf Gas Films of Spartina Anglica Enhance Rhizome and Root Oxygen during Tidal Submergence. *Plant Cell Environ.* **2011**, *34*, 2083–2092. [CrossRef]

45. Richards, J.H.; Kuhn, D.N.; Bishop, K. Interrelationships of Petiolar Air Canal Architecture, Water Depth, and Convective Air Flow in Nymphaea Odorata (*Nymphaeaceae*). *Am. J. Bot.* **2012**, *99*, 1903–1909. [CrossRef]

46. Dacey, J.W.H. Internal Winds in Water Lilies: An Adaptation for Life in Anaerobic Sediments. *Science* **1980**, *210*, 1017–1019. [CrossRef]

47. Grosse, W.; Bernhard Büchel, H.; Tiebel, H. Pressurized Ventilation in Wetland Plants. *Aquat. Bot.* **1991**, *39*, 89–98. [CrossRef]

48. Dacey, J.W.H.; Klug, M.J. Ventilation by Floating Leaves in Nuphar. *Am. J. Bot.* **1982**, *69*, 999. [CrossRef]

49. Grosse, W.; Jovy, K.; Tiebel, H. Influence of Plants on Redox Potential and Methane Production in Water-Saturated Soil. In *Management and Ecology of Freshwater Plants*; Springer: Dordrecht, The Netherlands, 1996; pp. 93–99.

50. Grosse, W.; Schröder, P. Pflanzenleben Unter Anaeroben Umweltbedingungen, Die Physikalischen Grundlagen Und Anatomischen Voraussetzungen. *Ber. Der Dtsch. Bot. Ges.* **1986**, *99*, 367–381. [CrossRef]

51. Schröder, P. Characterization of a Thermo-Osmotic Gas Transport Mechanism in *Alnus glutinosa* (L.) Gaertn. *Trees* **1989**, *3*, 38–44. [CrossRef]

52. Konnerup, D.; Sorrell, B.K.; Brix, H. Do Tropical Wetland Plants Possess Convective Gas Flow Mechanisms? *New Phytol.* **2011**, *190*, 379–386. [CrossRef]

53. Mevi-Scutz, J.; Grosse, W. A Two-Way Gas Transport System in *Nelumbo nucifera*. *Plant Cell Environ.* **1988**, *11*, 27–34. [CrossRef]

54. Matthews, P.G.D.; Seymour, R.S. Stomata Actively Regulate Internal Aeration of the Sacred Lotus *Nelumbo Nucifera*. *Plant Cell Environ.* **2014**, *37*, 402–413. [CrossRef]

55. Vogel, S. Contributions to the Functional Anatomy and Biology of Nelumbo Nucifera (*Nelumbonaceae*) I. Pathways of Air Circulation. *Plant Syst. Evol.* **2004**, *249*, 9–25. [CrossRef]

56. Matthews, P.G.D.; Seymour, R.S. Anatomy of the Gas Canal System of *Nelumbo nucifera*. *Aquat. Bot.* **2006**, *85*, 147–154. [CrossRef]

57. Gaberščik, A. Measurements of Apparent CO_2 Flux in Amphibious Plant *Polygonum amphibium* L. Growing over Environmental Gradient. *Photosynthetica* **1993**, *29*, 473–476.

58. Colmer, T.D.; Pedersen, O. Underwater Photosynthesis and Respiration in Leaves of Submerged Wetland Plants: Gas Films Improve CO_2 and O_2 Exchange. *New Phytol.* **2008**, *177*, 918–926. [CrossRef] [PubMed]

59. Hyvönen, T.; Ojala, A.; Kankaala, P.; Martikainen, P.J. Methane Release from Stands of Water Horsetail (*Equisetum Fluviatile*) in a Boreal Lake. *Freshw. Biol.* **1998**, *40*, 275–284. [CrossRef]

60. Armstrong, J.; Armstrong, W. Reasons for the Presence or Absence of Convective (Pressurized) Ventilation in the Genus *Equisetum*. *New Phytol.* **2011**, *190*, 387–397. [CrossRef] [PubMed]

61. Armstrong, J.; Armstrong, W. Record Rates of Pressurized Gas-flow in the Great Horsetail, Equisetum Telmateia. Were Carboniferous Calamites Similarly Aerated? *New Phytol.* **2009**, *184*, 202–215. [CrossRef]

62. Strand, V.V. The Influence of Ventilation Systems on Water Depth Penetration of Emergent Macrophytes. *Freshw. Biol.* **2002**, *47*, 1097–1105. [CrossRef]

63. Vretare Strand, V.; Weisner, S.E.B. Interactive Effects of Pressurized Ventilation, Water Depth and Substrate Conditions on Phragmites Australis. *Oecologia* **2002**, *131*, 490–497. [CrossRef]

64. Vretare, V.; Weisner, S.E.B. Influence of Pressurized Ventilation on Performance of an Emergent Macrophyte (*Phragmites australis*). *J. Ecol.* **2000**, *88*, 978–987. [CrossRef]

65. Armstrong, J.; Armstrong, W.; Beckett, P.M. Phragmites Australis: Venturi- and Humidity-Induced Pressure Flows Enhance Rhizome Aeration and Rhizosphere Oxidation. *New Phytol.* **1992**, *120*, 197–207. [CrossRef]

66. Faußer, A.C.; Dušek, J.; Čížková, H.; Kazda, M. Diurnal Dynamics of Oxygen and Carbon Dioxide Concentrations in Shoots and Rhizomes of a Perennial in a Constructed Wetland Indicate Down-Regulation of below Ground Oxygen Consumption. *AoB Plants* **2016**, *8*, plw025. [CrossRef]

67. Kludze, H.K.; DeLaune, R.D.; Patrick, W.H. Aerenchyma Formation and Methane and Oxygen Exchange in Rice. *Soil Sci. Soc. Am. J.* **1993**, *57*, 386–391. [CrossRef]

68. Bart, D.; Hartman, J.M. Environmental Determinants of Phragmites Australis Expansion in a New Jersey Salt Marsh: An Experimental Approach. *Oikos* **2000**, *89*, 59–69. [CrossRef]

69. Visser, E.J.W.; Colmer, T.D.; Blom, C.W.P.M.; Voesenek, L.A.C.J. Changes in Growth, Porosity, and Radial Oxygen Loss from Adventitious Roots of Selected Mono- and Dicotyledonous Wetland Species with Contrasting Types of Aerenchyma. *Plant Cell Environ.* **2000**, *23*, 1237–1245. [CrossRef]

70. Singer, A.; Eshel, A.; Agami, M.; Beer, S. The Contribution of Aerenchymal CO_2 to the Photosynthesis of Emergent and Submerged Culms of Scirpus Lacustris and Cyperus Papyrus. *Aquat. Bot.* **1994**, *49*, 107–116. [CrossRef]

71. Jermy, A.C.; Anthony, C.; Simpson, D.; Oswald, P.H. Botanical Society of the British Isles. "*Schoenoplectus lacustris* (L.) Palla". In *Sedges of the British Isles. BSBI Handbook*, 3rd ed.; Botanical Society of the British Isles: London, UK, 2007; Volume 1, pp. 105–107. ISBN 9780901158352.

72. Tulbure, M.G.; Ghioca-Robrecht, D.M.; Johnston, C.A.; Whigham, D.F. Inventory and Ventilation Efficiency of Nonnative and Native Phragmites Australis (Common Reed) in Tidal Wetlands of the Chesapeake Bay. *Estuaries Coasts* **2012**, *35*, 1353–1359. [CrossRef]

73. Tornberg, T.; Bendix, M.; Brix, H. Internal Gas Transport in *Typha latifolia* L. and *Typha angustifolia* L. 2. Convective Throughflow Pathways and Ecological Significance. *Aquat. Bot.* **1994**, *49*, 91–105. [CrossRef]

74. White, S.D.; Ganf, G.G. Influence of Stomatal Conductance on the Efficiency of Internal Pressurisation in Typha Domingensis. *Aquat. Bot.* **2000**, *67*, 1–11. [CrossRef]

75. Bansal, S.; Lishawa, S.C.; Newman, S.; Tangen, B.A.; Wilcox, D.; Albert, D.; Anteau, M.J.; Chimney, M.J.; Cressey, R.L.; DeKeyser, E.; et al. Typha (*Cattail*) Invasion in North American Wetlands: Biology, Regional Problems, Impacts, Ecosystem Services, and Management. *Wetlands* **2019**, *39*, 645–684. [CrossRef]

76. Duarte, V.P.; Pereira, M.P.; Corrêa, F.F.; de Castro, E.M.; Pereira, F.J. Aerenchyma, Gas Diffusion, and Catalase Activity in Typha Domingensis: A Complementary Model for Radial Oxygen Loss. *Protoplasma* **2021**, *258*, 765–777. [CrossRef]

77. Matsui, T.; Tsuchiya, T. Root Aerobic Respiration and Growth Characteristics of Three Typha Species in Response to Hypoxia. *Ecol. Res.* **2006**, *21*, 470–475. [CrossRef]

78. Raskin, I.; Kende, H. Mechanism of Aeration in Rice. *Science* **1985**, *228*, 327–329. [CrossRef]

79. Bailey-Serres, J.; Fukao, T.; Ronald, P.; Ismail, A.; Heuer, S.; Mackill, D. Submergence Tolerant Rice: SUB1's Journey from Landrace to Modern Cultivar. *Rice* **2010**, *3*, 138–147. [CrossRef]

80. Nishiuchi, S.; Yamauchi, T.; Takahashi, H.; Kotula, L.; Nakazono, M. Mechanisms for Coping with Submergence and Waterlogging in Rice. *Rice* **2012**, *5*, 2. [CrossRef] [PubMed]

81. Herzog, M.; Konnerup, D.; Pedersen, O.; Winkel, A.; Colmer, T.D. Leaf Gas Films Contribute to Rice (*Oryza Sativa*) Submergence Tolerance during Saline Floods. *Plant Cell Environ.* **2018**, *41*, 885–897. [CrossRef] [PubMed]

82. Konnerup, D.; Pedersen, O. Flood Tolerance of Glyceria Fluitans: The Importance of Cuticle Hydrophobicity, Permeability and Leaf Gas Films for Underwater Gas Exchange. *Ann. Bot.* **2017**, *120*, 521–528. [CrossRef]

83. Groot, T.T.; van Bodegom, P.M.; Meijer, H.A.J.; Harren, F.J.M. Gas Transport through the Root–Shoot Transition Zone of Rice Tillers. *Plant Soil* **2005**, *277*, 107–116. [CrossRef]

84. Yamauchi, T.; Tanaka, A.; Mori, H.; Takamure, I.; Kato, K.; Nakazono, M. Ethylene-Dependent Aerenchyma Formation in Adventitious Roots Is Regulated Differently in Rice and Maize. *Plant Cell Environ.* **2016**, *39*, 2145–2157. [CrossRef]

85. Steffens, B.; Geske, T.; Sauter, M. Aerenchyma Formation in the Rice Stem and Its Promotion by H_2O_2. *New Phytol.* **2011**, *190*, 369–378. [CrossRef]

86. Ni, X.-L.; Gui, M.-Y.; Tan, L.-L.; Zhu, Q.; Liu, W.-Z.; Li, C.-X. Programmed Cell Death and Aerenchyma Formation in Water-Logged Sunflower Stems and Its Promotion by Ethylene and ROS. *Front. Plant Sci.* **2019**, *9*, 1928. [CrossRef]

87. Parlanti, S.; Kudahettige, N.P.; Lombardi, L.; Mensuali-Sodi, A.; Alpi, A.; Perata, P.; Pucciariello, C. Distinct Mechanisms for Aerenchyma Formation in Leaf Sheaths of Rice Genotypes Displaying a Quiescence or Escape Strategy for Flooding Tolerance. *Ann. Bot.* **2011**, *107*, 1335–1343. [CrossRef]

88. Nagai, K.; Kurokawa, Y.; Mori, Y.; Minami, A.; Reuscher, S.; Wu, J.; Matsumoto, T.; Ashikari, M. SNORKEL Genes Relating to Flood Tolerance Were Pseudogenized in Normal Cultivated Rice. *Plants* **2022**, *11*, 376. [CrossRef]

89. Colmer, T.D.; Cox, M.C.H.; Voesenek, L.A.C.J. Root Aeration in Rice (*Oryza sativa*): Evaluation of Oxygen, Carbon Dioxide, and Ethylene as Possible Regulators of Root Acclimatizations. *New Phytol.* **2006**, *170*, 767–778. [CrossRef]

90. Colmer, T.D.; Pedersen, O. Oxygen Dynamics in Submerged Rice (*Oryza sativa*). *New Phytol.* **2008**, *178*, 326–334. [CrossRef] [PubMed]

91. Seago, J.L.; Peterson, C.A.; Enstone, D.E. Cortical Development in Roots of the Aquatic Plant Pontederia Cordata (*Pontederiaceae*). *Am. J. Bot.* **2000**, *87*, 1116–1127. [CrossRef] [PubMed]

92. Herzog, M.; Pedersen, O. Partial versus Complete Submergence: Snorkelling Aids Root Aeration in R Umex Palustris but Not in R. Acetosa. *Plant Cell Environ.* **2014**, *37*, 2381–2390. [CrossRef] [PubMed]

93. Ting, T.M.; Poulsen, A.D. Understorey Vegetation at Two Mud Volcanoes in North-East Borneo. *J. Trop. For. Sci.* **2009**, *21*, 198–209.

94. Bonde, S.D.; Kumaran, K.P.N. A Permineralized Species of Mangrove Fern Acrostichum L. from Deccan Intertrappean Beds of India. *Rev. Palaeobot. Palynol.* **2002**, *120*, 285–299. [CrossRef]

95. Fonini, A.M.; Barufi, J.B.; Schmidt, É.C.; Rodrigues, A.C.; Randi, Á.M. Leaf Anatomy and Photosynthetic Efficiency of Acrostichum Danaeifolium after UV Radiation. *Photosynthetica* **2017**, *55*, 401–410. [CrossRef]

96. Chomicki, G.; Bidel, L.P.R.; Baker, W.J.; Jay-Allemand, C. Palm Snorkelling: Leaf Bases as Aeration Structures in the Mangrove Palm (*Nypa fruticans*). *Bot. J. Linn. Soc.* **2014**, *174*, 257–270. [CrossRef]

97. Rozainah, M.Z.; Aslezaeim, N. A Demographic Study of a Mangrove Palm, Nypa Fruticans. *Sci. Res. Essays* **2010**, *5*, 3896–3902.

98. Srikanth, S.; Lum, S.K.Y.; Chen, Z. Mangrove Root: Adaptations and Ecological Importance. *Trees* **2016**, *30*, 451–465. [CrossRef]

99. McKee, K.L. Soil Physicochemical Patterns and Mangrove Species Distribution-Reciprocal Effects? *J. Ecol.* **1993**, *81*, 477–487. [CrossRef]

100. Scholander, P.F.; van Dam, L.; Scholander, S.I. Gas Exchange in the Roots of Mangroves. *Am. J. Bot.* **1955**, *42*, 92. [CrossRef]

101. Purnobasuki, H.; Suzuki, M. Aerenchyma Formation and Porosity in Root of a Mangrove Plant, Sonneratia Alba (*Lythraceae*). *J. Plant Res.* **2004**, *117*, 465–472. [CrossRef]

102. Purnobasuk, H. Primary Structure and Intercellular Space Formation of Feeding Root in *Sonneratia alba* J. Smith. *Int. J. Bot.* **2012**, *9*, 44–49. [CrossRef]

103. Zhang, Y.; Ding, Y.; Wang, W.; Li, Y.; Wang, M. Distribution of Fish among Avicennia and Sonneratia Microhabitats in a Tropical Mangrove Ecosystem in South China. *Ecosphere* **2019**, *10*, e02759. [CrossRef]

104. Cheng, H.; Wu, M.-L.; Li, C.-D.; Sun, F.-L.; Sun, C.-C.; Wang, Y.-S. Dynamics of Radial Oxygen Loss in Mangroves Subjected to Waterlogging. *Ecotoxicology* **2020**, *29*, 684–690. [CrossRef] [PubMed]

105. Cheng, H.; Wang, Y.-S.; Fei, J.; Jiang, Z.-Y.; Ye, Z.-H. Differences in Root Aeration, Iron Plaque Formation and Waterlogging Tolerance in Six Mangroves along a Continues Tidal Gradient. *Ecotoxicology* **2015**, *24*, 1659–1667. [CrossRef]

106. Curran, M.; Cole, M.; Allaway, W.G. Root Aeration and Respiration in Young Mangrove Plants (*Avicennia marina* (Forsk.) Vierh.). *J. Exp. Bot.* **1986**, *37*, 1225–1233. [CrossRef]

107. Hovenden, M. Horizontal Structures on Pneumatophores of *Avicennia marina* (Forsk.) Vierh.—A New Site of Oxygen Conductance. *Ann. Bot.* **1994**, *73*, 377–383. [CrossRef]

108. Curran, M.; James, P.; Allaway, W.G. The Measurement of Gas Spaces in the Roots of Aquatic Plants—Archimedes Revisited. *Aquat. Bot.* **1996**, *54*, 255–261. [CrossRef]

109. Rogers, G.K. Bald Cypress Knees, Taxodium Distichum (Cupressaceae): An Anatomical Study, with Functional Implications. *Flora* **2021**, *278*, 151788. [CrossRef]

110. Martin, C.E.; Francke, S.K. Root Aeration Function of Baldcypress Knees (*Taxodium distichum*). *Int. J. Plant Sci.* **2015**, *176*, 170–173. [CrossRef]

111. Wang, G.-B.; Cao, F.-L. Formation and Function of Aerenchyma in Baldcypress (*Taxodium distichum* (L.) Rich.) and Chinese Tallow Tree (*Sapium Sebiferum* (L.) Roxb.) under Flooding. *S. Afr. J. Bot.* **2012**, *81*, 71–78. [CrossRef]

112. Middleton, B. Carbon Stock Trends of Baldcypress Knees along Climate Gradients of the Mississippi River Alluvial Valley Using Allometric Methods. *For. Ecol. Manag.* **2020**, *461*, 117969. [CrossRef]

113. Sou, H.-D.; Masumori, M.; Kurokochi, H.; Tange, T. Histological Observation of Primary and Secondary Aerenchyma Formation in Adventitious Roots of *Syzygium kunstleri* (King) Bahadur and *R.C.Gaur Grown* in Hypoxic Medium. *Forests* **2019**, *10*, 137. [CrossRef]

114. Grosse, W.; Schulte, A.; Fujita, H. Pressurized Gas Transport in Two Japanese Alder Species in Relation to Their Natural Habitats. *Ecol. Res.* **1993**, *8*, 151–158. [CrossRef]

115. Takaishi, T.; Sensui, Y. Thermal Transpiration Effect of Hydrogen, Rare Gases and Methane. *Trans. Faraday Soc.* **1963**, *59*, 2503–2514. [CrossRef]

116. Denbigh, K.G. Thermo-Osmosis of Gases through a Membrane. *Nature* **1949**, *163*, 60. [CrossRef]

117. Denbigh, K.G.; Raumann, G. The Thermo-Osmosis of Gases through a Membrane. I. Theoretical. *Proc. R. Soc. London Ser. A Math. Phys. Sci.* **1952**, *210*, 377–387.

118. Buchel, H.B.; Grosse, W. Localization of the Porous Partition Responsible for Pressurized Gas Transport in *Alnus glutinosa* (L.) Gaertn. *Tree Physiol.* **1990**, *6*, 247–256. [CrossRef]

119. Dittert, K.; Wätzel, J.; Sattelmacher, B. Responses of Alnus Glutinosa to Anaerobic Conditions-Mechanisms and Rate of Oxygen Flux into the Roots. *Plant Biol.* **2006**, *8*, 212–223. [CrossRef] [PubMed]

120. Armstrong, W.; Armstrong, J. Stem Photosynthesis Not Pressurized Ventilation Is Responsible for Light-Enhanced Oxygen Supply to Submerged Roots of Alder (*Alnus glutinosa*). *Ann. Bot.* **2005**, *96*, 591–612. [CrossRef] [PubMed]

121. Batzli, J.M.; Dawson, J.O. Development of Flood-Induced Lenticels in Red Alder Nodules Prior to the Restoration of Nitrogenase Activity. *Can. J. Bot.* **1999**, *77*, 1373–1377. [CrossRef]

122. Huber, M.; Linder, H.P. The Evolutionary Loss of Aerenchyma Limits Both Realized and Fundamental Ecohydrological Niches in the Cape Reeds (*Restionaceae*). *J. Ecol.* **2012**, *100*, 1338–1348. [CrossRef]

123. Bedoya, A.M.; Madriñán, S. Evolution of the Aquatic Habit in Ludwigia (*Onagraceae*): Morpho-Anatomical Adaptive Strategies in the Neotropics. *Aquat. Bot.* **2015**, *120*, 352–362. [CrossRef]

124. Hoffmann, M.H. To the Roots of Carex: Unexpected Anatomical and Functional Diversity. *Syst. Bot.* **2019**, *44*, 26–31. [CrossRef]

125. Visser, E.J.W.; Bogemann, G.M.; van de Steeg, H.M.; Pierik, R.; Blom, C.W.P.M. Flooding Tolerance of Carex Species in Relation to Field Distribution and Aerenchyma Formation. *New Phytol.* **2000**, *148*, 93–103. [CrossRef]

126. Armstrong, W.; Armstrong, J. Plant Internal Oxygen Transport (Diffusion and Convection) and Measuring and Modelling Oxygen Gradients. In *Low-Oxygen Stress in Plants*; Springer: Berlin, Germany, 2014; Volume 21, pp. 267–297.

127. Schröder, P.; Grosse, W.; Woermann, D. Localization of Thermo-Osmotically Active Partitions in Young Leaves of Nuphar Lutea. *J. Exp. Bot.* **1986**, *37*, 1450–1461. [CrossRef]

128. Bazhin, N.M. Influence of Plants on the Methane Emission from Sediments. *Chemosphere* **2004**, *54*, 209–215. [CrossRef]

129. Bridgham, S.D.; Cadillo-Quiroz, H.; Keller, J.K.; Zhuang, Q. Methane Emissions from Wetlands: Biogeochemical, Microbial, and Modeling Perspectives from Local to Global Scales. *Glob. Chang. Biol.* **2013**, *19*, 1325–1346. [CrossRef]

130. Bansal, S.; Johnson, O.F.; Meier, J.; Zhu, X. Vegetation Affects Timing and Location of Wetland Methane Emissions. *J. Geophys. Res. Biogeosci.* **2020**, *125*, e2020JG005777. [CrossRef]

131. Bhullar, G.S.; Iravani, M.; Edwards, P.J.; Olde Venterink, H. Methane Transport and Emissions from Soil as Affected by Water Table and Vascular Plants. *BMC Ecol.* **2013**, *13*, 32. [CrossRef] [PubMed]
132. Turetsky, M.R.; Kotowska, A.; Bubier, J.; Dise, N.B.; Crill, P.; Hornibrook, E.R.C.; Minkkinen, K.; Moore, T.R.; Myers-Smith, I.H.; Nykänen, H.; et al. A Synthesis of Methane Emissions from 71 Northern, Temperate, and Subtropical Wetlands. *Glob. Chang. Biol.* **2014**, *20*, 2183–2197. [CrossRef]
133. Kosten, S.; Piñeiro, M.; de Goede, E.; de Klein, J.; Lamers, L.P.M.; Ettwig, K. Fate of Methane in Aquatic Systems Dominated by Free-Floating Plants. *Water Res.* **2016**, *104*, 200–207. [CrossRef] [PubMed]
134. Henneberg, A.; Sorrell, B.K.; Brix, H. Internal Methane Transport through *Juncus effusus*: Experimental Manipulation of Morphological Barriers to Test Above- and Below-ground Diffusion Limitation. *New Phytol.* **2012**, *196*, 799–806. [CrossRef] [PubMed]
135. Kim, J.; Lee, J.; Yun, J.; Yang, Y.; Ding, W.; Yuan, J.; Kang, H. Mechanisms of Enhanced Methane Emission Due to Introduction of *Spartina anglica* and *Phragmites australis* in a Temperate Tidal Salt Marsh. *Ecol. Eng.* **2020**, *153*, 105905. [CrossRef]
136. van den Berg, M.; van den Elzen, E.; Ingwersen, J.; Kosten, S.; Lamers, L.P.M.; Streck, T. Contribution of Plant-Induced Pressurized Flow to CH4 Emission from a Phragmites Fen. *Sci. Rep.* **2020**, *10*, 12304. [CrossRef]
137. Were, D.; Kansiime, F.; Fetahi, T.; Hein, T. Carbon Dioxide and Methane Fluxes from Various Vegetation Communities of a Natural Tropical Freshwater Wetland in Different Seasons. *Environ. Processes* **2021**, *8*, 553–571. [CrossRef]
138. Jones, M.B. Methane Production and Emission from Papyrus-Dominated Wetlands. *SIL Proc.* **2000**, *27*, 1406–1409. [CrossRef]
139. Van der Nat, F.J.W.A.; Middelburg, J.J.; van Meteren, D.; Wielemakers, A. Diel Methane Emission Patterns from *Scirpus lacustris* and *Phragmites australis*. *Biogeochemistry* **1998**, *41*, 1–22. [CrossRef]
140. Sebacher, D.I.; Harriss, R.C.; Bartlett, K.B. Methane Emissions to the Atmosphere Through Aquatic Plants. *J. Environ. Qual.* **1985**, *14*, 40–46. [CrossRef]
141. Inubushi, K.; Sugii, H.; Nishino, S.; Nishino, E. Effect of Aquatic Weeds on Methane Emission from Submerged Paddy Soil. *Am. J. Bot.* **2001**, *88*, 975–979. [CrossRef] [PubMed]
142. Iqbal, M.F.; Liu, S.; Zhu, J.; Zhao, L.; Qi, T.; Liang, J.; Luo, J.; Xiao, X.; Fan, X. Limited Aerenchyma Reduces Oxygen Diffusion and Methane Emission in Paddy. *J. Environ. Manag.* **2021**, *279*, 111583. [CrossRef] [PubMed]
143. Machacova, K.; Borak, L.; Agyei, T.; Schindler, T.; Soosaar, K.; Mander, Ü.; Ah-Peng, C. Trees as Net Sinks for Methane (CH_4) and Nitrous Oxide (N_2O) in the Lowland Tropical Rain Forest on Volcanic Réunion Island. *New Phytol.* **2021**, *229*, 1983–1994. [CrossRef] [PubMed]
144. Pangala, S.R.; Enrich-Prast, A.; Basso, L.S.; Peixoto, R.B.; Bastviken, D.; Hornibrook, E.R.C.; Gatti, L.v.; Marotta, H.; Calazans, L.S.B.; Sakuragui, C.M.; et al. Large Emissions from Floodplain Trees Close the Amazon Methane Budget. *Nature* **2017**, *552*, 230–234. [CrossRef] [PubMed]
145. Pangala, S.R.; Moore, S.; Hornibrook, E.R.C.; Gauci, V. Trees Are Major Conduits for Methane Egress from Tropical Forested Wetlands. *New Phytol.* **2013**, *197*, 524–531. [CrossRef]
146. Laanbroek, H.J. Methane Emission from Natural Wetlands: Interplay between Emergent Macrophytes and Soil Microbial Processes. A Mini-Review. *Ann. Bot.* **2010**, *105*, 141–153. [CrossRef]

Review

The Orchids of Wetland Vegetation in the Central Balkans

Vladan Djordjević [1],*, Svetlana Aćić [2], Eva Kabaš [1], Predrag Lazarević [1], Spyros Tsiftsis [3] and Dmitar Lakušić [1]

[1] Institute of Botany and Botanical Garden, Faculty of Biology, University of Belgrade, Takovska 43, 11000 Belgrade, Serbia

[2] Department of Botany, Faculty of Agriculture, University of Belgrade, Nemanjina 6, 11080 Belgrade, Serbia

[3] Department of Forest and Natural Environment Sciences, International Hellenic University, 1st km Dramas-Mikrochoriou, P.O. Box 172, 66100 Drama, Greece

* Correspondence: vdjordjevic@bio.bg.ac.rs

Abstract: Wetland ecosystems are important habitats for the growth and survival of numerous terrestrial orchids in Europe. This study reviews the current knowledge on the orchids of wetland vegetation in the Central Balkans. The orchid flora was analyzed from taxonomic, phytogeographical, ecological and conservation aspects. The most important taxa include the two Balkan endemics (*Dactylorhiza cordigera* subsp. *bosniaca* and *D. kalopissi* subsp. *macedonica*) and the three subendemics of the Balkans and the Carpathians (*Dactylorhiza cordigera* subsp. *cordigera*, *D. maculata* subsp. *transsilvanica* and *Gymnadenia frivaldii*), as well as a considerable number of Central European, Eurasian and boreal orchid representatives. Several orchid taxa occurring in the wet meadows and fens of the Central Balkans have a southern limit of their distribution in this part of Europe, suggesting that wetlands are important refuges for them. In total, 33 orchid taxa were recorded in plant communities from five classes, 10 orders and 17 alliances. Most orchid taxa grow in the following wetland vegetation types: wet meadows (class *Molinio-Arrhenatheretea*, order *Molinietalia caeruleae*, alliances *Molinion caeruleae*, *Deschampsion cespitosae* and *Calthion palustris*); fens (class *Scheuchzerio palustris-Caricetea fuscae*, order *Caricetalia fuscae*, alliance *Caricion fuscae*); tall-herb vegetation along mountain streams and springs (class *Mulgedio-Aconitetea*); marshes and herb-land vegetation of freshwater or brackish water bodies (class *Phragmito-Magnocaricetea*). This study highlights the importance of serpentine and silicate wetland vegetation types as important habitats for the survival of terrestrial orchids. In addition, detailed taxonomic, ecological and chorological studies of the wetland orchids of the Central Balkans need to be carried out in order to establish a successful plan for their conservation.

Keywords: Orchidaceae; species diversity; wet meadow; fens; mires; vegetation; ecology; Balkan Peninsula

Citation: Djordjević, V.; Aćić, S.; Kabaš, E.; Lazarević, P.; Tsiftsis, S.; Lakušić, D. The Orchids of Wetland Vegetation in the Central Balkans. *Diversity* **2023**, *15*, 26. https://doi.org/10.3390/d15010026

Academic Editors: Mateja Germ, Igor Zelnik, Matthew Simpson and Michael Wink

Received: 3 December 2022
Revised: 19 December 2022
Accepted: 20 December 2022
Published: 23 December 2022

1. Introduction

The orchid family (Orchidaceae) is one of the largest and most diverse families in the plant kingdom, with approximately 28,000 species and 880 genera [1]. Because of their germination limitation, mycorrhizal specificity and pollinator specialization, orchids are particularly vulnerable to changes in ecosystem balance, especially changes in moisture content, light regime, nutrient availability and competition levels [2,3]. Habitat changes or their complete destruction have led to the extinction or decline in abundance and distribution of many orchids and, consequently, many species are protected by laws and/or are included in Red Data Books [4]. Therefore, knowledge of the habitats and ecological preferences of orchids is a prerequisite for their appropriate conservation.

Orchids are known to occur in almost all terrestrial ecosystems, while they are absent or less abundant in extremely dry deserts, salt marshes and agricultural lands [5,6]. Terrestrial orchids in Europe occur in forests and scrubs, grasslands, meadows, heaths, tall-herb vegetation as well as in mires, bogs, fens, marshlands and even in anthropogenic vegetation [7]. Studies have often pointed out that orchid species occur in different vegetation types in different geographic regions and that the greatest differences occur when comparing the center and the edge of their range [8,9]. According to the "abundant-centre hypothesis", species at the edges of their range occur primarily in a limited number of plant communities, while species at the center of their range usually inhabit a wide variety of vegetation types [10].

Wetland ecosystems are important habitats for the growth and survival of numerous orchid species in Europe [6,7]. Due to climate change and global warming, these habitats are expected to decline or disappear in certain areas, especially near the Mediterranean. Therefore, the unique orchid flora of these areas can also be expected to decline or disappear. Knowledge of their spatial distribution and ecological characteristics is of great importance for conservation. So far, these habitats have been the main topic of several scientific papers. Some of them summarize the knowledge of habitats and phytocoenological affiliations of wetland orchids [11–16]. However, recent studies have examined the importance of wetland vegetation types as factors affecting the distribution and abundance of orchids [17,18]. In addition, over the past decades, numerous ecological studies have focused on the effects of wetland management (e.g., mowing) on orchid performance [19,20]. According to recent studies, orchids inhabiting wetlands in western and central Europe are most threatened [21–23]. On the other hand, knowledge about which wetland vegetation types are particularly rich in orchids, which orchids are specialists and which are generalists in wetland habitats, the impact of geological substrates as factors affecting the abundance and composition of orchids and the importance of these habitats for conservation in specific regions of Europe is still limited. Detailed insight into the preferences of wetland orchids will lead to a better understanding of conservation priorities and the application of conservation plans. In addition, this knowledge will allow predictions of species distribution and abundance in response to future changes in land cover and climate.

The Balkan Peninsula is an important center of orchid diversity, with the highest number of recorded species in the Mediterranean region, especially in the Aegean part of Greece [24,25]. Moreover, the Balkan Peninsula represents one of the most important centers of diversity of the genus *Dactylorhiza*, known for its numerous water-demanding representatives [26]. Although the area of the Central Balkans is insufficiently studied in terms of orchids, recent research indicates that wetland habitats in this area are important for many terrestrial orchids [15,18]. The orchids of the wetland vegetation in the Central Balkans have been studied mostly within the framework of extensive phytocoenological studies, which include a list of species with information on their abundance and sociality [27,28]. Given the strong influence of the humid climate in western Serbia, northern Montenegro, eastern Bosnia and Herzegovina, northern Albania and the northwestern part of North Macedonia, the significant presence of wet meadows, bogs, fens and marshes is understandable. In the Central Balkans, wetland vegetation can be divided into the following types: (a) submerged rooted herbaceous vegetation of brackish waters (*Ruppietea maritimae*); (b) salt marshes within the classes *Therosalicornietea* and *Festuco-Puccinellietea*; (c) freshwater aquatic vegetation (*Lemnetea*, *Charetea intermediae* and *Potamogetonetea*); (d) vegetation of freshwater springs, shorelines and marshes (*Montio-Cardaminetea*, *Isoëto-Nanojuncetea* and *Phragmito-Magnocaricetea*); (e) bogs and fens (*Oxycocco-Sphagnetea* and *Scheuchzerio palustris-Caricetea fuscae*); (f) wet meadows (some alliances within the class *Molinio-Arrhenatheretea*); and (g) tall-herb vegetation along mountain streams and water springs (some alliances within the class *Mulgedio-Aconitetea*) [27–32].

The present study represents a synthesis of knowledge on the orchids of the wetland vegetation of the Central Balkans, based on long-term personal field investigations, checking and revision of herbarium material and published sources. The study is based mainly on knowledge from Serbia, but also on some data from Bosnia, Montenegro and North Macedonia. The main objectives were: (i) to analyze the orchid flora of wetland vegetation from taxonomic, phytogeographical and life-form perspectives; (ii) to provide an overview of the classes, orders and associations of wetland vegetation in which orchids occur; (iii) to determine the richness of orchid taxa in relation to wetland vegetation; (iv) to demonstrate the importance of geological substrates as factors affecting the distribution, abundance and composition of orchids; and (v) to identify the main threats to orchids of wetland vegetation and to draw basic conclusions for orchid conservation.

2. Overview of the Orchid Flora of Wetland Vegetation in the Central Balkans

The overall survey of orchid taxa occurring in the wetland vegetation of the Central Balkans given here (Table 1) is based on long-term personal field investigations, herbarium material and relevant published sources. The material in the herbarium of the University of Belgrade (BEOU) and in that of the Museum of Natural History in Belgrade (BEO) was reviewed and revised. The nomenclature and taxonomy in this study follow Djordjević et al. [33] and Euro+Med [34].

Table 1. Overview of orchid taxa of wetland vegetation in the Central Balkans with indication of their degree of representation and chorological groups and life forms. BOR—boreal, CE—Central European, CEM—Central European mountainous, CE-EUX-CAUC—Central European-Euxine-Caucasian, EAS—Eurasian, MED-SUBMED—Mediterranean-Submediterranean; R—rhizomatous orchids, I—intermediate type of orchids, T—tuberous orchids.

Taxon	Degree of Representation	Chorological Group	Life Form
Anacamptis coriophora (L.) R.M.Bateman, Pridgeon & M.W.Chase subsp. *coriophora*	3	CE	T
Anacamptis laxiflora (Lam.) R.M.Bateman, Pridgeon & M.W.Chase subsp. *laxiflora*	1	MED-SUBMED	T
Anacamptis morio (L.) R.M.Bateman, Pridgeon & M.W.Chase subsp. *morio*	4	CE	T
Anacamptis palustris (Jacq.) R.M.Bateman, Pridgeon & M.W.Chase subsp. *palustris*	1	EAS	T
Anacamptis palustris subsp. *elegans* (Heuff.) R.M.Bateman, Pridgeon & M.W.Chase	1	EAS	T
Anacamptis pyramidalis (L.) Rich.	4	MED-SUBMED	T
Coeloglossum viride (L.) Hartm.	3	BOR	I
Dactylorhiza cordigera (Fr.) Soó subsp. *cordigera*	2	CEM	I
Dactylorhiza cordigera subsp. *bosniaca* (Beck) Soó	2	CEM	I
Dactylorhiza fuchsii (Druce) Soó subsp. *fuchsii*	3	BOR	I
Dactylorhiza incarnata (L.) Soó subsp. *incarnata*	1	BOR	I
Dactylorhiza kalopissii subsp. *macedonica* (J.Hölzinger & Künkele) Kreutz	1	MED-SUBMED	I
Dactylorhiza maculata (L.) Soó subsp. *maculata*	2	BOR	I
Dactylorhiza maculata subsp. *transsilvanica* (Schur) Soó	2	CE	I
Dactylorhiza majalis (Rchb.) P.F.Hunt & Summerh. subsp. *majalis*	1	CE	I
Dactylorhiza saccifera (Brongn.) Soó subsp. *saccifera*	3	MED-SUBMED	I
Dactylorhiza sambucina (L.) Soó	4	CE	I
Epipactis palustris (L.) Crantz	1	EAS	R
Gymnadenia conopsea (L.) R.Br.	3	EAS	I
Gymnadenia densiflora (Wahlenb.) A.Dietr.	2	EAS	I
Gymnadenia frivaldii Hampe ex Griseb.	1	CEM	I
Gymnadenia odoratissima (L.) Rich.	4	CE	I
Herminium monorchis (L.) R.Br.	2	EAS	T
Neotinea ustulata (L.) R.M.Bateman, Pridgeon & M.W.Chase	3	CE	T
Neottia ovata (L.) Bluff & Fingerh.	3	EAS	R
Nigritella rhellicani Teppner & E.Klein	4	CEM	I

Table 1. *Cont.*

Taxon	Degree of Representation	Chorological Group	Life Form
Orchis mascula subsp. *speciosa* (Mutel) Hegi	4	CE-EUX-CAUC	T
Orchis militaris L. subsp. *militaris*	4	EAS	T
Platanthera bifolia (L.) Rich.	3	EAS	I
Platanthera chlorantha (Custer) Rchb.	3	CE	I
Pseudorchis albida (L.) Á.Löve & D.Löve subsp. *albida*	3	BOR	I
Spiranthes spiralis (L.) Chevall.	4	CE	T
Traunsteinera globosa (L.) Rchb.	3	CEM	T

In the overview of the orchid flora of wetland vegetation in the Central Balkans, we have particularly emphasized the degree of orchid presence in wetland vegetation types (Table 1):

1. species occurs exclusively in wetland vegetation types (at 100% of its sites, it is found in wetland vegetation types);
2. species grows mainly in wetland vegetation types and rarely occurs in other vegetation types (at 50–100% of its sites, it is found in wetland vegetation types);
3. species occurs in wetland vegetation types but mostly inhabits other vegetation types (at 10–50% of its sites, it is found in wetland vegetation types);
4. species rarely occurs in wetland vegetation types and mostly inhabits other vegetation types (at < 10% of its sites, it is found in wetland vegetation types).

2.1. Richness of Orchid Taxa and Taxonomic Analysis

The floristic composition of the orchid flora of wetland vegetation in the Central Balkans includes 33 species and subspecies classified in 14 genera (Table 1). Eight taxa occur exclusively in these vegetation types, while six taxa grow mainly in wetland vegetation types and rarely occur in other vegetation types (Table 1). On the other hand, there are 11 taxa that grow in wetland vegetation types but occur more frequently and with greater abundance in other vegetation types, while eight taxa very rarely occur in wetland vegetation types (Table 1). The most taxon-rich genera are *Dactylorhiza* (ten taxa), *Anacamptis* (six taxa) and *Gymnadenia* (four taxa). The genera *Orchis* and *Platanthera* are represented by two taxa, while nine genera (*Coeloglossum, Epipactis, Herminium, Neotinea, Neottia, Nigritella, Pseudorchis, Spiranthes* and *Traunsteinera*) are represented by a single taxon (Table 1).

The genus *Dactylorhiza* has the highest number of taxa within the total orchid flora of wetland vegetation in the Central Balkans (Figure 1), which is not surprising considering that wet habitats (fens, bogs, marshes and wet meadows) are typical habitats for representatives of this genus [26,35]. The occurrence of a large number of *Dactylorhiza* taxa can also be explained by the significant presence of silicate geological substrates in the study area, known for their high water-storage capacity, which is favorable for the growth and survival of numerous representatives of this genus [36]. The presence of the two Balkan endemics (*D. cordigera* subsp. *bosniaca* and *D. kalopissi* subsp. *macedonica*) and two Carpathian-Balkan subendemics (*D. cordigera* subsp. *cordigera* and *D. maculata* subsp. *transsilvanica*) is particularly important (Figure 1). *Dactylorhiza* × *serbica* (H.Fleischm.) Soó, a natural hybrid between *D. incarnata* subsp. *incarnata* and *D. saccifera* subsp. *saccifera*, which also inhabits the wetlands, was described in Serbia [33]. In addition, *D. maculata* subsp. *maculata, D. maculata* subsp. *trassilvanica, D. cordigera* subsp. *bosniaca* and *D. majalis* have a southern limit of their distribution in the Central Balkans [33], making their habitats of high conservation value in this region. The area of the Central Balkans is also a contact zone where *D. fuchsii* and *D. maculata* subsp. *maculata* from the west, north and northwest and *D. saccifera* from the south and southeast meet [37,38], so there is potential for their future taxonomic and phylogeographic research. Due to their complicated taxonomy, the *D. maculata* and *D. majalis* groups require detailed taxonomic and phylogeographic research in the Central Balkans.

Figure 1. Some representatives of the genus *Dactylorhiza* of wetland vegetation in the Central Balkans: (**a**) *Dactylorhiza cordigera* subsp. *cordigera*, (**b**) *Dactylorhiza maculata* subsp. *maculata*, (**c**) *Dactylorhiza maculata* subsp. *transsilvanica*, (**d**) *Dactylorhiza incarnata* subsp. *incarnata*, (**e**) *Dactylorhiza saccifera* subsp. *saccifera*, (**f**) *Dactylorhiza sambucina* (photos V. Djordjević).

Three *Anacamptis* taxa that occur exclusively in wetland vegetation types are *Anacamptis laxiflora*, *A. palustris* subsp. *palustris* and *A. palustris* subsp. *elegans* (Figure 2). However, knowledge on the distribution of these taxa in the Central Balkans is insufficient, as these taxa are usually presented in the literature under their species rank for the flora of Serbia, i.e., as *Orchis laxiflora* Lam. [33]. Therefore, published data on the distribution and habitat preferences of these taxa in Serbia cannot be considered with any degree of certainty. Recent studies have shown that *A. palustris* subsp. *elegans* is the most widespread taxon, while *A. palustris* subsp. *palustris* is a rarer taxon, distributed mainly in the southern part of the Pannonian plain and very rarely in other parts of the Central Balkans [33,39–41]. Although *A. coriophora* subsp. *coriophora*, *A. morio* subsp. *morio* and *A. pyramidalis* occur in wet habitats, they are more common in other habitat types (mesophilous and xerophilous grasslands) [15].

Figure 2. Some representatives of the genus *Anacamptis* of wetland vegetation in the Central Balkans: (**a**) *Anacamptis palustris* subsp. *palustris*, (**b**) *Anacamptis palustris* subsp. *elegans*, (**c**) *Anacamptis laxiflora* ((**a**) photo I. Stevanoski; (**b**,**c**) photo S. Tsiftsis).

Among the *Gymnadenia* taxa, the Carpathian-Balkan subendemic *Gymnadenia frivaldii* (Figure 3a), which occurs exclusively in wetland vegetation types, should be emphasized.

The center of its distribution is on the mountains of the southeastern Dinaric Alps and on the mountains of the Scardo-Pindhian province, with a disjunction in the southeastern Carpathians [42–44]. This species has the southern and western limits of its distribution in the Central Balkans. Although *G. conopsea* is very common in wetland vegetation types in the Central Balkans, where it is often very abundant, this species also grows in other habitats such as mesophilous and xero-mesophilous meadows, as well as open woodlands [15].

Figure 3. Some representatives of orchids of wetland vegetation in the Central Balkans: (**a**) *Gymnadenia frivaldii*, (**b**) *Epipactis palustris*, (**c**) *Orchis mascula* subsp. *speciosa*, (**d**) *Platanthera bifolia*, (**e**) *Pseudorchis albida*, (**f**) *Traunsteinera globosa* (photos V. Djordjević).

The orchid flora of the wetland vegetation of the Central Balkans includes a small number of representatives of the genera *Epipactis* and *Neottia*, known for their typical forest representatives. The species *Epipactis palustris* (Figure 3b), which occurs exclusively in wetland vegetation, is widespread throughout the Central Balkans [43]. *Neottia ovata*, on the other hand, is an ecologically very plastic species that grows in wetland vegetation as well as in other vegetation types, including forests [6,41,45]. The genus *Orchis* is also represented by only a few representatives, which is not surprising knowing that the species of this genus tend to grow in xerophilous and mesophilous habitats and often in forest ecosystems. Among the representatives of orchids, *Pseudorchis albida* (Figure 3e) and *Traunsteinera globosa* (Figure 3f) are of great importance, because these species have the southern limit of their distribution in the Central Balkans.

2.2. Phytogeographical Analysis

Chorological analysis of the orchid flora of wetland vegetation in the Central Balkans revealed the presence of six chorological groups (Figure 4). The chorological types for phytogeographical analysis were determined according to the principles defined by Meusel et al. [46,47], Meusel and Jäger [48], Stevanović [49] and Djordjević et al. [33]. The occurrence of orchids from different chorological groups can be explained by the fact that the Central Balkans is located in an area influenced by different floristic-vegetation regions due to historical, geological, geomorphological and climatic reasons and the considerable altitude differences in the study area.

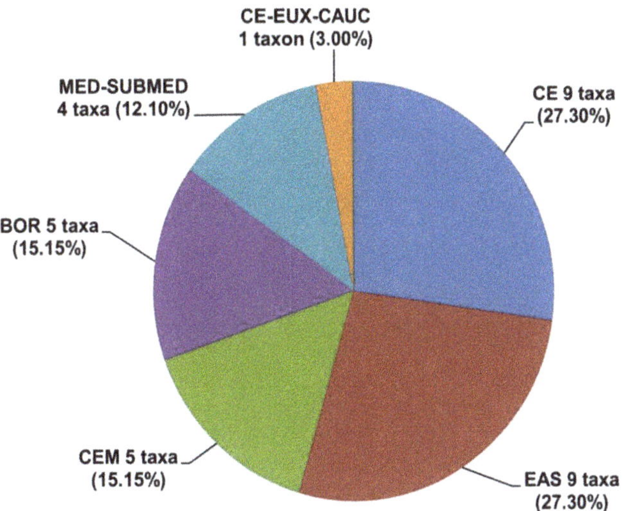

Figure 4. Spectrum of basic chorological groups of the orchid flora of wetland vegetation in the Central Balkans. CE—Central European; EAS—Eurasian; CEM—Central European mountainous; BOR—Boreal; MED-SUBMED—Mediterranean-Submediterranean; CE-EUX-CAUC—Central European-Euxine-Caucasian.

The chorological analysis of the orchid flora of wetland vegetation in Serbia indicates a pronounced dominance of orchids belonging to the Central European and Eurasian chorological groups (Figure 4). The Central European chorological group includes nine taxa from six genera (*Anacamptis, Dactylorhiza, Gymnadenia, Neotinea, Platanthera* and *Spiranthes*). The significant representation of Central European orchids is not surprising, considering that a large part of the Central Balkans has a temperate-continental climate and many different types of habitats where the majority of orchids of this chorological group occur. The Central European mountainous group is represented by five taxa from four genera (*Dactylorhiza, Gymnadenia, Nigritella* and *Traunsteinera*). Considering the numerous high-mountain areas in the Central Balkans, orchid representatives from this chorological group are expected.

The Eurasian chorological group includes nine taxa from seven genera (*Anacamptis, Epipactis, Gymnadenia, Herminium, Neottia, Orchis* and *Platanthera*) (Table 1). Many orchids of this chorological group are characterized by great ecological plasticity, which allows them to grow and survive in different habitats. *Gymnadenia conopsea, Neottia ovata* and *Platanthera bifolia* are among the least specialized and most widespread species [6,15,18].

The boreal chorological group has a significant presence—five orchid taxa from three genera (*Coeloglossum, Dactylorhiza* and *Pseudorchis*). Their occurrence in the Central Balkans can be explained not only by historical factors but also by favorable climatic conditions, adequate habitats as well as by the widespread presence of siliceous geological substrates. The fact is that most siliceous substrates, especially acidic and intermediate igneous rocks, have a high water-holding capacity, and siliceous substrates occupy large areas at higher altitudes suitable for many representatives of boreal orchids [36].

The Mediterranean-Submediterranean chorological group includes four taxa from two genera (*Anacamptis* and *Dactylorhiza*). While *A. laxiflora* is distributed mainly in the Mediterranean and Submediterranean regions and is less common in continental areas, *A. pyramidalis* and *D. saccifera* are species widely distributed throughout the Central Balkans [41]. However, the localities of *D. kalopissi* subsp. *macedonica* and *D. saccifera* in the Central Balkans represent their northern distribution limits in this part of Europe. In addition, *D. kalopissi* subsp. *macedonica* occurs in the Central Balkans only in North Macedonia but is also distributed in Albania, Greece and Bulgaria [50]. *Orchis mascula* subsp. *speciosa* is the only taxon belonging to the Central European-Euxine-Caucasian chorological group.

2.3. Life Forms

The orchid representatives of wetland vegetation in the Central Balkans are terrestrial orchids that display characteristics of the geophyte life form [51,52]. However, we classified orchids according to the concept presented by Tsiftsis et al. [24], Averyanov [26], Štípková et al. [53] and Djordjević et al. [54]: (1) rhizomatous orchids; (2) "intermediate orchids" (intermediate in evolutionary history between rhizomatous orchids and orchids with spheroid tubers), i.e., orchids with palmate, fusiform, or stoloniferous tubers; and (3) tuberous orchids, i.e., orchids with spheroid tubers. The structure of life forms of orchids of wetland vegetation in the Central Balkans is shown in Figure 5.

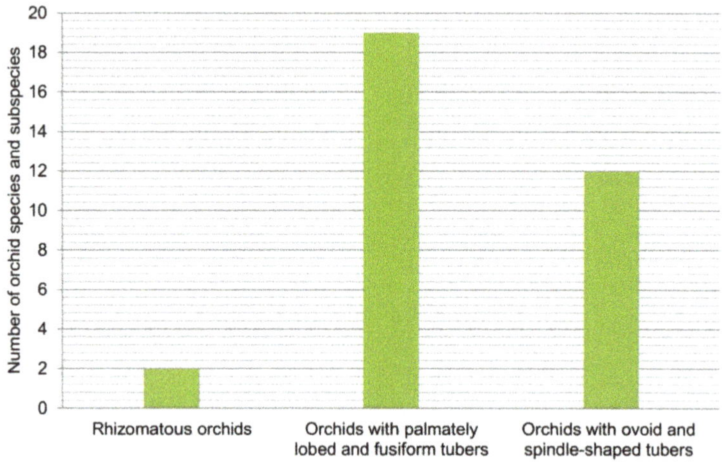

Figure 5. Structure of life forms of orchids of wetland vegetation in the Central Balkans.

The orchid flora of wetland vegetation in the Central Balkans is dominated by "intermediate orchids" (orchids with palmately lobed and fusiform tubers) (Figure 5). This group includes 19 orchid taxa from six genera (*Coeloglossum, Dactylorhiza, Gymnadenia, Nigritella, Platanthera* and *Pseudorchis*). Among these orchids, taxa of the genera *Coeloglossum, Dactylorhiza, Nigritella* and *Gymnadenia* have palmately lobed (finger-like) tubers, whereas species of the genus *Platanthera* are characterized by fusiform tubers. The significant presence of these orchids in the wetland vegetation of the Central Balkans is not surprising considering the origin and evolutionary development of orchids of this life form. The first occurrences of "intermediate orchids" have been associated with Alpine orogeny, i.e., the emergence of lower-temperature mountain habitats [26]. These orchids significantly expanded their range as a result of cooling at the end of the Neogene and in the Pleistocene and were able to colonize areas with plains where the degradation of the Tertiary thermophilic flora took place [26]. Therefore, they can be considered to have well-developed adaptations to the cold and wet conditions of the habitats.

The group having ovoid and spindle-shaped tubers includes 12 orchid species and subspecies classified into six genera (*Anacamptis, Herminium, Neotinea, Orchis, Spiranthes* and *Traunsteinera*). A lower proportion of orchids with spherical tubers in wetland vegetation is to be expected since these orchids usually inhabit dry and semi-dry habitats. Their tubers represent the final stage in the development of the underground organs of orchids, which enable many representatives to survive in habitats with dry and warm conditions [24]. However, *A. laxiflora, A. palustris* subsp. *palustris* and *A. palustris* subsp. *elegans* are taxa that represent exceptions to the rule and grow exclusively in wetland vegetation types.

There are only two orchids with rhizomes (*Epipactis palustris* and *Neottia ovata*). The smaller number of representatives of rhizomatous orchids is understandable because it is known that these orchids occur mainly in forest ecosystems [6,15].

3. Wetland Vegetation

Terrestrial orchids are widely represented in various types of wetland vegetation, including wet meadows, as well as bogs, fens and marshes [15,18,55,56]. In this section, an overview of the main wetland vegetation types with terrestrial orchids and literature sources is presented (Table 2). A total of 33 orchid species and subspecies were recorded in plant communities from five classes, 10 orders and 17 alliances (Table 2). The syntaxonomic nomenclature follows Mucina et al. [57] and Peterka et al. [58].

Table 2. Overview of wetland vegetation types with terrestrial orchids in the Central Balkans.

Vegetation Class	Vegetation Order	Vegetation Alliance	Orchid Species and Subspecies	Literature Sources
Molinio-Arrhenatheretea Tx. 1937	*Molinietalia caeruleae* Koch 1926	*Calthion palustris* Tx. 1937	*Anacamptis palustris* subsp. *elegans*, *Anacamptis morio* subsp. *morio*, *Dactylorhiza incarnata* subsp. *incarnata*, *Dactylorhiza maculata* subsp. *maculata*, *Dactylorhiza maculata* subsp. *transsilvanica*, *Dactylorhiza cordigera* subsp. *cordigera*, *Dactylorhiza cordigera* subsp. *bosniaca*, *Dactylorhiza saccifera*, *Epipactis palustris*, *Gymnadenia conopsea*, *Neottia ovata*, *Platanthera bifolia*, *Traunsteinera globosa*	[15,18,41,59–62]
Molinio-Arrhenatheretea Tx. 1937	*Molinietalia caeruleae* Koch 1926	*Molinion caeruleae* Koch 1926	*Anacamptis coriophora* subsp. *coriophora*, *Anacamptis morio* subsp. *morio*, *Anacamptis palustris* subsp. *palustris*, *Anacamptis palustris* subsp. *elegans*, *Anacamptis pyramidalis*, *Dactylorhiza fuchsii*, *Dactylorhiza incarnata* subsp. *incarnata*, *Dactylorhiza maculata* subsp. *maculata*, *Dactylorhiza maculata* subsp. *transsilvanica*, *Dactylorhiza majalis* subsp. *majalis*, *Dactylorhiza saccifera*, *Dactylorhiza sambucina*, *Epipactis palustris*, *Gymnadenia conopsea*, *Herminium monorchis*, *Neotinea ustulata*, *Neottia ovata*, *Orchis mascula* subsp. *speciosa*, *Orchis militaris*, *Platanthera bifolia*, *Platanthera chlorantha*, *Pseudorchis albida*, *Spiranthes spiralis*, *Traunsteinera globosa*	[15,18,41,60,63–70]
Molinio-Arrhenatheretea Tx. 1937	*Molinietalia caeruleae* Koch 1926	*Deschampsion cespitosae* Horvatić 1930	*Anacamptis coriophora* subsp. *coriophora*, *Anacamptis morio* subsp. *morio*, *Anacamptis palustris* subsp. *palustris*, *Anacamptis palustris* subsp. *elegans*, *Dactylorhiza cordigera* subsp. *cordigera*, *Dactylorhiza incarnata* subsp. *incarnata*, *Dactylorhiza maculata* subsp. *maculata*, *Dactylorhiza maculata* subsp. *transsilvanica*, *Dactylorhiza saccifera*, *Epipactis palustris*, *Gymnadenia conopsea*, *Neotinea ustulata*, *Neottia ovata*, *Platanthera bifolia*, *Traunsteinera globosa*	[15,18,40,41,60,62,66,71–74]
Molinio-Arrhenatheretea Tx. 1937	*Filipendulo ulmariae-Lotetalia uliginosi* Passarge 1975	*Mentho longifoliae-Juncion inflexi* T. Müller et Görs ex de Foucault 2009	*Dactylorhiza incarnata* subsp. *incarnata*, *Dactylorhiza saccifera*, *Epipactis palustris*, *Gymnadenia conopsea*, *Gymnadenia densiflora*, *Platanthera bifolia*	[15,18,41,75]
Molinio-Arrhenatheretea Tx. 1937	*Trifolio-Hordeetalia* Horvatić 1963	*Trifolion pallidi* Ilijanić 1969	*Anacamptis morio* subsp. *morio*, *Anacamptis palustris* subsp. *palustris*, *Anacamptis palustris* subsp. *elegans*	[15,18,41,76–78]
Molinio-Arrhenatheretea Tx. 1937	*Trifolio-Hordeetalia* Horvatić 1963	*Trifolion resupinati* Micevski 1957	*Anacamptis coriophora* subsp. *coriophora*, *Anacamptis laxiflora*, *Anacamptis morio* subsp. *morio*, *Anacamptis palustris* subsp. *elegans*, *Gymnadenia conopsea*	[74,79–86]
Molinio-Arrhenatheretea Tx. 1937	*Trifolio-Hordeetalia* Horvatić 1963	*Molinio-Hordeion secalini* Horvatić 1934	*Anacamptis laxiflora*	[87]
Molinio-Arrhenatheretea Tx. 1937	*Potentillo-Polygonetalia avicularis* Tx. 1947	*Potentillion anserinae* Tx. 1947	*Anacamptis coriophora* subsp. *coriophora*, *Anacamptis palustris* subsp. *palustris*, *Anacamptis palustris* subsp. *elegans*, *Dactylorhiza incarnata* subsp. *incarnata*, *Orchis militaris*	[68,71,88]
Mulgedio-Aconitetea Hadač et Klika in Klika et Hadač 1944c 1944	*Adenostyletalia alliariae* Br.-Bl. 1930	*Cirsion appendiculati* Horvat et al. 1937	*Dactylorhiza maculata* subsp. *maculata*, *Dactylorhiza sambucina*, *Dactylorhiza cordigera* subsp. *cordigera*, *Dactylorhiza saccifera*, *Gymnadenia frivaldii*, *Gymnadenia odoratissima*, *Gymnadenia conopsea*, *Nigritella rhellicani*	[74,89,90]
Phragmito-Magnocaricetea Klika in Klika et Novák 1941	*Phragmitetalia* Koch 1926	*Phragmition communis* Koch 1926	*Anacamptis palustris* subsp. *palustris*, *Anacamptis palustris* subsp. *elegans*, *Dactylorhiza incarnata* subsp. *incarnata*, *Dactylorhiza kalopissii* subsp. *macedonica*, *Epipactis palustris*, *Gymnadenia conopsea*	[15,18,41,60,91–93]

Table 2. *Cont.*

Vegetation Class	Vegetation Order	Vegetation Alliance	Orchid Species and Subspecies	Literature Sources
Phragmito-Magnocaricetea Klika in Klika et Novák 1941	*Magnocaricetalia* Pignatti 1953	*Magnocaricion elatae* Koch 1926	*Anacamptis coriophora* subsp. *coriophora*, *Anacamptis morio* subsp. *morio*, *Dactylorhiza incarnata* subsp. *incarnata*, *Gymnadenia conopsea*, *Traunsteinera globosa*	[15,41,60,94]
Phragmito-Magnocaricetea Klika in Klika et Novák 1941	*Magnocaricetalia* Pignatti 1953	*Magnocaricion gracilis* Géhu 1961	*Anacamptis palustris* subsp. *elegans*	[86,95]
Scheuchzerio palustris-Caricetea fuscae Tx. 1937	*Caricetalia davallianae* Br.-Bl. 1950	*Caricion davallianae* Klika 1934	*Dactylorhiza cordigera* subsp *cordigera*, *Dactylorhiza cordigera* subsp. *bosniaca*, *Epipactis palustris*, *Gymnadenia frivaldii*	[15,27,41,69]
Scheuchzerio palustris-Caricetea fuscae Tx. 1937	*Caricetalia fuscae* Koch 1926	*Caricion fuscae* Koch 1926	*Anacamptis coriophora* subsp. *coriophora*, *Anacamptis morio* subsp. *morio*, *Dactylorhiza cordigera* subsp. *bosniaca*, *Dactylorhiza cordigera* subsp. *cordigera*, *Dactylorhiza incarnata* subsp. *incarnata*, *Dactylorhiza kalopissii* subsp. *macedonica*, *Dactylorhiza maculata* subsp. *maculata*, *Dactylorhiza maculata* subsp. *transsilvanica*, *Dactylorhiza majalis* subsp. *majalis*, *Dactylorhiza saccifera*, *Epipactis palustris*, *Gymnadenia conopsea*, *Gymnadenia frivaldii*, *Herminium monorchis*, *Neottia ovata*, *Nigritella rhellicani*, *Platanthera bifolia*, *Pseudorchis albida*, *Traunsteinera globosa*	[15,18,27,41,59,62,74, 89,93,96–103]
Scheuchzerio palustris-Caricetea fuscae Tx. 1937	*Caricetalia fuscae* Koch 1926	*Narthecion scardici* Horvat ex Lakušić 1968	*Gymnadenia frivaldii*, *Gymnadenia conopsea*, *Dactylorhiza cordigera* subsp. *cordigera*, *Dactylorhiza cordigera* subsp. *bosniaca*, *Pseudorchis albida*	[27,90,102,104]
Scheuchzerio palustris-Caricetea fuscae Tx. 1937	*Caricetalia fuscae* Koch 1926	*Sphagno-Caricion canescentis* Passarge (1964) 1978	*Dactylorhiza cordigera* subsp. *cordigera*, *Dactylorhiza incarnata* subsp. *incarnata*, *Dactylorhiza maculata* subsp. *maculata*, *Dactylorhiza maculata* subsp. *transsilvanica*, *Epipactis palustris*	[15,27,41,103]
Montio-Cardaminetea Br.-Bl. et Tx. ex Klika et Hadač 1944	*Montio-Cardaminetalia* Pawłowski et al. 1928	*Cardamino-Montion* Br.-Bl. 1926	*Dactylorhiza cordigera* subsp. *cordigera*	[89]

Orchid richness in relation to vegetation classes, orders and alliances in the Central Balkans is presented in Figures 6–8. The greatest number of orchids was recorded in the class *Molinio-Arrhenatheretea* (28 taxa or 84.9% of the total analyzed orchid flora), followed by *Scheuchzerio palustris-Caricetea fuscae* (19 taxa), *Phragmito-Magnocaricetea* (nine taxa), *Mulgedio-Aconitetea* (eight taxa) and *Montio-Cardaminetea* (one taxon) (Figure 6).

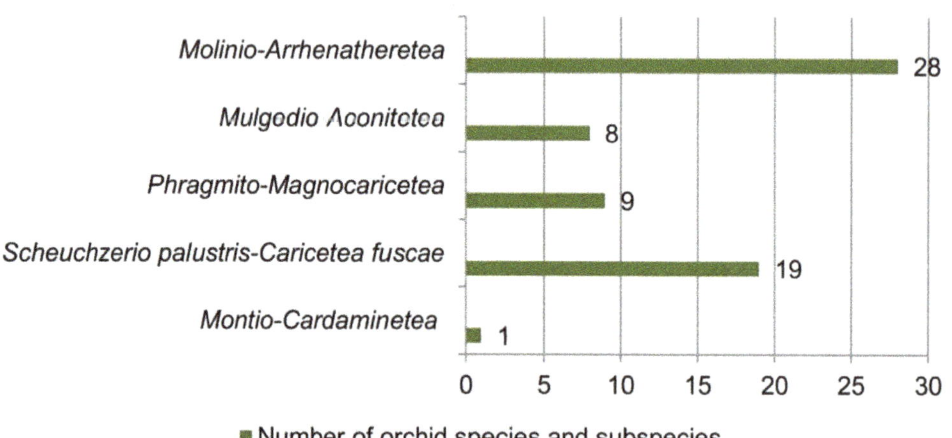

Figure 6. Richness of orchid species and subspecies in relation to vegetation classes.

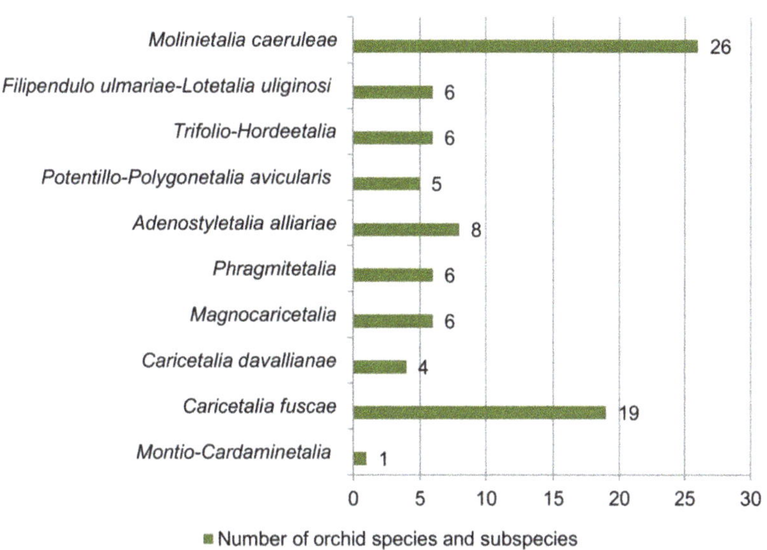

Figure 7. Richness of orchid species and subspecies in relation to vegetation orders.

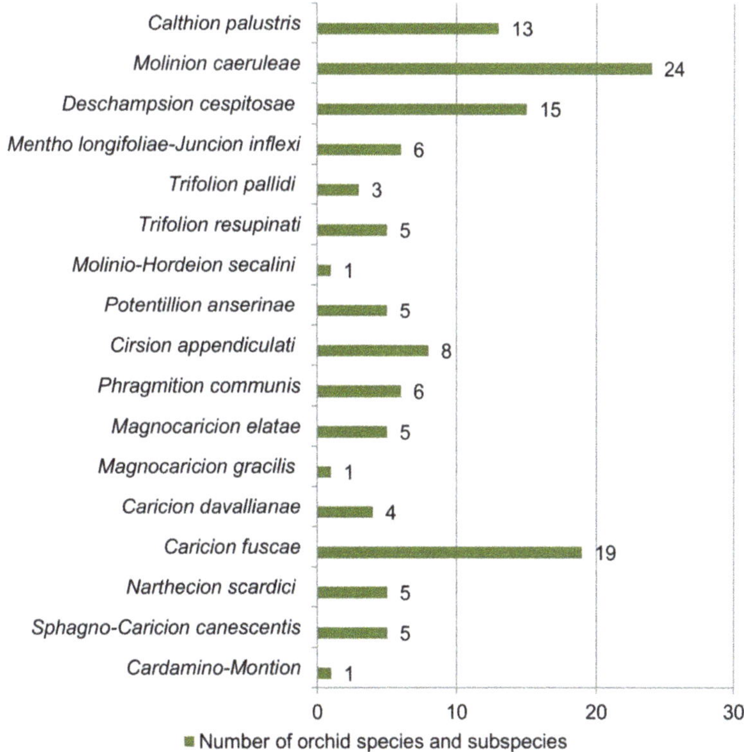

Figure 8. Richness of orchid species and subspecies in relation to vegetation alliances.

Concerning vegetation orders, the greatest number of orchids was recorded in the *Molinietalia caeruleae* (26 taxa or 78.8% of the total analyzed orchid flora), followed by

Caricetalia fuscae (19 taxa), *Adenostyletalia alliariae* (eight taxa), *Filipendulo ulmariae-Lotetalia uliginosi*, *Trifolio-Hordeetalia*, *Phragmitetalia*, *Magnocaricetalia* (six taxa each) and *Potentillo-Polygonetalia avicularis* (five taxa) (Figure 7). The smallest number of orchid taxa was found in the orders *Caricetalia davallianae* and *Montio-Cardaminetalia* (Figure 7).

Regarding the affiliation to vegetation alliances, the greatest number of orchid taxa was recorded in the *Molinion caeruleae* (24 taxa or 72.7%), followed by *Caricion fuscae* (19 taxa), *Deschampsion cespitosae* (15 taxa), *Calthion palustris* (13 taxa), *Cirsion appendiculati* (eight taxa), *Mentho longifoliae-Juncion inflexi*, *Phragmition communis* (six taxa each), and *Trifolion resupinati*, *Potentillion anserinae*, *Magnocaricion elatae*, *Narthecion scardici* and *Sphagno-Caricion canescentis* (five taxa each). The smallest number of orchid taxa was found in the alliances *Trifolion pallidi*, *Molinio-Hordeion secalini*, *Magnocaricion gracilis*, *Caricion davallianae* and *Cardamino-Montion* (Figure 8).

3.1. Wet Meadows

Orchids of the Central Balkans, which require hygrophilous and hygro-mesophilous habitat conditions, are particularly abundant in communities of the vegetation class *Molinio-Arrhenatheretea*. Many orchid taxa, including numerous taxa of the genus *Dactylorhiza*, were found in communities of the order *Molinietalia caeruleae* (mown meadows on mineral and peaty soils), especially in the alliances *Molinion caeruleae*, *Deschampsion cespitosae* and *Calthion palustris* (Table 2).

Recent studies from Serbia have shown that in the case of the alliance *Molinion caeruleae*, orchids are most abundant in stands of the communities *Molinietum caeruleae* W. Koch 1926 (Figure 9), *Molinio caeruleae-Deschampsietum cespitosae* Pavlović 1951 and *Lathyro pannonici-Molinietum caeruleae* Tatić et al. ex Aćić et al. 2013 [41]. Moreover, it has been shown that the composition of orchids in this vegetation type largely depends on the bedrock types. For example, *Molinion caeruleae* communities in Serbia on Quaternary sediments and carbonate clastites are particularly suitable for *Dactylorhiza incarnata* and *Anacamptis palustris* subsp. *elegans*, while *Molinion caeruleae* communities on serpentine support significant populations of *Platanthera bifolia* and *Dactylorhiza maculata* subsp. *transsilvanica* [18,105]. The importance of the order *Molinietalia caeruleae* as an important vegetation type for orchids has also been recognized in other European regions. The following orchid taxa have been recorded in *Molinion caerulae* communities in Europe: *Epipactis palustris*, *Dactylorhiza majalis*, *D. maculata* subsp. *maculata*, *Neotinea ustulata*, *Gymnadenia conopsea*, *G. densiflora* and *Neottia ovata* [11,13,45,106].

Figure 9. The association *Molinietum caeruleae* W. Koch 1926 (Serbia, photos V. Djordjević).

In the Central Balkans, a significant occurrence of orchids has been noted within the alliance *Calthion palustris*, which represents wet grasslands and tall herb communities that are often unmanaged and found on flat lands along streams or on saturated soils near headwaters. Orchids are most frequently recorded in communities of *Equiseto palustris-Eriophoretum latifolii* Petković ex Aćić et al. 2013, *Scirpetum sylvatici* Ralski 1931 (Figure 10), *Calthaetum palustris* s.l. and *Cirsietum rivularis* Nowiński 1927 (Figure 11). Previous studies in Europe have shown that *Dactylorhiza incarnata, D. maculata, D. majalis, D. praetermissa, D. saccifera, D. cordigera, Epipactis palustris, Gymnadenia conopsea* and *Neottia ovata* have significant populations in *Calthion palustris* communities [11,13,19,45,107–109].

Figure 10. The association *Scirpetum sylvatici* Ralski 1931 (Serbia, photo V. Djordjević).

Figure 11. The association *Cirsietum rivularis* Nowiński 1927 (Serbia, photo V. Djordjević).

In addition, many orchids occur in the Central Balkans in communities of the alliance *Deschampsion cespitosae*, which are mown temporarily wet meadows on heavy soils on floodplains in the forest and forest-steppe zones of (sub)continental Central and Eastern Europe. The orchids were most frequently recorded in the Central Balkans within the following communities: *Deschampsietum cespitosae* Horvatić 1930, *Agrostio stoloniferae-Juncetum effusi* Cincović 1959, *Junco articulati-Deschampsietum cespitosae* Petković ex Aćić et al. 2013 and *Rhinantho borbasii-Festucetum pratensis* Gajić ex Aćić et al. 2013 (Table 2). According to earlier published data from Europe, *Dactylorhiza incarnata, D. saccifera, Epipactis palustris, Gymnadenia conopsea* and *Platanthera bifolia* are orchids commonly found in communities of this alliance [11,13,107].

Some orchids in the Central Balkans have significant representation within the vegetation order *Filipendulo ulmariae-Lotetalia uliginosi* (tall-herb wet meadow fringe vegetation on mineral soils) (Table 2). Within this order, *Epipactis palustris* and *Dactylorhiza incarnata* are among the most common species, especially abundant in *Mentho longifoliae-Juncion inflexi* communities [18,41].

Orchids are less prevalent in communities of the vegetation order *Trifolio-Hordeetalia* (Table 2). This vegetation type represents the wet meadows of the humid continental regions of the north-central Balkans, occurring on clayey, mesotrophic to eutrophic soils on riverside terraces and gentle slopes along the rivers [32,57,110]. *Anacamptis palustris* subsp. *elegans* is one of the most common taxa that have been recorded both in communities of the alliance *Trifolion resupinati* (vegetation of wet meadows of the subarid continental regions of the Southern Balkans) and in communities of the alliance *Trifolion pallidi* (vegetation of wet meadows of the humid continental regions of the north-central Balkans). *Anacamptis laxiflora* is especially common in communities of the alliances *Trifolion resupinati* (*Cynosuro-Caricetum hirtae* K. Micevski 1957, *Hordeo-Caricetum distantis* K. Micevski 1957 and *Trifolietum nigrescentis-subterranei* K. Micevski 1957) and *Molinio-Hordeion secalini* [79].

Some orchid taxa in the Central Balkans have been recorded in communities of the order *Potentillo-Polygonetalia avicularis* Tx. 1947 and the alliance *Potentillion anserinae* Tx. 1947 (Table 2). These are temporarily flooded and heavily grazed nutrient-rich pastures experiencing variable wet-dry or brackish-freshwater alternating conditions of temperate Europe [57,110,111].

3.2. Tall-Herb Vegetation along Mountain Streams and Springs

Representatives of the family Orchidaceae are less abundant in communities of the vegetation class *Mulgedio-Aconitetea* in the Central Balkans (Table 2). This vegetation represents tall-herb vegetation in nutrient-rich habitats moistened and fertilized by percolating water at high altitudes in Europe, Siberia and Greenland [57]. Within this vegetation class, certain orchid species were recorded in communities of the order *Adenostyletalia alliariae* (tall-herb vegetation on fertile soils at high altitudes of temperate and Mediterranean Europe) and the alliance *Cirsion appendiculati* (tall-herb vegetation on acidic soils along mountain streams and springs at high altitudes of the Eastern and Central Balkans) (Table 2).

3.3. Marshland Vegetation

In the Central Balkans, orchids also inhabit marsh communities of the class *Phragmito-Magnocaricetea* (reed, sedge bed and herb-land vegetation of freshwater or brackish water bodies and streams of Eurasia) (Table 2). Based on recent studies in the Central Balkans, it can be stated that especially *Dactylorhiza incarnata* and *Epipactis palustris* are significantly represented in the communities of *Magnocaricion elatae* (*Magnocaricetalia*) and *Phragmition communis* (*Phragmitetalia*) (Figure 12), whereas *Anacamptis palustris* subsp. *elegans* is recorded in the community of *Magnocaricion gracilis* (*Magnocaricetalia*). In Germany, *E. palustris* has also been recorded in communities of *Magnocaricion elatae* (marsh vegetation on oligotrophic to mesotrophic organic sediments of temperate Europe) [107]. In addition, *Dactylorhiza incarnata, D. majalis, Epipactis palustris, Hammarbya paludosa* and *Liparis loeselii* were found in the Czech Republic, Hungary and Germany in communities with *Phragmites australis* as a strongly represented species [109,112].

Figure 12. The association *Phragmitetum australis* Savič 1926 (Serbia, photo V. Djordjević).

3.4. Vegetation of Bogs and Fens

The vegetation class *Scheuchzerio palustris-Caricetea fuscae* (fens, transitional mires and bog hollows in the temperate, boreal and Arctic zones of the Northern Hemisphere) represents important vegetation types for many moisture-demanding orchid taxa in the Central Balkans (Table 2). This vegetation type has been estimated to occupy less than 0.001% of the total Serbian territory [18], so the existence of 19 orchid taxa in these wetland communities in the Central Balkans indicates its great conservation value. Moreover, recent studies in western Serbia indicated that four orchids (*Dactylorhiza cordigera* subsp. *cordigera, D. maculata* subsp. *maculata, D. saccifera* and *Gymnadenia frivaldii*) were significantly correlated with this vegetation class [18]. Orchids in the Central Balkans were recorded in communities of the order *Caricetalia fuscae* (sedge-moss vegetation of acidic fens in the boreal and temperate zones and in the supra-Mediterranean belt of mountains in Southern Europe) (Table 2). Within the order *Caricetalia fuscae*, orchids are significantly represented in the following communities: *Carici-Sphagno-Eriophoretum* R. Jovanović 1978, *Eriophoro-Caricetum paniculatae* R. Jov. 1983 (Figure 13), *Eriophoro-Caricetum echinatae* V. Randjelović 1998 (within the alliance *Caricion fuscae*), and *Sphagno-Caricetum nigrae* P. Lazarević 2016, *Molinio-Sphagnetum fusci* P. Lazarević 2016, *Sphagno-Caricetum rostratae* P. Lazarević 2016 (within the alliance *Sphagno-Caricion canescentis*) (Table 2).

Orchids belonging to the alliance *Carici-Nardion* V. Ranđelović 1998 at the national level are assigned to the alliance *Caricion fuscae* [57,58]. These are wet communities dominated by *Nardus stricta*, which are not well defined and for which research is still needed, not only in ecological terms but also in terms of nomenclature and classification. Some of the typical orchid taxa in these communities are *Anacamptis coriophora* subsp. *coriophora, Dactylorhiza sambucina, D. maculata* subsp. *maculata, D. maculata* subsp. *transsilvanica, D. cordigera* subsp. *bosniaca, D. cordigera* subsp. *cordigera, Gymnadenia conopsea, Platanthera bifolia* and *Traunsteinera globosa*. In addition, it should be noted that the separation of the alliances *Sphagno-Caricion canescentis* and *Caricion fuscae* in the area of the Central Balkans requires additional studies.

Figure 13. The association *Eriophoro-Caricetum paniculatae* R. Jov. 1983 (Serbia, photo V. Djordjević).

The specificity of the Central Balkans is the presence of orchids in the alliance *Narthecion scardici*, which represents relic oro-Mediterranean moderately-rich fens of the Balkans. Within this alliance, orchids are significantly represented in the community *Carici-Narthecietum scardici* Ht. 1953 [102]. The communities of this alliance have great conservation value, hosting significant populations of *Gymnadenia frivaldii, Dactylorhiza cordigera* subsp. *cordigera, D. cordigera* subsp. *bosniaca* and *Pseudorchis albida*. Orchids in the Central Balkans are less prevalent in communities of the vegetation alliance *Caricion davallianae* (sedgemoss calcareous mineral-rich fen vegetation of Europe and Western Asia) within the order *Caricetalia davallianae* (Table 2).

Communities of the class *Scheuchzerio palustris-Caricetea fuscae* are considered important for the growth and survival of numerous orchids in Europe. The following orchids have significant representation within this vegetation class in other European countries: *Anacamptis palustris* subsp. *palustris, Liparis loeselii, Dactylorhiza cordigera, D. maculata, D. majalis, D. fuchsii, D. incarnata, D. lapponica, D. russowii, D. traunsteineri, Epipactis palustris, Gymnadenia densiflora, G. frivaldii, G. conopsea, Malaxis monophyllos, Herminium monorchis, Hammarbya paludosa, Neottia ovata, Platanthera bifolia, Pseudorchis albida, Spiranthes aestivalis* and *S. sinensis* [11–13,19,45,57,112–117].

3.5. Vegetation of Springs

In the Central Balkans, orchids are less common in communities of the vegetation class *Montio-Cardaminetea* (vegetation of springs of Europe, the European Arctic archipelagos and Greenland) (Table 2). Within this vegetation class, only *Dactylorhiza cordigera* subsp. *cordigera* was found in communities of the order *Montio-Cardaminetalia* and the alliance *Cardamino-Montion* (vegetation of springs with cold and nutrient-poor water in the subalpine and alpine belts of mountains of central and southwestern Europe).

4. Geological Substrates

The geological substrates and soil properties represent important factors influencing the diversity patterns of terrestrial orchids [7,17,118,119]. Recent studies in the Central Balkans have shown that the bedrock type significantly affects the distribution, abundance and composition of orchids of wetland vegetation and that the greatest differences occur when comparing orchids in habitats on serpentine, carbonate and silicate bedrocks [18,36,105]. These studies underline the important role of bedrock types in separating niches of orchid taxa. Differences in the chemical and physical composition of geological substrates and soils also affect the size of orchid populations [105].

The carbonate geological substrates and soils are the most important for the growth and development of orchids in Central Europe [6,36,120,121]. The great representation of orchid taxa on carbonates in the Central Balkans is explained not only by the physical and chemical properties of the substrate but also by the considerable surface area of this substrate, considering that carbonate substrates are represented from lowlands to high-mountain areas [36]. However, many orchids, known to be characteristic species of carbonate habitats, have also been found on non-carbonate geological substrates in the study area. For example, *Epipactis palustris*, *Dactylorhiza fuchsii*, *Gymnadenia conopsea*, *Nigritella rhellicani* and *Neotinea ustulata* were found to grow in the Central Balkans on limestone-dolomite and carbonate clastites, as well as on various types of silicate substrates, whereas previous studies indicated that these species occur mainly or exclusively on carbonate substrates [6,11,13,122].

Recent studies on orchid ecological preferences suggest that wet habitats on serpentines are particularly important to the survival of numerous orchid species [18,36,105]. Orchids with large population sizes that are common in wet habitats on serpentines in the Central Balkans include *Gymnadenia conopsea*, *Platanthera bifolia*, *Dactylorhiza maculata* subsp. *transsilvanica*, *D. maculata* subsp. *maculata*, *D. sambucina*, *D. incarnata*, *Anacamptis coriophora* and *A. morio*, whereas *Coeloglossum viride*, *Traunsteinera globosa*, *Spiranthes spiralis*, *D. saccifera* and *A. pyramidalis* occur somewhat less frequently on these substrates [36,41]. The surprisingly large number of orchid taxa found in wet habitats can be explained by the physical and chemical properties of serpentine soils, especially their low nutrient content, as most orchid species are sensitive to increased phosphorus, nitrogen and potassium content in the soil [118,123–126]. It is known that serpentine substrates allow the development of open habitats with a generally low level of competition between plants, which enables the survival of low-competitive orchid taxa that have high light requirements [105]. In addition, mycorrhizal fungi are thought to play a key role in increasing tolerance to high levels of heavy metals in serpentine soils. Although serpentine soils are characterized by high concentrations of Ni, Cr and Co, an unfavorable ratio of Ca to Mg, and low content of macronutrients (N, P and K) [127], the impact of these specific characteristics is much lower when soils are moist and well developed, which is usually the case in wetlands on serpentine bedrock. This is one possible reason why many species characteristic of carbonate substrates are abundant in serpentine wetlands.

Orchids growing in wetland vegetation in the Central Balkans are very common on ophiolitic mélanges and sandstones from the Carboniferous and Permian periods, which include diabase, gabbro, spilite, cherts, sandstones, shales and marls of the Jurassic period, and sandstones from the Carboniferous and Permian periods. The great abundance of orchids on these geological substrates is due to their heterogeneous composition since these volcanogenic-sedimentary formations (the old name is "diabase-chert formation") usually contain diabase and cherts [128–130]. Orchids highly represented on ophiolitic mélanges and sandstones from the Carboniferous and Permian periods include the Carpathian-Balkan subendemics (*Dactylorhiza cordigera* subsp. *cordigera*, *D. maculata* subsp. *transsilvanica*, *Gymnadenia frivaldii*), as well as *Anacamptis morio*, *A. coriophora*, *A. pyramidalis*, *Epipactis palustris*, *Traunsteinera globosa*, *Neotinea ustulata*, *Platanthera bifolia*, *Pseudorchis albida*, *Dactylorhiza incarnata*, *D. maculata* subsp. *maculata*, *D. fuchsii* and *D. saccifera*. On the Stara planina mountain (eastern Serbia), on the substrate of the "red sandstone formation": conglomerates, sandstones and siltstones from the Permian period, significant populations of the following orchids were found in wetland vegetation: *Gymnadenia frivaldii, G. conopsea, Pseudorchis albida, Dactylorhiza cordigera* subsp. *cordigera, Dactylorhiza saccifera* and *Traunsteinera globosa* [131].

Many species inhabiting wetland vegetation in the Central Balkans have been found on metamorphic rocks (schists, gneisses and phyllites) [18,36]. Among them, those that occur mainly in high-altitude areas and are rare in the study area stand out. For example, *Gymnadenia frivaldii* and *Nigritella rhellicani* are very common on phyllites on Golija Mountain in western Serbia [41]. A recent study revealed that these two orchids are indicator species of schists, gneisses and phyllites [18]. Orchids are found very often on this bedrock type in western Serbia and on Kopaonik mountain, Mts Šar-Planina and Vlasina Plateau [62,99,132]. Other orchid taxa that have a large presence on schists, gneisses and phyllites in the Central Balkans are *Dactylorhiza maculata* subsp. *maculata, D. incarnata, D. saccifera, Gymnadenia conopsea, Anacamptis morio, Platanthera bifolia, Traunsteinera globosa* and *Epipactis palustris* [18,36,41,62,99,132].

Some orchid taxa of wetland vegetation in the Central Balkans have been recorded on acidic igneous rocks [18,36,41]. Orchids that occur to a considerable extent on quartz latites are *Dactylorhiza incarnata, D. maculata* subsp. *maculata, D. saccifera, D. sambucina, Nigritella rhellicani, Gymnadenia conopsea, Coeloglossum viride* and *Traunsteinera globosa*, while the orchids that are particularly abundant on granodiorites are *D. cordigera* subsp. *cordigera, D. cordigera* subsp. *bosniaca, D. saccifera, G. conopsea, G. frivaldii* and *T. globosa* [18,36,41,62,99,132]. Furthermore, orchid taxa of herbaceous wetlands were found growing on intermediate igneous rocks (andesite, dacite and porphyrite) [18,36,41]. Among these species, the following should be highlighted: *Anacamptis morio, A. laxiflora, Dactylorhiza incarnata, Epipactis palustris, Gymnadenia conopsea, Platanthera bifolia* and *Traunsteinera globosa*.

In the Central Balkans, numerous water-demanding orchids have been found on Quaternary sediments that include proluvial and alluvial deposits, eluvial-deluvial sediments and fluvial terraces. *Anacamptis palustris* subsp. *palustris, Anacamptis palustris* subsp. *elegans, A. morio, A. pyramidalis, Dactylorhiza incarnata, D. saccifera, Epipactis palustris, Gymnadenia conopsea, Neottia ovata, Orchis militaris* and *Traunsteinera globosa* grow on this type of substrate [18,36,41]. The lowest number of orchid species in the wetlands of the Central Balkans was recorded on flysch, which is a series of sedimentary rocks where marls, clay shales, sandstones, conglomerates and limestones are the most common [133]. Among the species that occur on this type of substrate, the following are noteworthy: *Dactylorhiza maculata* subsp. *maculata, Anacamptis morio, Gymnadenia conopsea, Platanthera bifolia* and *Traunsteinera globosa* [18,36,41].

5. Threat Factors and Conservation Priorities

5.1. Threat Factors

Factors threatening orchids of wetland vegetation in the Central Balkans can be classified into several groups: (a) hydrologic regime alteration; (b) pollution; (c) uncontrolled urbanization, industrialization and construction of transport infrastructures; (d) grazing intensity, mowing time and frequency; (e) agriculture; (f) tourism; (g) invasive and non-native species; (h) collection of orchids; and (i) climate change.

Hydromelioration works, soil drainage, creation of hydroaccumulations, capture of springs, channelization of natural runoff, deepening and straightening of river courses and other forms of hydrologic regime alterations are the main factors threatening water-demanding orchids in the Central Balkans. In Peštersko polje (southwestern Serbia), significant changes in the hydrological regime were made when a system of canals, dams and levees was built to divert the water basin into the Uvac hydropower system [27]. It is assumed that due to the drainage of the central part of Peštersko polje, the groundwater level has decreased and the vegetation has developed from fen communities of the class *Scheuchzerio palustris-Caricetea fuscae* to wet and mesophilous meadows [27,41]. It is important to emphasize that uncontrolled water use at springs and in the headwaters of rivers, especially in mountainous regions, affects the water balance of entire regions and poses a potential threat to many orchid species.

Various forms of physical, chemical and biological pollution, directly and indirectly, threaten orchids of wetland vegetation in the Central Balkans. A direct negative impact can be seen in the vicinity of agricultural land, rural households, transport routes, industrial plants and tourist facilities. Orchids are particularly threatened by wastewater discharge, municipal waste disposal and soil nitrification. Since orchids are particularly sensitive to increased levels of nitrogen and phosphorus in the soil [118], their lower occurrence has been observed in wet meadows and fens near farms that use artificial or natural fertilizers and pesticides. One of the examples is Divčibare (northwestern Serbia), where waste oil, fuel oil and fecal water are occasionally discharged into the upper reaches of the river from restaurants and hotels. This polluted the soil, surface and groundwater and directly affected the degradation of this part of the mire area [41].

Uncontrolled urbanization, industrialization and construction of transport infrastructures without ecologically oriented spatial planning pose a significant threat to orchids of the Central Balkans. Urbanization and road construction have destroyed many wet habitats of orchids, especially in lowland areas and near tourist centers. Roads cut through natural ecosystems, disrupt or prevent communication between coenobionts, increase the erosion process and affect water and soil pollution, threatening orchids directly and indirectly. Incidentally, based on studies that included over 8,000 plant species, it was found that representatives of the family Orchidaceae have the highest risk of disappearing from the immediate vicinity of cities [134]. Among the many negative consequences of urbanization, the above authors pointed out in particular the decline of orchid populations and competition with invasive species. A particularly sharp decline in orchid populations due to a high degree of urbanization was observed in the northern areas of Western Europe (northern France, Belgium and Luxembourg) [5,22].

Intesive grazing and mowing in lowland, mountain and high-mountain areas of the Central Balkans negatively affect orchid taxa in wetland vegetation. Extensive animal husbandry leads to the intensification of erosion processes, damage to soil structure and quality, and thus to negative zoo-anthropogenic selection of plant cover. Grazing by cattle and sheep leads to the spread of the species *Nardus stricta* L. and the degradation of many mires and wet meadow ecosystems. Livestock management is the greatest threat to the survival of endangered plant species in Europe [135]. The negative consequences are not only due to direct grazing, but the impoverishment of the floristic composition of plant species is mainly due to nitrification and soil compaction. Early mowing of wet meadows has a negative impact, especially, on orchids that complete their reproductive phase (seed formation) by the time of mowing. The negative effects of this factor are mainly seen in the reduction of cross-pollination. It is important to note that the complete abandonment of traditional activities such as mowing or grazing would threaten the survival of many orchid species, as open habitats would thus be threatened by the development of forest and shrub vegetation [135]. Without the above-mentioned traditional activities, mires and wet meadows are particularly at risk due to the establishment of forest vegetation [27]. Previous studies have shown that mowing to some extent reduces competition between plants in the habitat and thus has a beneficial effect on the development of orchid populations. Regular annual mowing, when carried out in seasons when orchids do not appear above ground, has been shown to be beneficial to the optimal development of many species of the genus *Dactylorhiza* [19,20].

The spreading of arable land at the expense of natural ecosystems (wet meadows, fens and marshes) threatens orchids in the Central Balkans. In addition to the direct loss of natural habitats where orchids grow, the negative effects of the formation of agroecosystems can be seen in the fragmentation of habitats, fertilization of the soil and pollution of the soil with chemical substances, especially pesticides.

Uncontrolled tourism development is another important factor threatening orchids in the wetlands of the Central Balkans. Tourism has a negative impact on the status of orchid populations, especially in the mountain tourist areas of the Central Balkans. The negative effects of this factor are manifested in the fragmentation and destruction of wet habitats where orchids grow, disruption of the water balance, ruderalization of ecosystems and pollution of air, water and soil [136]. The most severe impacts of tourism have been found in the mountains of Zlatibor and Kopaonik (Serbia), where many wet habitats have been destroyed or degraded. Previous research has shown that orchids in tourist areas are threatened primarily by habitat loss, picking by individuals, trampling of areas and ecosystem disturbance by motor vehicles and bicycles, as well as horseback riding [136–138].

Invasive and non-native species also threaten orchids in the Central Balkans. The negative impacts due to the disruption of cenotic relationships and reduction of biodiversity are most evident in lowland areas, near roads, agricultural lands, rural households and tourist centers, where many wet meadows are ruderalized and under the strong influence of invasive and non-native species (*Erigeron annuus*, *Conyza canadensis*, *Ambrosia artemisiifolia*, *Ailanthus altissima*, *Robinia pseudoacacia* and others) [41]. However, the negative impacts of non-native and invasive species in high-altitude areas have not been observed. In North America, invasive and non-native species threaten especially orchids of mire habitats [5].

Although orchid collecting is the most important threat to orchids on the global IUCN Red List [139], this factor does not pose a significant threat to wetland orchids in the Central Balkans. The consequence of picking orchids is a decrease in reproductive success, considering that they are prevented from cross-pollination and reproduction by seed [139]. The aboveground parts of orchids are harvested for their decorative flowers, especially in tourist areas. The use of orchid tubers for the production of the drink salep has been noted in the Pešter region (southwestern Serbia) and North Macedonia [140]. From a survey in the Pešter area, it appears that the locals use mainly *Anacamptis morio* and *Gymnadenia conopsea* for the production of the drink salep [41]. The production of salep threatens the survival of many orchid species, especially in the eastern Mediterranean countries, where salep is traditionally used as a food, tonic and aphrodisiac [141].

Climate change is another factor threatening the survival of orchids of wetland vegetation in the Central Balkans. Considering that the global temperature has increased in the last century (1.1 °C warming since 1850–1900) and that an average warming of 1.5 °C or more is expected in the next 20 years [142], drought is expected to lead to a decrease in the distribution of wetlands and consequently of orchids, and many species will be restricted altitudes. Thus, orchids in lowland and mid-altitude areas are more at risk than at higher altitudes due to higher temperatures. Orchids that grow exclusively in wetland vegetation types are most at risk, while orchids that are generalists, i.e., orchids that inhabit other habitats (dry and semi-dry grasslands and forest habitats), are less at risk. From a recent study on the effects of climate change on the distribution of *Traunsteinera globosa* and its pollinators, it appears that the distribution of *T. globosa* may decline significantly as a result of global warming, and pollinators of this orchid will also face a loss of habitat [143]. As a warmer climate makes growing seasons longer and warmer, increases productivity and decreases water levels, these effects increase the duration and intensity of interspecific competition, encourage competing species and force the niches of specialized wetland species towards narrower pH ranges [144]. This means that orchids known to be weakly competitive will face stronger competition, and the question for future research is how climate change will affect orchids with different ecological preferences for soil pH.

In addition to the factors already mentioned, the internal factors affecting the distribution and abundance of orchids are natural factors that operate during the belowground (need for mycorrhizal association) and aboveground (need for successful pollination) stages of orchid development [2,3]. It is important to emphasize that for most terrestrial orchids, the presence and effectiveness of mycorrhizae in the soil have a greater influence on survival than other factors.

5.2. Conservation Priorities

The conservation priorities defined in this study are based on the degree of representation of orchids of wetland vegetation (Table 1), the marginality and breadth of the species' niches [18], the size of their populations, the rarity and conservation status of their habitats, as well as the extent of their geographical distribution.

Special attention should be paid to orchid taxa that occur exclusively or mainly in wet habitats (Table 1). These orchids have the highest level of habitat specialization [18], the highest requirements for soil moisture, and their habitats (especially the fens) are the rarest and most threatened habitats in the study area [27,41]. Moreover, both the study area and the entire Balkan Peninsula represent one of the most important centers of evolution, diversity and endemicity of the genus *Dactylorhiza*, to which most specialists belong [17,18,26,35]. Conservation of these orchids requires ensuring adequate water supply in wet meadows, fens and marshes, while the optimal performance of many *Dactylorhiza* taxa can be achieved by regular annual mowing [19,20].

Special priority should be given to orchids whose southernmost range falls within the study area (e.g., *Dactylorhiza maculata* subsp. *maculata*, *D. maculata* subsp. *transsilvanica*, *D. cordigera* subsp. *bosniaca*, *D. fuchsii*, *D. majalis* and *Traunsteinera globosa*). Biomonitoring of these taxa in light of global warming is necessary because they are expected to respond rapidly to climatic changes [35].

One of the conservation priorities relates to wet habitats occurring on serpentine in the study area. Because the wet serpentine habitats of the Central Balkans are less used for agriculture, they could be considered as potential orchid reserves, especially considering that orchids are very common in these habitats and have large population sizes. In addition, wetland vegetation types at higher altitudes, occurring on silicate rocks known for their water-holding capacity, harbor numerous representatives of orchids and can be considered the most important habitats for specialized orchids, including an endemic taxon of the Balkans (*Dactylorhiza cordigera* subsp. *bosniaca*) and subendemic taxa of the Balkans and Carpathians (*Dactylorhiza cordigera* subsp. *cordigera*, *D. maculata* subsp. *transsilvanica* and *Gymnadenia frivaldii*).

6. Conclusions

The presence of 33 orchid species and subspecies was established in the wetland vegetation types of the Central Balkans. *Dactylorhiza* is the most taxon-rich genus (with ten taxa), followed by *Anacamptis* (six taxa) and *Gymnadenia* (four taxa). The phytogeographical analysis shows that representatives of the Central European and Eurasian chorological groups dominate, followed by orchids of the Central European mountainous and boreal groups. The analysis of life forms revealed that representatives with palmately lobed and fusiform tubers are dominant, followed by orchids with ovoid and spindle-shaped tubers and orchids with rhizomes.

According to the degree of occurrence in wetland vegetation types, eight taxa were found to occur exclusively in these vegetation types, six taxa grow mainly in wetland vegetation types and rarely occur in other vegetation types, 11 taxa grow in wetland vegetation types but occur more frequently in other vegetation types, while eight taxa occur very rarely in wetland vegetation types. Most of the orchid taxa were found in communities of the classes *Molinio-Arrhenatheretea* and *Scheuchzerio palustris-Caricetea fuscae*; the orders *Molinietalia caeruleae* and *Caricetalia fuscae* and the alliances *Molinion caeruleae*, *Caricion fuscae*, *Deschampsion cespitosae* and *Calthion palustris*.

Serpentine and silicate bedrock types and their wet habitats in the Central Balkans are important for many orchids, suggesting that they may play an important role in orchid conservation. The study highlights the importance of establishing biomonitoring for orchids that have southern limits of their distribution in the Central Balkans, in the face of global warming. Future detailed taxonomic, chorological and ecological studies of orchids of wetland vegetation in the Central Balkans are necessary to conduct their successful conservation.

Author Contributions: Conceptualization, V.D.; data collection was performed by V.D., S.A., E.K., P.L. and D.L.; writing—original draft preparation, V.D.; writing—review and editing, V.D., S.A., E.K., S.T. and D.L. All authors have read and agreed to the published version of the manuscript.

Funding: This research was supported by the Science Fund of the Republic of Serbia, grant number 7750112—Balkan biodiversity across spatial and temporal scales—patterns and mechanisms driving vascular plant diversity (BalkBioDrivers).

Institutional Review Board Statement: Not applicable.

Informed Consent Statement: Not applicable.

Data Availability Statement: The data presented in this study are available on request from the corresponding author.

Acknowledgments: The authors thank Ivana Stevanoski for providing a photograph of *Anacamptis palustris* subsp. *palustris*. The authors thank Slavčo Hristovski (North Macedonia) and Đorđije Milanović and Elvedin Šabanović (Bosnia and Herzegovina) for their help in finding specific literary data. We greatly thank and appreciate the two anonymous reviewers and the editor for their useful suggestions and comments on a previous version of the manuscript.

Conflicts of Interest: The authors declare no conflict of interest.

References

1. Givnish, T.J.; Spalink, D.; Ames, M.; Lyon, S.P.; Hunter, S.J.; Zuluaga, A.; Doucette, A.; Caro, G.G.; McDaniel, J.; Clements, M.A.; et al. Orchid historical biogeography, diversification, Antarctica and the paradox of orchid dispersal. *J. Biogeogr.* **2016**, *43*, 1905–1916. [CrossRef]
2. Waterman, R.J.; Bidartondo, M.I. Deception above, deception below: Linking pollination and mycorrhizal biology of orchids. *J. Exp. Bot.* **2008**, *59*, 1085–1096. [CrossRef]
3. Swarts, N.D.; Dixon, K.W. Terrestrial orchid conservation in the age of extinction. *Ann. Bot.* **2009**, *104*, 543–556. [CrossRef] [PubMed]
4. Whigham, D.F.; Willems, J.H. Demographic studies and life-history strategies of temperate terrestrial orchids as a basis for conservation. In *Orchid Conservation*; Dixon, K.W., Kell, S.P., Barrett, R.L., Cribb, P.J., Eds.; Natural History Publications: Kota Kinabaluk, Malaysia, 2003; pp. 137–158.
5. Hágsater, E.; Dumont, V. (Eds.) *Orchids: Status, Survey and Conservation Action Plan*; IUCN: Gland, Switzerland; Cambridge, UK, 1996.
6. Delforge, P. *Orchids of Europe, North Africa and the Middle East*; A. & C. Black: London, UK, 2006.
7. Djordjević, V.; Tsiftsis, S. The Role of Ecological Factors in Distribution and Abundance of Terrestrial Orchids. In *Orchids Phytochemistry, Biology and Horticulture*; Reference Series in Phytochemistry; Mérillon, J.-M., Kodja, H., Eds.; Springer Nature: Cham, Switzerland, 2022; pp. 3–72.
8. Duffy, K.J.; Scopece, G.; Cozzolino, S.; Fay, M.F.; Smith, R.J.; Stout, J.C. Ecology and genetic diversity of the dense-flowered orchid, *Neotinea maculata*, at the centre and edge of its range. *Ann. Bot.* **2009**, *104*, 507–516. [CrossRef] [PubMed]
9. Pfeifer, M.; Passalacqua, N.G.; Bartram, S.; Schatz, B.; Croce, A.; Carey, P.D.; Kraudelt, H.; Jeltsch, F. Conservation priorities differ at opposing species borders of a European orchid. *Biol. Conserv.* **2010**, *143*, 2207–2220. [CrossRef]
10. Sagarin, R.D.; Gaines, S.D. The 'abundant centre' distribution: To what extent is it a biogeographical rule? *Ecol. Lett.* **2002**, *5*, 137–147. [CrossRef]
11. Jacquemyn, H.; Brys, R.; Hutchings, M.J. Biological flora of the British Isles: *Epipactis palustris*. *J. Ecol.* **2014**, *102*, 1341–1355. [CrossRef]
12. Jersáková, J.; Malinová, T.; Jeřábková, K.; Dötteri, S. Biological Flora of the British Isles: *Pseudorchis albida* (L.) Á. & D. Löve. *J. Ecol.* **2011**, *99*, 1282–1298.
13. Meekers, T.; Hutchings, M.J.; Honnay, O.; Jacquemyn, H. Biological Flora of the British Isles: *Gymnadenia conopsea* s.l. *J. Ecol.* **2012**, *100*, 1269–1288. [CrossRef]
14. Jersáková, J.; Traxmandlová, I.; Ipser, Z.; Matthias, K.; Pellegrino, G.; Schatz, B.; Djordjević, V.; Kindlmann, P.; Renner, S.S. Biological flora of Central Europe: *Dactylorhiza sambucina* (L.) Soó. *Perspect. Plant Ecol. Evol. Syst.* **2015**, *17*, 318–329. [CrossRef]
15. Djordjević, V.; Tsiftsis, S.; Lakušić, D.; Jovanović, S.; Stevanović, V. Orchid species richness and composition in relation to vegetation types. *Wulfenia* **2020**, *27*, 183–210.
16. Kirillova, I.A.; Dubrovskiy, Y.A.; Degteva, S.V.; Novakovskiy, A.B. Ecological and habitat ranges of orchids in the northernmost regions of their distribution areas: A case study from Ural Mountains, Russia. *Plant Divers.* **2022**. [CrossRef]
17. Tsiftsis, S.; Tsiripidis, I.; Karagiannakidou, V.; Alifragis, D. Niche analysis and conservation of the orchids of east Macedonia (NE Greece). *Acta Oecol.* **2008**, *33*, 27–35. [CrossRef]
18. Djordjević, V.; Tsiftsis, S.; Lakušić, D.; Jovanović, S.; Stevanović, V. Factors affecting the distribution and abundance of orchids in grasslands and herbaceous wetlands. *Syst. Biodivers.* **2016**, *14*, 355–370. [CrossRef]

19. Wotavová, K.; Balounová, Z.; Kindlmann, P. Factors affecting persistence of terrestrial orchids in wet meadows and implications for their conservation in a changing agricultural landscape. *Biol. Conserv.* **2004**, *118*, 271–279. [CrossRef]
20. Janečková, P.; Wotavová, K.; Schödelbauerová, I.; Jersáková, J.; Kindlmann, P. Relative effects of management and environmental conditions on performance and survival of populations of a terrestrial orchid, *Dactylorhiza majalis*. *Biol. Conserv.* **2006**, *129*, 40–49. [CrossRef]
21. Jacquemyn, H.; Brys, R.; Hermy, M.; Willems, J.H. Does nectar reward affect rarity and extinction probabilities of orchid species? An assessment using historical records from Belgium and the Netherlands. *Biol. Conserv.* **2005**, *121*, 257–263. [CrossRef]
22. Vogt-Schilb, H.; Munoz, F.; Richard, F.; Schatz, B. Recent declines and range changes of orchids in Western Europe (France, Belgium and Luxembourg). *Biol. Conserv.* **2015**, *190*, 133–141. [CrossRef]
23. Kull, T.; Selgis, U.; Pecina, M.V.; Metsare, M.; Ilves, A.; Tali, K.; Shefferson, R.P. Factors influencing IUCN threat levels to orchids across Europe on the basis of national red lists. *Ecol. Evol.* **2016**, *6*, 6245–6265. [CrossRef]
24. Tsiftsis, S.; Štípková, Z.; Kindlmann, P. Role of way of life, latitude, elevation and climate on the richness and distribution of orchid species. *Biodivers. Conserv.* **2019**, *28*, 75–96. [CrossRef]
25. Tsiftsis, S. The complex effect of heterogeneity and isolation in determining alpha and beta orchid diversity on islands in the Aegean archipelago. *Syst. Biodivers.* **2020**, *18*, 281–294. [CrossRef]
26. Averyanov, L. A review of the genus *Dactylorhiza*. In *Orchid Biology—Reviews and Perspectives*; Arditti, J., Ed.; V. Timber Press Inc.: Portland, OR, USA, 1990; pp. 159–206.
27. Lazarević, P. Fens of the class *Scheuchzerio-Caricetea fuscae* (Nordh. 1936) R. Tx. 1937. in Serbia—Floristic and Vegetation Characteristics, Threats and Protection. Ph.D. Thesis, University of Belgrade, Belgrade, Serbia, 2016. (In Serbian)
28. Aćić, S. Synecological and Phytocoenological Study of Grassland Vegetation of Serbia. Ph.D. Thesis, University of Belgrade, Belgrade, Serbia, 2018.
29. Kojić, M.; Popović, R.; Karadžić, B. *Syntaxonomic Review of the Vegetation of Serbia*; Institute for Biological Research Siniša Stanković: Belgrade, Serbia, 1998. (In Serbian)
30. Aćić, S.; Šilc, U.; Lakušić, D.; Vukojičić, S.; Dajić-Stevanović, Z. Typification and correction of syntaxa from the class *Molinio-Arrhenatheretea* Tx. 1937 in Serbia. *Hacquetia* **2013**, *12*, 39–54. [CrossRef]
31. Tomović, G.; Niketić, M.; Lakušić, D.; Ranđelović, V.; Stevanović, V. Balkan endemic plants in Central Serbia and Kosovo regions: Distribution patterns, ecological characteristics and centres of diversity. *Bot. J. Linn. Soc.* **2014**, *176*, 173–202. [CrossRef]
32. Čarni, A.; Ćuk, M.; Zelnik, I.; Franjić, J.; Igić, R.; Ilić, M.; Krstonošić, D.; Vukov, D.; Škvorc, Z. Wet Meadow Plant Communities of the Alliance *Trifolion pallidi* on the Southeastern Margin of the Pannonian Plain. *Water* **2021**, *13*, 381. [CrossRef]
33. Djordjević, V.; Niketić, M.; Stevanović, V. Orchids of Serbia: Taxonomy, Life Forms, Pollination Systems, and Phytogeographical Analysis. In *Orchidaceae: Characteristics, Distribution and Taxonomy*; Djordjević, V., Ed.; Nova Science Publishers Inc.: New York, NY, USA, 2021; pp. 57–163.
34. Euro+Med. The Euro+Med PlantBase—The Information Resource for Euro-Mediterranean Plant Diversity. Available online: http://ww2.bgbm.org/EuroPlusMed/query.asp (accessed on 18 October 2022).
35. Pillon, Y.; Fay, M.F.; Shipunov, A.B.; Chase, M.W. Species diversity versus phylogenetic diversity: A practical study in the taxonomically difficult genus *Dactylorhiza* (Orchidaceae). *Biol. Conserv.* **2006**, *129*, 4–13. [CrossRef]
36. Djordjević, V.; Tsiftsis, S. Patterns of orchid species richness and composition in relation to geological substrates. *Wulfenia* **2019**, *26*, 1–21.
37. Djordjević, V.; Jovanović, S.; Stevanović, V. *Dactylorhiza fuchsii* (Orchidaceae), a new species in the flora of Serbia. *Arch. Biol. Sci.* **2014**, *66*, 1227–1232. [CrossRef]
38. Šabanović, E.; Djordjević, V.; Milanović, Đ.; Boškailo, A.; Šarić, Š.; Huseinović, S.; Randjelović, V. Checklist of the *Orchidaceae* of Bosnia and Herzegovina. *Phyton-Ann. Rei Bot.* **2021**, *61*, 83–95.
39. Niketić, M.; Tomović, G.; Perić, R.; Zlatković, B.; Anačkov, G.; Djordjević, V.; Jogan, N.; Radak, B.; Duraki, Š.; Stanković, M.; et al. Material on the Annotated Checklist of Vascular Flora of Serbia. Nomenclatural, taxonomic and floristic notes I. *Bull. Nat. Hist. Mus. Belgr.* **2018**, *11*, 101–180. [CrossRef]
40. Radak, B. Morphological Variability of Species of the Genus *Anacamptis* Rich. (Orchidoideae, Orchidaceae) in the Balkan Peninsula and the Pannonian Plain. Ph.D. Thesis, University of Novi Sad, Novi Sad, Serbia, 2019; pp. 1–284.
41. Djordjević, V. *The Orchid Flora (Orchidaceae) of Western Serbia*; Serbian Academy of Sciences and Art: Belgrade, Serbia, 2021; pp. 1–467. (In Serbian)
42. Bartók, A.; Csergő, A.-M.; Balázs, Ö.; Hurdu, B.-I.; Jakab, G. A *Gymnadenia frivaldii* Hampe ex Griseb. Újrafelfedezése areája északihatáran (Keleti Kárpátok, Románia). *Kitaibelia* **2016**, *21*, 213–220. [CrossRef]
43. Djordjević, V.; Lakušić, D.; Jovanović, S.; Stevanović, V. Distribution and conservation status of some rare and threatened orchid taxa in the central Balkans and the southern part of the Pannonian Plain. *Wulfenia* **2017**, *24*, 143–162.
44. Milanović, Đ.; Stupar, V.; Šabanović, E.; Djordjević, V.; Brujić, J.; Boškailo, A.; Ranđelović, V. On the distribution and conservation status of some rare orchid taxa (Orchidaceae) in Bosnia and Herzegovina (Western Balkans). *Hacquetia* **2022**, *21*, 327–346. [CrossRef]
45. Kotilínek, M.; Těšitelová, T.; Jersáková, J. Biological Flora of the British Isles: *Neottia ovata*. *J. Ecol.* **2015**, *103*, 1354–1366. [CrossRef]
46. Meusel, H.; Jäger, E.; Weinert, E. *Comparative Chorology of Central European Flora 1 (Vergleichende Chorologie der Zentraleuropäischen Flora 1)*; Gustav Fischer: Jena, Germany, 1965.

47. Meusel, H.; Jäger, E.; Weinert, E. *Comparative Chorology of Central European flora 2 (Vergleichende Chorologie der Zentraleuropäischen Flora 2)*; Gustav Fischer: Jena, Germany, 1978.

48. Meusel, H.; Jäger, E. *Vergleichende Chorologie der Zentraleuropäischen Flora 3*; Gustav Fischer: Stuttgart, Germany, 1992.

49. Stevanović, V. Floristic division of the territory of Serbia with an overview of higher chorion and appropriate floral elements. In *The Flora of Serbia I*; Sarić, M.R., Ed.; Serbian Academy of Sciences and Arts: Belgrade, Serbia, 1992; pp. 49–65.

50. Tsiftsis, S.; Antonopoulos, Z. *Atlas of the Greek Orchids*; Mediterraneo Editions: Rethymno, Greece, 2017; Volume 1.

51. Raunkiaer, C. *The Life Forms of Plants and Statistical Plant Geograph*; Clarendon: Oxford, UK, 1934.

52. Ellenberg, H.; Mueller-Dambois, D. A key to Raunkiaer plant life froms with revised subdivisions. *Ber. Des Geobot. Inst. Der Eidg. Techn. Hochsch. Stift. Rübel* **1967**, *37*, 56–73.

53. Štípková, Z.; Tsiftsis, S.; Kindlmann, P. Distribution of Orchids with Different Rooting Systems in the Czech Republic. *Plants* **2021**, *10*, 632. [CrossRef] [PubMed]

54. Djordjević, V.; Tsiftsis, S.; Kindlmann, P.; Stevanović, V. Orchid diversity along an altitudinal gradient in the central Balkans. *Front. Ecol. Evol.* **2022**, *10*, 929266. [CrossRef]

55. Bernardos, S.; García-Barriuso, M.; Sánchez-Anta, M.A.; Amich, F. Composition, geographical affinities and endemism of the Iberian Peninsula orchid flora. *Nord. J. Bot.* **2007**, *25*, 227–237. [CrossRef]

56. Vakhrameeva, M.G.; Tatarenko, I.V.; Varlygina, T.I.; Torosyan, G.K.; Zagulski, M.N. *Orchids of Russia and Adjacent Countries (Within the Borders of the Former USSR)*; Gantner Verlag: Ruggell, Liechtenstein, 2008.

57. Mucina, L.; Bültmann, H.; Dierßen, K.; Theurillat, J.-P.; Raus, T.; Čarni, A.; Šumberová, K.; Willner, W.; Dengler, J.; Gavilán García, R.; et al. Vegetation of Europe: Hierarchical floristic classification system of plant, lichen, and algal communities. *Appl. Veg. Sci.* **2016**, *19*, 3–264. [CrossRef]

58. Peterka, T.; Hájek, M.; Jiroušek, M.; Jiménez-Alfaro, B.; Aunina, L.; Bergamini, A.; Dítě, D.; Felbaba-Klushyna, L.; Graf, U.; Hájková, P.; et al. Formalized classification of European fen vegetation at the alliance level. *Appl. Veg. Sci.* **2017**, *20*, 124–142. [CrossRef]

59. Mišić, V.; Jovanović-Dunjić, R.; Popović, M.; Borisavljević, L.; Antić, M.; Dinić, A.; Danon, J.; Blaženčić, Ž. *Plant Communities and Habitats of Stara Planina*; Serbian Academy of Sciences and Arts: Belgrade, Serbia, 1978; pp. 1–389.

60. Petković, B. Marsh vegetation in the area of Tutin. *Bull. Inst. Jard. Bot. Univ. Belgrade* **1983**, *17*, 61–102. (In Serbian)

61. Ranđelović, V.; Zlatković, B. Vegetation of the alliance *Calthion* in southeastern Serbia. *Ekologija* **1994**, *28–29*, 19–31. (In Serbian)

62. Ranđelović, V.; Zlatković, B. *Flora and Vegetation of the Vlasina Plateau*; Faculty of Science, University of Niš: Niš, Serbia, 2010; pp. 1–448. (In Serbian)

63. Košanin, N. Vegetation of Jakupica mountain in Macedonia. *Glas. Srp. Kralj. Akad.* **1911**, *85*, 184–242. (In Serbian)

64. Pavlović, Z. Vegetation of the Mountain Zlatibor. *Zb. Rad. Inst. Ekol. I Biogeogr. SANU* **1951**, *2*, 115–182. (In Serbian)

65. Stjepanović-Veseličić, L. Vegetation of Deliblato Sands. *Inst. Ecol. Biogeogr. SANU* **1953**, *4*, 1–113. (In Serbian)

66. Tatić, B.; Veljović, V.; Petković, B.; Stefanović, M.; Radotić, S. Ass. *Lathyreto-Molinietum Coerulae*—A new community of meadow vegetation from the Pešter plateau—southwestern Serbia. *Bull. Inst. Jard. Bot. Univ. Belgrade* **1988**, *12*, 31–38.

67. Butorac, B.; Hulo, I. Phytocoenological, floristic and ornithological values of the area Selevenjska pustara as the basis for protection. *Zaštita Prir.* **1992**, *45*, 65–76. (In Serbian)

68. Butorac, B.; Hulo, I. *Contribution to Knowledge of Marsch Meadows around the Kereš River*; Book of Abstracts of XXIV Tiszakutató Ankét; Szegedi Ökologógiai Napok: Segedin, Hungary, 1993; p. 6.

69. Lazarević, P. *Floristic-Ecological Study of the Peštersko Polje Mire in Southwestern Serbia*; University of Belgrade: Belgrade, Serbia, 2009. (In Serbian)

70. Šabanović, E. Family Orchidaceae in the flora of Bosnia and Herzegovina. Ph.D. Thesis, University of Tuzla, Tuzla, Bosnia and Herzegovina, 2022; pp. 1–301. (In Bosnian).

71. Babić, N. Lowland meadows in Podunavlje. *Rad Vojvođanskih Muz.* **1955**, *4*, 155–156. (In Serbian)

72. Cincović, T. Types of meadows in Posavina. *Proc. Fac. Agric. Belgrade* **1956**, *4*, 1–25. (In Serbian)

73. Ranđelović, N. Phytocenological-Ecological Characteristics of the Mountain Grasslands of Southeastern Serbia. Ph.D. Thesis, University of Zagreb, Zagreb, Croatia, 1978. (In Serbian)

74. Ranđelović, V. Flora and Vegetation of the Vlasin Plateau. Ph.D. Thesis, University of Belgrade, Belgrade, Serbia, 2002. (In Serbian)

75. Radak, B.; Hristovski, S.; Matevski, V.; Anačkov, G. New orchid taxa for North Macedonia. In *Proceedings of the 6th Congress of Ecologists of the Republic of North Macedonia*; Ohrid, North Macedonia, 15–18 October 2022, Macedonian Ecological Society: Ohrid, North Macedonia, 2022; p. 82.

76. Cincović, T. Meadow Vegetation in River Valleys of Western Serbia. Ph.D. Thesis, University of Belgrade, Belgrade, Serbia, 1959; pp. 1–62. (In Serbian)

77. Parabućski, S.; Stojanović, S. *Oenanthe (banatica)-Alopecuretum pratensis* ass. nova. *Matica Srp. J. Nat. Sci.* **1988**, *74*, 71–78. (In Serbian)

78. Gajić, M.; Karadžić, D. *Flora of flat Srem with Special Reference to Obedska Bara*; Šumarski Fakultet: Belgrade, Serbia; Šumsko Gazdinstvo: Sremska Mitrovica, Serbia, 1991; pp. 1–437.

79. Micevski, K. Typological classification of lowland meadow and swamp vegetation in Macedonia. *Folia Balc.* **1957**, *1*, 29–33.

80. Jovanović, R. Types of valley meadows of Jasenica. *Arch. Biol. Sci.* **1957**, *9*, 1–14. (In Serbian)

81. Jovanović-Dunjić, R. Typology, Ecology and Dynamics of Swamp and Meadow Vegetation in the Velika Morava Valley. Ph.D. Thesis, University of Belgrade, Belgrade, Serbia, 1965; pp. 1–399. (In Serbian)
82. Jovanović, V. Kukavica Mountain in southeastern Serbia and the vegetation of its northern part. *Leskovački Zb.* **1977**, *17*, 271–299. (In Serbian)
83. Ranđelović, N.; Rexhepi, F.; Jovanović, V. Plant communities of Southeast Kosovo. In Proceedings of the 2nd Congress of Ecologists of Yugoslavia, Zadar, Plitvice, Yugoslavia, 1–7 October 1979; pp. 957–995.
84. Jovanović, V. Meadow Vegetation of Southeastern Serbia—Mount Radan, Goljak, Part of Kukavica and Their Surroundings. Ph.D. Thesis, University of Novi Sad, Novi Sad, Serbia, 1979. (In Serbian)
85. Hundozi, B. Vegetation of Lowland Meadows in Kosovo. Ph.D. Thesis, University of Zagreb, Zagreb, Croatia, 1980; pp. 1–173. (In Serbian)
86. Jovanović-Dunjić, R.; Stefanović, K.; Popović, R.; Dimitrijević, J. A contribution to the knowledge of meadow ecosystems in the Veliki Jastrebec area. *Bull. Inst. Jard. Bot. Univ. Belgrade* **1986**, *20*, 7–31. (In Serbian)
87. Milanović, Đ.; (Faculty of Forestry, University of Banja Luka, Banja Luka, Bosnia and Herzegovina). Personal communication, 2022.
88. Danon, J. *Phytocoenological Investigations of Meadows in the Vicinity of Krivi Vir with Special Reference to Nutritional Value of the Hay*; Faculty of Natural Sciences, University in Belgrade: Belgrade, Serbia, 1960. (In Serbian)
89. Mišić, V.; Dinić, A. The vegetation of nature preserves on the Stara Planina Mt. In Proceedings of the 5th Symposium on the Flora of Southeast Serbia, Zaječar, Serbia, 5–9 June 1997; pp. 129–138.
90. Berisha, N.; Rizani, K.L.; Kadriaj, B.; Millaku, F. Notes on the distribution, ecology, associated vegetation and conservation status of *Gymnadenia* (Orchidaceae) in Kosovo. *Ital. Bot.* **2021**, *12*, 1–27. [CrossRef]
91. Petković, B.; Tatić, B. Ass. *Scirpeto-Phragmitetum* Koch. W. 1926, around the Ubavac stream on Fruška Gora. *Bull. Nat. Hist. Mus. Belgrade* **1978**, *33B*, 55–68. (In Serbian)
92. Ranđelović, V. Marsh Vegetation along the Upper Reaches of the South Morava. Diploma Thesis, University of Novi Sad, Novi Sad, Serbia, 1988. (In Serbian)
93. Ponert, J. Contributions to the orchids of Republic of Macedonia and Serbia. *J. Eur. Orch.* **2014**, *46*, 561–577.
94. Obratov-Petković, D.; Popović, I.; Dajić-Stevanović, Z. Diversity of the vascular flora of Mt. Zlatar (Southwest Serbia). *Eurasia J. Biosci.* **2007**, *5*, 35–47.
95. Perišić, S. Flora and Vegetation of Blačko Lake. Master's Thesis, University of Belgrade, Belgrade, Serbia, 2002; pp. 1–201. (In Serbian)
96. Čolić, D. New localities of round-leaved sundew (*Drosera rotundifolia* L.) on Mt. Stara Planina—Eastern Serbia. *Zaštita Prir.* **1965**, *29–30*, 523. (In Serbian)
97. Petković, B. The new community of matgrass, ass. *Carici-Nardetum strictae*, from the area of southwestern Serbia. *Bull. Nat. Hist. Mus. Belgrade* **1985**, *40*, 89–95.
98. Stevanović, V.; Niketić, M.; Lakušić, D. Chorological additions to the flora of eastern Yugoslavia. *Flora Mediterr.* **1991**, *1*, 121–142.
99. Lakušić, D. The high mountain flora of Kopaonik—an ecological-phytogeographical study. Master's Thesis, University of Belgrade, Belgrade, Serbia, 1993; pp. 1–230. (In Serbian)
100. Zlatković, B.; Ranđelović, V.; Ranđelović, N. Material for the flora of southeastern Serbia. In *Flora and Vegetation, Proceedings of the 3rd Symposium on the Flora of Southeast Serbia, Leskovac-Pirot, Serbia, 1–4 June 1993*; Ranđelović, N., Ed.; University of Niš: Niš, Serbia; Faculty of Technology in Leskovac: Leskovac, Serbia, 1993; pp. 95–110.
101. Petković, B.; Krivošej, Z.; Veljić, M. *Selaginello-Eriophoretum latifoli*—ass. nova from Mount Ošljak (Serbia, Kosovo). *Bull. Inst. Jard. Bot. Univ. Belgrade* **1996**, *30*, 89–95. (In Serbian)
102. Ranđelović, V.; Zlatković, B.; Amidžić, L. Flora and vegetation of the high mountain mires of the Mts. Šar planina. *Zaštita Prir.* **1998**, *50*, 377–397. (In Serbian)
103. Todorović, T. Phytocenological analysis of endangered plant species of the Vlasina plateau. Master's Thesis, University of Niš, Niš, Serbia, 2013; pp. 1–58.
104. Rudski, I. Plant communities in the high mountains of southern Serbia. *Šumarski List.* **1938**, *61*, 611–623. (In Serbian)
105. Djordjević, V.; Tsiftsis, S.; Lakušić, D.; Stevanović, V. Niche analysis of orchids of serpentine and non-serpentine areas: Implications for conservation. *Plant Biosyst.* **2016**, *150*, 710–719. [CrossRef]
106. Budzhak, V.V.; Chorney, I.I.; Tokariuk, A.I.; Kuzemko, A.A. Numeric syntaxonomical analysis of the communities with participation of species from *Molinia caerulea* complex in the southwest of Ukraine. *Hacquetia* **2016**, *15*, 63–78. [CrossRef]
107. Oberdorfer, E. *Suddeutsche Pflanzengesellschaften (Plant Communities of Southern Germany). III [Secalietea–Molinio-Arrhenatheretea]*; Gustav Fischer: Stuttgart, Germany, 1983. (In German)
108. Dijk, E.; Grootjans, A.P. Performance of four *Dactylorhiza* species over a complex trophic gradient. *Acta Bot. Neerl.* **1998**, *47*, 351–368.
109. Schrautzer, J.; Fichtner, A.; Huckauf, A.; Rasran, L.; Jensen, K. Long-term population dynamics of *Dactylorhiza incarnata* (L.) Soó after abandonment and re-introduction of mowing. *Flora* **2011**, *206*, 622–630. [CrossRef]
110. Šilc, U.; Aćić, S.; Škvorc, Ž.; Krstonošić, D.; Franjić, J.; Dajić Stevanović, Z. Grassland vegetation of the *Molinio-Arrhenatheretea* class in the NW Balkan Peninsula. *Appl. Veg. Sci.* **2014**, *17*, 591–603. [CrossRef]
111. Škvorc, Ž.; Ćuk, M.; Zelnik, I.; Franjić, J.; Igić, R.; Ilić, M.; Krstonošić, D.; Vukov, D.; Čarni, A. Diversity of wet and mesic grasslands along a climatic gradient on the southern margin of the Pannonian Basin. *Appl. Veg. Sci.* **2020**, *23*, 676–697. [CrossRef]

112. Illyés, Z.; Halász, K.; Rudnóy, S.; Ouanphanivanh, N.; Garay, T.; Bratek, Z. Changes in the diversity of the mycorrhizal fungi of orchids as a function of the water supply of the habitat. *J. Appl. Bot. Food.* **2009**, *83*, 28–36.
113. Hrivnák, R.; Gömöry, D.; Cvachová, A. Inter-annual variabilty of the abundance and morphology of *Dactylorhiza majalis* (Orchidaceae-Orchideae) in two permanent plots of a mire in Slovakia. *Phyton-Ann. Rei Bot.* **2006**, *46*, 27–44.
114. Molnár, A. (Ed.) *Atlas of Hungarian Orchids*; Kossuth Kiadó: Budapest, Hungary, 2011. (In Hungarian)
115. Urban, D. Characteristics of the locality of *Hammarbya paludosa* (L.) O. Kuntze on the Łęczna-Włodawa Plain (West Polesie). *Teka Komis Ochr. Środow. Przyr.* **2013**, *10*, 448–454.
116. Blinova, I.V. Spatial population structure of rare orchid species in rich fens in the central part of Murmansk oblast. *Russ. J. Ecol.* **2016**, *47*, 234–240. [CrossRef]
117. Jermakowicz, E.; Brzosko, E.; Kotowicz, J.; Wróblewska, A. Genetic diversity of orchid *Malaxis monophyllos* over European range as an effect of population properties and postglacial colonization. *Pol. J. Ecol.* **2017**, *65*, 69–86.
118. Dijk, E.; Willems, J.H.; van Andel, J. Nutrient responses as a key factor to the ecology of orchid species. *Acta Bot. Neerl.* **1997**, *46*, 339–363. [CrossRef]
119. Bowles, M.; Zettler, L.; Bell, T.; Kelsey, P. Relationship between soil characteristics, distribution and restoration potential of the federal threatened eastern prairie fringed orchid, *Platanthera leucophaea* (Nutt.) Lindl. *Am. Midl. Nat.* **2005**, *154*, 273–285. [CrossRef]
120. Leuschner, C.; Ellenberg, H. *Ecology of Central European Non-Forest Vegetation: Coastal to Alpine, Natural to Man-Made Habitats: Vegetation Ecology of Central Europe*; Springer International Publishing: Cham, Switzerland, 2017.
121. Hrivnák, M.; Slezák, M.; Galvánek, D.; Vlčko, J.; Belanová, E.; Rízová, V.; Senko, D.; Hrivnák, R. Species Richness, Ecology, and Prediction of Orchids in Central Europe: Local-Scale Study. *Diversity* **2020**, *12*, 154. [CrossRef]
122. Tali, K.; Foley, M.J.Z.; Kull, T. Biological flora of the British Isles, 232. *Orchis ustulata* L. *J. Ecol.* **2004**, *92*, 174–184.
123. Beyrle, H.; Penningsfeld, F.; Hock, B. The role of nitrogen concentration in determining the outcome of the interaction between *Dactylorhiza incarnata* (L.) Soó and *Rhizoctonia* sp. *New Phytol.* **1991**, *117*, 665–672. [CrossRef]
124. Dijk, E.; Olff, H. Effects of nitrogen, phosphorus and potassium fertilization on field performance of *Dactylorhiza majalis*. *Acta Bot. Neerl.* **1994**, *43*, 383–392. [CrossRef]
125. Silvertown, J.; Wells, D.A.; Gillman, M.; Dodd, M.E.; Robertson, H.; Lakhani, K.H. Short-term effects and long-term aftereffects of fertilizer application on the flowering population of green-winged orchid *Orchis morio*. *Biol. Conserv.* **1994**, *69*, 191–197. [CrossRef]
126. Hornemann, G.; Michalski, S.G.; Durka, W. Short-term fitness and long-term population trends in the orchid *Anacamptis morio*. *Plant Ecol.* **2012**, *213*, 1583–1595. [CrossRef]
127. Stevanović, V.; Tan, K.; Iatrou, G. Distribution of the endemic Balkan flora on serpentine I. Obligate serpentine endemics. *Plant Syst. Evol.* **2003**, *242*, 149–170. [CrossRef]
128. Chiari, M.; Djerić, N.; Garfagnoli, F.; Hrvatović, H.; Krstić, M.; Levi, N.; Malasoma, A.; Marroni, M.; Menna, F.; Nirta, G.; et al. The geology of the Zlatibor-Maljen area (western Serbia): A geotraverse accross theophiolites of the Dinaric-Hellenic collisional belt. *Ofioliti* **2011**, *36*, 139–166.
129. Gawlick, H.-J.; Sudar, M.; Missoni, S.; Suzuki, H.; Lein, R.; Jovanović, D. Triassic-Jurassic geodynamic evolution of the Dinaridic Ophiolite Belt (Inner Dinarides, SW Serbia). *J. Alpine Geol.* **2017**, *55*, 1–167.
130. Perry, E.C.; Lefticariu, L. Formation and geochemistry of Precambrian cherts. *Treatise Geochem.* **2003**, *7*, 99–113.
131. Djordjević, V.; University of Belgrade, Faculty of Biology, Institute of Botany and Botanical Garden, Belgrade, Serbia. Unpublished work. 2022.
132. Duraki, Š. Vascular flora of the high mountain ridge of Kobilica on Šar mountain. Master's Thesis, University of Belgrade, Belgrade, Serbia, 2008. (In Serbian)
133. Dzulynski, S.; Walton, E.K. *Sedimentary Features in Flysch and Greywackes. Developments in Sedimentology*; Elsevier: Amsterdam, The Netherlands, 1965; Volume 7.
134. Duncan, R.P.; Clemants, S.E.; Corlett, R.T.; Hahs, A.K.; McCarthy, M.A.; McDonnell, M.J.; Schwartz, M.W.; Thompson, K.; Vesk, P.A.; Williams, N.S.G. Plant traits and extinction in urban areas: A meta-analysis of 11 cities. *Glob. Ecol. Biogeogr.* **2011**, *20*, 509–519. [CrossRef]
135. Bilz, M.; Kell, S.P.; Maxted, N.; Lansdown, R.V. *European Red List of Vascular Plants*; Publications Office of the European Union: Luxembourg, 2011.
136. Ballantyne, M.; Pickering, C. Ecotourism as a threatening process for wild orchids. *J. Ecotourism.* **2012**, *11*, 34–47. [CrossRef]
137. Light, M.; MacConaill, M. Effects of trampling on a terrestrial orchid environment. *Lankesteriana* **2007**, *7*, 294–298. [CrossRef]
138. Pickering, C.M.; Hill, W.; Newsome, D.; Leung, Y.-F. Comparing hiking, mountain biking and horse riding impacts on vegetation and soils in Australia and the United States of America. *J. Environ. Manag.* **2010**, *91*, 551–562. [CrossRef]
139. Wraith, J.; Pickering, C. Quantifying anthropogenic threats to orchids using the IUCN Red List. *Ambio* **2018**, *47*, 307–317. [CrossRef] [PubMed]
140. Grlić, L. *Encyclopedia of Wild Edible Plants*; August Cesarec: Zagreb, Croatia, 1986. (In Serbian)
141. Arditti, J. *Fundamentals of Orchid Biology*; Wiley Interscience: New York, NY, USA, 1992.
142. IPCC. *The Physical Science Basis*; Contribution of Working Group I to the Sixth Assessment Report of the Intergovernmental Panel on Climate Change; Cambridge University Press: Cambridge, MA, USA, 2021.

143. Kolanowska, M. The future of a montane orchid species and the impact of climate change on the distribution of its pollinators and magnet species. *Glob. Ecol. Conserv.* **2021**, *32*, e01939. [CrossRef]
144. Hájek, M.; Těšitel, J.; Tahvanainen, T.; Peterka, T.; Jiménez-Alfaro, B.; Jansen, F.; Pérez-Haase, A.; Garbolino, E.; Carbognani, M.; Kolari, T.H.M.; et al. Rising temperature modulates pH niches of fen species. *Glob. Change Biol.* **2022**, *28*, 1023–1037. [CrossRef] [PubMed]

 *diversity*

Article

Three New Diatom Species from Spring Habitats in the Northern Apennines (Emilia-Romagna, Italy)

Marco Cantonati [1,*], **Olena Bilous** [2], **Nicola Angeli** [1], **Liesbeth van Wensen** [1] **and Horst Lange-Bertalot** [3]

1. Limnology & Phycology Section, MUSE—Museo delle Scienze, Corso del Lavoro e della Scienza 3, I-38123 Trento, Italy; nicola.angeli@muse.it (N.A.); lhvanwensen@gmail.com (L.v.W.)
2. Institute of Hydrobiology, National Academy of Sciences of Ukraine, Geroiv Stalingrada Ave., 12, 04210 Kyiv, Ukraine; bilous_olena@ukr.net
3. Institute for Ecology, Evolution and Diversity, Biologicum, University of Frankfurt, Max-von-Laue Straße 13, 60438 Frankfurt am Main, Germany; lange-bertalot@web.de
* Correspondence: marco.cantonati@muse.it; Tel.: +39-0-461-270-342

Abstract: Using light (LM, including plastid characterization on fresh material) and scanning electron microscopy (SEM), as well as a thorough morphological, physical, chemical, and biological characterization of the habitats, the present study aims at describing three species new to science. They belong to the genera *Eunotia* Ehrenb., *Planothidium* Round and L. Bukht., and *Delicatophycus* M.J. Wynne, and were found in two contrasting spring types in the northern Apennines. The three new species described differ morphologically from the most similar species by: less dense striae and areolae, and the absence of a ridge at the valve face-mantle transition (SEM feature) [*Eunotia crassiminor* Lange-Bert. et Cantonati sp. nov.; closest established species: *Eunotia minor* (Kütz.) Grunow]; narrower and shorter cells [*Planothidium angustilanceolatum* Lange-Bert. et Cantonati sp. nov.; most similar species: *Planothidium lanceolatum* (Bréb. ex Kütz.) Lange-Bert.]; barely-dorsiventral symmetry, set off ends, and lower density of the central dorsal striae [*Delicatophycus crassiminutus* Lange-Bert. et Cantonati sp. nov.; most similar species: *Delicatophycus minutus* M.J.Wynne]. Two of the three species we described are separated from the closest species by dimensions. Their description improved knowledge on two taxa (*Eunotia minor* s.l. and *Planothidium lanceolatum* s.l.) likely to be only partially resolved species complexes. We could also refine knowledge on the ecological profiles of the three newly-described species. *Eunotia crassiminor* sp. nov., as compared to *Eunotia minor*, appears to occur in colder inland waters with a circumneutral pH and a strict oligotrophy as well with respect to nitrogen. The typical habitat of *Planothidium angustilanceolatum* sp. nov. appears to be oligotrophic mountain flowing springs with low conductivity. *Delicatophycus crassiminutus* sp. nov. was observed only in limestone-precipitating springs, and is therefore likely to be restricted to hard water springs and comparable habitats where CO_2 degassing leads to carbonate precipitation. Springs are a unique but severely threatened wetland type. Therefore, the in-depth knowledge of the taxonomy and ecology of characteristic diatom species is important, because diatoms are excellent indicators of the quality and integrity of these peculiar ecosystems in the face of direct and indirect human impacts.

Keywords: diatom; springs; size; *Eunotia*; *Planothidium*; *Delicatophycus*; taxonomy; ecology; plastids

Citation: Cantonati, M.; Bilous, O.; Angeli, N.; van Wensen, L.; Lange-Bertalot, H. Three New Diatom Species from Spring Habitats in the Northern Apennines (Emilia-Romagna, Italy). *Diversity* **2021**, *13*, 549. https://doi.org/10.3390/d13110549

Academic Editors: Mateja Germ, Igor Zelnik, Matthew Simpson and Michael Wink

Received: 3 October 2021
Accepted: 26 October 2021
Published: 29 October 2021

Publisher's Note: MDPI stays neutral with regard to jurisdictional claims in published maps and institutional affiliations.

1. Introduction

Springs are characterized by rich species pools at the landscape level (ɣ diversity, [1]). They are unique habitats: multiple ecotones and extremely heterogeneous environments, offering to the sheltered organisms a wide range of environmental conditions [2]. They are also the systems where the utility of a deep integration of hydrogeological and ecological approaches becomes obvious (ecohydrogeology, [3]). However, these ecosystems are menaced by many threats, the main ones being the water-resource exploitation, and the reduction of precipitation and recharge due to climate change [4].

Springs have been classified in many ways, and a number of spring types have been recognized (e.g., [3]). Those relevant for the present paper are the two following, contrasting spring types: rheocrenic mountain springs with low conductivity, and limestone precipitating springs.

Rheocrenic mountain springs with low conductivity are typically high-ecological-integrity, oligotrophic systems with relevant discharge and current velocity. They provide a habitat to many threatened-Red-List and recently discovered species (e.g., [5]), but they are sensitive to disturbance from human activities and climate and environmental change (e.g., [6,7]).

Limestone precipitating springs (LPS) are a very peculiar kind of spring where hard water and CO_2 degassing lead to the precipitation of carbonates. They host relatively low-diversity assemblages that however include many highly-adapted and characteristic taxa. This is the only widespread spring type clearly recognized by nature-protection legislation in the EU, but these springs are nevertheless often affected by many impacts (e.g., [8]).

Diatoms are the most diverse groups of algae in springs, where they can be excellent indicators of environmental features and ecosystem integrity [2,9]. Many rare and Red List diatom species occur in springs. Many diatom species were described from springs, and it is easy to provide examples also with reference to one of the genera discussed in the present paper, namely *Eunotia*: *E. arcofallax* Lange-Bert., *E. braendlei* Lange-Bert. et Werum, *E. kruegeri* Lange-Bert., *E. palatina* Lange-Bert. et W. Krüger, and *E. pexii* Lange-Bert. [10], *E. glacialispinosa* Lange-Bert. et Cantonati [11], *E. cisalpina* Lange-Bert. and Cantonati, *E. fallacoides* Lange-Bert. and Cantonati, and *E. insubrica* Lange-Bert. and Cantonati [12].

The importance of size in diatom species delimitation has been stressed in classical diatom literature (e.g., [13]) and current articles combining molecular and morphological approaches (e.g., [14]). Krammer [13] proposes the ratio of maximal and minimal width as a reliable means to test the quality of taxa: if >1.5, this ratio would point to an unresolved species complex whilst it is <1.5 in well-defined species. In this context, Krammer [13] also recalls Geitler's [15] first rule on life-cycle form changes: over the population developmental cycle the apical axis is shortened not only absolutely but also relatively more strongly than the transapical axis, leading to smaller specimens that are comparatively wider than larger specimens. Many diatom species were separated from the most similar species mainly on the basis of dimensions, and it is easy to provide examples as well with reference to one of the genera discussed in the present paper, namely *Eunotia*: *E. nanopapilio* Lange-Bert., and *E. superpaludosa* Lange-Bert. [16].

The great biogeographic and conservation importance of the Apennines is confirmed by the uniqueness of subalpine and alpine belts forming these mountains, peculiar climatic characteristics, and complex paleogeographic and paleoclimatic history of the region, combined with high geodiversity [1]. The Northern Apennines differ from the central and southern ones historically, geographically, and morphologically. They have some scattered stands of alpine vegetation, dominated by the orophilous central-European, boreal and Eurasiatic species, as well as a few limited endemics. Owing to its floristic similarity to the Alps, the summit area of the Northern Apennines has been considered as the southernmost part of a larger phytogeographic unit that also includes the main central-European massifs [17]. The described factors formed the autonomy of the Apennine communities from the Alps and central European mountains. This situation might also have favoured the discovery of new diatom species in this geographic area [6,18]. The focus of this paper is on the uniqueness of some representatives from three genera found in the Northern Apennines: *Eunotia* Ehrenb., *Planothidium* Round et L. Bukht., and *Delicatophycus* M.J.Wynne.

Eunotia Ehrenb. [19] is one of the most diverse diatom genera, consisting of more than 800 species [20] and more than 150 taxa known for Europe [16]. The morphology of the *Eunotia* species is characterized by dorsiventral outline, and at least one rimoportula at one apex. The species in the genus have striae which are punctate externally and interrupted near the ventral portion of the valve, and most representatives are characterized by a "rudimentary" raphe system [16,21]. Although some fossil *Eunotia* have been described from New Zealand marine sediments [20], the extant species of the genus are restricted to freshwater environments [16,22]. The genus includes many species from tropical and subtropical areas as well, inhabiting mainly oligotrophic waterbodies in the epiphyton and metaphyton [22–25]. The large majority of the *Eunotia* species have an ecological optimum in acidic, low conductivity, and oligo-dystrophic conditions [12,25–27]. New species of *Eunotia* are continuously described. Recent examples are Ruwer et al. [28]; *E. nupeliana* D.T.Ruwer, L.Rodrigues), and Luo et al. [21]; *E. mugecuo* F. Luo, Q.-M.You and Q.-X.Wang, *E. filiformis* F. Luo, Q.-M.You and G.-X.Wang), who worked on high elevation aquatic habitats.

The genus *Planothidium* F.E. Round and L. Bukhtiyarova [29] includes more than 110 names flagged as accepted taxonomically on the basis of the literature listed under the species name, according to Guiry and Guiry [30]. A search in DiatomBase [31] yielded 142 matching extant records, 67 of which have been verified by a taxonomic editor. The species of the genus have heterovalvar frustules that are usually solitary, with elliptic to lanceolate valves. *Planothidium* taxa are characterized by slightly concave raphe valves and have an asymmetrical central area and convex rapheless sternum valves with continuous ('delicatulum' type) or interrupted striae on one side showing a clear space in the central area [32,33]. Along with other morphological characteristics, the central area serves as a distinguishing feature for the taxa of the genus presenting a shallow depression (named sinus), a hood (also known as cavum), or the lack of both of these structures [34]. The genus is formed of species with a wide geographical distribution, from South and North America [33,35,36], Africa [37,38], Europe [39,40], Asia [34,41], to the Antarctic region (e.g., [42]). Most species are known from freshwater environments, although there are some representatives reported from brackish and marine environments, and also from aerial environments (e.g., [43,44]). The species belonging to this genus are predominantly epilithic, epipsammic, and epiphytic on aquatic plants and algae [33,34]. The species inhabit flowing and standing waters, with low to high conductivity, occur from circumneutral to alkaline waters, and seem to be tolerant up to mesotrophic conditions [33,34,45]. Examples of recently described *Planothidium* species are: *Planothidium hinzianum* C.E.Wetzel, Van de Vijver and L.Ector [34], *P. potapovae* C.E.Wetzel and L.Ector [34], *P. sheathii* Stancheva [33], *P. tujii* C.E.Wetzel and L.Ector [34], *P. californicum* Stancheva and N. Kristan [46], *P. nanum* Bąk, Kryk et Halabowski [47,48], and *P. marganaiensis* Lai, L.Ector and C.E.Wetzel [40].

Delicatophycus M.J.Wynne [49] is the correct name for the genus known as *Delicata* Krammer [50]. This name was invalid because it is a technical term and was amended by Wynne [49], who also noted that names ending in -phycus (φύκος, phykos), ought to be neutral, but were treated as masculine in accordance with tradition (International Code of Nomenclature for algae, fungi, and plants, Shenzhen Code) [51]. The current circumscription of the genus accounts for eight accepted species names [31], and 28 have been flagged as accepted taxonomically on the basis of the literature listed under the species name by Guiry and Guiry [30]. The morphology of the taxa belonging to *Delicatophycus* is characterized by dorsiventral valves with a lateral structure of the raphe, the presence of pseudosigmoids, the absence of apical pore fields and of stigmata, and foramina with a tendency to form undulated transapical structures externally. The species have strongly ventrally curved proximal raphe branches, distal raphe fissures deflected dorsally, with combination of the partly zig–zag shaped striae [50,52]. The representatives of the genus are found across a large climatic and geographical range: Europe [50], Asia [53,54], Africa [50], and South and North America [55,56]. The ecological preferences of the genus are still insufficiently known. However, common species of the genus are found in aerial habitats (dripping wet moss, wet rocks) and in the littoral of oligotrophic lakes [50]. Even within this relatively small genus new species are continuously described, a very recent example being *Delicatophycus liuweii* Y.-L. Li [57].

Using light microscopy (LM, both fresh and prepared materials) and scanning electron microscopy (SEM) observations, as well as a thorough morphological, physical, chemical, and biological characterization of the habitats, the present study aims to describe in detail three new species from the genera *Eunotia*, *Planothidium* and *Delicatophycus* found in two contrasting spring types in the Northern Apennines.

2. Materials and Methods

The samples on which this study is based were collected during surveys for the EBERs (Exploring the Biodiversity of Emilia-Romagna springs, 2011–2013) project [1]. Samples were collected by scraping 8–10 stones, and by collecting specimens of the dominant bryophyte species in three points of the spring area [9], and then digested using hydrogen peroxide (EN 13,946 2003 [58]). The cleaned material was mounted in Naphrax (refractive index of 1.74). Relative abundances were determined by identifying and counting a total of at least 450 valves using a Zeiss Axioskop 2 (Zeiss, Jena, Germany) and x1000 magnification.

Materials (slides, prepared material, and aliquots of the original samples), including the holotype of the new species, are held at the Diatom Collection of the MUSE—Museo delle Scienze of Trento (TR) (Northern Italy). Isotype slides and aliquots of prepared material from the same locality and substratum were deposited at the Diatom Collection of the Botanical Garden and Botanical Museum of the Freie University of Berlin (B) (Germany) and at the Diatom Collection of the Academy of Natural Sciences of Drexel University (PH) (PA, USA).

If not otherwise stated, measurements on 30 different specimens representative of the size-diminution series were made to obtain ranges and averages of the morphological and ultrastructural features.

SEM observations were carried out at the University of Frankfurt using a Hitachi S-4500 (Hitachi Ltd., Tokyo, Japan) and at the MUSE—Museo delle Scienze (Trento) using a Zeiss-EVO40XVP, Carl Zeiss SMT Ltd., Cambridge, UK at high vacuum on gold-coated stubs.

Plastid characteristics and type were assessed using Cox [59]. Terminology to describe valve morphology is based on Round et al. [22].

As concerns the typification of the new species, we chose to use the entire slide as the holotype following article 8.2 of the International Code for Botanical Nomenclature [51]. The choice for the entire population on the slide is, in our opinion, more consistent with the fact that most diatom species show an extensive variability during their population cell cycle.

In an attempt to increase data on the distribution of the three new species, both diatom and environmental data of selected springs from a comprehensive dataset of the southeastern Alps (CRENODAT Project, Biodiversity assessment and integrity evaluation of springs of Trentino—Italian Alps—and long-term ecological research, 2004–2008 [9]) were used. We looked for *Planothidium angustilanceolatum* sp. nov. in the epibryon samples from ten CRENODAT springs that were selected because counts included at least 100 valves of *P. lanceolatum* s.l. and because they had an ecomorphology/hydrochemistry consistent with the type locality of this species. We also looked for *Delicatophycus crassiminutus* sp. nov. in the epibryon slides from the five LPS included in the CRENODAT Project.

All the statistical analyses were performed within the R statistical environment [60]. To find out more about the ecological preferences of *Eunotia crassiminor* sp. nov. as compared to *Eunotia minor* (Kütz.) Grunow, we considered the 12 CRENODAT sites and the 4 EBERs sites in which both species occur. If necessary, by revisiting the slide, we carefully checked the relative abundances of the two species in each site. We then calculated weighted average, mode, percentiles, minimum, and maximum for each environmental parameter, tested differences between the two species for statistical significance using *t*-tests, and illustrated the preferences of the two species for each factor for which there was a significant difference with box plots (Figure 3a–l). The R packages used for these *Eunotia* analyses were *corrplot*, weights, and *ENmisc*. For the biometry part of the study, all relevant morphological parameters listed in Table 2 were measured on 100 raphe-valves and 100 rapheless-valves of *Planothidium angustilanceolatum* sp. nov. and *P. lanceolatum* (Bréb. ex Kütz.) Lange-Bert., respectively. The R package *plotrix* was used for this *Planothidium* analysis.

3. Results

Eunotia crassiminor Lange-Bert. et Cantonati sp. nov. (Figures 1 and 2)

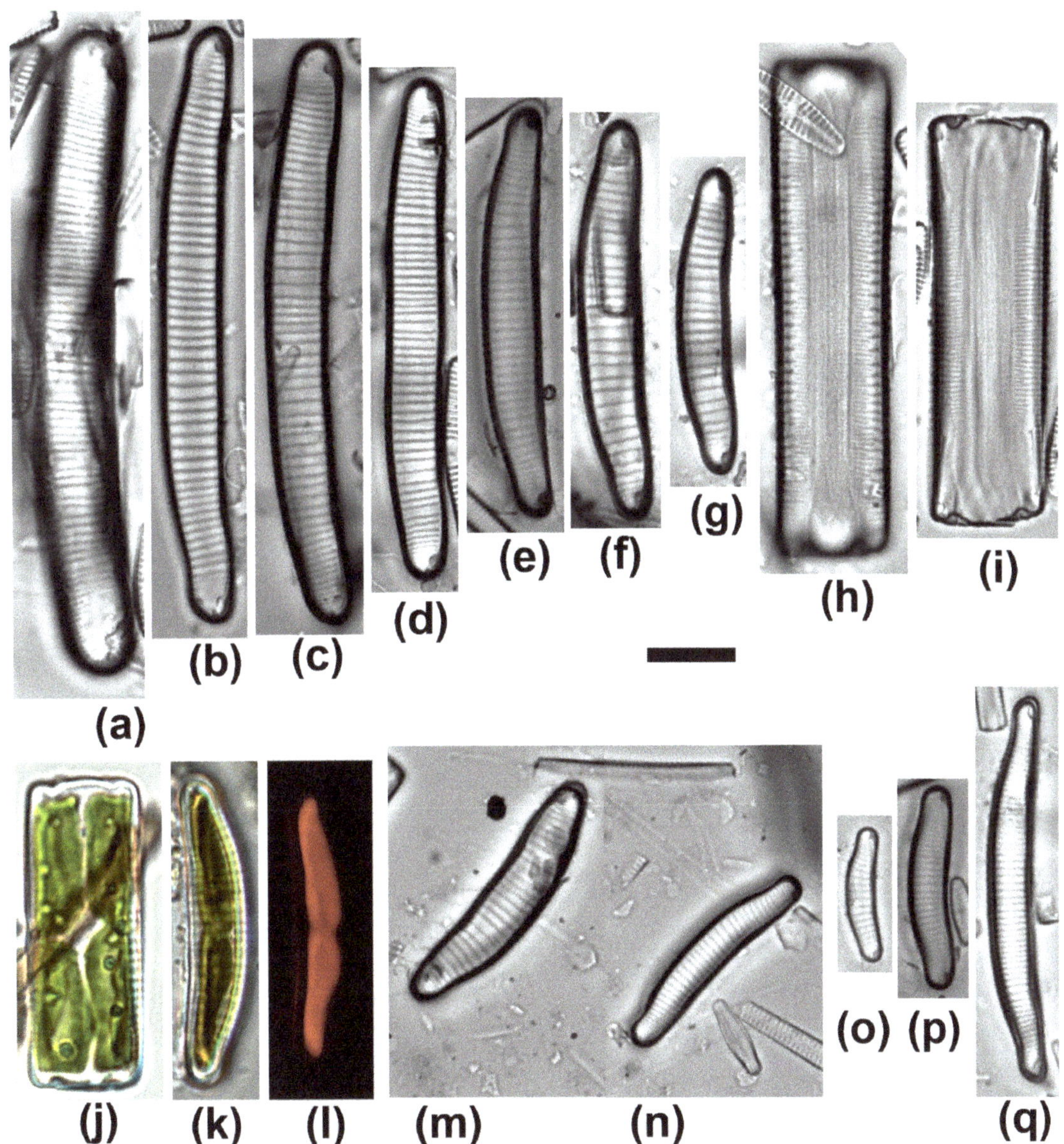

Figure 1. LM morphology of *Eunotia crassiminor* Lange-Bert. et Cantonati sp. nov. (**a–g,k–q**): valve views. (**h–j**): girdle views. (**a**): initial cell. (**j,k**): Chromoplast morphology. (**i,m–q**): *Eunotia minor* specimens shown for comparison. All micrographs bright field, with the exception of 12 which is based on chlorophyll autofluorescence. Scale bar 10 µm.

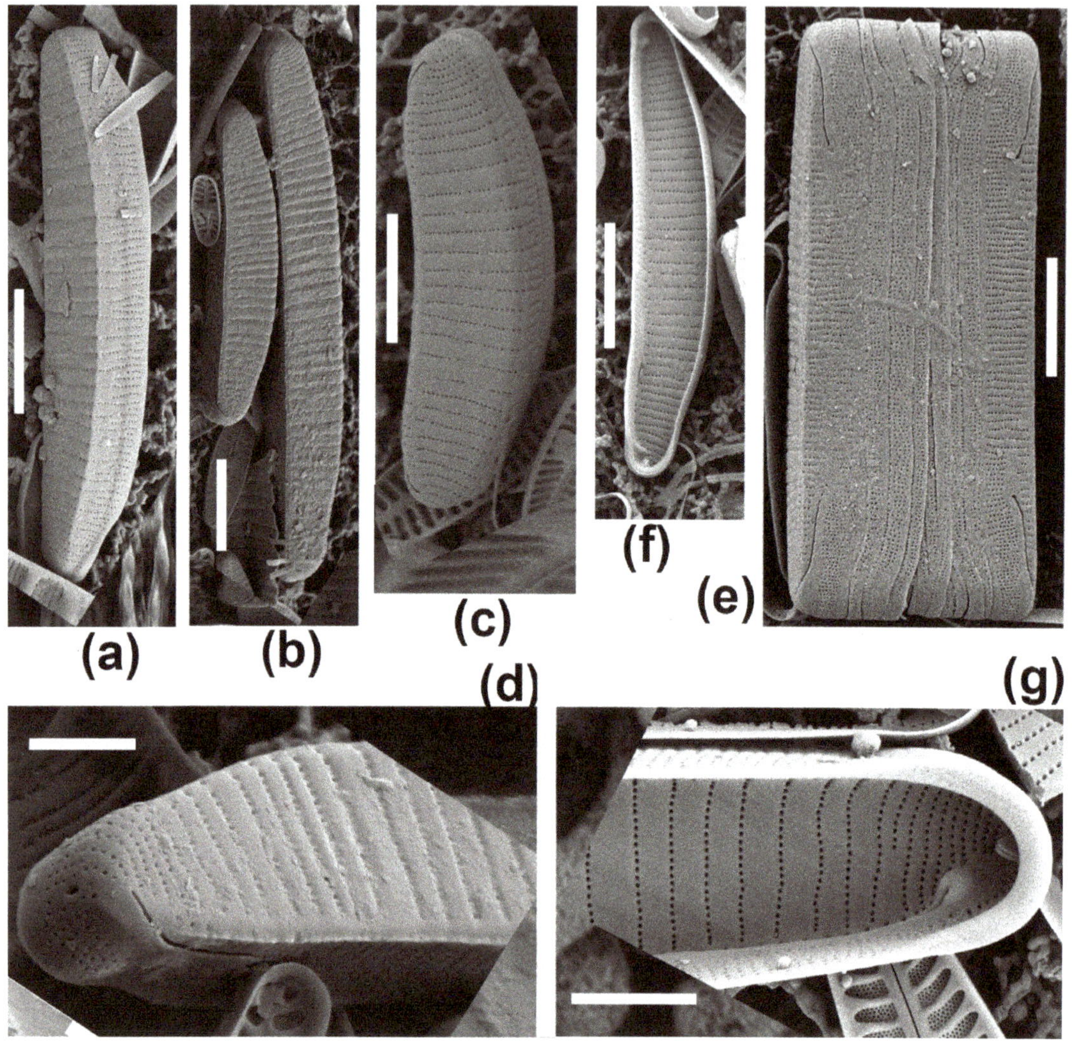

Figure 2. (a–g). SEM images of *Eunotia crassiminor* Lange-Bert. et Cantonati sp. nov. (a–e): External views. (f,g): internal views. Scale bars 10 μm (a–c,e,f), 4 μm (g), 3 μm (a–c,e,f).

Synonymy. *Eunotia minor* sensu Lange-Bert. et al. [16], Figure 159: 1–7.

To exclude from synonymy: *Eunotia minor* sensu Lange-Bert. et al. [16], Figure 159: 12–27.

Differential diagnosis versus *Eunotia minor* (Kütz.) Grunow [referred to *Himanthidium minus* Kütz., 1844, Material Kützing N. 30 from Jever, leg. Koch (=B.M. 17863), see [16], Figure 158: 14–17. Frustule morphology as in *E. minor* but specimens on average larger, appearing more strongly silicified. (Figure 1h,i provides a comparison for the girdle view, and Figure 1m,n for the valve view). Valve outline and shape variability during the cell cycle barely different. Length 28–63 (vs. 14–44) μm, breadth 6.0–8.0 (vs. 3.5–5.0) μm. Raphe course with terminal fissures not different. Transapical striae proximally 6–10 (vs. 10–17) in 10 μm, becoming rather abruptly much more densely spaced at the ends, 16–20 (vs.

becoming gradually denser up to 18–20) in 10 μm. Areolae precisely to count only with electron microscopical techniques.

As is typical for the genus, two elongate chromoplasts lying on the ventral side of the cell and extending onto the valve faces (Figure 1j,k).

SEM (Figure 2)

External view: Areolae 33–36 (vs. 39–41) in 10 μm. Sternum (=ventral area) broader in comparison. Most remarkable distinguishing character a delicate ridge on both valve margins at the junction between face and ventral/dorsal mantles (e.g., Figure 2a,b,d). Ridges lacking in *E. minor* (e.g., Figure 160: 1–2 in Lange-Bertalot et al. [16]) but present even in small cell-cycle stages of *E. crassiminor* (Figure 19).

Only one valve pole with a rimoportula in both taxa (Figure 2d,g). A faint pseudoseptum sometimes developed at the poles (Figures 23 and 24).

Other characters seen with light microscopy (e.g., striae becoming abruptly much denser towards the poles, e.g., Figure 2b) could be confirmed.

Type material. HOLOTYPE. Diatom collection of the MUSE—Museo delle Scienze, Trento, Italy, TR, slide cLIM007 DIAT 1971 (Mt. Penna spring, bryophytes). Collected by M. Cantonati on the 25th of July 2011. The holotype material is shown in Figure 1a–h,j–m and Figure 2a–g.

ISOTYPES. Diatom Collection of the Academy of Natural Sciences of Drexel University, Philadelphia, PA, USA: ANSP GC14462 (slide), ANSP GCM15149 (cleaned material), ANSP GCM15150 (raw material); -Botanical Museum of the University of Berlin, Germany: B 40 0,041,535 (slide), B 40 0,041,536 (cleaned material), B 40 0,041,537 (raw material).

REGISTRATION.—http://phycobank.org/102929

Type locality. Monte Penna spring (EBERs Project code: MtPe_ShFS-Hi, [1]). Shaded (Sh) Flowing Spring (FS) with the crustose red alga Hildenbrandia (Hi). Coordinates: Longitude: 9°30′29.493″ E, Latitude 44°29′6.029″ N. 1324 m a.s.l. Lithology: ophiolite (basalts) hard rock aquifer.

Etymology. Resembles *E. minor* but is larger and with a more robust structure.

Distribution. As yet critically observed with SEM and distinguished from *E. minor* in several springs with low conductivity in the south-eastern Alps and in the Northern Apennine but probably occurring elsewhere under appropriate conditions, waiting for critical differentiation from other morphodemes of *E. minor* sensu lato. At the type locality, the new species was more abundant in the epibryon than in the epilithon (relative abundance: 4.9 vs. 2.3%, respectively).

Ecology, co-occurring diatom species, and associated photoautotrophs. Environmental conditions at the type locality: Discharge (L s^{-1}): 3.5, Temperature (°C): 5.3, conductivity (μS cm^{-1}): 62, alkalinity (μeq L^{-1}): 311, pH: 6.6, nitrate (mg L^{-1}): 1.2, TP (μg L^{-1}): 7 (see [1] for more details). As concerns photoautotrophs, in this very shaded source the competitive balance between large groups (algae, lichens, bryophytes, and vascular plants) is clearly favorable to the mosses, which cover almost all the lithic substrata [dominance of *Brachythecium rivulare* W.P. Schimper, both submerged and, in large portions, emerged, and a certain relevance of *Plagiomnium undulatum* (Hedw.) T.J. Kop. and *Rhizomnium punctatum* (Hook.) T.J. Kop.]. Vascular plants are not abundant (as cover), and *Adenostyles glabra* (Miller) DC. and *Saxifraga rotundifolia* L. can be mentioned among them. In terms of cover, bryophytes are followed by lichens. These include two species which are rarely reported in Italy: *Verrucaria madida* Orange, an amphibious species in frequently flooded sites on siliceous rocks, often in association with other aquatic lichens and bryophytes, and *Verrucaria aquatilis* Mudd., common both in springs and along streams, in conditions of perennial/frequent submersion. Benthic macroalgae are rare and mainly represented by the red freshwater alga *Hildenbrandia rivularis* (Liebmann) J. Agardh, which is characteristic of shaded springs with well-buffered waters and medium-high conductivity.

The main co-occurring diatom species at the type locality (at least 5% relative abundance in one of the slides): *Achnanthidium minutissimum* sp. gr., *Amphora inariensis* Krammer, *Amphora indistincta* Levkov, *Brachysira exilis* (Kütz.) Round and D.G.Mann *Cocconeis pseu-*

dolineata (Geitler) Lange-Bert., *Gomphonema elegantissimum* E.Reichardt and Lange-Bert., *Humidophila perpusilla* (Grunow) Lowe, Kociolek, J.R.Johansen, Van de Vijver, Lange-Bert. and Kopalová, *Planothidium angustilanceolatum* sp. nov., *P. frequentissimum* (Lange-Bert.) Lange-Bert., *P. lanceolatum*, *Psammothidium grischunum* Bukht. and Round.

Ecology (Table 1, Figure 3a–l). With reference to temperature, *E. crassiminor* has a lower optimum weighted average than *E. minor* (Table 1); consistently, *E. crassiminor* also seems to prefer sites which are more shaded (Figure 3b). As concerns pH (Figure 3e), interestingly, *E. crassiminor* appears to prefer circumneutral values whilst *E. minor* occurs at slightly acidic ones. *E. crassiminor* has a higher weighted average for sulphates whilst *E. minor* has a higher optimum for manganese (Table 1). In particular, with reference to nitrogen, *E. crassiminor* appears to be associated with more strict oligotrophy than *E. minor*.

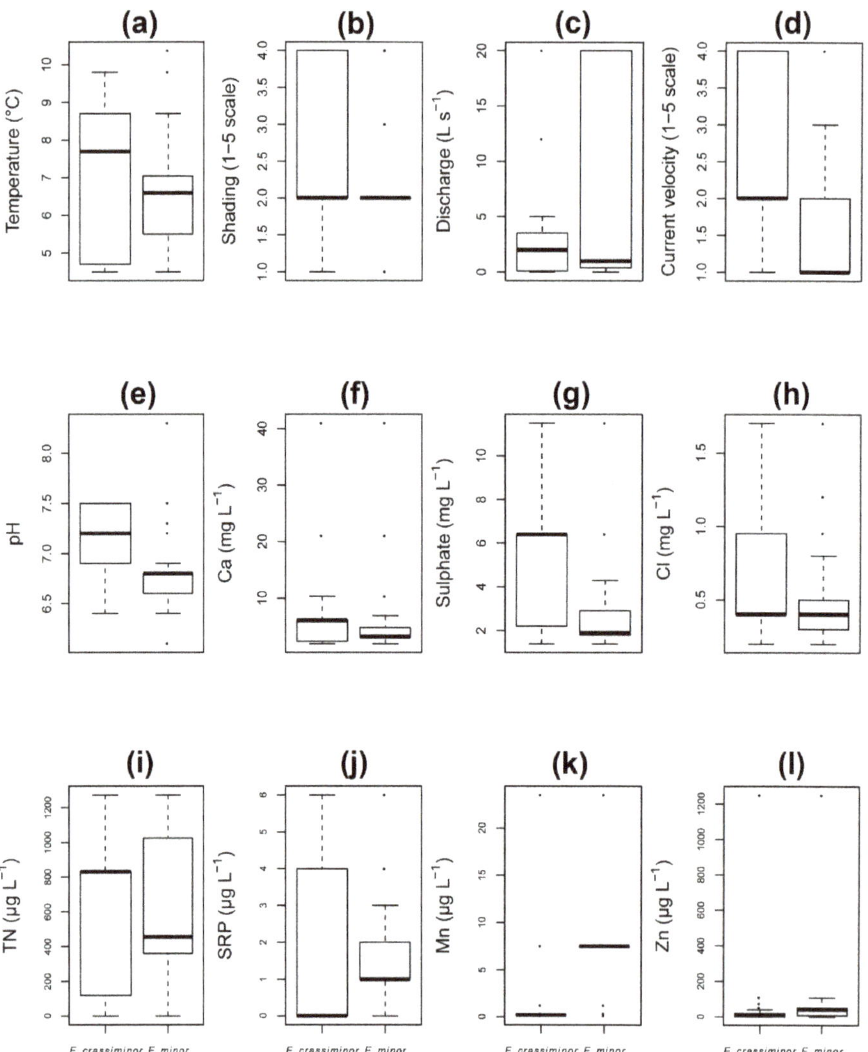

Figure 3. (**a–l**). Box and whisker plots showing the ecological preferences of *Eunotia crassiminor* as compared to *E. minor*. Only environmental factors/parameters for which statistically significant differences could be found are shown.

Table 1. Ecological preferences of *Eunotia crassiminor* as compared to *E. minor*. Only environmental factors/parameters for which statistically significant differences could be found are shown.

Factor	Eunotia crassiminor			Eunotia minor			t-Tests	
	Weighted Average	Min	Max	Weighted Average	Min	Max	t-Value	p-Value
Temperature (°C)	5.6	4.5	9.8	7.8	4.5	8.7	7.58	5.99×10^{-11}
Shading (1–5 scale)	3	1	4	2	2	2	−9.08	1.20×10^{-14}
Discharge (L s^{-1})	2.0	0.0	5.0	4.3	0.0	20.0	2.0	4.85×10^{-2}
Current velocity (1–5 scale)	3	1	4	2	1	3	−6.07	2.33×10^{-8}
pH	6.8	6.4	7.5	7.0	6.4	6.9	2.8	6.11×10^{-3}
Ca^{2+} (mg L^{-1})	4.9	2.0	10.3	8.5	2.0	6.9	2.20	3.11×10^{-2}
Sulphates (mg L^{-1})	5.1	1.4	11.5	2.2	1.4	4.3	−7.39	1.45×10^{-11}
Cl^{-1} (mg L^{-1})	0.4	0.2	1.7	0.8	0.2	0.8	4.53	2.28×10^{-5}
TN (µg L^{-1})	340	0	1272	649	0	1272	4.29	7.69×10^{-5}
SRP (µg L^{-1})	2	0	6	2	0	3	−2.25	2.69×10^{-2}
Mn (µg L^{-1})	0.6	0.2	0.2	9.8	7.5	7.5	4.95	2.59×10^{-5}
Zn (µg L^{-1})	101	0	40	37	0	107	−2.12	3.56×10^{-2}

Taxonomic comments. The obvious heterogeneity of various morphodemes and problems of identity concerning type and typification of *Eunotia minor* (Kütz.) Grunow have been discussed at length by Lange-Bertalot et al. ([16], pp. 157–160, Figure captions of plates 158–164). Whilst the true identity of *Himanthidium minus* Kütz. is not yet quite clear, *E. crassiminor* from the south-eastern Alps can be defined taxonomically and is well characterized from the morphological and ecological standpoints. It was possible to find associated in a single sample from a helocrenic spring in the Apennines (Elocrena Lago Scuro) *E. minor* and *E. crassiminor*, both never converging and easy to distinguish with the light microscope. *E. crassiminor* roughly resembles *E. pomeranica* Lange-Bert., Bąk et Witkowski [16] from peat bogs in north-western Poland. However, the latter differs by almost evenly spaced striae in the proximal and distal parts of the valve, 11–15 and 16–18 in 10 µm, respectively. The amount of areolae and striae in the new described species is less than in similarly compared species. Marginal ridges in SEM view are missing.

Planothidium angustilanceolatum Lange-Bert. et Cantonati sp. nov. (Figures 4 and 5)

Differential diagnosis compared with an associated population of *Planothidium lanceolatum* (Bréb. ex Kütz.) Lange-Bert. Specimens with conspicuously narrower valves, concerning in particular medium-sized and smaller cell-cycle stages. Valves linear-elliptic with rounded ends (vs. broadly elliptic to elliptic-lanceolate with broadly rounded ends). Largest stages rhombic-lanceolate). Length 5–24 µm, breadth 2.5–4 µm (vs. 13–32 and 5–7.5 µm, respectively). Length-to-breadth ratio 2.8–5.8 (vs. 1.6–4.7). Areae, raphe, and striae are barely different in both taxa, stria density 12–16 in 10 µm, considerably variable in both taxa.

As is typical for the genus, one chromoplast, in girdle view lying against the more convexly curved valve but extending under the other one; only moderate indentation between the two shorter lobes (Figure 4m–o; *Planothidium lanceolatum* plastid shown for comparison in Figure 4a′,b′).

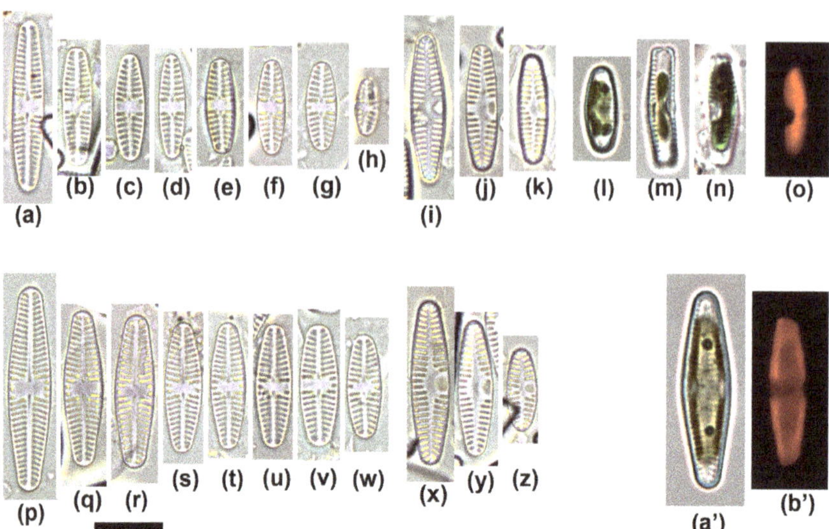

Figure 4. (**a–h**). LM morphology of *Planothidium angustilanceolatum* sp. nov. (**a–o**) as compared to a population of *Planothidium lanceolatum* co-occurring at the type locality (**p–b′**). All bright-field micrographs with the exception of (**o,b′**) which are based on chlorophyll autofluorescence. All valve views, with the exception of (**m,n**) that are girdle views. (**a–h,p–w**): Raphe valves. (**i–k,x–z**): Rapheless valves. (**l–o,a′,b′**): Chromoplast morphology. Scale bar 10 μm.

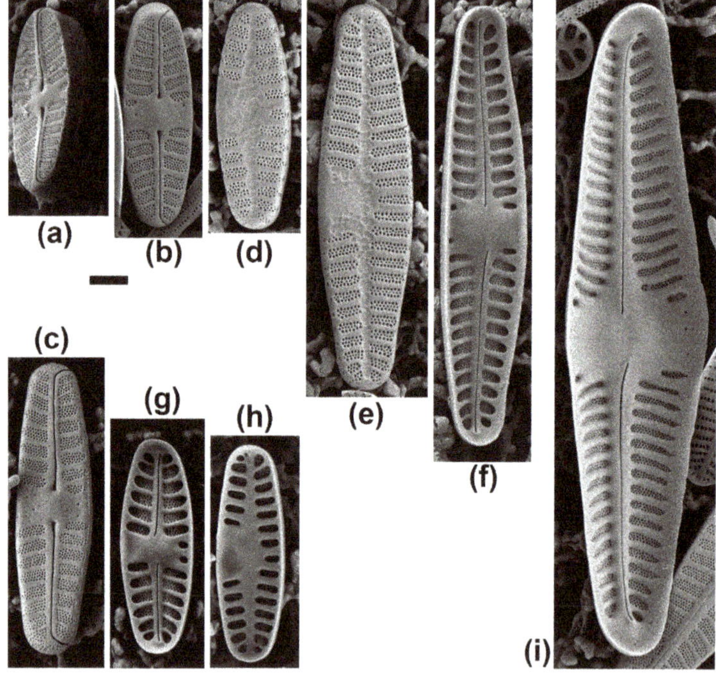

Figure 5. (**a–i**). SEM images of *Planothidium angustilanceolatum* Lange-Bert. et Cantonati sp. nov. (**a–h**), and of *P. lanceolatum* for comparison (**i**). (**a–e**): External views. (**f–i**): Internal views. (**a–c,f,g,i**): Raphe valves. (**d,e,h**): Rapheless valves. Scale bar 1 μm.

SEM (Figure 5).

Externally the areae of raphid valves are smooth (Figure 5b), whereas the araphid valves are covered by shallow irregular grooves on both the axial and the central area (Figure 5d,e), the latter more extended but not restricted unilaterally. The pluriseriate areolae (3–5 series) extend more or less clearly over the valve face margins onto the mantle. More in rapheless (Figure 5d) and less in raphid (Figure 5c) valves. The characteristic unilateral depression, "sinus", of rapheless valves is very shallow comparatively (Figure 5d,e).

Type material. HOLOTYPE. Diatom collection of the MUSE—Museo delle Scienze, Trento, Italy, TR, slide cLIM007 DIAT 1971. The holotype material is shown in Figures 4a–o and 5a–h.

ISOTYPES. Diatom Collection of the Academy of Natural Sciences of Drexel University, Philadelphia, PA, USA: ANSP GC14463 (slide), ANSP GCM15151 (cleaned material), ANSP GCM15152 (raw material); Botanical Museum of the University of Berlin, Germany: B 40 0,041,538 (slide), B 40 0,041,539 (cleaned material), B 40 0,041,540 (raw material).

REGISTRATION. http://phycobank.org/102930

Type locality. Monte Penna spring (EBERs Project code: MtPe_ShFS-Hi, [1]). Shaded (Sh) Flowing Spring (FS) with the crustose red alga *Hildenbrandia* (Hi) (see the description of *Eunotia crassiminor* for complete information).

Etymology. Resembles *P. lanceolatum* but is narrower.

Distribution. As yet critically observed exclusively at the type location in the Northern Apennines and in a spring with very similar hydrochemistry in the southeastern Alps (Belvedere spring epibryon: 3.7% *Planothidium angustilanceolatum* sp. nov., 15.6% *Planothidium lanceolatum*). At the type locality, the new species was clearly more abundant in the epibryon than in the epilithon (relative abundance: 13.6% vs. 5.4%, respectively).

Ecology, co-occurring diatom species, and associated photoautotrophs. See the description of *Eunotia crassiminor* sp. nov. for complete information.

The search for this new species in ten comparable CRENODAT springs allowed us to find it in the low-conductivity high-mountain (2056 m a.s.l.) spring Belvedere (CRENODAT Project code: OC2056). Environmental conditions at Belvedere spring: Discharge (L s^{-1}): 2, Temperature (°C): 4.5, conductivity (μS cm^{-1}): 60, alkalinity (μeq L^{-1}): 360, pH: 6.9, nitrate (mg L^{-1}): 0.48, TP (μg L^{-1}): 5.

Taxonomic comments. Obviously *Planothidium lanceolatum* sensu stricto is the closest related taxon. A population, probably identical with the *P. lanceolatum* type, is associated in the samples from the type locality. Geitler [15] described the entire cell cycle of *Achnanthes lanceolata* (syn. *Planothidium lanceolatum*): length of the auxospores (apical axis) 32–36, rarely up to 40 μm; copulating cells (gametes) 11–16, rarely up to 20 μm; smallest cells length: 7 μm; breadth, transapical axis, of post-initial cells 8–10 μm, of copulating cells 5–7 μm, of smallest specimens 4.5–5 μm, resulting in a length-to-breadth ratio of 4.1, 2.5, and 1.6, respectively. The cultured clones originate from the calcium-carbonate-rich, oligotrophic Lake Lunz in the Austrian northern Alps. The valve outlines documented by line drawings are broadly elliptical in smallest stages and elliptic-lanceolate to rhombic-lanceolate. All with broadly rounded ends. Valve outlines conforming to *P. angustilanceolatum* do not occur. Stria density 13–14 in 10 μm. On the other hand, various photographically documented specimens from all continents conform to Geitler's description as far as *P. lanceolatum* sensu stricto is concerned, excluding many misidentified similar taxa. Examples are given by Rumrich et al. ([61], Figure 28: 11–16) from the Andes in Chile, 4000 m a.s.l., Sonneman et al. ([62], p. 15, Figure 10a–d) from Australia, Dorofeyuk and Kulikovskiy ([63], Figure 41: 1–8, 14–29, Figure 45: 1–4, 46: 1–6 from Mongolia, Metzelin et al. ([64], Figure 28: 1–8) from Mongolia, Blanco ([65], Figure 44: 1–22) from Spain, Wojtal ([66], Figure 138: 11–22) from Poland, Van de Vijner et al. ([67], Figure 23: 7–14, 24: 8–a) from the Ile Crozet Archipelago, Sub Antarctica.

Other, just roughly similar taxa with comparable size and outlines are: *P. frequentissimum* (Lange-Bert.) Lange-Bert. in Metzelin et al. ([64], Figure 28: 9–11), *P. aff. fragilarioides* sensu Lange-Bert. et Krammer ([68], Figure 88: 16–21), both distinguished mainly by the presence of a horseshoe-shaped "cavum", *P. aueri* (Krasske) Lange-Bert. distinguished by striae consisting of biseriate areolae and a "cavum".

Geitler ([15], p.41, Figure 64a–d) mentioned and displayed by line drawings smaller specimens that occurred together with *P. lanceolatum*, 8–10 µm long, 4 µm broad, resembling but not identical with *P. angustilanceolatum* sp. nov. Later on, [69] validly established *Achnanthes lanceolata* var. *minor* (Schulz) Lange-Bert. as an infraspecific taxon, with a type originating from the Botanical Garden of Vienna, Austria, distinguished by conspicuously lower dimensions. Length was 6.5–21 µm, breadth 3.2–6 µm; copulating cells (gametes) were 6.5–9.5 µm long, ca. 3.9 µm broad, primary (initial) cells 18.5–21 µm long and 5.5 µm broad. Very likely this taxon is not a synonym of *P. angustilanceolatum* sp. nov. The valves of primary cells possessed rhombic outlines, i.e., inflated in the central part, whereas 22 µm long stages of *P. angustilanceolatum* sp. nov. are distinguished by non-inflated linear to slightly linear-lanceolate outlines. Moreover, the distinctly contoured "horseshoe"-shaped depressions of the rapheless valves of *A. lanceolata* var. *minor* point to a covered "cavum" rather than to a shallow open depression ("sinus"). Photographically documented specimens identified by Reichard ([70], Figure 7: 13–15) as Geitler's taxon display that character clearly. Thus, it appears to be more closely related to *P. frequentissimum* and *P. frequentissimum* var. *minus* (Shulz) Lange-Bert. rather than to *P. lanceolatum* sensu stricto.

Biometry of *Planothidium angustilanceolatum* sp. nov. at the type locality, compared with a co-occurring population of *P. lanceolatum* (Table 2, Figure 6a–e). By means of the biometry part of the study we could confirm that width (Table 2, Figure 6b) and length (Table 2, Figure 6a) differ in a statistically significant way between the *Planothidium angustilanceolatum* sp. nov. and *P. lanceolatum* population, being larger in the latter species. This is underlined also by the length—width relationships plotted in Figure 6e.

Delicatophycus crassiminutus Lange-Bert. et Cantonati sp. nov. (Figures 7 and 8)

Light microscopy. Valves rather weakly dorsiventral, lanceolate, ends distinctly protracted rostrate-subcapitate (Figure 7a–i). Length 18–27 µm, breadth 3.5–5 µm proximally, becoming ca. 2 µm below the expanded ends, maximum length-to-breadth ratio 6.2. Axial area narrow, curved, extended gradually to the valve centre, slightly displaced to the ventral side. No set off central area developed, since dorsal striae commonly not, occasionally very little shortened gradually, never abruptly and barely wider spaced than the subproximal striae (Figure 7d–g). Occasionally the area system even appears slightly constricted in the centre. Raphe strongly reverse-lateral towards proximal ends. Terminal fissures comma-shaped and dorsally bent (see SEM). Dorsal striae radiate throughout, 15–16 in 10 µm. Ventral striae are distinctly more narrowly spaced, 18–21 in 10 µm, the median ones at most very little radiate. Areolation of the striae is indistinctly resolvable in LM.

As is typical for the genus, one highly-lobed chromoplast lying with its centre against the ventral side of the girdle but extending under each valve with longitudinal indentations (H-shaped); pyrenoid on the dorsal side of the cell (Figure 7k,l).

SEM. External and internal views see Figure 8a,b. Basic pattern of the fine structures is principally the same as in other taxa of the genus, particularly *D. minutus* M.J.Wynne. However, the wavy appearance of fusing areolae externally is less expressed in our materials probably due to corrosion effects, see likewise Figure 8c for *D. minutus*. Areola density 35–44 in 10 µm. Proximal raphe ends deflected to the ventral side, distal ends strongly bent onto the dorsal mantle, sickle-shaped. Striae in the subcapitate ends become subparallel on the dorsal side, even convergent on the ventral side.

Table 2. Morphological characteristics of *Planothidium angustilanceolatum* sp. nov. at the type locality, compared with a co-occurring population of *P. lanceolatum*.

	P. angustilanceolatum						P. lanceolatum						t-Tests					
	R-Valves			RL-Valves			R-Valves			RL-Valves			R-Valves			RL-Valves		
	Mean	Min	Max	Mean	Min	Max	Mean	Min	Max	Mean	Min	Max	t-Value	df	p-Value	t-Value	df	p-Value
Length (μm)	14.9	7.9	24.5	14.0	7.9	24.1	18.8	12.8	29.1	18.9	11.7	27.4	5.12	99	1.51×10^{-6}	6.86	101	5.44×10^{-10}
Width (μm)	4.1	3.1	5.2	3.9	3.1	4.8	5.2	4.1	7.2	5.1	4.1	7.4	11.09	98	$<2.2 \times 10^{-16}$	13.41	100	$<2.2 \times 10^{-16}$
Length/Width	3.6	2.6	5.7	3.6	2.4	5.8	3.6	2.8	4.9	3.7	2.6	5.4	−0.09	75	9.28×10^{-1}	0.78	117	$4.38 \times 10^{-}$
Striae in 10 μm	13	12	14	13	12	14	13	12	16	13	12	14	1	88	1.84×10^{-1}	0.21	114	$8.37 \times 10^{-}$

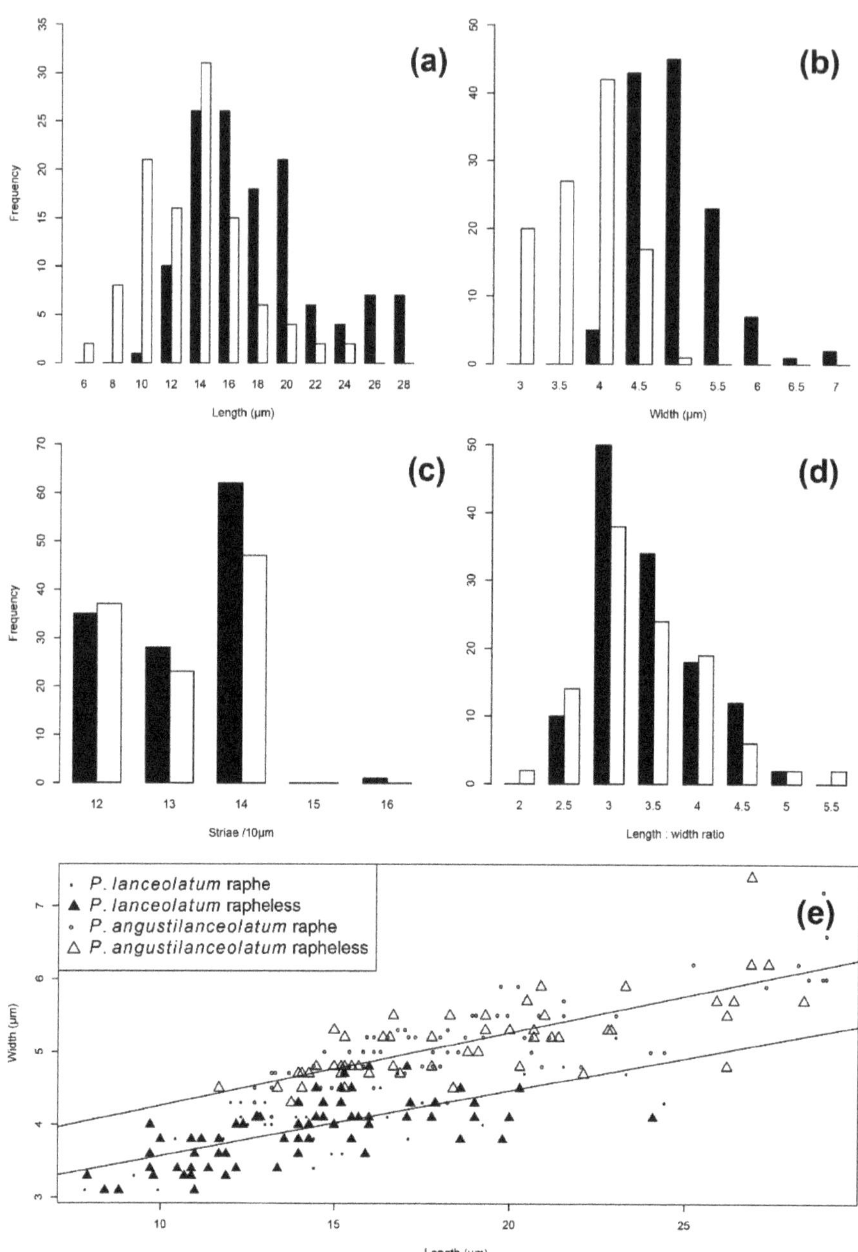

Figure 6. (**a–e**) Morphological characteristics of *Planothidium angustilanceolatum* sp. nov. at the type locality, compared with a co-occurring population of *P. lanceolatum*.

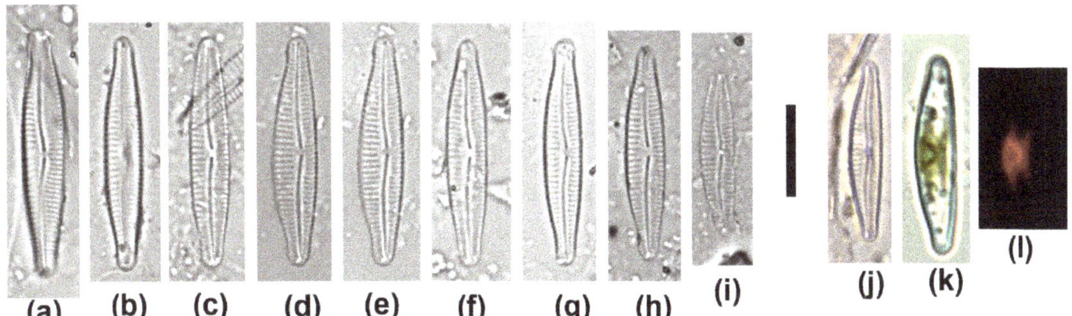

Figure 7. (**a–l**). LM morphology of *Delicatophycus crassiminutus* Lange-Bert. et Cantonati sp. nov. (**a–i,k,l**), and of *D. minutus* (**j**) for comparison. All bright-field micrographs with the exception of (**l**) which is based on chlorophyll autofluorescence. (**k,l**): Chromoplast morphology. Scale bar 10 μm.

Figure 8. (**a–c**). SEM images of *Delicatophycus crassiminutus* Lange-Bert. & Cantonati sp. nov. (**a,b**), and of *D. minutus* for comparison (**c**). (**a,c**): External views. (**b**): Internal view. Scale bars 5 μm.

Type material.

HOLOTYPE. Diatom collection of the MUSE—Museo delle Scienze, Trento, Italy, TR, slide cLIM007 DIAT 1962 (Carameto spring, epilithon). Collected by M. Cantonati on the 28th of July 2011. The holotype material is shown in Figure 7a–i,k,l and Figure 8a,b.

ISOTYPES. Diatom Collection of the Academy of Natural Sciences of Drexel University, Philadelphia, PA, USA: ANSP GC14464 (slide), ANSP GCM15153 (cleaned material), ANSP GCM15154 (raw material); -Botanical Museum of the University of Berlin, Germany: B 40 0,041,541 (slide), B 40 0,041,542 (cleaned material), B 40 0,041,543 (raw material).

REGISTRATION.—http://phycobank.org/102931

Type locality. Carameto spring (EBERs Project code: Cara_LPS-sn, [1]). Small (s), near-natural (n) limestone-precipitating spring (LPS). Coordinates: Longitude: 9°45′35.165″ E, Latitude 44°39′50.840″ N. 758 m a.s.l. Lithology: limestones, flyschs; calcarenites.

Distribution. At the type locality, the new species was found only in the epilithon (relative abundance: 2.2%). Observations on CRENODAT materials confirmed that the species is restricted to LPS springs.

Ecology, co-occurring diatom species, and associated photoautotrophs. Environmental conditions at the type locality: Discharge (L s^{-1}): 0.07, Temperature (°C): 12.8, conductivity (µS cm^{-1}): 462, HCO$_3^-$ (mg L^{-1}): 145, pH: 7.7, nitrate (mg L^{-1}): 0.12, TP (µg L^{-1}): 6 (see [1] for more details). The vegetation occurring in the Carameto spring belongs to the *Adiantion* alliance with dominance of the characteristic bryophyte species *Eucladium verticillatum* Bruch and W.P. Schimper, *Hymenostylium recurvirostrum* (Hedw.) Dixon, and *Pellia endiviifolia* (Dicks.) Dumort. Other taxa occurring at the type locality, but all with low cover, were the bryophytes *Bryum pseudotriquetrum* (Hedw.) Gaertn., Meyer and Scherb. and *Palustriella commutata* Ochyra, and the vascular plants *Carex flacca* Schreb. and *Molinia caerulea* (L.) Moench. Species diversity was low (13 taxa), as expected in this kind of community [71].

Main co-occurring diatom species at the type locality (at least 5% relative abundance in one of the slides): *Achnanthidium minutissimum* sp. gr., *Achnanthidium trinode* Ralfs, *Delicatophycus minutus*, *Denticula tenuis* Kütz., *Encyonopsis lange-bertalotii* Krammer, *Gomphonema tenoccultum* E.Reichardt.

Taxonomic comments. Among the moderately few taxa of the genus, only *D. minutus* is actually similar (see [50], summarizing table of the taxa with illustrations (pp. 112–113). It differs mainly by more distinctly dorsiventral symmetry, missing set off ends, and presence of a central area dorsally. In SEM, external view ([50], Figure 137: 16), striae are distinctly radiate throughout; convergent striae on the dorsal side of the ends are lacking due to the simply rounded, not subcapitate ends. Moreover, *D. minutus* has a higher density of the central dorsal striae (16–21 vs. 15–16 in 10 µm).

4. Discussion

Springs are a unique but severely threatened wetland type. Therefore, the in-depth knowledge of the taxonomy and ecology of characteristic diatom species is important because diatoms are excellent indicators of the quality and integrity of these peculiar ecosystems in the face of direct and indirect human impacts.

The three new species described differ morphologically from the most similar existing species by: less dense striae and areolae, and absence of a ridge at the valve face-mantle transition (SEM feature) (*Eunotia crassiminor* Lange-Bert. et Cantonati sp. nov.; closest established species: *Eunotia minor*); narrower and shorter cells (*Planothidium angustilanceolatum* Lange-Bert. et Cantonati sp. nov.; closest established species: *Planothidium lanceolatum*); barely-dorsiventral symmetry, set off ends, and lower density of the central dorsal striae (*Delicatophycus crassiminutus* Lange-Bert. et Cantonati sp. nov.; closest established species: *Delicatophycus minutus*).

Two of the three species we described are separated from the most similar established species by dimensions. By applying Krammer's [13] ratio of maximal and minimal width as a reliable means to test the quality of taxa, we find that we have contributed to improve knowledge on two taxa (*Eunotia minor* s.l. and *Planothidium lanceolatum* s.l.), which are likely to be only partially resolved complexes of species because they have a max-min width ratio of 2 and 2.2, respectively (data of minimum and maximum width taken from [72].

We also could contribute to ameliorate knowledge on the ecological profiles of the three newly described species. In the case of *Eunotia crassiminor* sp. nov. data were sufficient to allow testing for statistical significance. *Eunotia crassiminor* sp. nov. as compared to *Eunotia minor*, appears to occur in colder inland waters with circumneutral pH, and strict oligotrophy also with respect to nitrogen. In the face of global warming, diffuse airborne nitrate pollution, and acidification risk, this realized ecological niche singles out *Eunotia crassiminor* sp. nov. as a species which is clearly more threatened than *E. minor*. This information is very useful as well for the generation and updating of diatom Red Lists that can provide excellent metrics for the conservation value of inland waters [5]. As far as the other two newly described species are concerned, data gained on the distribution were too few to allow for statistical treatment. However, they occur in spring types which are so peculiar that some generalization, though with caution, can be made. *Planothidium angustilanceolatum* sp. nov. was found only in two springs, one in the Northern Apennines and one in the south-eastern Alps, that however have almost identical morphological, physical, and chemical characteristics. It can thus be stated that the typical habitat of this species is oligotrophic mountain flowing springs with low conductivity (approximately 60 μS cm^{-1}). *Delicatophycus crassiminutus* sp. nov. was observed only in LPS, and is therefore likely to be restricted to hard water springs and comparable habitats where CO_2 degassing leads to carbonate precipitation.

The correct knowledge of the taxonomy and ecology of the species occurring in mountain aquatic habitats is of great importance to use diatoms as reliable indicators of environmental and climate change. Mountain ecosystems are sensitive and reliable indicators of climate change (e.g., [73]). There are many good reasons for protecting these freshwater habitats [74]: they are relatively scarce, provide clean water for many uses, harbour a large number of Red List taxa [compare [7,74], and they are sensitive to disturbance from human activities [7,12].

Author Contributions: Conceptualization, M.C. and H.L.-B.; methodology, M.C., N.A. and L.v.W.; validation, M.C. and H.L.-B.; formal analysis, L.v.W. and M.C.; investigation, M.C. and N.A.; resources, M.C.; data curation, M.C. and N.A.; writing—original draft preparation, M.C. and O.B.; writing—review and editing, M.C., O.B., H.L.-B. and L.v.W.; visualization, N.A. and M.C.; supervision, M.C. and H.L.-B.; project administration, M.C.; funding acquisition, M.C. All authors have read and agreed to the published version of the manuscript.

Funding: The study (EBERs Project) generating the data on which the descriptions of the three new species are based was funded by the Geological Survey of the Emilia-Romagna Region (Italy).

Institutional Review Board Statement: Not applicable.

Informed Consent Statement: Not applicable.

Data Availability Statement: The data presented in this study are available on request from the corresponding author.

Acknowledgments: Authors other than H.L.B. are most glad to use the occasion of the publication of this paper to celebrate the 85th birthday of Horst Lange-Bertalot acknowledging his outstanding contributions to the taxonomy and ecology of diatoms. We are grateful to the Geological Survey of the Emilia-Romagna Region (in particular Stefano Segadelli) for fostering the EBERs (Exploring the Biodiversity of Emilia-Romagna springs; 2011–2013) Project, in the frame of which samples on which the three species described as new in this paper were collected. We thank Daniel Spitale and Juri Nascimbene for useful information on the bryophytes + higher plants and the lichens of the type localities, respectively.

Conflicts of Interest: The authors declare no conflict of interest.

References

1. Cantonati, M.; Segadelli, S.; Spitale, D.; Gabrieli, J.; Gerecke, R.; Angeli, N.; De Nardo, M.T.; Ogata, K.; Wehr, J.D. Geological and hydrochemical prerequisites of unexpectedly high biodiversity in spring ecosystems at the landscape level. *Sci. Total Environ.* **2020**, *740*, 140157. [CrossRef]
2. Cantonati, M.; Füreder, L.; Gerecke, R.; Jüttner, I.; Cox, E.J. Crenic habitats, hotspots for freshwater biodiversity conservation: Toward an understanding of their ecology. *Freshw. Sci.* **2012**, *31*, 463–480. [CrossRef]
3. Cantonati, M.; Stevens, L.; Segadelli, S.; Springer, A.; Goldscheider, N.; Celico, F.; Filippini, M.; Ogata, K.; Gargini, A. Ecohydrogeology: The interdisciplinary convergence needed to improve the study and stewardship of springs and other groundwater-dependent habitats, biota, and ecosystems. *Ecol. Indic.* **2020**, *110*, 1. [CrossRef]
4. Cantonati, M.; Fensham, R.J.; Stevens, L.E.; Gerecke, R.; Glazier, D.S.; Goldscheider, N.; Knight, R.L.; Richardson, J.S.; Springer, A.E.; Tockner, K. Urgent plea for global protection of springs. *Conserv. Biol.* **2021**, *35*, 378–382. [CrossRef]
5. Cantonati, M.; Hofmann, G.; Spitale, D.; Werum, M.; Lange-Bertalot, H. Diatom Red Lists: Important tools to assess and preserve biodiversity and habitats in the face of direct impacts and environmental change. *Biodivers. Conserv.* **2021**. Accepted.
6. Cantonati, M.; Angeli, N.; Lange-Bertalot, H. Three new *Fragilaria* species (Bacillariophyta) from low-conductivity mountain freshwaters (Alps and Apennines). *Phytotaxa* **2019**, *404*, 261–274. [CrossRef]
7. Bahls, L. Diatom indicators of climate change in Glacier National Park. *Intermt J. Sci.* **2007**, *13*, 99–109.
8. Cantonati, M.; Segadelli, S.; Ogata, K.; Tran, H.; Sanders, D.; Gerecke, R.; Rott, E.; Filippini, M.; Gargini, A.; Celico, F. A global review on ambient Limestone-Precipitating Springs (LPS): Hydrogeological setting, ecology, and conservation. *Sci. Total. Environ.* **2016**, *568*, 624–637. [CrossRef]
9. Cantonati, M.; Angeli, N.; Bertuzzi, E.; Spitale, D.; Lange-Bertalot, H. Diatoms in springs of the Alps: Spring types, environmental determinants, and substratum. *Freshw. Sci.* **2012**, *31*, 499–524. [CrossRef]
10. Werum, M.; Lange-Bertalot, H. Diatoms in springs, from Central Europe and elsewhere under the influence of hydrogeology and anthropogenic impages. *Iconogr. Diatomol.* **2004**, *13*, 1–417.
11. Cantonati, M.; Lange-Bertalot, H. Diatom biodiversity of springs in the berchtesgaden national park (north-eastern alps, germany), with the ecological and morphological characterization of two species new to science. *Diatom Res.* **2010**, *25*, 251–280. [CrossRef]
12. Cantonati, M.; Lange-Bertalot, H. Diatom monitors of close-to-pristine, very-low alkalinity habitats: Three new *Eunotia* species from springs in Nature Parks of the south-eastern Alps. *J. Limnol.* **2011**, *70*, 209–221. [CrossRef]
13. Krammer, K. *Cymbella Diatoms of Europe, Diatoms of the European Inland Waters and Comparable Habitats*; Lange-Bertalot, H., Ed.; A.R.G. Gantner Verlag K.G.: Rugell, Liechtenstein, 2002; Volume 3, pp. 1–584.
14. Maltsev, Y.; Maltseva, S.; Kociolek, J.P.; Jahn, R.; Kulikovskiy, M. Biogeography of the cosmopolitan terrestrial diatom Hantzschia amphioxys sensu lato based on molecular and morphological data. *Sci. Rep.* **2021**, *11*, 1–19. [CrossRef]
15. Geitler, L. Der Formwechsel der pennaten Diatomeen (Kieselalgen) *Arch. Protistenkunde.* **1932**, *78*, 1–226.
16. Lange-Bertalot, H.; Bąk, M.; Witkowski, A. *Eunotia* and some related genera. In *Diatoms of Europe. Diatoms of the European in-Land Water and Comparable Habitats*; Lange-Bertalot, H., Ed.; A.R.G. Gantner Verlag K.G.: Rugell, Liechtenstein, 2011; Volume 6, pp. 1–747.
17. Blasi, C.; Del Vico, E. High mountain vegetation of the Apennines. *Ber. d. Reinh.-Tüxen-Ges.* **2012**, *24*, 179–194.
18. Cantonati, M.; Angeli, N.; Lange-Bertalot, H.; Levkov, Z. New *Amphora* and *Halamphora* (Bacillariophyta) species from springs in the northern Apennines (Emilia-Romagna, Italy). *Plant Ecol. Evol.* **2019**, *152*, 285–292. [CrossRef]
19. Ehrenberg, C.G. *Über ein aus Fossilen Infusorien Bestehendes, 1832 zu Brod Verbacknes Bergmehl von den Grenzen Lapplands in Schweden. Bericht über die zur Bekanntmachung Geeigneten Verhandlungen der Königl*; Preuß. Akademie der Wissenschaften zu Berlin Erster Jarhrgang: Berlin, Germany, 1837; pp. 43–45.
20. Novitski, L.; Kociolek, P. Preliminary Light and Scanning Electron Microscope Observations of Marine Fossil *Eunotia* Species with Comments on the Evolution of the Genus *Eunotia*. *Diatom Res.* **2005**, *20*, 137–143. [CrossRef]
21. Luo, F.; You, Q.; Yu, P.; Pang, W.; Wang, Q. Eunotia (Bacillariophyta) biodiversity from high altitude, freshwater habitats in the Mugecuo Scenic Area, Sichuan Province, China. *Phytotaxa* **2019**, *394*, 133–147. [CrossRef]
22. Round, F.E.; Crawford, R.M.; Mann, D.G. *The Diatoms: Biology and Morphology of the Genera*; Cambridge University Press: Cambridge, CA, USA, 1990; pp. 1–747.
23. Kociolek, J.P.; Spaulding, S. Eunotioid and asymmetrical naviculoid diatoms. In *Freshwater Algae of North America. Ecology and classification*; Wehr, J.D., Sheath, R.G.F., Eds.; Academic Press Elsevier Science: London, UK, 2003; pp. 655–668.
24. Furey, P.C.; Lowe, R.L.; Johansen, R.J. Eunotia Ehrenberg (Bacillariophyta) of the Great Smoky Mountains National Park, USA. *Biblioth. Diatomol.* **2011**, *56*, 1–133.
25. Cox, E. Coscinodiscophyceae, Mediophyceae, Fragilariophyceae, Bacillariophyceae (Diatoms). In *Syllabus of Plant Families. Adolf Engler's Syllabus der Pflanzenfamilien. 2/1 Photoautotrophic Eukaryotic Algae*; Frey, W., Ed.; Borntraeger Science Publishers: Stuttgart, Germany, 2015; pp. 64–103.
26. Metzeltin, D.; Lange-Bertalot, H. Tropical diatoms of South America I: About 700 predominantly rarely known or new taxa representative of the neotropical flora. In *Iconographia Diatomologica. Annotated Diatom Micrographs*; Lange-Bertalot, H., Ed.; Diversity-Taxonomy-Geobotany, Koeltz Scientific Books: Königstein, Germany, 1998; Volume 5, pp. 1–695.
27. Liu, Y.; Wang, Q.; Fu, C. Taxonomy and distribution of diatoms in the genus Eunotia from the Da'erbin Lake and Surrounding Bogs in the Great Xing'an Mountains, China. *Nova Hedwig* **2011**, *92*, 205–232. [CrossRef]

28. Ruwer, D.T.; Blanco, S.; Rodrigues, L. *Eunotia* Ehrenberg (Eunotiaceae, Bacillariophyta) in a subtropical floodplain: A new species and taxonomic contributions. *Phytotaxa* **2021**, *505*, 157–175. [CrossRef]
29. Round, F.E.; Bukhtiyarova, L. Four new genera based on *Achnanthes* (*Achnanthidium*) together with a re-definition of *Achnanthidium*. *Diatom Res.* **1996**, *11*, 345–361. [CrossRef]
30. Guiry, M.D.; Guiry, G.M. AlgaeBase. World-Wide Electronic Publication, National University of Ireland, Galway. 2021. Available online: https://www.algaebase.org (accessed on 1 October 2021).
31. Kociolek, J.P.; Blanco, S.; Coste, M.; Ector, L.; Liu, Y.; Karthick, B.; Kulikovskiy, M.; Lundholm, N.; Ludwig, T.V.; Potapova, M.; et al. DiatomBase. Available online: http://www.diatombase.org (accessed on 27 October 2021).
32. Jahn, R.; Abarca, N.; Gemeinholzer, B.; Mora, D.; Skibbe, O.; Kulikovskiy, M.; Gusev, E.; Kusber, W.-H.; Zimmermann, J. *Planothidium lanceolatum* and *Planothidium frequentissimum* reinvestigated with molecular methods and morphology: Four new species and the taxonomic importance of the sinus and cavum. *Diatom Res.* **2017**, *32*, 75–107. [CrossRef]
33. Stancheva, R. Planothidium sheathii, a new monoraphid diatom species from rivers in California, USA. *Phytotaxa* **2019**, *393*, 131–140. [CrossRef]
34. Wetzel, C.E.; Van de Vijver, B.; Blanco, S.; Ector, L. On some common and new cavum-bearing *Planothidium* (Bacillariophyta) species from freshwater. *Fottea* **2019**, *19*, 50–89. [CrossRef]
35. Potapova, M.G. New species and combinations in monoraphid diatoms (family Achnanthidiaceae) from North America. *Diatom Res.* **2012**, *27*, 29–42. [CrossRef]
36. Wetzel, C.E.; Van De Vijver, B.; Hoffmann, L.; Ector, L. *Planothidium incuriatum* sp. nov. a widely distributed diatom species (Bacillariophyta) and type analysis of *Planothidium biporomum*. *Phytotaxa* **2013**, *138*, 43–57. [CrossRef]
37. Compère, P.; Riaux–Gobin, C. Diatomées de quelques biotopes marins, saumâtres et dulçaquicoles de Guinée (Afrique occidentale) [Diatoms from some marine, brackish and freshwater biotopes of Guinea (West Africa)]. *Syst. Geogr. Plants* **2009**, *79*, 33–66.
38. N'Guessan, K.R.; Wetzel, C.E.; Ector, L.; Coste, M.; Cocquyt, C.; Van de Vijver, B.; Yao, S.S.; Ouattara, A.; Kouamelan, E.P.; Tison-Rosebery, J. *Planothidium comperei* sp. nov. (Bacillariophyta), a new diatom species from Ivory Coast. *Plant Ecol. Evol.* **2014**, *147*, 455–462. [CrossRef]
39. Blanco, S.; Álvarez–Blanco, I.; Cejudo–Figueiras, C.; Recio Espejo, J.M.; Borja Barrera, C.; Bécares, E.; Díaz del Olmo, F.; Cámara Artigas, R. The diatom flora in temporary ponds of Doñana National Park (southwest Spain): Five new taxa. *Nord. J. Bot.* **2013**, *31*, 489–499. [CrossRef]
40. Lai, G.G.; Ector, L.; Padedda, B.M.; Wetzel, C.E. *Planothidium marganaiensis* sp. nov. (Bacillariophyta), a new cavum-bearing species from a karst spring in south-western Sardinia (Italy). *Phytotaxa* **2021**, *489*, 140–154. [CrossRef]
41. Kulikovskiy, M.S.; Lange–Bertalot, H.; Kuznetsova, I.V. Lake Baikal: Hotspot of endemic diatoms, 2. *Iconogr. Diatomol.* **2015**, *26*, 1–656.
42. Van de Vijver, B.; Wetzel, C.; Kopalová, K.; Zidarova, R.; Ector, L. Analysis of the type material of *Achnanthidium lanceolatum* Brébisson ex Kützing (Bacillariophyta) with the description of two new *Planothidium* species from the Antarctic Region. *Fottea* **2013**, *13*, 105–117. [CrossRef]
43. Wetzel, C.E.; Ector, L. *Planothidium lagerheimii* comb. nov. (Bacillariophyta, Achnanthales) a forgotten diatom from South America. *Phytotaxa* **2014**, *188*, 261. [CrossRef]
44. Riaux-Gobin, C.; Witkowski, A.; Igersheim, A.; Lobban, C.S.; Al-Handal, A.Y.; Compère, P. *Planothidium juandenovense* sp. nov. (Bacillariophyta) from Juan de Nova (Scattered Islands, Mozambique Channel) and other tropical environments: A new addition to the Planothidium delicatulum complex. *Fottea* **2018**, *18*, 106–119. [CrossRef]
45. Cooper, J.T.; Steinitz-Kannan, M.; Kallmeyer, D.E. Bacillariophyta: The Diatoms. In *Algae: Source of Treatment*; Ripley, M.G., Ed.; American Water Works Association: Denver, CO, USA, 2010; pp. 207–249.
46. Stancheva, R.; Kristan, N.V.; Iii, W.B.K.; Sheath, R.G. Diatom genus *Planothidium* (Bacillariophyta) from streams and rivers in California, USA: Diversity, distribution and autecology. *Phytotaxa* **2020**, *470*, 1–30. [CrossRef]
47. Bąk, M.; Halabowski, D.; Kryk, A.; Lewin, I.; Sowa, A. Mining salinisation of rivers: Its impact on diatom (Bacillariophyta) assemblages. *Fottea* **2020**, *20*, 1–16. [CrossRef]
48. Halabowski, D.; Bąk, M.; Lewin, U. Distribution and ecology of two interesting diatom species *Navicula flandriae* Van de Vijver et Mertens and *Planothidium nanum* Bąk, Kryk et Halabowski in rivers of Southern Poland and their spring areas. *Oceanol. Hydrobiol. Stud.* **2021**, *50*, 137–149. [CrossRef]
49. Wynne, M.J. *Delicatophycus* gen. nov.: A validation of "Delicata Krammer" inval. (Gomphonemataceae, Bacillariophyta). *Not. Algarum* **2019**, *97*, 1–3.
50. Krammer, K. Cymbopleura, Delicata, Navicymbula, Gomphocymbellopsis, Afrocymbella. In *Diatoms of Europe, Diatoms of the Euro-Pean Inland Waters and Comparable Habitats*; Lange-Bertalot, H., Ed.; A.R.G. Gantner Verlag, K.G.: Rugell, Liechtenstein, 2003; Volume 4, pp. 1–529.
51. Turland, N.J.; Wiersema, J.H.; Barrie, F.R.; Greuter, W.; Hawksworth, D.L.; Herendeen, P.S.; Knapp, S.; Kusber, W.-H.; Li, D.-Z.; Marhold, K. (Eds.) *International Code of Nomenclature for Algae, Fungi, and Plants (Shenzhen Code) Adopted by the Nineteenth International Botanical Congress, Shenzhen, China, 23–29 July 2017*; Koeltz Botanical Books: Glashütten, Germany, 2018; Volume 159, pp. 1–253.

52. Cohu, R.L.; Lange-Bertalot, H.; Van de Viver, B.; Tudesque, L. Analysis and critical evaluation of structural features in four Cymbellaceae taxa from New Caledonia. *Fottea* **2020**, *20*, 75–85. [CrossRef]

53. Shi, Z.X. *Flora algarum Sinicarum Aquae Dulcis, Vol. XVI, Bacillariophyta Cymbellaceae*; Science Press: Beijing, China, 2013; pp. 1–217.

54. Liu, B.; Blanco, S.; Qing-Yan, L. Ultrastructure of *Delicata sinensis* Krammer et Metzeltin and *D. williamsii* sp. nov. (Bacillariophyta) from China. *Fottea* **2018**, *18*, 30–36. [CrossRef]

55. Metzeltin, D.; Lange-Bertalot, H. Tropical Diatoms of South America II. Special remarks on biogeography disjunction. *Iconogr. Diatomol.* **2007**, *18*, 1–877.

56. Bahls, L.L. *Diatoms from Western North America 1. Some New and Notable Biraphid Species*; Edited and Published by the Author: Helena, MT, USA, Montana Diatom Collection; 2017; pp. 1–52.

57. Liu, W.; Li, Y.-L.; Wu, H.; Kociolek, J.P. *Delicatophycus liuweii* sp. nov., a new cymbelloid diatom (Bacillariophyceae) from an upper tributary of the Liujiang River, Guangxi, China. *Phytotaxa* **2021**, *505*, 63–70. [CrossRef]

58. EN 13946. Water Quality. Guidance Standard for the Routine Sampling and Pretreatment of Benthic Diatoms from Rivers. 2003. Available online: http://http://www.safrass.com/partners_area/BSI%20Benthic%20diatoms.pdf (accessed on 1 October 2021).

59. Cox, E.J. *Identification of Freshwater Diatoms from Live Material*; Chapman and Hall: London, UK, 1996; p. 158.

60. Team, R.R. A language and Environment for Statistical Computing, Vienna, R Foundation for Statistical Computing. 2009. Available online: http://www.R-project.org (accessed on 27 October 2021).

61. Rumrich, U.; Lange-Bertalot, H.; Rumrich, M. Diatomeen der Anden von Venezuela bis Patagonien/Feuerland und zwei weitere Beiträge. Diatoms of the Andes from Venezuela to Patagonia/Tierra del Fuego and two additional contributions. *Iconogr. Diatomol.* **2009**, *9*, 1–673.

62. Sonneman, J.A.; Sincock, A.; Fluin, J.; Reid, M.A.; Newall, P.; Tibby, J.; Gell, P. *All Illustrated Guide to Common Stream Di-atom Species from Temperate Australia*; Cooperative Research Centre for Freshwater Ecology: Thurgoona, Australia, 2000; pp. 1–168.

63. Dorofeyuk, N.I.; Kulikovskiy, M.S. *Diatoms of Mongolia*; Inst. Ecol. Evol. RAN: Moscow, Russia, 2012; pp. 1–367.

64. Metzeltin, D.; Lange-Bertalot, H.; Soninkhishig, N. Diatoms in Mongolia. *Icon. Diatom.* **2009**, *20*, 1–686.

65. Blanco, S.; Cejudo-Figueiras, C.; Álvarez-Blanco, I.; Bécares, E.; Hoffmann, L.; Ector, L. *Atlas de las Diatomeas de la cuenca del Duero. Diatom atlas of the Duero Basin*; Universidad de León: León, Spain, 2010; pp. 1–382.

66. Wojtal, A.Z.; Łukasz, S. The influence of substrates and physicochemical factors on the composition of diatom assemblages in karst springs and their applicability in water-quality assessment. *Hydrobiol* **2012**, *695*, 97–108. [CrossRef]

67. Van de Vijver, B.; Frenot, Y.; Beyens, L. Freshwater diatoms from Ile de la Possession (Crozet Archipelago, Subantarctica). *Biblioth. Diatomol.* **2002**, *46*, 1–412.

68. Lange-Bertalot, H.; Krammer, K. *Achnanthes* Eine Monographie der Gattung. *Biblioth. Diatomol.* **1989**, *18*, 1–393.

69. Geitler, L. Beiträge zur Entwicklungsgeschichte und Taxonomie einiger *Achnanthes*-Arten, Subgenus *Microneis* (Bacillariophyceae). *Plant Syst. Evol.* **1980**, *134*, 1–10. [CrossRef]

70. Reichardt, E. Die Diatomeen der Altmühl (Beitragë zur Diatomeenflora der Altmühl 2). *Biblioth. Diatomol.* **1984**, *6*, 1–169.

71. Zechmeister, H.; Mucina, L. Vegetation of European springs: High-rank syntaxa of the Montio-Cardaminetea. *J. Veg. Sci.* **1994**, *5*, 385–402. [CrossRef]

72. Cantonati, M.; Kelly, M.G.; Lange-Bertalot, H. *Freshwater Benthic Diatoms of Central Europe: Over 800 Common Species Used in Ecological Assessment*; Koeltz Botanical Books: Schmitten-Oberreifenberg, Germany, 2017; pp. 1–942.

73. Rogora, M.; Frate, L.; Carranza, M.L.; Freppaz, M.; Stanisci, A.; Bertani, I.; Bottarin, R.; Brambilla, A.; Canullo, R.; Carbognani, M.; et al. Assessment of climate change effects on mountain ecosystems through a cross-site analysis in the Alps and Apennines. *Sci. Total. Environ.* **2018**, *624*, 1429–1442. [CrossRef]

74. Bahls, L.; Pierce, J.; Apfelbeck, R.; Olsen, L. *Encyonema droseraphilum* sp. nov. (Bacillariophyta) and other rare diatoms from undisturbed floating-mat fens in the northern Rocky Mountains, USA. *Phytotaxa* **2013**, *127*, 32. [CrossRef]

Article

Relations between Benthic Diatom Community and Characteristics of Karst Ponds in the Alpine Region of Slovenia

Katarina Novak [1,2] and Igor Zelnik [1,*]

[1] Department of Biology, Biotechnical Faculty, University of Ljubljana, Jamnikarjeva 101, 1000 Ljubljana, Slovenia; novakkatarina7@gmail.com
[2] Environmental Agency of the Republic of Slovenia, Vojkova 1b, 1000 Ljubljana, Slovenia
* Correspondence: igor.zelnik@bf.uni-lj.si; Tel.: +386-1-3203339

Abstract: The aim of this research was to investigate the structure of the benthic diatom community and its relations to selected environmental parameters. We collected samples in 16 karst ponds in the alpine region of Slovenia, where the Alpine karst is found. Since the predominating substrate in these ponds was clay, the epipelic community was analyzed. Hydromorphological characteristics, and physical and chemical conditions were also measured at each site. We found 105 species of diatoms, which belonged to 32 genera. The most frequent taxa were *Gomphonema parvulum* (Kützing) Kützing, *Navicula cryptocephala* Kützing, *Sellaphora pupula* (Kützing) Mereschkowsky (species group) and *Achnanthidium pyrenaicum* (Hustedt) Kobayasi. The pond with the lowest diversity was found at the highest altitude, while, on the other hand, the most species-rich pond was found at the lowest altitude. Regarding the ecological types, the most common were motile species. We confirmed a positive correlation between the number of diatom species and the saturation of water with oxygen, while correlation between species richness and NH_4-N was negative. The content of NO_3-N and NH_4-N explained almost 20% of the total variability of diatom community. Unlike our expectations, we calculated a negative correlation between the diversity of macroinvertebrates and diatoms, which is probably a consequence of different responses to environmental conditions.

Keywords: epipelon; diatoms; karst ponds; wetlands; southeast Alps

Citation: Novak, K.; Zelnik, I. Relations between Benthic Diatom Community and Characteristics of Karst Ponds in the Alpine Region of Slovenia. *Diversity* **2021**, *13*, 531. https://doi.org/10.3390/d13110531

Academic Editor: Michael Wink

Received: 28 September 2021
Accepted: 21 October 2021
Published: 25 October 2021

Publisher's Note: MDPI stays neutral with regard to jurisdictional claims in published maps and institutional affiliations.

1. Introduction

Ponds are water bodies ranging from 1 m^2 to 2 hectares, of natural or anthropogenic origin, with permanent or seasonal water [1]. Researchers used to treat them as lakes, but ponds differ from lakes due to several characteristics [2]: (a) smaller surface area and depth, (b) smaller ratio between the volume of water and the shore area, and therefore more direct contact with the terrestrial environment making them more susceptible to various influences; (c) smaller drainage basin and therefore bigger isolation [1]; (d) relatively small volume and water intake, which increases the connection between the sediments and water column and a more significant impact of sediment on the nutrient content in water, (e) due to the low water depth, the surface of the entire waterbody could be covered with macrophytes [3,4]. This is also the main reason why we consider ponds as a type of wetland. It is characteristic that their conditions change faster than in larger water bodies [5], which is reflected in large daily and seasonal fluctuations [1,5].

Ponds as a habitat have been neglected in ecological studies [6]. Today, we recognize them as an important carbon sink, pollution filter, and source of biodiversity, hosting several specialized and rare species [2,6]. For organisms living in the aquatic environment, ponds are refuges in degraded and inhospitable areas [1,7].

Karst ponds were made in areas with no surface water bodies (e.g., Karst), where people had problems with water supply [8]. Although they were used to water livestock and gardens, they lost their importance when water pipelines were constructed. However, today they represent an important source of biodiversity, like all other types of ponds [1,7,9,10]. Smol and Stoermer [11] suggest that Karstic aquatic habitats are the most interesting environments in which to study algae, especially diatoms.

With their distribution, they form a network of aquatic ecosystems, which increases γ diversity [1,8]. Biodiversity and abundance of the biota in Alpine ponds significantly correlate with altitude—with it, the average air temperature decreases, the amount of local precipitation increases, and UV radiation is more intense. In addition, the organisms in these environments face high daily and annual temperature differences and have a short period suitable for growth, which gives cold stenothermic species a better chance of survival [12–14].

The substrate consisting of clay and silt mostly covers the entire bottom of these ponds. On such a substrate, an epipelic biofilm develops, which is dominated by diatoms constituting the basal trophic levels for extensive food webs [15]. Diatoms are present in different aquatic environments and their sensitivity to various environmental factors, makes them a good bioindicator of water quality [16]. Recent studies have highlighted the high level of cryptic diversity of diatoms [17]. The diatom community is influenced by several factors such as water chemistry (pH, nutrient concentration, and organic load), physical (electrical conductivity, temperature, light) hydromorphological characteristics (substrate, water regime), and biotic pressures such as grazing, competition, and parasitism [17–20].

Benthic diatoms are important primary producers in shallow waters where light penetrates to the bottom [21]. On a fine substrate, a specific epipelic diatom community usually forms, which is adapted to low light conditions, consisting mainly of motile taxa that can move through interstitial waters to avoid newly deposited sediments [22]. Due to their location between substrate and water, they play a fundamental role in various biogeochemical cycles and dynamics of aquatic ecosystems [23].

The biological characteristics of diatoms, such as cell size class and ecological types, give us information about the structure of the community [17,24], as well as environmental conditions. Low-profile diatoms are well adapted to physical disturbances and are more abundant in waters with low nutrient content [17,24,25]. For high-profile diatoms, the formation of colonies allows exploiting nutrients that are not available to other groups but are therefore more exposed to grazing [24,25]. Motile diatoms are fast-growing species. Their abundance increases with a higher concentration of nutrients and organic load. They are also well adapted to high physical disorders [24]. Planktic species are present in lentic water, where they float in the water column [25], but due to sinking they can also be abundant in phytobenthos [26].

Despite their import roles, karst ponds are disappearing due to the abandonment of their original use. In addition to natural processes such as overgrowing with plants, they are also threatened by anthropogenic factors, especially intensification of agriculture, abandonment of livestock farming, backfilling, the input of non-native species and chemical pollution [1–3]. Pollutants cannot be sufficiently diluted [27], and nutrients are retained and potentially recycled by internal processes, which is difficult in the affected ecosystem [4]. All this can be significantly reflected in the structure of the diatom community.

However, we have not found any published work on the epipelic diatom community in karst ponds. Even the studies of periphytic diatom communities in ponds are rare, which had been discovered by Šumberova et al. [28]. In central Europe, we have found one paper about epipelic diatoms in ponds [29], while in southern Europe there are some papers that analyze epipelic diatoms (e.g., [16,30–33]).

We measured physical, hydromorphological, and chemical factors in 16 ponds at various locations in the Alpine region and sampled the epipelon. In this paper, we focused primarily on their response to various environmental characteristics. The study aimed to determine the species composition of the benthic diatom community in the Alpine karst ponds, determine the relationships between the structure of diatom community and the studied parameters, and find out the significant correlation between them.

We hypothesized that: (a) the diatom's species diversity correlates with the diversity of macroinvertebrates; (b) the diversity of species will decline with altitude and declining of ponds size; (c) the species composition will be significantly affected by the pH and electrical conductivity of the water and the land use in the drainage basin.

2. Materials and Methods

2.1. Study Sites and Sampling

We chose 16 karst ponds in the alpine region of Slovenia, which is a part of the South—Eastern calcareous Alps. Since limestone and dolomite are predominating rocks in this area, the Alpine karst is found there [34]. These water bodies are found in the area of the Julian Alps (Pokljuka, Jelovica, Ratitovec) and the Kamnik Alps (Krvavec, Velika planina, and Menina) (Figure 1). During the sample preparation we realized, that there were almost no frustules in samples from four ponds.

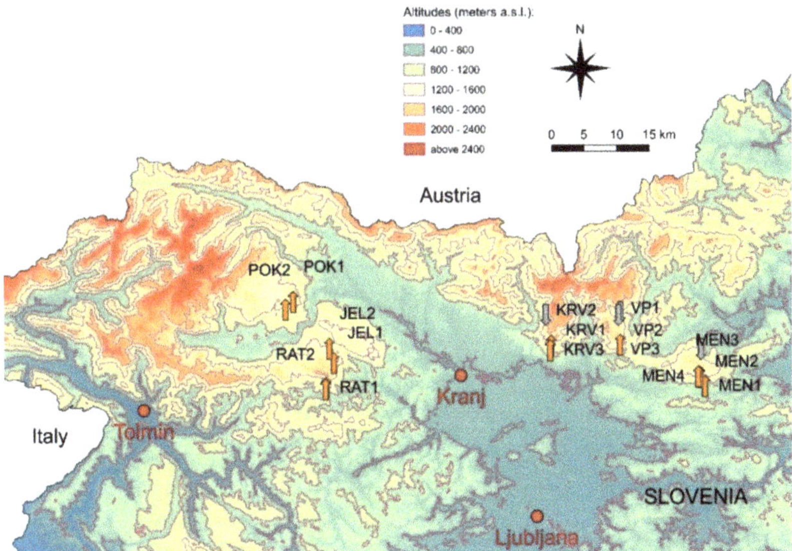

Figure 1. Map of sampled karst ponds. The arrow-tips indicate the localities of the studied ponds. Gray arrows represent ponds where samples contained low number of frustules. POK1, POK2—Pokljuka; JEL1, JEL2—Jelovica; RAT1, RAT2—Ratitovec; KRV1, KRV2, KRV3—Krvavec; VP1, VP2, VP3—Velika planina; MEN1, MEN2, MEN3, MEN4—Menina.

Mountain climate prevails in the area, where the average temperature of the coldest month is lower than $-3\ °C$, and the average temperature of the warmest month depends on the altitude and location [35]. Macrophyte and macroinvertebrate communities were studied before in the same ponds and results were published in Zelnik et al. [10].

Sampling took place in August of 2016, during the peak pasture season. Argilal and clay, respectively, was the only type of substrate present in all sites, so we decided to sample epipelon. Since we experienced difficulties with cleaning the samples from four ponds as well as very poor presence of diatom frustules in them, the samples from 12 sites were studied only (Table 1).

Basic physical and chemical factors were measured with a portable multimeter (EU-TECH, PCD 650). For each pond, we measured the pH and T of water ($°C$), electrical conductivity ($\mu S/cm$), total dissolved solids (mg/L), saturation with O_2 (%), and O_2 concentration (mg/L). For laboratory analyzes, a water sample (1 L) was taken at each site.

In the laboratory, the concentrations of NO_3-N (LCK 339), NH_4-N (LCK 304), TN (LCK 138), and orthophosphates (LCK 349) were determined using HACH Lange cuvette tests. Values were measured in individual samples with a HACH Lange LT 200 spectrophotometer. Dry mass and total suspended solids content (TSS) were determined by filtration and drying at 105 °C.

Table 1. Information about sampling sites.

Code	Karst Pond	Altitude [m]	Gauß-Krüger Coordinates		Precipitation per Year [mm]
			Y	X	
POK1	Pokljuka 1	1201	425202	134889	2200
POK2	Pokljuka 2	1302	424023	133737	2200
JEL1	Jelovica 1	1129	431399	125787	1900
JEL2	Jelovica 2	1138	430695	127923	1900
KRV1	Krvavec 1	1724	464378	128300	1650
	Krvavec 2	1509	463564	128355	1600
KRV3	Krvavec 3	1445	464227	127589	1600
RAT1	Ratitovec 1	1577	430192	122130	2100
RAT2	Ratitovec 2	1620	430104	121849	2100
	Velika planina 1	1434	475035	128689	1700
	Velika planina 2	1481	474750	128275	1700
VEL3	Velika planina 3	1454	474958	128408	1700
MEN1	Menina 1	1318	488084	122280	1250
MEN2	Menina 2	1403	487335	123639	1500
	Menina 3	1360	487473	123194	1500
MEN4	Menina 4	1419	487053	123695	1500

2.2. Biotic Analyses

Due to the absence of a firm substrate, diatom samples were taken from the surface of the loamy substrate. We scraped the top layer of argilal with an area of approximately 2 cm^2, with a spoon, at a 20–25 cm water depth. The samples were placed into bottles and 37% formaldehyde was added for fixation, in a ratio of 1:9.

Each sample was first homogenized with magnetic stirrer at a rate of 1200 rpm. We put 2 mL of the sample into a test tube and added 2.5 mL of 65% nitric (V) acid (HNO_3). The samples were heated over a fire until the smoke turned white to remove organic matter from the sample. After cooling the tube contents were centrifuged with a SIGMA 2-16PK centrifuge, 4 min at 4000 rpm, and the supernatant was discarded. The sample was further washed with distilled water. The resulting pellet was added to 2 mL of distilled water and mixed. We put single drops onto slides, dried them, and fixed them with Naphrax® mountant.

The prepared preparations were examined with an Olympus CX41 microscope under 1000× magnification, and the first 400 frustules of each sample were determined. Identification was performed using the keys of Hoffman et al. [36], Lange-Bertalot et al. [37], and in some cases Krammer and Lange-Bertalot [38–41].

2.3. Data Analysis

Correlation analysis was performed with PAST program [42]. Some data (land use, number of habitat types, turbidity) were of the interval type and thus not normally distributed, so we used Kendall correlation coefficients (tau).

Similarity in taxonomic composition of diatom community between the ponds was calculated using Sørensen similarity index. Diversity was calculated as Shannon-Wiener diversity index (S-WI) and Margalef diversity index. The trophic index (TI) was calculated according to Rott et al. [43].

The influence of individual factors on the composition of the diatom community was checked by direct gradient analyzes. First, we performed a detrended correspondence analysis (DCA) to determine whether the distribution of the diatom species along potential gradients is unimodal or linear. We found that the mentioned distribution was unimodal (Length of gradient: 9.7 S.D.), so we used Canonical Correspondence Analysis (CCA). All analyzes were performed with the Canoco 4.5 software package [44].

Environmental parameters were grouped into spatial variables (coordinates, altitude, annual precipitation, a distance from the next pond or road), substrate (inorganic and organic), chemical and physical variables, hydromorphological data, drainage basin etc. We used the method of forward selection to check the effect of individual environmental factors on the taxonomic composition. The program made 999 permutations in each round, three rounds were performed. In each next round, we considered only factors with p less than 0.1. In the last round, we considered the two most statistically significant factors, that were in fact marginally significant ($p = 0.06$ and 0.07). Based on two factors that had a marginally statistically significant effect on the structure of the diatom community, we also created an ordination diagram in which the ponds are distributed along gradients of environmental factors.

3. Results

3.1. Structure of the Benthic Diatom Community

A total of 105 species of diatoms were identified in 12 ponds (Table A1). Of these, most species-rich was JEL1 (43 species) and POK1 (30 species) (Figure 2). The pond with the lowest number of species was KRV1 (14 species). Dominant species and their proportions vary significantly between ponds (Table 2). *Navicula cryptocephala* Kützing was the most dominant in four ponds (POK1, JEL2 MEN2 and MEN4), and it was also present in a large proportion in RAT1. The pioneer complex *Achnanthidium minutissimum* (Kützing) Czarnecki was the most common taxon in three ponds (JEL1, RAT2 and MEN2). The highest dominance index is in POK2 and KRV1, where two dominant taxa represent 77% of the identified species (Table 2).

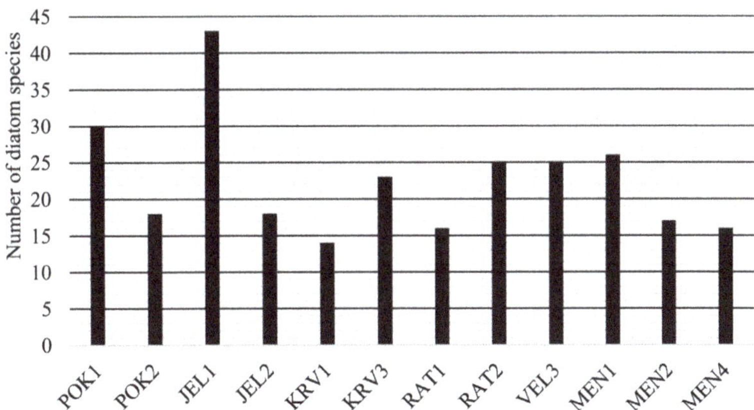

Figure 2. Number of diatom species in individual karst ponds.

Table 2. Dominance index (proportion in %) of the two most common species (highlighted in gray) in studied ponds. Diatoms that are not dominant in the sample but have a proportion ≥10% are also shown.

Species	POK1	POK2	JEL1	JEL2	KRV1	KRV3	RAT1	RAT2	VEL3	MEN1	MEN2	MEN4
Achnanthidium minutissimum			13					24			35	
Achnanthidium pyrenaicum				13				38			18	
Craticula accomoda						16						
Eucoconeis alpestris										10		
Eunotia bilunaris										19		
Eunotia tenella		41										
Gomphonema angustum									10	13		

Diversity **2021**, 13, 531

Table 2. *Cont.*

Species	POK1	POK2	JEL1	JEL2	KRV1	KRV3	RAT1	RAT2	VEL3	MEN1	MEN2	MEN4
Gomphonema parvulum				11					13			19
Navicula cryptocephala	26			40			14				19	14
Navicula exilis							45					
Nitzschia acicularis					53							
Nitzschia adamata						16						
Nitzschia palea												
Nitzschia perminuta											16	
Nitzschia supralitorea					19							
Pinnularia interrupta		36										
Sellaphora pseudopupula				10								
Sellaphora pupula	28							17				10
Tabellaria flocculosa			15									
Dominance index	53.3	77.3	27.5	52.5	71.5	32.4	61.3	62.7	23.1	34.9	53	32.6

Ponds with the highest similarity of diatom community are POK1 and VEL3, although the huge distance between them (see Figure 1). On the other side there was POK2, which stood out the most in rare species—with four ponds (KRV1, KRV3, MEN2, and MEN4) had no species in common (Table 3).

Table 3. Similarity of diatom community between the studied ponds according to Sørenson index. The similarity indices >0.5 are in bold.

POK1	POK2	JEL1	JEL2	KRV1	KRV3	RAT1	RAT2	VEL3	MEN1	MEN2	MEN4	
	0.29	0.44	0.46	0.23	0.30	0.26	0.40	**0.62**	0.43	0.47	0.44	POK1
		0.13	0.11	0	0	0.06	0.05	0.23	0.18	0	0	POK2
			0.33	0.25	0.27	0.24	0.41	0.44	0.32	0.40	0.31	JEL1
				0.31	0.49	0.29	**0.56**	0.37	0.27	0.46	0.29	JEL2
					0.22	0.27	0.26	0.15	0.20	0.39	0.33	KRV1
						0.26	0.38	0.29	0.25	0.40	0.31	KRV3
							0.39	0.34	0.19	0.30	0.31	RAT1
								0.36	0.28	0.52	0.39	RAT2
									0.55	0.52	0.44	VEL3
										0.37	0.48	MEN1
											0.55	MEN2

Figure 3 shows the proportion of diatoms according to their ecological type. Motile and high-profile diatoms are present in all samples. Low-profile diatoms are absent in one pond, while in four ponds (KRV3, RAT1, MEN1, and MEN4) they are very rare. Their largest proportion is in RAT2 (68%) and MEN2 (52%). Planktic diatoms are present with a negligible proportion (JEL1, KRV3, and RAT1), except for KRV1, representing half of the specimens. The most common are motile diatoms. In POK1, JEL2, KRV3, RAT1, VEL3 and MEN4, they represent the majority proportion of diatoms.

Figure 4 shows the size classes of diatoms. The most common size class is 3, followed by 2 and 4. Members of size classes 1 and 5 are infrequent. Smaller diatoms (size classes 1 and 2) are dominant in RAT2, POK2, and MEN2. Data for POK2 are not representative, as 77% of specimens were not determined a size class due to lack of data in the literature. There is also a considerable proportion of unknown size classes in KRV1 and KRV3 (22% and 28%).

3.2. Effects of Environmental Factors on the Diatom Community Composition

The concentration of NO_3-N and NH_4-N in water explains almost 20% of the total variability of the diatom community in ponds (Table 4). The concentration of NO_3-N explains 10% of the variability, and the NH_4-N concentration in water 9.6%. The content of these two nutrients or nitrogen species is probably mainly due to the higher load in ponds and their basin area with livestock. The same shows the ordination diagram based on CCA (Figure 5), where ponds are arranged according to the diatom taxonomic composition along the gradients of NO_3-N and NH_4-N concentration in water.

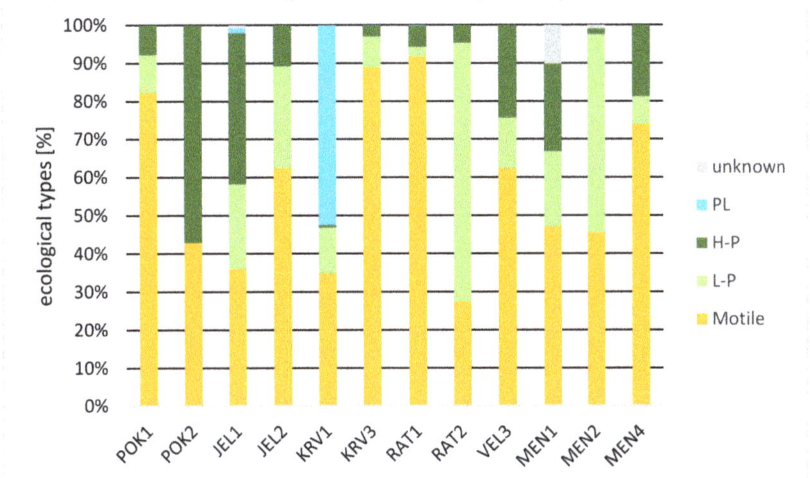

Figure 3. Diatoms according to their ecological type [in %]. (PL—planktic, H-P—high-profile, L-P—low-profile).

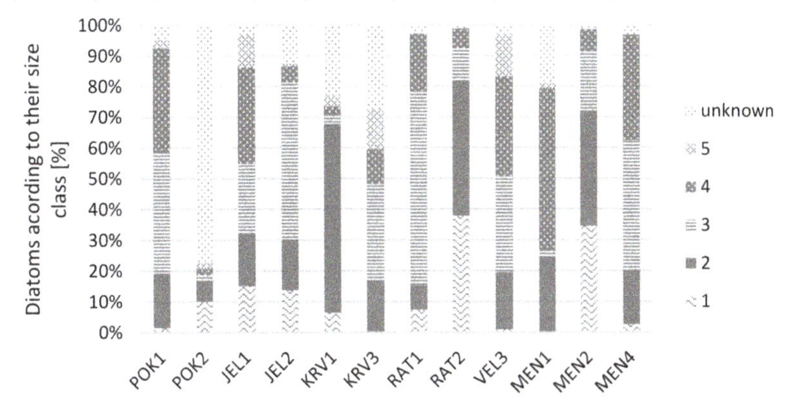

Figure 4. Diatoms according to their size class [%].

Table 4. Results of Canonical correspondence analysis (CCA) and forward selection. (% TVE-proportion of the explained variability by specific variable).

Variable	P	% TVE
NO_3-N	0.064	10.0
NH_4-N	0.072	9.6

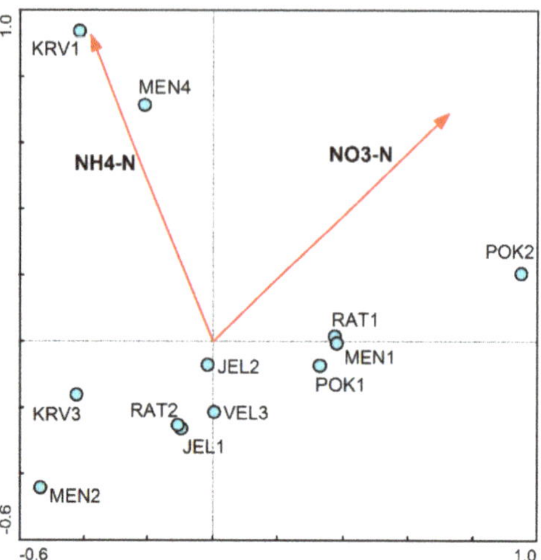

Figure 5. A CCA-based ordination diagram in which karst ponds are distributed along environmental gradients.

According to the S-WI index (Figure 6), the highest diversity is in JEL1, VEL 3 is next. The lowest diversity is in POK2, the lower diversity is also in KRV1 and RAT1. The Margalef index (Figure 7) showed a different assessment of diversity than S-WI.

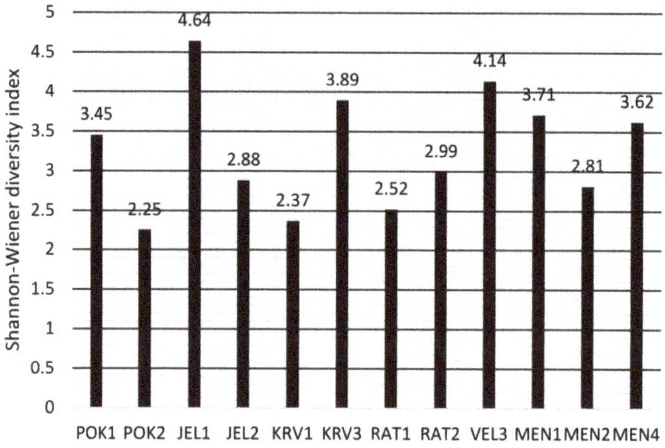

Figure 6. Shannon-Wiener diversity index values of diatoms in karst ponds.

JEL1 still has the highest diversity value (7.01), but the ponds with the lowest diversity are RAT1 and MEN4.

3.3. Environmental Factors and Diversity of Diatom Community

Kendall correlation coefficients showed that the number of diatom species is in a statistically significant positive correlation with oxygen saturation and a negative correlation with the concentration of NH_4-N (Table 5). The Margalef index was also positively correlated with oxygen saturation and negatively with NH_4-N concentration. A negative statistically significant correlation ($p = 0.05$) was calculated between altitude and the Margalef index.

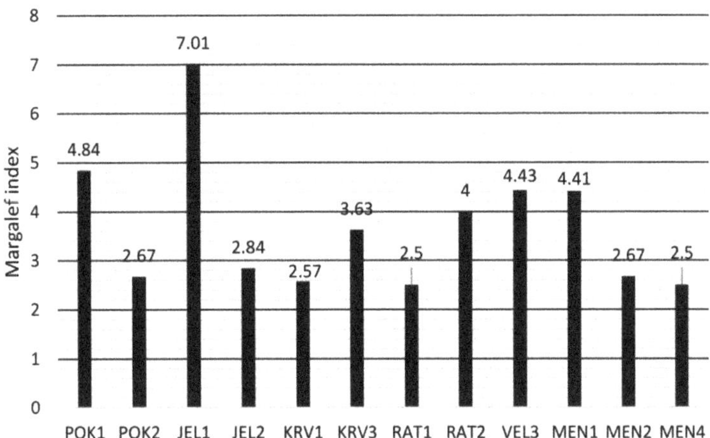

Figure 7. Margalef index values of diatoms in karst ponds.

Table 5. Kendall (tau) correlation coefficients between environmental factors and diversity parameters of diatom communities in ponds. Only statistically significant correlations (*—$p < 0.05$) and marginally statistically significant correlations ($p = 0.05$) are shown.

	No. of Diatom Species	Margalef Index
altitude [m]	n.s.	−0.431
O_2 saturation [%]	0.531 *	0.543 *
NH_4-N [mg/L]	−0.481 *	−0.492 *
SW_I h.taxa macoinvertebrates	−0.481 *	−0.492 *

We also found a negative correlation between the number of diatom species and S-WI and the Margalef index calculated based on the composition of the invertebrate community, which was contrary to our expectations.

Great differences in TI values were found between the ponds (Figure 8). The lowest TI value was in POK1 (ultraoligotrophic) and the highest in JEL2, KRV3, KRV1, POK1, and MEN4 (polytrophic).

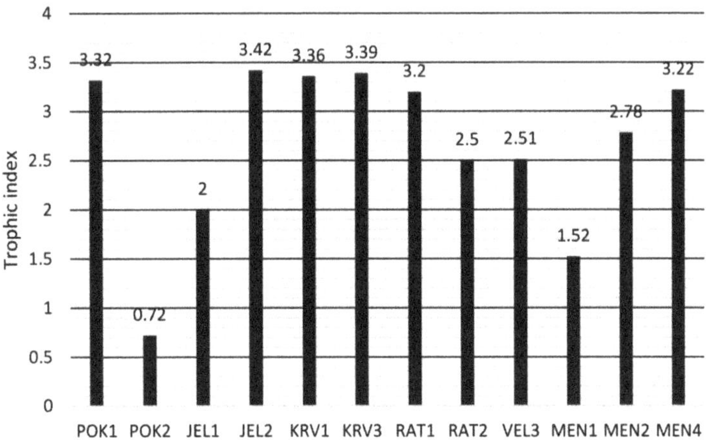

Figure 8. Trophic index values for sampled karst ponds.

4. Discussion

4.1. Structure of the Benthic Diatom Community

In total, 105 diatom species belonging to 32 genera were identified. The most common taxa were *Gomphonema parvulum* (Kützing) Kützing, *Navicula cryptocephala* Kützing, species group *Sellaphora pupula* (Kützing) Mereschkowsky (present in 10 sites). Almost half of the species (52) were present in only one site, from which we can assume that the composition of diatom communities differs much between the ponds. The genera with the highest number of species were *Nitzschia*, *Pinnularia*, *Navicula* and *Neidium*. The highest number of species was identified in the JEL1, whereas in KRV1, we found the lowest number of species, of which *Nitzschia acicularis* (Kützing) W. Smith represented more than half of the identified frustules. We expected lower diversity as well as variability of epipelic diatom community, as karst ponds are small water bodies with frequent disturbances, which make the conditions unfavorable. The number of species varied from 14 to 43, which is much higher than 11–26 taxa from ponds in South-eastern Alps reported by Cantonati et al. [29]. However, the mentioned researchers studied different type of ponds in alpine region.

Among the ecological types, the motile diatoms were the most common. They dominated in four ponds (POK1, KRV3, RAT1, and MEN4) and were codominant in another four ponds (Figure 2). Sites where deposition occurs are advantageous for motile diatoms [45–47] as well as nutrient-rich sites [48–50]. Typical representatives from genera *Navicula*, *Nitzschia*, *Sellaphora*, and *Surirella* [24] were also present in our samples. However, we did not calculate any significant correlation between environmental factors and the share of motile species. In ponds with higher trophic index values motile species dominated, which are well adapted to higher nutrient content. We expected that high-profile (H-P) diatoms would also be present here with higher proportion. However, they were probably not present in such high proportion due to physical disturbances.

High-profile diatoms, which are also common in nutrient-rich water but with fewer disturbances [48] are less common in our samples. The proportion of H-P negatively correlated with TSS ($p = 0.009$), which negatively influence light conditions with turbidity and deposition. On the other hand, we calculated positive correlation between proportions of H-P diatoms and argilal ($p = 0.029$). The typical genera of this group, which were also present in our samples, were *Eunotia*, *Fragilaria* and *Gomphonema*. The proportions of H-P diatoms were lower than motile, except POK2, where H-P represent two-thirds of the community. Disturbances and grazing, made motile species more efficient than H-P ones.

Low-profile (L-P) diatoms were rare, but in two samples (RAT2 and MEN2) they were dominant. Both ponds are fenced, so with no access of the cattle. Proportions of L-P diatoms negatively correlated with NO_3-N ($p = 0.023$) and positively with habitat diversity in the catchment area ($p = 0.039$), which actually means low density of the cattle. Typical representatives are from the genera *Achnantes*, *Achnanthidium*, *Amphora*, *Cocconeis*, and *Meridion* [24]. *Achnantidium minutissiumum* (Kützing) Czarnecki was the most dominant taxon in RAT2 and MEN2, as well as in JEL1. It seems that cattle cause problem for L-P diatoms due to high input of nutrients to ponds, to which L-P species are not adapted [48]. In some samples (JEL1, KRV1, KRV3, and RAT1), planktic diatoms were also present.

In ponds with higher concentrations of orthophosphates, we find mainly motile and H-P diatoms adapted to higher concentrations of nutrients [24,51,52] (Figure 3). In POK2 (0.3 mg/L of ortophosphate), MEN1 (0.92 mg/L) and VEL3 (0.23 mg/L) motile and H-P diatoms represent almost the entire sample, L-P diatoms are almost absent. However, the significant correlation between P and ecological types was not calculated. There was also no correlation between P concentration and diatom size classes, which also report Lavoie et al. [53].

The concentration of NO_3-N and NH_4-N in water explained almost 20% of the total variability of diatom community (Table 4). The concentration of NO_3-N explains 10% of the variability of the diatom community, and the concentration of NH_4-N 9.6% (Table 4, Figure 5). The ponds are arranged according to the taxonomic composition of diatom communities along the gradients of NO_3-N and NH_4-N concentration in water.

The results did not show statistically significant correlations between the composition of diatom community and concentrations of either orthophosphate or TP as expected, which is consistent with Soininen et al. [54]. This is probably because absorption rate for phosphorus from the water column by epipelon is lower than in other groups of primary producers [55].

Haubois et al. [56] report that large and small species do coexist within the epipelon. We found that size-class three had the highest proportion in five ponds, while size-class 2 and 4 in three ponds each (Figure 4). However, most of the identified frustules belonged to the middle-size class (3), which also report Lavoie et al. [53]. In ponds with higher biodiversity (JEL1, KRV3, VEL3, MEN1, and MEN4), size-classes 4 and 3 dominated.

4.2. Diversity of Benthic Diatom Community and Environmental Factors

In general, altitude affects biota in ponds as it affects temperature, precipitation, and radiation [12]. The results showed a negative correlation between altitude and the Margalef index, which is in line with our hypothesis and with the general rules in ecology [57]. The diatom species richness did not correlate with altitude, but pond at the highest altitude (KRV1) had the lowest number of species, while pond at the lowest altitude (JEL1) had the highest diversity. On the contrary for mountain ponds in Spain Blanco et al. [31] report positive correlation of diatom diversity with altitude.

The water depth in these shallow ponds is important mainly because of poor light conditions in turbid water. One of the dominant species was also *Nitzschia perminuta* (Grunow) M. Peragallo, which dominates in low light conditions [58]. Due to shallowness, there is no stratification during the summer [59].

We calculated no significant correlation between pH and diversity indices. The most extreme values were measured at POK2 (pH = 3.8) and MEN1 (pH = 9.6) (Table A2). The first is located in a coniferous forest and is a dystrophic system. Therefore, diatom species in this pond differed from others the most (Table 3). As reported in DeNicola [60] and Della Bella [16], we found there mainly species from the genera *Neidium*, *Eunotia*, *Pinnularia*, *Stauroneis*, and *Sellaphora*, which occurred in small numbers or were absent in other ponds. Diatom community from this pond had no species in common with four other ponds. This pond was more similar to the shallow ponds on mires presented in [29,61]. The lowest value of the electrical conductivity was also measured there (16 μS/cm), which coincides with the trophic index, which defines it as ultraoligotrophic.

We found a positive correlation between the number of diatom species and water saturation with oxygen and the Margalef index and water saturation with oxygen. The highest oxygen saturation was in MEN1 (almost 250%) due to intense photosynthetic activity of the phytoplankton, making the water very turbid.

In KRV1 and MEN4, a large proportion of N is in the form of NH_4-N, which can be explained by the high density of cattle in their catchments. Correlation coefficients showed a negative correlation between the Margalef index and the NH_4-N concentration. In ponds with a higher concentration (KRV1 and MEN4), the diversity was lower, while it was higher in ponds with lower NH_4-N concentrations (POK1, JEL1, RAT1, and VEL3). In contrast to NH_4-N concentrations, NO_3-N concentrations did not differ much between ponds. Values were 0.2–0.5 mg/L. NO_3-N and NH_4-N concentrations classify our ponds as eutrophic (POK2, JEL2, KRV3, RAT1), mesotrophic (POK1 and MEN1), or oligotrophic (JEL1, RAT2, and MEN2) [54]. In KRV1 and MEN4, the values of NH_4-H and NO_3-N were so high that they can be classified as hypereutrophic.

Cattle can have a substantial negative effect on the diversity of communities in ponds [62]. Trampling the bottom and the shore presents physical disturbances. In ponds with moderate intensity of trampling, the diatom diversity was higher than in those without trampling, which is consistent with the intermediate-disturbance hypothesis [63]. More important is the influence of the cattle as the source of nutrients and organic matter from their excrements. Smaller water bodies in the agricultural landscape are highly exposed to influences from nearby agricultural areas, since they can be strongly affected by nutrient accumulation [4].

Based on the trophic index (TI), ponds vary from ultraoligotrophic to polytrophic. Della Bella et al. [30] report that trophic diatom index highly correlated with nutrient content, especially orthophosphate and NO_3-N in wetlands in central Italy. However, in our case orthophosphate concentrations were the highest where the TI values were low (POK2 and MEN1). According to TP concentrations and nutrient estimates for lakes [58], both ponds were hypertrophic, but TI classified them as ultraoligotrophic (POK2) and mesotrophic (MEN1). Due to the pH = 3.8, there were probably not enough basic ions in POK2, despite the high concentration of TP and NO_3^-. Insufficient amount of HCO_3^- was present at pH = 9.6, which reduced primary production and thus nutrient uptake, which was probably the explanation for the condition in the MEN1.

4.3. Correlations between Diatoms and Macroinvertebrates

We found a negative correlation between the diatom species richness and the S-WI, and Margalef index calculated on the base of the macroinvertebrate community, which was contrary to our expectations. Similar findings report also Gascón et al. [64], which found out that different aquatic communities respond differently to the environmental factors, so we could not generalize relations between parameters and diversity patterns. Due to the larger size of macroinvertebrates, they might be more susceptible to physical destruction of the littoral zone, and loss of mesohabitats due to trampling of the bottom compared to diatoms, whereas diatoms, as primary producers, are particularly sensitive to water chemistry and light conditions [16,65]. Another reason is probably grazing [66]. We should not neglect the fact that on the same substrate on which diatoms thrive, Chironomidae dominate, which graze on epipelon.

5. Conclusions

We found a negative correlation between species-richness and diversity of the diatom community and diversity of the macroinvertebrate community (S-WI, Margalef index).

Despite relatively small differences in altitude, the results showed a marginal statistical correlation between altitude and Margalef Index. No effect of the pond size on the diversity of diatom community was observed.

We did not calculate significant correlations between pH and diversity. Half of the species in most acidic pond POK2 were present only in this pond. Correlations between electrical conductivity, land use, and diversity of diatom community were not significant.

Motile diatoms were most common. They are adapted to high nutrient concentrations and disturbances and can migrate to the site with sufficient light or nutrients when the re-suspended substrate is depositing.

We found a positive correlation between the number of diatom species and O_2 saturation and the Margalef index and O_2 saturation. The pond with the lowest oxygen saturation value (KRV1) had the lowest species diversity.

The results also showed a negative correlation between the number of diatoms and NH_4-N concentration and the Margalef index and NH_4-N concentration. NH_4-N is probably present in the ponds due to the cattle grazing in the area in the summer. The concentrations of NO_3-N and NH_4-N explain almost 20% of the total variability of the diatom community.

Author Contributions: Conceptualization, I.Z.; methodology, I.Z. and K.N.; validation, K.N.; formal analysis, I.Z. and K.N.; investigation, K.N.; data curation, I.Z. and K.N.; writing—original draft preparation, K.N.; writing—review and editing, I.Z. and K.N.; visualization, I.Z. and K.N.; supervision, I.Z.; funding acquisition, I.Z. All authors have read and agreed to the published version of the manuscript.

Funding: This research and the APC were partly funded by the Slovenian Research Agency, Research program Biology of plants, grant number P1-0212.

Institutional Review Board Statement: Not applicable.

Informed Consent Statement: Not applicable.

Data Availability Statement: Data are stored within the documentation of Master Study program theses and P1-0212 Research program.

Acknowledgments: Authors thank to Matej Holcar for creation of the Figure 1.

Conflicts of Interest: The authors declare no conflict of interest.

Appendix A

Table A1. The list of the names of diatom taxa found in studied karst ponds.

Achnanthidium pyrenaicum (Hustedt) Kobayasi
Achnanthidium minutissimum (Kützing) Czarnecki
Adlafia minuscula (Grunow) Lange-Bertalot var. *minuscula*
Amphora copulata (Kützing) Schoeman et Archibald
Amphora pediculus (Kützing) Grunow
Brachysira neoexilis Lange-Bertalot
Caloneis tenuis (Gregory) Krammer
Chamaepinnularia mediocris (Krasske) Lange-Bertalot
Chamaepinnularia muscicola (Petersen) Kulikovskiy, Lange-Beralot et Witkowski
Chamaepinnularia soehrensis (Krasske) Lange-Bertalot et Krammer
Cocconeis pediculus Ehrenberg
Craticula accomoda (Hustedt) D.G. Mann
Craticula ambigua (Ehrenberg) D.G. Mann
Craticula halophila (Grunow) D.G. Mann
Craticula molestiformis (Hustedt) Lange-Bertalot
Cyclotella stelligera Cleve & Grunow
Cymbopleura amphicephala (Nägeli) Krammer
Cymbopleura naviculiformis (Auerswald) Krammer
Diploneis krammeri Lange-Bertalot et Reichardt
Encyonema hebridicum Grunow ex Cleve
Encyonema minutum (Hilse) D.G. Mann
Encyonema silesiacum (Bleisch) D.G. Mann
Eucocconeis alpestris (Brun) Lange-Bertalot
Eunotia arcus Ehrenberg
Eunotia bilunaris (Ehrenberg) Schaarschmidt
Eunotia exigua (Brébisson) Rabenhorst
Eunotia minor (Kützing) Grunow
Eunotia paludosa Grunow
Eunotia pseudogroenlandica Lange-Bertalot et Tagliaventi
Eunotia subarcuatoides Alles, Nörpel et Lange-Bertalot
Eunotia tenella (Grunow) Hustedt
Fragilaria radians (Kützing) Williams et Round
Fragilaria tenera (W. Smith) Lange-Bertalot
Frustulia crassinervia (Brébisson) Lange-Bertalot et Krammer
Gomphonema acuminatum Ehrenberg
Gomphonema angustum (Kützing) Rabenhorst
Gomphonema calcifugum Lange-Bertalot et Reichardt
Gomphonema exilissimum (Grunow) Lange-Bertalot et Reichardt
Gomphonema occultum Reichardt et Lange-Bertalot
Gomphonema parvulum (Kützing) Kützing
Gomphonema sarcophagus Gregory
Hantzschia abundans Lange-Bertalot
Luticola nivalis (Ehrenberg) D.G. Mann
Luticola mutica (Kützing) D.G. Mann
Meridion circulare (Gréville) C. Agardh
Navicula antonii Lange-Bertalot

Table A1. *Cont.*

Navicula cryptocephala Kützing
Navicula cryptotenella Lange-Bertalot
Navicula exilis Kützing
Navicula menisculus Schumann
Navicula reichardtiana Lange-Bertalot
Navicula trivialis Lange-Bertalot
Navicula veneta Kützing
Navicula wildii Lange-Bertalot
Neidium affine (Ehrenberg) Pfitzer
Neidium alpinum Hustedt
Neidium ampliatum (Ehrenberg) Krammer
Neidium bergii (Cleve-Euler) Krammer
Neidium binodeforme Krammer
Neidium bisulcatum (Lagerstedt) Cleve var. bisulcatum
Neidium dubium (Ehrenberg) Cleve
Neidium iridis (Ehrenberg) Cleve
Neidium productum (W. Smith) Cleve
Nitzschia acicularis (Kützing) W. Smith
Nitzschia adamata Hustedt
Nitzschia angustata (W. Smith) Grunow
Nitzschia communis Rabenhorst
Nitzschia dissipata (Kützing) Grunow ssp. dissipata
Nitzschia fonticola Grunow
Nitzschia gisela Lange-Bertalot
Nitzschia palea (Kützing) W. Smith
Nitzschia perminuta (Grunow) M. Peragallo
Nitzschia pura Hustedt
Nitzschia pusilla Grunow
Nitzschia supralitorea Lange-Bertalot
Nitzschia umbonata (Ehrenberg) Lange-Bertalot
Pinnularia borealis Ehrenberg
Pinnularia gibba Ehrenberg
Pinnularia grunowii Krammer
Pinnularia interupta W. Smith
Pinnularia marchica I. Schönfelder ex Krammer
Pinnularia microstauron (Ehrenberg) Cleve
Pinnularia rupestris Hantzsch
Pinnularia sinistra Krammer
Pinnularia subcapitata Gregory var. subcapitata
Pinnularia viridiformis Krammer
Placoneis ignorata (Schimanski) Lange-Bertalot
Placoneis paraelginensis Lange-Bertalot
Planothidium lanceolatum (Brébisson ex Kützing) Lange-Bertalot
Psammothidium grischunum (Wunthrich) Bukhtiyarova et Round
Psammothidium helveticum (Hustedt) Bukhtiyarova & Round
Sellaphora pseudopupula (Krasske) Lange-Bertalot
Sellaphora pupula (Kützing) Mereschkowsky (species group)
Sellaphora stroemii (Hustedt) D.G.Mann
Sellaphora verecundiae Lange-Bertalot
Stauroneis acidoclinata Lang-Bertalot et Werum
Stauroneis anceps Ehrenberg
Stauroneis gracilis Ehrenberg
Stauroneis kriegeri Patrick
Stauroneis smithii Grunow
Stauroneis thermicola (Petersen) Lund
Stephanodiscus alpinus Hustedt
Surirella angusta Kützing
Surirella minuta Brébisson ex Kützing
Tabellaria flocculosa (Roth) Kützing

Table A2. Characteristics of karst ponds in the year 2016. * Secchi depth in most transparent ponds is the same as water depth; the bottom of the pond MEN2 was covered with plastic layer on which fine substrate deposited. + represents presence of substrate, cover <5%.

Sample	POK1	POK2	JEL1	JEL2	KRV1	KRV3	RAT1	RAT2	VEL3	MEN1	MEN2	MEN4
date	23.8.	23.8.	23.8.	23.8.	19.8.	19.8.	23.8.	23.8.	19.8.	18.8.	18.8.	18.8.
pH	5.9	3.8	6.5	6.4	6.7	8.3	7.4	6.5	5.9	9.6	7.2	6.2
T [°C]	17.5	12.2	14.1	9.8	14.9	15.3	7.7	10.2	17.3	17.7	17.9	16.0
Conductivity [μS/cm]	37	16	149	47	242	92	95	256	36	158	55	90
O_2 saturation [%]	75	53	56	62	10	69	56	74	100	244	90	25
O_2 [mg/L]	6.6	4.7	5.0	6.0	0.9	5.9	4.9	7.5	8.1	19.4	7.3	2.0
Secchi depth [cm]	25 *	30 *	60 *	55 *	30 *	13.0	20 *	30 *	35	10	56	36
depth [cm]	25	30	60	55	30	100	20	30	40	20	100	48
Turbidity [1–3]	1	1	1	3	3	3	1	1	3	3	1	3
Clay, silt [%]	100	100	100	90	80	5	100	100	100	95	-	100
Sand, gravel [%]	0	0	0	10	20	65	0	0	0	0	-	0
Pebbles [%]	0	0	0	+	0	30	0	0	0	5	-	0
Stones [%]	0	0	+	0	0	0	0	0	0	0	-	0
CPOM [%]	0	20	0	0	+	5	+	+	+	0	1	0
FPOM [%]	0	80	0	0	0	1	100	80	100	100	80	0
[%] of trampled shore	1	1	0	45	70	70	20	0	50	100	0	80
Intensity of trampled shores (0–5)	1	1	0	3	5	2	3	0	4	5	0	4
TP [mg/L]	0.17	0.34	0.03	0.05	0.28	0.07	0.07	0.06	0.23	0.92	0.08	0.15
PO_4^{3-} [mg/L]	0.17	0.30	0.02	0.02	0.07	0.03	0.01	0.001	0.23	0.92	0.05	0.02
TN [mg/L]	1.35	0.82	0.59	0.84	5.91	1.21	1.62	0.56	1.53	6.56	0.95	16.0
NO_3-N [mg/L]	0.39	0.52	0.30	0.34	0.42	0.26	0.41	0.30	0.32	0.40	0.21	0.42
NH_4-N [mg/L]	0.08	0.14	0.03	0.51	4.0	0.73	0.28	0.07	0.06	0.21	0.03	3.08
TDS [mg/l]	72	70	96	50	120	78	94	58	80	226	74	92
TSS [mg/L]	3	8	17	58	151	98	49	93	30	201	257	39

References

1. Biggs, J.; Williams, P.; Whitfield, M.; Nicolet, P.; Weatherby, A. 15 years of pond assessment in Britain: Results and lessons learned from the work of Pond Conservation. *Aquat. Conserv. Mar. Freshw. Ecosyst.* **2005**, *6*, 693–714. [CrossRef]
2. Oertli, B.; Biggs, J.; Céréghino, R.; Grillas, P.; Joly, P.; Lachavanne, J.-B. Conservation and monitoring of pond biodiversity: Introduction. *Aquat. Conserv. Mar. Freshw. Ecosyst.* **2005**, *6*, 535–540. [CrossRef]
3. Declerck, S.; De Bie, T.; Ercken, D.; Hampel, H.; Schrijvers, S.; Van Wichelen, J.; Gillard, V.; Mandiki, R.; Losson, B.; Bau-wens, D.; et al. Ecological characteristics of small farmland ponds: Associations with land use practices at multiple spatial scales. *Biol. Conserv.* **2006**, *131*, 523–532. [CrossRef]
4. Søndergaard, M.; Jeppesen, E.; Jensen, J.P. Pond or lake: Does it make any difference? *Fundam. Appl. Limnol.* **2005**, *162*, 143–165. [CrossRef]
5. Davies, B.; Biggs, J.; Williams, P.; Whitfield, M.; Nicolet, P.; Sear, D.; Bray, S.; Maund, S. Comparative biodiversity of aquatic habitats in the European agricultural landscape. *Agric. Ecosyst. Environ.* **2008**, *125*, 1–8. [CrossRef]
6. Hassall, C.; Hollinshead, J.; Hull, A. Environmental correlates of plant and invertebrate species richness in ponds. *Biodivers. Conserv.* **2011**, *20*, 3189–3222. [CrossRef]
7. Céréghino, R.; Biggs, J.; Oertli, B.; Declerck, S. The ecology of European ponds: Defining the characteristics of a neglected freshwater habitat. *Hydrobiologia* **2008**, *597*, 1–6. [CrossRef]
8. Čelik, T.; Zelnik, I.; Babij, V.; Vreš, B.; Pirnat, A.; Seliškar, A.; Drovenik, B. Inventory of karstic ponds and their importance for biotic diversity. In *Kras: Water and Life in a Rocky Landscape*; Mihevc, A., Ed.; ZRC: Ljubljana, Slovenia, 2005; pp. 72–82.
9. Zelnik, I.; Potisek, M.; Gaberščik, A. Environmental Conditions and Macrophytes of Karst Ponds. *Pol. J. Environ. Stud.* **2012**, *21*, 1911–1920.
10. Zelnik, I.; Gregorič, N.; Tratnik, A. Diversity of macroinvertebrates positively correlates with diversity of macrophytes in karst ponds. *Ecol. Eng.* **2018**, *117*, 96–103. [CrossRef]
11. Smol, J.P.; Stoermer, E.F. (Eds.) *The Diatoms: Applications for the Environmental and Earth Sciences*, 2nd ed.; Cambridge University Press: Cambridge, UK, 2010; ISBN 978-1-107-56496-1.
12. Hinden, H.; Oertli, B.; Menetrey, N.; Sager, L.; Lachavanne, J.-B. Alpine pond biodiversity: What are the related environmental variables? *Aquat. Conserv. Mar. Freshw. Ecosyst.* **2005**, *15*, 613–624. [CrossRef]

13. Ilg, C.; Oertli, B. How can we conserve cold stenotherm communities in warming Alpine ponds? *Hydrobiologia* **2014**, *723*, 53–62. [CrossRef]

14. Frisbie, M.P.; Lee, R.E. Inoculative Freezing and the Problem of Winter Survival for Freshwater Macroinvertebrates. *J. N. Am. Benthol. Soc.* **1997**, *16*, 635–650. [CrossRef]

15. Ocón, C.S.; López van Oosterom, M.V.; Munoz, M.I.; Rodrigues-Capítulo, A. Macroinvertebrate trophic responses to nutrient addition in a temperate stream in South America. *Arch. Hydrobiol.* **2013**, *182*, 17–30. [CrossRef]

16. Bella, V.D.; Mancini, L. Freshwater diatom and macroinvertebrate diversity of coastal permanent ponds along a gradient of human impact in a Mediterranean eco-region. *Hydrobiologia* **2009**, *634*, 25–41. [CrossRef]

17. Berthon, V.; Bouchez, A.; Rimet, F. Using diatom life-forms and ecological guilds to assess organic pollution and trophic level in rivers: A case study of rivers in south-eastern France. *Hydrobiologia* **2011**, *673*, 259–271. [CrossRef]

18. Sayer, C.D. Problems with the application of diatom-total phosphorus transfer functions: Examples from a shal-low English lake. *Freshw. Biol.* **2001**, *46*, 743–757. [CrossRef]

19. Wu, N.; Faber, C.; Sun, X.; Qu, Y.; Wang, C.; Ivetic, S.; Riis, T.; Ulrich, U.; Fohrer, N. Importance of sampling fre-quency when collecting diatoms. *Sci. Rep.* **2016**, *6*, 36950. [CrossRef] [PubMed]

20. Vis, C.; Hudon, C.; Cattaneo, A.; Pinel-Alloul, B. Periphyton as an indicator of water quality in the St Lawrence River (Québec, Canada). *Environ. Pollut.* **1998**, *101*, 13–24. [CrossRef]

21. Revsbech, N.P.; Nielsen, J.; Hansen, P.K. Benthic Primary Production and Oxygen Profiles. In *Nitrogen Cycling in Coastal Marine Environments*; Blackburn, T.H., Sørensen, J., Eds.; John Wiley & Sons Ltd.: Hoboken, NY, USA, 1988; pp. 69–81.

22. Battarbee, R.W.; Jones, V.J.; Flower, R.J.; Cameron, N.G.; Bennion, H.; Carvalho, L.; Juggins, S. Diatoms. In *Tracking Environmental Change Using Lake Sediments: Terrestrial, Algal, and Siliceous Indicators, Developments in Paleoenvironmental Research*; Smol, J.P., Birks, H.J., Last, W.M., Eds.; Springer: Dordrecht, The Netherlands, 2001; pp. 155–202.

23. Kröpfl, K.; Vladár, P.; Szabó, K.; Ács, É.; Borsodi, A.K.; Szikora, S.; Caroli, S.; Záray, G. Chemical and biological characterisation of biofilms formed on different substrata in Tisza river. *Environ. Pollut.* **2006**, *144*, 626–631. [CrossRef]

24. Rimet, F.; Bouchez, A. Life-forms, cell-sizes and ecological guilds of diatoms in European rivers. *Knowl. Manag. Aquat. Ecosyst.* **2012**, *406*, 01. [CrossRef]

25. Rimet, F.; Berthon, V.; Bouchez, A. *Formes de vie, Guildes Écologiques et Classes de Tailles des Diatomées d'eau Douce*; INRA, Station d'hydrobiologie lacustre: Thonon, France, 2010; p. 10.

26. Zelnik, I.; Sušin, T. Epilithic Diatom Community Shows a Higher Vulnerability of the River Sava to Pollution during the Winter. *Diversity* **2020**, *12*, 465. [CrossRef]

27. De Marco, P.; Nogueira, D.S.; Correa, C.C.; Vieira, T.B.; Silva, K.D.; Pinto, N.S.; Bichsel, D.; Hirota, A.S.V.; Vieira, R.R.S.; Carneiro, F.M.; et al. Patterns in the organization of Cerrado pond biodiversity in Brazilian pasture landscapes. *Hydrobiologia* **2014**, *723*, 87–101. [CrossRef]

28. Šumberová, K.; Vild, O.; Ducháček, M.; Fabšičová, M.; Potužák, J.; Fránková, M. Drivers of Macrophyte and Diatom Diversity in a Shallow Hypertrophic Lake. *Water* **2021**, *13*, 1569. [CrossRef]

29. Cantonati, M.; Lange-Bertalot, H.; Decet, F.; Gabrieli, J. Diatoms in very-shallow pools of the site of community importance Danta di Cadore Mires (south-eastern Alps), and the potential contribution of these habitats to diatom biodiversity conservation. *Nova Hedwig.* **2011**, *93*, 475–507. [CrossRef]

30. Bella, D.V.; Puccinelli, C.; Marcheggiani, S.; Mancini, L. Benthic diatom communities and their relationship to water chemistry in wetlands of central Italy. *J. Limnol.* **2007**, *43*, 89–99. [CrossRef]

31. Blanco, S.; Olenici, A.; Ortega, F.; Jiménez-Gómez, F.; Guerrero, F. Identifying environmental drivers of benthic diatom diversity: The case of Mediterranean mountain ponds. *PeerJ* **2020**, *8*, e8825. [CrossRef]

32. Kochoska, H.; Zaova, D.; Videska, A.; Mitic-Kopanja, D.; Naumovska, H.; Wetzel, C.E.; Ector, L.; Levkov, Z. Sellaphora pelagonica (Bacillariophyceae), a new species from dystrophic ponds in the Republic of North Macedonia. *Phytotaxa* **2021**, *496*, 2. [CrossRef]

33. Vidaković, D.; Levkov, Z.; Hamilton, P.B. *Neidiopsis borealis* sp. nov., a new diatom species from the mountain Shar Planina, Republic of North Macedonia. *Phytotaxa* **2019**, *402*, 21. [CrossRef]

34. Mihevc, A.; Gabrovšek, F.; Knez, M.; Kozel, P.; Mulec, J.; Otoničar, B.; Petrič, M.; Pipan, T.; Prelovšek, M.; Slabe, T.; et al. Karst in Slovenia. *Boletín Geológico y Minero* **2016**, *127*, 79–97.

35. Ogrin, D. Podnebni tipi v Sloveniji. *Acta Geogr. K* **1996**, *68*, 39–56.

36. Hofmann, G.; Werum, M.; Lange-Bertalot, H. *Diatomeen im Süßwasser-Benthos von Mitteleuropa: Bestimmungsflora Kieselalgen für die Ökologische Praxis*; Koeltz Scientific Books: Königstein, Germany, 2013; p. 908.

37. Lange-Bertalot, H.; Hofmann, G.; Werum, M.; Cantonati, M. *Freshwater Benthic Diatoms of Central Europe. Over 800 Common Species Used in Ecological Assessment*; English Edition with Updated Taxonomy and Added Species; Koeltz Scientific Books: Oberreifenberg, Germany, 2017; p. 942.

38. Krammer, K.; Lange-Bertalot, H. Bacillariophyceae. 1. Teil: Naviculaceae. In *Süßwasserflora von Mitteleuropa*; Ettl, H., Gerloff, J., Heynig, H., Mollenhauer, D., Eds.; Fischer: Jena, Germany, 1986; p. 876.

39. Krammer, K.; Lange-Bertalot, H. Bacillariophyceae. 2. Teil: Epithemiaceae, Surirellaceae. In *Süßwasserflora von Mitteleuropa*; Ettl, H., Gerloff, J., Heynig, H., Mollenhauer, D., Eds.; Fischer: Jena, Germany, 1988; p. 596.

40. Krammer, K.; Lange-Bertalot, H. Bacillariophyceae—Teil 3: Centrales, Fragilariaceae, Eunotiaceae. In *Süßwasserflora von Mitteleuropa*; Ettl, H., Gerloff, J., Heynig, H., Mollenhauer, D., Eds.; Fischer: Jena, Germany, 1991; p. 576.

41. Krammer, K.; Lange-Bertalot, H. Bacillariophyceae. 4. Teil: Achnanthaceae, Kritische Ergänzungen zu Navicula (Lineolatae) und Gomphonema Gesamtliteraturverzeichnis. In *Süßwasserflora von Mitteleuropa*; Ettl, H., Gerloff, J., Heynig, H., Mollenhauer, D., Eds.; Fischer: Jena, Germany, 1991; p. 437.

42. Hammer, Ø.; Harper, D.A.T.; Ryan, P.D. PAST: Paleontological Statistics Software Package for Education and Data Analysis. *Palaeontol. Electron.* **2001**, *4*, 1–9.

43. Rott, E.; Pipp, E.; Pfister, P.; Van Dahm, H.; Ortler, K.; Binder, N.; Pall, K. *Indikationslisten fur Aufwuchsalgen in Östereichen Fließgevessern, Teil 2: Trophienindikation so vie geochemische Präferenz, Taxonomische und toxicologische Anmerkungen*; Bundesministerium für Land und Forstwirtschaft: Wien, Austria, 1999; p. 248.

44. Braak, C.J.F.; Šmilauer, P. *CANOCO Reference Manual and CanoDraw for Windows User's Guide: Software for Canonical Community Ordination*; Microcomputer Power: Ithaca, NY, USA, 2002; p. 500.

45. Pan, Y.; Hughes, R.; Herlihy, A.; Kaufmann, P. Non-wadeable river bioassessment: Spatial variation of benthic diatom assemblages in Pacific Northwest rivers, USA. *Hydrobiologia* **2012**, *684*, 241–260. [CrossRef]

46. Heine-Fuster, I.; López-Allendes, C.; Aránguiz-Acuña, A.; Véliz, D. Differentiation of Diatom Guilds in Extreme Environments in the Andean Altiplano. *Front. Environ. Sci.* **2021**, *9*, 266. [CrossRef]

47. Licursi, M.; Gómez, N.; Sabater, S. Effects of nutrient enrichment on epipelic diatom assemblages in a nutrient-rich lowland stream, Pampa Region, Argentina. *Hydrobiologia* **2016**, *766*, 135–150. [CrossRef]

48. Passy, S.I. Diatom ecological guilds display distinct and predictable behavior along nutrient and disturbance gradients in running waters. *Aquat. Bot.* **2007**, *86*, 171–178. [CrossRef]

49. Gottschalk, S.; Kahlert, M. Shifts in taxonomical and guild composition of littoral diatom assemblages along environmental gradients. *Hydrobiologia* **2012**, *694*, 41–56. [CrossRef]

50. Béres, V.; Török, P.; Kókai, Z.; Krasznai, E.T.; Tóthmérész, B.; Bácsi, I. Ecological diatom guilds are useful but not sensitive enough as indicators of extremely changing water regimes. *Hydrobiologia* **2014**, *738*, 191–204. [CrossRef]

51. Zelnik, I.; Balanč, T.; Toman, M.J. Diversity and Structure of the Tychoplankton Diatom Community in the Limnocrene Spring Zelenci (Slovenia) in Relation to Environmental Factors. *Water* **2018**, *10*, 361. [CrossRef]

52. Peszek, Ł.; Zgrundo, A.; Noga, T.; Kochman-Kędziora, N.; Poradowska, A.; Rybak, M.; Puchalski, C.; Lee, J. The influence of drought on diatom assemblages in aemperate climate zone: A case study from the Carpathian Mountains, Poland. *Ecol. Indic.* **2021**, *125*, 107579. [CrossRef]

53. Lavoie, I.; Lento, J.; Morin, A. Inadequacy of size distributions of stream benthic diatoms for environmental monitoring. *J. N. Am. Benthol. Soc.* **2010**, *29*, 586–601. [CrossRef]

54. Soininen, J.; Jamoneau, A.; Rosebery, J.; Passy, S.I. Global patterns and drivers of species and trait composition in diatoms: Global compositional patterns in stream diatom. *Glob. Ecol. Biogeogr.* **2016**, *25*, 940–950. [CrossRef]

55. Scinto, L.J.; Reddy, K.R. Biotic and abiotic uptake of phosphorus by periphyton in a subtropical freshwater wetland. *Aquat. Bot.* **2003**, *77*, 203–222. [CrossRef]

56. Haubois, A.G.; Sylvestre, F.; Guarini, J.M.; Richard, P.; Blanchard, G.F. Spatio-temporal structure of the epipelic diatom assemblage from an intertidal mudflat in Marennes-Oléron Bay, France. *Estuar. Coast. Shelf Sci.* **2005**, *64*, 385–394. [CrossRef]

57. Körner, C. The use of 'altitude' in ecological research. *Trends. Ecol. Evol.* **2007**, *22*, 569–574. [CrossRef] [PubMed]

58. van der Grinten, E.; Janssen, A.P.H.; de Mutsert, K.; Barranguet, C.; Admiraal, W. Temperature- and Light-Dependent Performance of the Cyanobacterium Leptolyngbya Foveolarum and the Diatom Nitzschia Perminuta in Mixed Biofilms. *Hydrobiologia* **2005**, *548*, 267–278. [CrossRef]

59. Jurczak, T.; Wojtal-Frankiewicz, A.; Kaczkowski, Z.; Oleksińska, Z.; Bednarek, A.; Zalewski, M. Restoration of a shady urban pond—The pros and cons. *J. Environ. Manag.* **2018**, *217*, 919–928. [CrossRef] [PubMed]

60. DeNicola, D.M. A review of diatoms found in highly acidic environments. *Hydrobiologia* **2000**, *433*, 111–122. [CrossRef]

61. Krivograd-Klemenčič, A.; Smolar-Žvanut, N.; Istenič, D.; Griessler-Bulc, T. Algal community patterns in Slovenian bogs along environmental gradients. *Biologia* **2010**, *65*, 422–437. [CrossRef]

62. Schmutzer, A.C.; Gray, M.J.; Burton, E.C.; Miller, D.L. Impacts of cattle on amphibian larvae and the aquatic environment. *Freshw. Biol.* **2008**, *53*, 2613–2625. [CrossRef]

63. Connell, J.H. Diversity in tropical rainforests and coral reefs. *Science* **1978**, *199*, 1302–1310. [CrossRef]

64. Gascón, S.; Boix, D.; Sala, J. Are different biodiversity metrics related to the same factors? A case study from Mediter-ranean wetlands. *Biol. Conserv.* **2009**, *11*, 2602–2612. [CrossRef]

65. Feio, M.; Almeida, S.; Craveiro, S.; Calado, A. Diatoms and macroinvertebrates provide consistent and complementary information on environmental quality. *Fundam. Appl. Limnol.* **2007**, *169*, 247–258. [CrossRef]

66. Lange, K.; Liess, A.; Piggott, J.J.; Townsend, C.R.; Matthaei, C.D. Light, nutrients and grazing interact to determine stream diatom community composition and functional group structure: Diatom responses to light, nutrients and grazing. *Freshw. Biol.* **2011**, *56*, 264–278. [CrossRef]

Article

Spatial Distribution of the Taxonomic Diversity of Phytoplankton and Bioindication of the Shallow Protected Lake Borovoe in the Burabay National Natural Park, Northern Kazakhstan

Sophia Barinova [1,*], **Elena Krupa** [2,3] **and Elena Khitrova** [3]

1   Institute of Evolution, University of Haifa, Mount Carmel, 199 Abba Khoushi Ave., Haifa 3498838, Israel
2   Institute of Zoology, Ministry of Science and Higher Education of the Republic of Kazakhstan, 93 Al Farabi Str., Almaty 050060, Kazakhstan
3   Kazakh Agency of Applied Ecology, 42 Amangeldy Str., Almaty 050038, Kazakhstan
*   Correspondence: sophia@evo.haifa.ac.il; Tel.: +972-4824-97-99

Abstract: The Burabay National Natural Park unites six lakes located among the steppe landscape, with Lake Borovoe being the most visited among them. The phytoplankton of the protected Lake Borovoe was examined in the summer season of 2019, at eight stations, which were defined for the first time as the monitoring sites on the lake surface. Altogether, 72 algae and cyanobacteria species from seven taxonomic phyla were found in the Lake Borovoe phytoplankton during the study period. The most species-rich were three phyla: diatoms, green algae, and cyanobacteria. The average phytoplankton abundance was 3012.6 cells L^{-1}, and biomass was 2383.41 mg L^{-1}. The ecological status of the lake in 2019 was assessed based on the species richness, abundance, biomass, and calculated indices of organic pollution and toxic impact. The statistical mapping, calculated community similarity, correlation, and Redundancy Analysis (RDA) revealed zones affected by human impact. These were located in the lake shores and low-alkaline water with the saprobity index of 1.63–2.00. This is typical for naturally clean lakes, indicating the oligotrophic-to-mesotrophic status of the lake during the study period. The increase in cyanobacteria species in coastal communities can be associated with an increase in the biogenic load on the lake ecosystem in recent times. Therefore, our multivariate analysis allowed us to assess the ecological state of Lake Borovoe, which can be the result of the interaction of many external environmental factors, such as climatic conditions, long-term accumulation of organic substances, the intensity and duration of anthropogenic press, and internal lake processes such as the development of algae communities. The results suggest a tendency for the eutrophication of Lake Borovoe to increase because of pollution coming from the human impact zones on the lake shores.

Keywords: phytoplankton; species richness; abundance; biomass; bioindicators; statistics; ecological mapping; Lake Borovoe; Burabay National Natural Park; Kazakhstan

Citation: Barinova, S.; Krupa, E.; Khitrova, E. Spatial Distribution of the Taxonomic Diversity of Phytoplankton and Bioindication of the Shallow Protected Lake Borovoe in the Burabay National Natural Park, Northern Kazakhstan. *Diversity* **2022**, *14*, 1071. https://doi.org/10.3390/d14121071

Academic Editors: Jun Sun, Mateja Germ, Igor Zelnik and Matthew Simpson

Received: 31 October 2022
Accepted: 2 December 2022
Published: 5 December 2022

Publisher's Note: MDPI stays neutral with regard to jurisdictional claims in published maps and institutional affiliations.

1. Introduction

The shallow Lake Borovoe is located on the Schuchinsk-Borovsk resort area in Northern Kazakhstan as a part of the territory of the National Natural Park "Burabay" [1]. International Union for Conservation of Nature and Natural Resources, IUCN [2], classified the Burabay wetlands as Category II—a national park that can be managed in a way that may contribute to local economies through promoting to educational and recreational tourism on a scale that will not reduce the effectiveness of conservation efforts. Lake Borovoe is not only the site most visited by tourists; it is also occupied by a few balneological resorts. The enrichment of the lake water with nutrients has been going on since the foundation of the first resort 100 years ago [3]. Increased anthropogenic pressure on the lake is causing serious problems in regard to the water quality. High abundances of phytoplankton and zooplankton in the lake in the last decades have been documented, along with the blooming of Cyanobacteria [4–6].

Phytoplankton and zooplankton communities in Lake Borovoe demonstrate high heterogeneity, but the species richness has not been regularly studied. Nevertheless, research into the algae in Lake Borovoe was started in 1947 by N. Voronikhin [7–9], and it continued sporadically up until 2013; only a few species were mentioned in the previous reports [6,10–15]. The monitoring of the protected waterbodies has started revealing the species richness of its communities. All monitoring programs usually included the research of biodiversity and quantitative variables of the biological part of ecosystem, as well as the definition of major environmental variables [16,17]. Assessing the water quality of the protected waterbody is also an important task to identify the main sources of its pollution [18,19]. Floristic studies in aquatic systems are very important because the flora are environmental indicators that can be used to infer the environmental impacts on the natural, climatic, and economic conditions of the Burabay National Natural Park, as was revealed during the hydrochemical and hydrobiological study of five other lakes in the park territory [20]. Studies of algae are of special interest since the formation of their floras occurs under conditions of water flowing from the catchment basin and thus represents an accumulative result of natural and anthropogenic conditions throughout the entire catchment area over many years. So, the first step for the monitoring program is the screening of major chemical and biological variables on the net of sampling stations on the lake surface. Despite the occasional hydrochemical and hydrobiological studies of Lake Borovoe, a comprehensive assessment of its ecological status has not been performed yet.

Statistical data mapping is an effective tool for solving this application problem [21,22]. The effectiveness of this method in monitoring studies of water bodies in Kazakhstan has been shown previously [23,24]. The mapping of environmental and biological variables helps to reveal human impact zones in aquatic objects, as well as the ecosystem damage by them.

The purpose of this work was to study the species composition, abundance, and biomass of phytoplankton in Lake Borovoe in the protected area of the Burabay National Natural Park. We hypothesized that the distribution of algae indicators over the surface of the lake and the comparison of it to environmental variables, using statistical methods can, for the first time for Northern Kazakhstan, show the zones and main factors of human impact on the shallow lake.

2. Materials and Methods

2.1. Description of Study Site

Lake Borovoe is located at an altitude of 315.0 m above sea level. Borovoe is separated from the nearby lakes by ridges 400–900 m high (Figure 1a). It has an almost circular shape. The maximum length of the lake is about 4 km, with a width of 3.27 km. It is a shallow lake with a maximum depth of about 5 m. Two bays are placed in the western part (Figure 1b). The catchment area is 164.0 km^2. The east coast is sandy, sloping. The banks are overgrown with pine and birch. The Sary-Bulak River is inputted into the lake. The lake is fed by this small river, temporary streams, and precipitation. Only one small river, Kyrkyruek, flows from the lake. Sediments are represented by black silts. The coastal zone and northwestern bays are overgrown with *Potamogeton lucens* L. Macrophyte overgrowth is no more than 35–40% of the lake's surface and shoreline. The bottom is overgrown with Charophyta algae, except for the central part of the lake [25]. The position of Northern Kazakhstan in the depths of the mainland causes a sharp continental climate. Its characteristic features are a long cold winter with strong winds and snowstorms, and a short but hot summer [20,26]. Average long-term temperatures of the coldest month of January are about −17.6 °C to −18.5 °C, with a minimum of −45.0 °C. In July, the air temperature reaches an average of +19.0 °C to +19.5 °C, with an absolute maximum of +41.0 °C. The average annual precipitation varies from 290–295 to 425–435 mm. The snow cover lasts for about 5 months, from November to March. The Lake Borovoe area has been a resort of national importance since 1920. The lake received conservation status in 2000 when the Burabay National Natural Park was created. The Borovoye village is located on the eastern shore of the lake, with numerous rest houses and balneological resorts. The two largest resorts are located on the southeast and northwest shores of the lake.

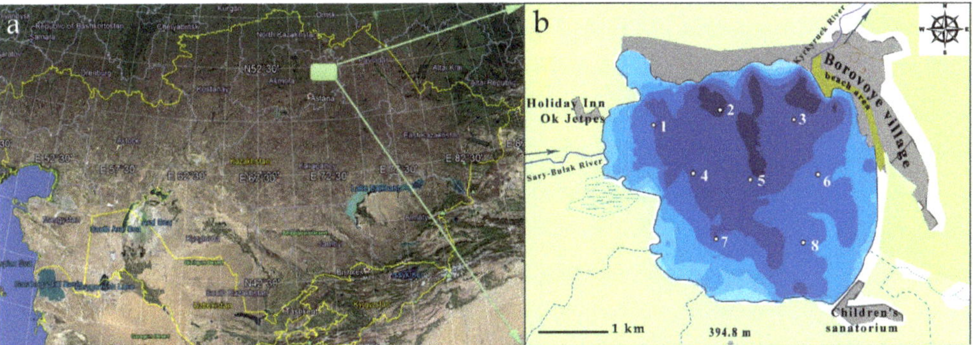

Figure 1. Map of study site on Lake Borovoe, Northern Kazakhstan, in July of 2019. Position of Lake Borovoe in Kazakhstan (**a**). Position of the phytoplankton sampling stations 1–8 on the lake surface (**b**). Map creating in ESRI ArcGIS 10.8. USA program on the base of our measurements of GIS coordinates and the lake depth on each sampling station.

2.2. Sampling and Laboratory Analysis

The Lake Borovoe was examined in the summer season (29 July of 2019). It was the first trip with sampling and measurements that covered the entire lake's surface as a screening stage for preliminary assessment for the purpose of monitoring. A total of eight sampling stations have been identified for the first time as monitoring sites on the surface of the lake. Coordinate referencing of the stations was performed by Garmin eTrex GPS-navigator. The temperature, Total Dissolved Solids (TDS), and pH values of the surface layers of water were measured in parallel with sampling with HANNA HI 9813-0, and N-NO$_3$ with HANNA HI 93728 (HANNA Instruments, Vensokit, RI, USA), with three repetitions. The transparency of the water was measured by using a Secchi disk.

Phytoplankton samples were taken from the surface water layers about 0–0.5 m into 1 L plastic containers [27], fixed in 3% neutral formaldehyde, and transported to the lab in an icebox. The sedimentary method was used to process phytoplankton samples, with the final volume of the concentrated sample being 5–10 mL [28,29]. Fixed phytoplankton samples were also studied in the lab in three repetitions, from wet and permanent slides [30], under light microscopes A Nikon ECLIPSE E200 (Nikon Instruments Inc., Melville, NY, USA) was used, with a magnification of ×100–×2000. Cells' abundance was calculated in the Nageotte counting chamber (Hausser Scientific, Horsham, PA, USA). Wet biomass was calculated from the volume of the cell in mg L^{-1}. Species identification was performed by using standard methods with relevant guides to the species identifications [31–36]. Modern taxonomy was adopted with [37].

The Shannon Diversity Index [38] was calculated with the Primer 6 program, using the following Formula (1):

$$H = -\sum_{i=1}^{n} p_i \times log2\, p_i \tag{1}$$

where H is the Shannon Index (bits/ind.), p_i is the share of the i-th species in the total abundance, $log2$ is the logarithm to base 2, n is the number of species in the sample, and $\sum$ is the sum of values for the sample.

Index saprobity S was calculated for each algal community, according to V. Sládeček [39], as a function of the number of saprobic species and their relative abundances (2):

$$S = \sum_{i=1}^{n}(s_i h_i) / \sum_{i=1}^{n}(h_i) \tag{2}$$

where S is the index of saprobity for algal community (unitless), s is species-specific saprobity index [40,41], and h is the cell density of each species, n is the number of species in the sample, and $\sum$ is the sum of values for the sample.

Bioindicator analysis was performed with species-specific ecological preferences of planktonic algae found at each sampling station [40–42] for revealing influencing external factors such as temperature, salinity, pH, oxygen conditions, organic pollution level, and trophic status of a water body.

Statistical maps [22] that reflect the probability of mapped variable distribution over the lake surface were built in the Statistica 12.0 Software based on the GPS coordinates of sampling points for each measured biological and chemical variable. Cluster analysis of the Bray–Curtis similarity of phytoplankton communities was carried out by using the Biodiversity-Pro program. A correlation analysis of the revealed data was carried out by using the JASP 0.16.4.0 program [43]. A heat map was constructed in the ExStatR program [44]. The linear ordination method *Redundancy Detrended Analysis* (RDA) was processed in the CANOCO 4.5 program in order to recognize the species–environment relationships [45].

3. Results

3.1. Environmental Characteristics of the Lake Borovoe

The lake was shallow, as can be seen in Appendix Table A1 and Figure 2c. During the study period, the temperature of the surface water reached 22.0–22.86 °C (Appendix Table A1) and was lowest in the central part of the lake (Figure 2b). The water was fresh, with a TDS of about 200 mg L^{-1}, and highest ionic content was found in the central part of the lake (Figure 2a). The water was slightly alkaline, with a pH of about 8.3 and transparency of about 0.5–0.8 m (Figure 2d). So, major environmental variables can demonstrate the uniform condition in two parts of the Lake Borovoe surface—northwestern and southeastern parts—as represented in the maps in Figure 2a,b. The distribution of TDS and water temperature was controversial (R = –0.93, p = 0.001), especially in the central part with no correlations with the depth (p = 0.53) (Figure 2c). The water transparency was not correlated with temperature (p = 0.39), but visually, we can see a similar distribution near the wetlands of Station 3 and Borovoe village (Figure 2b,d).

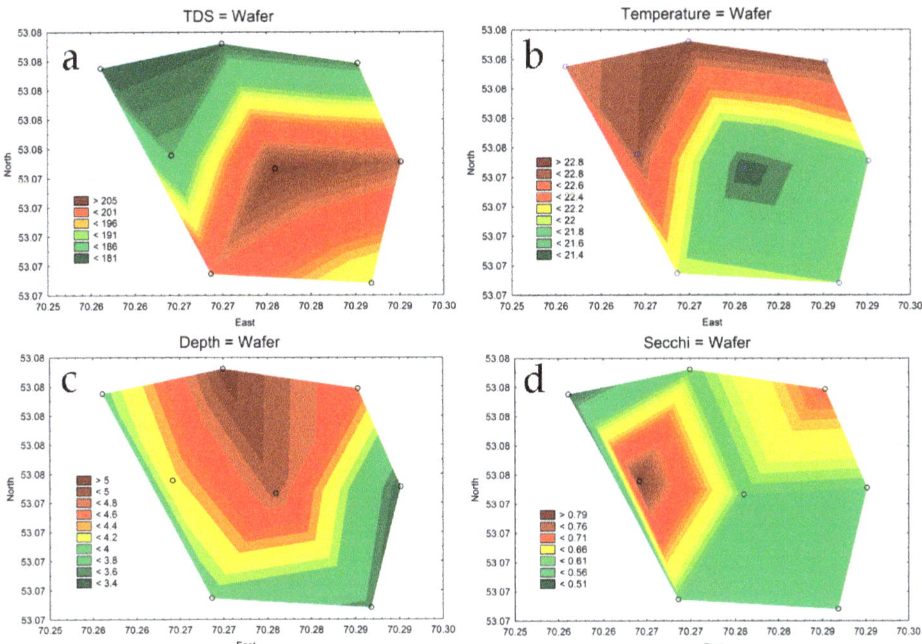

Figure 2. Distribution of major environmental variables in the sampling stations of Lake Borovoe's surface, 2019. TDS (**a**), water temperature (**b**), lake depth (**c**), and transparency by Secchi disk (**d**).

3.2. Phytoplankton in the Lake Borovoe

Altogether, 72 algae and cyanobacteria taxa from seven taxonomic phyla were found in Lake Borovoe during the sampling period (Appendix Table A2). The diatom species was the richest, with 29 taxa, followed green algae, with 18, and cyanobacteria, with 15 taxa (Table 1). Species richness at sampling stations varies within a small range, from 21 to 29 taxa (Table 1). The lowermost species number was in Station 4, where water transparency was maximal (Figure 2d).

Table 1. Species richness of phytoplanktonic algae and cyanobacteria in Lake Borovoe, summer 2019.

Phylum	1	2	3	4	5	6	7	8	Percent
Bacillariophyta	7	11	8	10	9	11	9	7	40
Chlorophyta	7	6	7	2	8	8	8	10	25
Cyanobacteria	8	4	7	5	5	7	6	5	21
Euglenozoa	0	0	1	1	2	0	1	1	3
Miozoa	1	2	3	1	1	1	2	3	7
Ochrophyta	1	1	1	1	1	1	1	1	1
Charophyta	2	1	1	1	1	1	1	1	3
Total	26	25	28	21	27	29	28	28	100

The Bray–Curtis similarity analysis divided the species richness into two clusters, the first of which included the three most species-rich phyla–diatoms, green algae, and cyanobacteria (Figure 3); the second cluster included four other phyla, which contained a small number of species (Table 1).

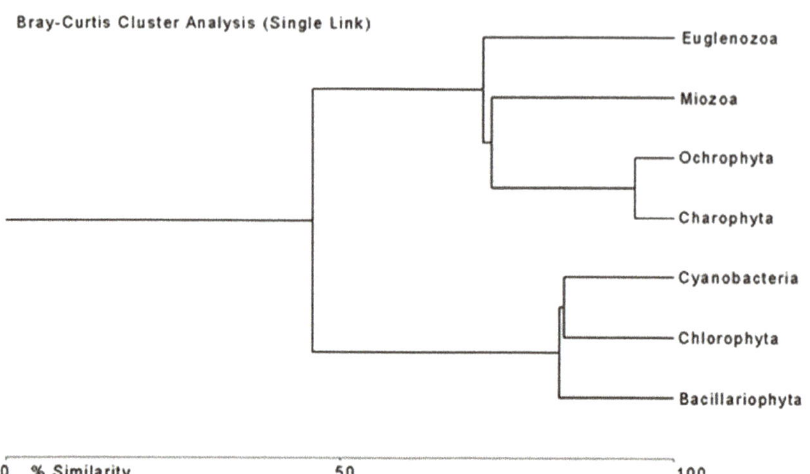

Figure 3. Cluster dendrogram of phytoplankton community species richness in the Lake Borovoe, 2019.

Phytoplankton species occupy the first level of the food pyramid and depend on the concentration of essential nutrients such as nitrate and phosphate. Thus, phytoplankton species are rich where nutrients can be available. We tried to reveal the relationship between phytoplankton species richness in sampling stations by using statistical methods to reveal some homogeneity of the lake environment. Figure 4 shows a graph of JASP analysis of the correlation of phytoplankton species number at sampling stations. At a similarity level of more than 50%, the graph divides the phytoplankton species richness into three groups. The most similar communities are at Stations 1, 2, and 6 (Group 1); 3 and 7 (Group 2); and 4, 5, and 8 (Group 3).

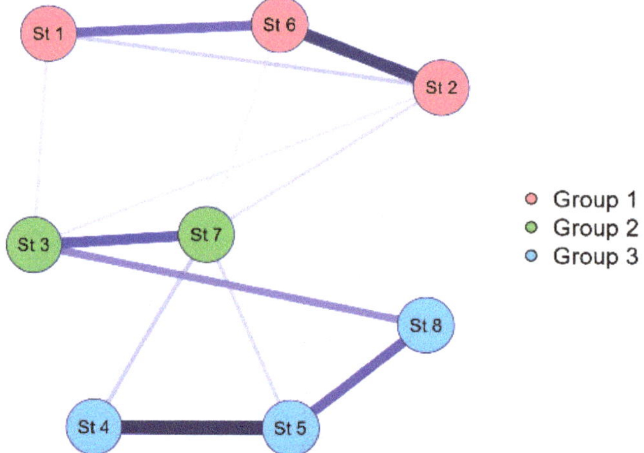

Figure 4. JASP correlation plot of the phytoplankton species richness in Lake Borovoe, 2019. Bold line shows largest similarity on type of analysis, "Huge" correlation > 0.5.

The averages of the parameter values (Table 2) were calculated to indicate the differences in the stations grouped in Figure 4. Group 1 includes Stations 1, 3, and 6, where, compared to other groups, there is less depth, transparency, TDS, and biomass, but higher water warming, which indicates not only the high ecosystem activity work of the lake, but also greater stress for phytoplankton, when the WESI index decreases. The second group of stations (Stations 3 and 7) is characterized by a higher abundance of phytoplankton, mainly at Station 7. Both stations are located near actively visited or populated lake shores. Group 3 consists of three stations (Stations 4, 5, and 8) that are deeper, transparent, cool, mineralized, and alkaline, where there are fewer nitrates, and the WESI index was higher. The stations of Group 3 are located closer to the middle of the lake and are the most remote from anthropogenic influence.

Table 2. Distribution of averaged environmental and biological variables of phytoplankton over groups of stations in Figure 4.

Station Group	1	2	3
Depth, m	4.13	4.22	4.21
Secchi, m	0.57	0.65	0.67
T, °C	22.44	22.43	22.00
pH	8.17	8.17	8.24
TDS, mg L^{-1}	188.47	191.80	194.43
N-NO3, mg L^{-1}	0.56	0.58	0.04
Index S	1.81	1.89	1.81
Index WESI	1.11	1.17	3.33
Abundance, cells L^{-1}	2516.5	4603.7	2448.0
Biomass mg L^{-1}	2087.4	3237.2	2110.2
Shannon Index	0.751	0.530	0.738

Therefore, to understand the importance of species dynamics at the phylum level, we constructed statistical maps for each taxa distribution in the sampling stations. Figure 5a shows that the total number of species was minimal at Station 4, which had the highest water transparency and temperature, as well as lower TDS values (Appendix Table A1). Figure 5b shows that the total number of phytoplankton species was determined by the number of green algae species that take a part of its composition. The distribution of diatoms is inversely relative to the distribution of green algae (Figure 5c). Cyanobacteria are most represented offshore, where the water temperature and anthropogenic load were higher, and the TDS was lower (Figure 5d). Euglenoids were mainly in the central part (Figure 5e), but charophyte planktonic species were concentrated in the northwestern bay of the lake (Figure 5f).

Spatial maps of phytoplankton abundance and biomass show an opposite distribution (Figure 6a,b). The abundance of phytoplankton was greatest where an influx of nitrates was found (Figure 6c). At the same time, the distribution of the WESI index (Figure 6d), which shows stress zones for algae, outlines the entire coast, including the points of nitrate inflow, as vulnerable, and the middle of the lake as the most preferable.

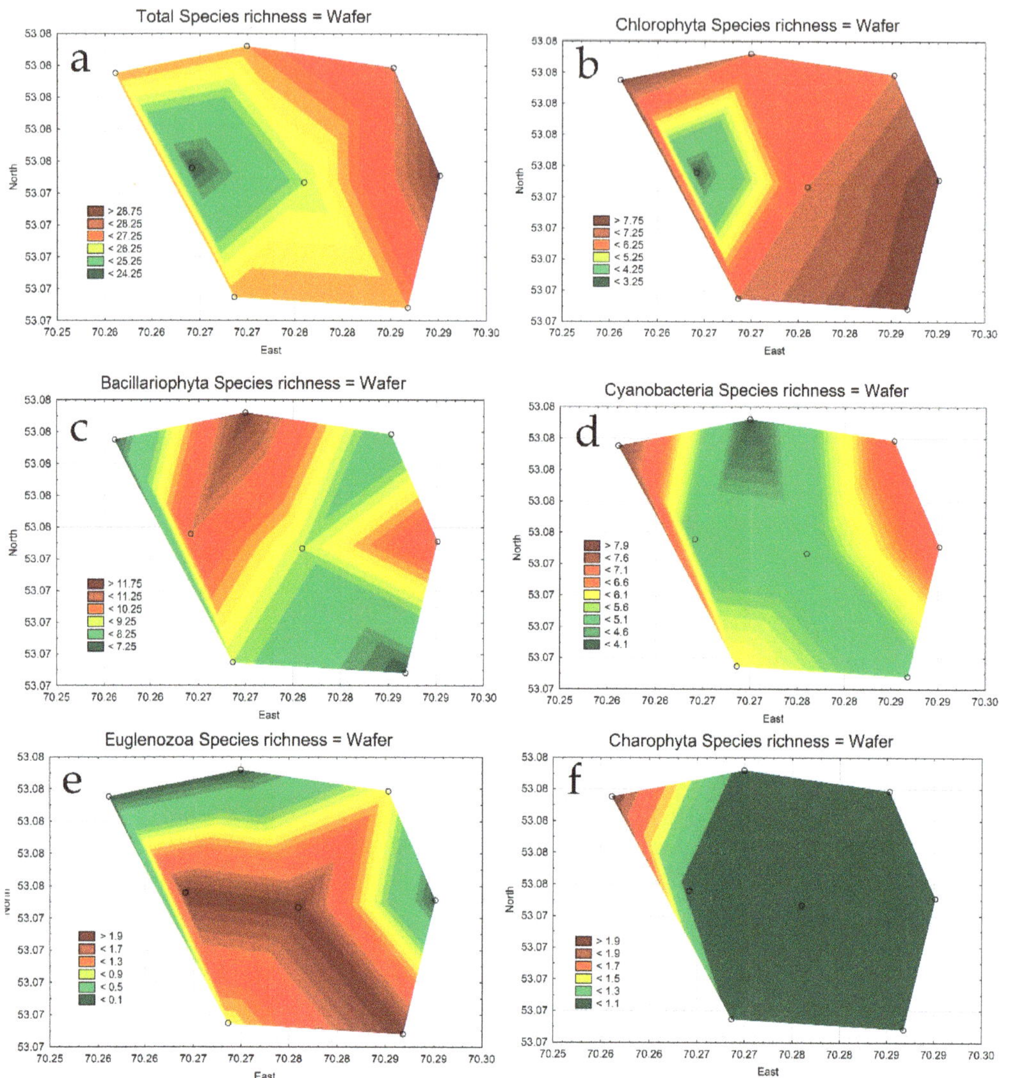

Figure 5. Statistical maps of distribution of phyla species richness in phytoplankton communities on the sampling stations of the Lake Borovoe surface, 2019. Total species richness (**a**), Chlorophyta (**b**), Bacillariophyta (**c**), Cyanobacteria (**d**), Euglenozoa (**e**), and Charophyta (**f**).

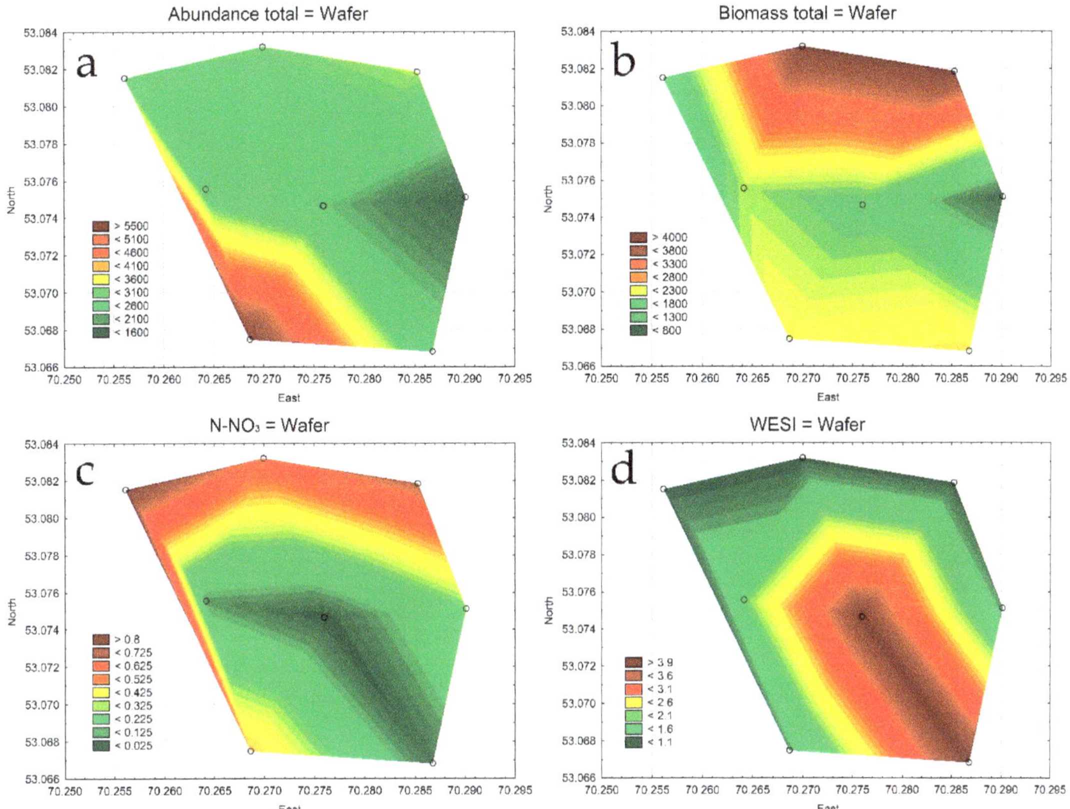

Figure 6. Maps of the surface phytoplankton abundance (**a**), biomass (**b**), nitrates (**c**), and WESI index (**d**) in Lake Borovoe, summer 2019.

3.3. Phytoplankton Indicators of Water Quality

The bioindication results on the basis of phytoplankton species' ecological preferences (Appendix Table A2) are represented in Appendix Table A3 and visualized in Figure 7. Even though the indicator values for stations represent a mosaic, it is possible to single out the main characteristic groups: Bacillariophyta and Chlorophyta species, planktonic and plankto-benthic inhabitants, and middle oxygenated water indicators of Class 3 of water quality.

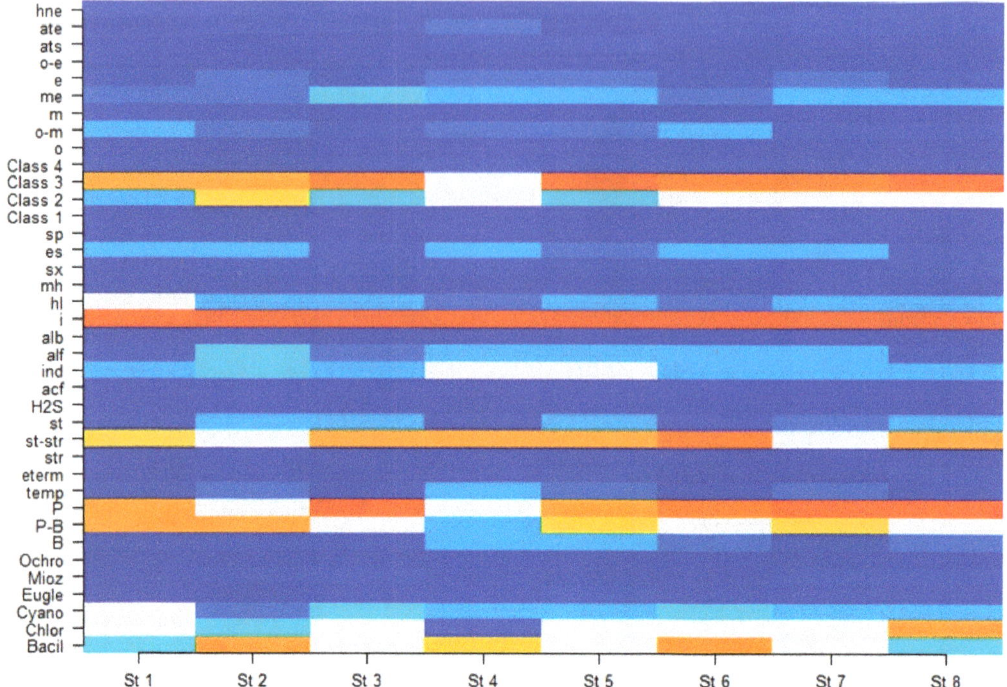

Figure 7. Heat map for distribution of indicators in phytoplankton communities over the sampling stations at Lake Borovoe, 2019. Abbreviations: Bacil, Bacillariophyta; Chlor, Chlorophyta; Cyano, Cyanobacteria; Eugle, Euglenozoa; Mioz, Miozoa; Ochro, Ochrophyta; B, benthic; P-B, plankto-benthic; P, planktonic; temp, inhabitants of temperate-temperature waters; eterm, eurythermic; str, streaming well-oxygenated waters inhabitants; st-str, inhabitants of standing to streaming middle-oxygenated waters; st, inhabitants of standing low-oxygenated waters; H_2S, sulfides anoxia indicators; acf, acidophilic; ind, pH indifferent; alf, alkaliphilic; alb, alkalibiontes; i, chlorides indifferent; hl, halophiles; mh, mesohalobes; sx, saproxenes; es, eurysaprobes; sp, saprophiles; Classes 1–4 of water-quality indicators according to Index S; o, oligotraphentes; o-m, oligo-mesotraphentes; m, mesotraphentes; me, meso-eutraphentes; e, eutraphentes; o-e, oligotraphentes to eutraphentes; ats, autotrophes-inhabited low-nitric-nitrogen waters; ate, autotrophes-inhabited high-nitric-nitrogen waters; hne, facultative-heterotrophes-inhabited low organically enriched waters. In the *x*-axis are station numbers, and in the *y*-axis are the same abbreviations as in Appendix Table A3. The color of the cells varies from white to blue and then to red, according to the proportion of the number in the entire distribution.

An RDA triplot was constructed on the base of environmental data (Appendix Table A1) and species richness in phyla (Table 1). Figure 8 shows that nitrate-nitrogen-stimulated Charophyta and Cyanobacteria grow in the waters of Stations 1 and 8 with less transparency (Figure 2d) and depth (Figure 2c). Diatom, green algae, and euglenoids were diverse in Stations 5, 6, and 7, where there was the highest TDS (Figure 2a) and total species richness (Figure 5a,b). Miozoa species (*Ceratium hirundinella*) were richest in Stations 2 and 3, which had highest temperature (Figure 2b) and lowest TDS (Figure 2a).

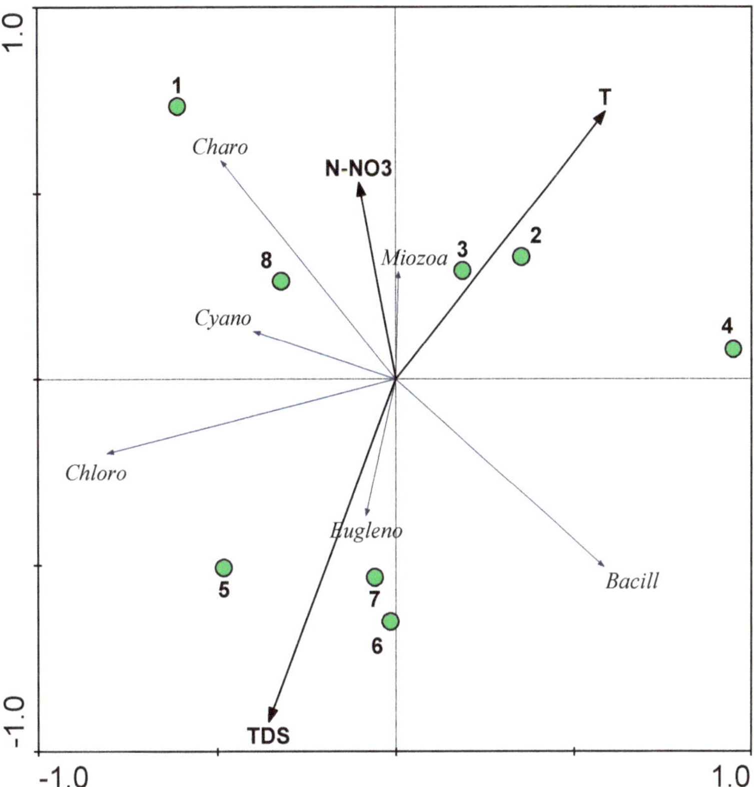

Figure 8. RDA plot of the relationships between environmental variables: temperature, nitrates-nitrogen concentration, TDS, depth, and species richness in phyla of phytoplankton in the Lake Borovoe, 2019. Monte Carlo test summary for 945 permutations: eigenvalue = 0.421; *p*-value = 0.136.

4. Discussion

Northern Kazakhstan abounds in shallow lakes, most of which are objects of WWF protection, as they are located on the migratory routes of birds [46]. Therefore, it is especially important not only to identify the diversity of organisms at protected sites, but also to determine the main influencing anthropogenic or natural factors. Algae, as the first level of the trophic pyramid, are the first to respond to changes in the lake's ecosystem, with the most noticeable impact on the example of phytoplankton [47]. Phytoplankton in the lakes of Northern Kazakhstan, including Lake Borovoe, has been studied sporadically [6–15]. In connection with the projects of the WWF on the status of protection, we undertook a large-scale study of phytoplankton during 1999–2000 in 34 North Kazakhstan lakes that have a conservation status or are under preparation [48]; with the help of bioindication and statistical analysis, it was possible to identify the main factor affecting the diversity of communities in the lakes of this region, i.e., salinity [49]. A subsequent comparison with lakes of similar size in the semi-arid climate zone in the Eastern Mediterranean not only confirmed the conclusion made, but also expanded the understanding of the importance of using bioindication to identify the main factors affecting phytoplankton [48]. The mentioned studies on protected lakes have shown that salinity is a regional natural factor related to climate. Our studies at Lake Borovoe also revealed a close inverse relationship between temperature and TDS (*p* = 0.001). A study of phytoplankton in the nearby Lake Zerenda showed that, for lakes whose shores are visited by tourists, the anthropogenic factor is the influx of nitrates [50]. At the same time, nitrates in all the studied lakes of Northern Kazakhstan turned out to be a factor stimulating the diversity of algae, and in Lake Borovoe, also its abundance and biomass. Thus, saprobity indices were higher where nitrate concentrations were higher [48]. Lake Borovoe, where the saprobity indices were in the range of 1.52–2.00, is comparable with freshwater lakes in terms of this indicator and the number of species in the community.

Our calculations of the index toxicity WESI were comparable with its values in the lakes of Northern Kazakhstan for those lakes where salinity (TDS) was higher, which indicates a long history of evaporation and a small inflow of surface water, that is, the natural state of the lakes [48]. For Lake Borovoe, the WESI index was higher (better condition) in the center of the lake, away from coastal pollution, where TDS is higher and water temperature is lower. This distribution shows that the phytoplankton of the lake was affected by organic pollution coming from the shore zone, which does not reach the center of the lake. Therefore, the lake retains sufficient self-purification capacity.

The spatial distribution of phytoplankton suggests that the lake water is influenced by pollution of the communal services of the Borovoye village and two resorts (Figure 1). The main contribution of the village is associated not only with wastewater discharges but also with surface runoff from the beach area along all east coasts. As can be seen from Figures 2d and 5, in the eastern and southeastern parts of the water surface, the signs of eutrophication of the lake can be recognized. Here, the high species richness of phytoplankton due to Chlorophyta taxa was recorded, as well as minimum values of water transparency and maximal value of TDS show that phytoplankton communities are formed under the conditions of a constant influx of fresh nutrients. As is known, the number of phytoplankton species decreases with an increase in their abundance and biomass [16], which we observe on spatial statistical maps for Lake Borovoe.

The spatial distribution of phytoplankton and environmental parameters in the lakes of Northern Kazakhstan were studied for the first time on Lake Borovoe. However, for more southerly lakes, this method has been applied with effective results. For a fairly large regional lake Balkhash, the zones of influence of both climatic (salinity) and anthropogenic (organic pollution) factors were clearly shown, using statistical mapping [51]. The phytoplankton of smaller lakes in Kazakhstan revealed the anthropogenic factor of organic pollution coming with the tourist flow to the lakes of the Kolsay cascade [52] or with runoff from fields and inflow to the Shardara reservoir [53]. Thus, statistical mapping, statistical methods, and bioindication can serve together as effective tools for identifying factors and zones of natural and anthropogenic impact in this region.

According to the results of the analysis of the main environmental variables in the summer of 2019 (Table 1), the water of Lake Borovoe can be assessed as fresh and low alkaline, corresponding to the level of clean, slightly polluted waters of Classes 2 and 3 of water quality [39]. This may be due to historically low nutrient content [4,6], even during the period of fish mortality in 1974 [3]. It is known that the best period for studying phytoplankton is the middle of summer, as in our case, when all processes in the lake give the maximum diversity and biomass of plankton communities [52–54]. The total species richness of phytoplankton in Lake Borovoe is currently determined mainly by the number of Chlorophyta species. This and phytoplankton abundance and biomass are typical for naturally clean lakes and may indicate the oligotrophic status of the lake during the study period. At the same time, the increase in cyanobacteria species in coastal communities may be associated with an increase in the biogenic load on the lake ecosystem in recent times [1]. The analysis shows that pollutants entering the lake water are associated with villages and resort areas located on the coast. However, they are utilized by algae almost completely, which can be seen in the distribution of pollution indices, showing cleaner water in the middle of the lake.

The phytoplankton is a part of total autotrophic organisms that, together with macrophytes, settled the lake and therefore can use only part of the total value of nutrient inflow to the lake. Since the IUCN protected Charophyte species [55] grow in the coastal area, it is important to monitor water quality and the state of their populations during monitoring, since the disappearance of charophytes will indicate the loss of a protected object in the IUCN system. The Lake Borovoe shores are inhabited by five species of *Chara*, *Nitellopsis obtusa*, and *Nitella flexilis* [25]. There is a unique diversity of charophytes in this one lake. Therefore, the role of macrophytes and charophytes in the trophic condition of the lake is very important. The anthropogenic load may be one of the reasons for the eutrophication of Lake Borovoe, leading to degradation of the macrophyte communities. The Lake Fund monitoring showed that macrophytes occupied 65–80% of the bottom surface in 1964–2002 [3], while their density had decreased to approximately 30–35% by 2019. Therefore, a decrease in the role of macrophytes consuming nutrients [56] has created favorable conditions for planktonic algae [19] and caused changes in the trophic conditions of the lake.

The differences in environmental and biological variables distribution were revealed with the help of statistical mapping, even in the case of low amplitude of variables [21], as in Lake Borovoe. The preliminary hydrochemical and hydrobiological assessment of the lakes of the Shchuchinsko-Borovsk resort zone [20] confirm that this approach can be recommended in the monitoring of different protected lakes' sustainability.

5. Conclusions

Our study showed that the ecological state of Lake Borovoe can be the result of the interaction of many environmental factors, such as climatic conditions, long-term accumulation of organic matter, the intensity and duration of anthropogenic pressure, as well as intralake processes, such as the development of a community of macrophytes, algae, and invertebrates. An assessment of the phytoplankton communities showed a trend towards an increase in eutrophication of the lake, as revealed during statistical mapping, because of organic pollution from populated and resort areas. The assessment of water quality by the bioindication of plankton species revealed weakly alkaline and slightly organically polluted water of Classes 2 and 3. With the help of indications and statistics, the salinity was determined as a climatic factor and organic pollution as an anthropogenic factor affecting the ecosystem of Lake Borovoe. Thus, the indication of phytoplankton and environmental parameters in Lake Borovoe reflect the state of the lake as oligotrophic with a transition to the mesotrophic stage, subject to organic pollution in the coastal part, but coping with anthropogenic impact.

Author Contributions: Conceptualization, S.B. (Sophia Barinova) and E.K. (Elena Krupa); methodology, S.B.; software, S.B.; validation, S.B., E.K. (Elena Krupa), and E.K. (Elena Khitrova); formal analysis, E.K. (Elena Khitrova) and S.B.; investigation, E.K. (Elena Krupa); resources, E.K. (Elena Krupa); data curation, S.B. and E.K. (Elena Khitrova); writing—original draft preparation, S.B. and E.K. (Elena Krupa); writing—review and editing, S.B.; visualization, S.B.; supervision, E.K. (Elena Krupa); project administration, E.K. (Elena Krupa). All authors have read and agreed to the published version of the manuscript.

Funding: This research received no external funding.

Institutional Review Board Statement: Not applicable.

Informed Consent Statement: Not applicable.

Data Availability Statement: Not applicable.

Acknowledgments: The work was carried out under project No. BR05236529 for the Committee of Science, Ministry of Education and Science, Republic of Kazakhstan, "A comprehensive assessment of the ecosystems of the Shchuchinsko-Borovsk resort area with the determination of the environmental load for the sustainable use of the recreational potential", as well as partly supported by the Israeli Ministry of Aliyah and Integration.

Conflicts of Interest: The authors declare no conflict of interest.

Appendix A

Table A1. Averaged environmental variables with standard deviation and coordinates of sampling stations on the Lake Borovoe, 2019.

Station	1	2	3	4	5	6	7	8	Average
North	53.08152	53.08321	53.08186	53.07557	53.07467	53.07512	53.06748	53.06682	
East	70.25613	70.26995	70.28529	70.26416	70.27596	70.29014	70.26864	70.28679	
Depth, m	3.92 ± 0.06	5.06 ± 0.05	4.55 ± 0.06	4.25 ± 0.06	4.85 ± 0.04	3.4 ± 0.02	3.89 ± 0.04	3.53 ± 0.02	4.18
Secchi, m	0.5 ± 0.09	0.6 ± 0.09	0.7 ± 0.08	0.8 ± 0.06	0.6 ± 0.05	0.6 ± 0.05	0.6 ± 0.05	0.6 ± 0.06	0.63
T $^\circ$C	22.61 ± 0.6	22.92 ± 0.5	22.86 ± 0.6	22.8 ± 0.6	21.3 ± 0.5	21.8 ± 0.5	22 ± 0.3	21.9 ± 0.5	22.27
pH	8.17 ± 0.01	8.16 ± 0.01	8.18 ± 0.02	8.16 ± 0.02	8.43 ± 0.01	8.19 ± 0.01	8.16 ± 0.01	8.12 ± 0.01	8.20
TDS, mg L^{-1}	180 ± 11.51	180.8 ± 11.2	182.2 ± 10.8	183.8 ± 9.58	207.8 ± 11.0	204.6 ± 10.0	201.4 ± 9.56	191.7 ± 9.12	191.5
N-NO$_3$, mg L^{-1}	0.816 ± 0.32	0.599 ± 0.12	0.700 ± 0.24	0.090 ± 0.01	0.000 ± 0.0	0.272 ± 0.02	0.466 ± 0.10	0.039 ± 0.01	0.373
Index S	1.82 ± 0.18	1.98 ± 0.15	1.77 ± 0.17	1.94 ± 0.18	1.99 ± 0.16	1.63 ± 0.13	2.00 ± 0.20	1.52 ± 0.11	1.83
Index WESI	1.00	1.00	1.00	2.00	4.00	1.33	1.33	4.00	1.96
Abundance, cells L^{-1}	2982.9 ± 193.3	3026.5 ± 220.7	3238.9 ± 245.9	2642.7 ± 222.8	2226.9 ± 137.9	1540.2 ± 80.1	5968.4 ± 575.0	2474.3 ± 141.5	3012.6
Biomass mg L^{-1}	1403.2 ± 78.4	4086.6 ± 270.0	4087.5 ± 215.0	2044.2 ± 95.6	1732.5 ± 101.6	772.5 ± 38.8	2386.8 ± 129.9	2554 ± 149.0	2383.4
Shannon Index	0.753 ± 0.17	0.614 ± 0.15	0.659 ± 0.14	0.552 ± 0.11	0.797 ± 0.17	0.885 ± 0.18	0.401 ± 0.10	0.864 ± 0.16	0.691

Table A2. Diversity of algae and cyanobacteria on the sampling stations (1–8) of the Lake Borovoe, 2019. Abbreviations: **Hab**, substrate preferences: B, benthic; P-B, plankto-benthic; and P, planktonic. **T**, temperature indicators: temp, inhabitants of temperate-temperature waters; and eterm, eurythermic. **Oxy**, Dissolved oxygen and water mobility indicators: str, inhabitants of streaming well-oxygenated waters; st-str, inhabitants of standing to streaming middle-oxygenated waters; st, inhabitants of standing low-oxygenated waters; H$_2$S, sulfides anoxia indicators. **pH**, water acidity indicators: acf, acidophilic; ind, pH indifferent; alf, alkaliphilic; alb, alkalibiontes. **Sal**, salinity indicators: i, chlorides indifferent; hl, halophiles; mh, mesohalobes; oh, broad spectrum oligohalobes. **D**, Watanabe diatom indicators for organic pollution: sx, saproxenes; es, eurysaprobes; sp, saprophiles. **Sap**, saprobity indicator categories with species-specific index S: b-a, 2.4–beta-alpha-mesosaprobiont; b-o, 1.6–beta-oligosaprobiont; o, 1.0–oligosaprobiont; o-a, 1.8–oligo-alpha-mesosaprobiont; o-b, 1.4–oligo-beta-mesosaprobiont; o-x, 0.6–oligo-xenosaprobiont; x, 0.0–xenosaprobiont; x-b, 0.8–xeno-beta-mesosaprobiont; x-o, 0.4–xeno-oligosaprobiont. **Index S**, species-specific index saprobity S. **Tro**, trophic-state indicators: o, oligotraphentes; o-m, oligo-mesotraphentes; m, mesotraphentes; me, meso-eutraphentes; e, eutraphentes; o-e, oligotraphentes to eutraphentes. **Aut-Het**, indicators of autotrophy-heterotrophy nutrition type: ats, autotrophes-inhabited low-nitric-nitrogen waters; ate, autotrophes-inhabited high-nitric-nitrogen waters; hne, facultative-heterotrophes-inhabited low organically enriched waters; "-", no data.

Taxa	1	2	3	4	5	6	7	8	Hab	T	Oxy	pH	Sal	D	Sap	Index S	Tro	Aut-Het
Cyanobacteria																		
Anabaena contorta Bachmann	0	0	0	0	0	1	0	0	P	-	st-str	-	-	-	-	-	-	-
Anagnostidinema amphibium (C. Agardh ex Gomont) Strunecký, Bohunická, J.R. Johansen and J. Komárek	1	0	0	0	0	0	0	0	P-B, S	-	st-str, H$_2$S	-	hl	-	a-o	2.6	m	-
Anathece clathrata (West and G.S. West) Komárek, Kastovsky and Jezberová	1	1	1	1	1	1	1	1	P	-	-	-	hl	-	o-a	1.8	me	-
Aphanocapsa holsatica (Lemmermann) G. Cronberg and Komárek	0	0	1	0	0	0	0	0	P	-	-	-	i	-	o-b	1.4	me	-

Table A2. *Cont.*

Taxa	1	2	3	4	5	6	7	8	Hab	T	Oxy	pH	Sal	D	Sap	Index S	Tro	Aut-Het
Cyanobacteria																		
Aphanocapsa incerta (Lemmermann) G. Cronberg and Komárek	0	0	0	0	0	1	1	0	P-B	-	-	-	-	i	-	b	2.2	me	-
Aphanocapsa planctonica (G.M. Smith) Komárek and Anagnostidis	1	1	1	0	0	1	0	0	P	-	-	-	-	i	-	-	-	o	-
Chroococcus minimus (Keissler) Lemmermann	1	0	0	0	0	0	0	0	P-B	-	-	-	-	hl	-	-	-	o-m	-
Chroococcus minutus (Kützing) Nägeli	0	0	0	1	1	0	0	0	P-B	-	-	ind	i	-	o-a	1.8	o-m	-
Merismopedia tenuissima Lemmermann	1	0	0	0	0	0	0	0	P-B	-	-	-	hl	-	b-a	2.4	e	-
Microcystis aeruginosa (Kützing) Kützing	1	1	1	1	1	1	1	1	P	-	-	-	hl	-	b	2.1	e	-
Planktolyngbya contorta (Lemmermann) Anagnostidis and Komárek	0	0	1	0	0	0	0	0	P	-	-	-	-	-	o-a	1.8	me	-
Planktolyngbya limnetica (Lemmermann) Komárková-Legnerová and Cronberg	1	1	0	1	0	0	0	0	P-B, S	-	st-str	-	hl	-	o-b	1.5	e	-
Radiocystis geminata Skuja	1	0	1	0	1	0	1	1	P	-	-	-	-	-	-	-	me	-
Rhabdoderma lineare Schmidle and Lauterborn	0	0	0	0	0	1	1	1	P	-	-	-	-	-	b	2.2	-	-
Snowella atomus Komárek and Hindák	0	0	1	1	1	1	1	1	P	-	-	-	-	-	-	-	me	-
Bacillariophyta																		
Achnanthidium minutissimum (Kützing) Czarnecki	0	0	0	0	0	1	0	0	P-B	eterm	st-str	ind	i	es	x-b	0.95	o-e	ate
Amphora ovalis (Kützing) Kützing	0	0	0	1	0	0	0	0	B	temp	st-str	alf	i	sx	o-b	1.5	me	ate
Aulacoseira granulata (Ehrenberg) Simonsen	1	1	0	1	0	0	0	0	P-B	temp	st-str	ind	i	es	b	2	me	Ate
Caloneis bacillum (Grunow) Cleve	0	0	0	1	1	0	0	0	B	temp	st-str	ind	i	es	o	1.3	me	ats
Cyclotella meneghiniana Kützing	0	1	1	0	1	0	1	0	P-B	temp	st	alf	hl	sp	a-o	2.8	e	hne
Cymbella cistula (Ehrenberg) O. Kirchner	0	0	0	0	1	0	0	0	B	-	st-str	alf	i	sx	o	1.2	e	ats
Cymbella helvetica Kützing	0	0	0	0	0	1	0	0	B	-	str	ind	i	-	o-x	0.6	o-m	-
Diatoma vulgaris Bory	0	0	1	0	0	0	0	0	P-B	-	st-str	ind	i	sx	b	2.2	me	ate
Discostella stelligera (Cleve and Grunow) Houk and Klee	1	1	1	1	1	1	1	1	P	-	-	ind	i	-	o-b	1.4	o-m	-
Epithemia adnata (Kützing) Brébisson	0	0	0	0	0	0	0	1	B	temp	st	alb	i	sx	o	1.2	me	ats
Eunotia arcus Ehrenberg	0	0	0	1	0	0	0	0	B	-	st-str	acf	i	-	x-o	0.5	ot	ats
Fragilaria capucina Desmazières	0	1	0	0	0	0	0	0	P-B	-	-	ind	i	es	b-o	1.6	m	-
Fragilaria radians (Kützing) D.M.Williams and Round	1	1	1	1	1	1	1	1	P-B	-	st-str	alf	i	sx	b-o	1.7	o-m	-
Gomphonella olivacea (Hornemann) Rabenhorst	0	1	1	1	1	1	1	1	B	-	st-str	alf	i	es	o-b	1.45	e	ate

Table A2. *Cont.*

Taxa	1	2	3	4	5	6	7	8	Hab	T	Oxy	pH	Sal	D	Sap	Index S	Tro	Aut-Het
Bacillariophyta																		
Gomphonema acuminatum Ehrenberg	1	0	0	0	0	0	0	0	B	-	st	ind	i	es	o-b	1.4	o-m	ats
Eunotia arcus Ehrenberg	0	0	0	1	0	0	0	0	B	-	st-str	acf	i	-	x-o	0.5	ot	ats
Fragilaria capucina Desmazières	0	1	0	0	0	0	0	0	P-B	-	-	ind	i	es	b-o	1.6	m	-
Fragilaria radians (Kützing) D.M. Williams and Round	1	1	1	1	1	1	1	1	P-B	-	st-str	alf	i	sx	b-o	1.7	o-m	-
Gomphonella olivacea (Hornemann) Rabenhorst	0	1	1	1	1	1	1	1	B	-	st-str	alf	i	es	o-b	1.45	e	ate
Gomphonema acuminatum Ehrenberg	1	0	0	0	0	0	0	0	B	-	st	ind	i	es	o-b	1.4	o-m	ats
Gomphonema gracile Ehrenberg	0	1	0	0	0	0	0	0	B	temp	st	alf	i	es	x-b	0.8	m	ats
Gyrosigma strigilis (W. Smith) J.W. Griffin and Henfrey	0	0	1	0	0	0	0	0	B	-	-	-	mh	-	-	-	-	-
Halamphora veneta (Kützing) Levkov	0	0	0	1	0	1	0	0	B	-	st-str	alf	i	es	a-o	2.6	e	ate
Lindavia comta (Kützing) Nakov, Gullory, Julius, Theriot, and Alverson	0	1	0	0	0	1	0	0	P	-	st	alf	i	sx	o	1.2	o-m	-
Melosira varians C.Agardh	1	1	1	1	1	1	1	1	P-B	temp	st-str	ind	hl	es	b	2.1	me	hne
Pauliella taeniata (Grunow) Round and Basson	0	0	0	0	1	0	1	0	B	-	-	alf	mh	-	b	2.0	-	-
Pinnularia viridis (Nitzsch) Ehrenberg	0	0	0	0	0	0	1	0	P-B	temp	st-str	ind	i	es	x	0.3	o-e	ate
Sellaphora pupula (Kützing) Mereschkovsky	0	0	0	0	0	0	0	1	B	eterm	st	ind	hl	sx	o-a	1.9	me	ate
Staurosira leptostauron (Ehrenberg) Kulikovskiy and Genkal	0	0	0	0	0	1	0	0		-	-	-	-	-	-	1.1	-	-
Stephanodiscus hantzschii Grunow	1	0	1	1	1	1	1	1	P	temp	st	alf	i	es	a-o	2.7	o-m	hne
Surirella elegans Ehrenberg	0	1	0	0	0	0	0	0	P-B	-	str	alf	i	-	o	1	me	-
Ulnaria acus (Kützing) Aboal	1	0	0	0	0	1	0	0	P-B	-	st-str	alf	i	es	o-a	1.85	o-m	-
Ulnaria amphirhynchus (Ehrenberg) Compère and Bukhtiyarova	0	1	0	0	0	0	0	0	P-B	-	-	alf	i	es	b	2	o-m	-
Ulnaria capitata (Ehrenberg) Compère	0	0	0	0	0	0	1	0	P-B	-	st-str	alf	i	es	o-b	1.5	e	ats
Euglenozoa																		
Lepocinclis ovum (Ehrenberg) Lemmermann	0	0	0	1	1	0	1	1	P	eterm	st	ind	i	-	b-a	2.4	-	-
Trachelomonas hispida (Perty) F.Stein	0	0	1	0	1	0	0	0	P-B	eterm	st-str	-	i	-	b	2.2	-	-
Miozoa																		
Ceratium hirundinella (O.F. Müller) Dujardin	1	1	1	0	1	1	1	1	P	-	st-str	-	i	-	o	1.3	-	-
Gymnodinium variabile E.C. Herdman	0	0	0	0	0	0	1	0	P	-	-	-	-	-	o-b	1.5	-	-
Naiadinium polonicum (Woloszynska) Carty	0	1	1	1	0	0	0	1	P	-	st	-	-	-	o	1.3	-	-

Table A2. *Cont.*

Taxa	1	2	3	4	5	6	7	8	Hab	T	Oxy	pH	Sal	D	Sap	Index S	Tro	Aut-Het
Miozoa																		
Peridinium bipes F. Stein	0	0	0	0	0	0	0	1	P	-	st-str	-	oh	-	o	1.3	-	-
Peridinium cinctum (O.F. Müller) Ehrenberg	0	0	1	0	0	0	0	0	P-B	-	st-str	-	i	-	b-o	1.6	-	-
Ochrophyta																		
Dinobryon divergens O.E. Imhof	1	1	1	1	1	1	1	1	P	-	st-str	ind	i	-	o-b	1.45	-	-
Chlorophyta																		
Binuclearia lauterbornii (Schmidle) Proschkina-Lavrenko	1	0	0	0	0	0	0	0	-	-	-	-	-	-	o-a	1.8	-	
Chlorella vulgaris Beyerinck [Beijerinck]	1	1	1	0	1	1	1	1	P-B, pb,S	-	-	-	hl	-	a	3.1	-	
Desmodesmus brasiliensis (Bohlin) E.Hegewald	0	0	0	0	1	1	1	1	P-B	-	st-str	-	-	-	b	2	-	
Monoraphidium contortum (Thuret) Komárková-Legnerová	1	0	1	0	0	0	0	0	P-B	-	st-str	-	i	-	b	2.2	-	
Monoraphidium convolutum (Corda) Komárková-Legnerová	0	0	0	0	0	1	0	1	P-B	-	st-str	-	-	-	b	2.3	-	
Monoraphidium minutum (Nägeli) Komárková-Legnerová	0	0	0	0	0	0	0	1	P-B	-	st-str	-	i	-	b-a	2.5	-	
Mucidosphaerium pulchellum (H.C. Wood) C.Bock, Proschold and Krienitz	0	0	0	0	1	0	0	0	P-B	-	st-str	ind	i	-	b	2.3	-	
Myrmecia irregularis (J.B. Petersen) Ettl and Gärtner	1	0	1	0	1	1	1	1	P	-	-	-	oh	-	-	-	-	
Neglectella solitaria (Wittrock) Stenclová and Kastovsky	0	0	1	0	1	0	0	0	P	-	st	ind	i	-	b-o	1.7	-	
Oocystis borgei J.W. Snow	0	1	0	0	0	0	0	0	P-B	-	st-str	ind	i	-	o-a	1.9	-	
Oocystis pusilla Hansgirg	0	0	0	0	0	0	1	1	P	-	-	-	oh	-	o-b	1.4	-	
Pseudodidymocystis planctonica (Korshikov) E.Hegewald and Deason	0	1	0	0	0	1	1	0	-	-	-	-	-	-	o-a	1.8	-	
Scenedesmus quadricauda (Turpin) Brébisson	1	1	0	1	1	1	1	1	P	-	-	ind	i	-	b	2.1	-	
Schroederia setigera (Schröder) Lemmermann	1	0	0	0	0	0	0	0	P	-	st-str	-	i	-	b-o	1.7	-	
Tetradesmus obliquus (Turpin) M.J.Wynne	0	1	1	0	1	0	1	1	P-B, S	-	st	-	i	-	b-a	2.4	-	
Tetraëdron minimum (A. Braun) Hansgirg	1	1	1	1	0	1	1	1	P-B	-	st-str	-	i	-	b	2.1	-	
Tetraëdron minutissimum Korshikov	0	0	1	0	1	0	0	1	P-B	-	st-str	-	i	-	b	2.1	-	
Tetrastrum staurogeniiforme (Schröder) Lemmermann	0	0	0	0	0	1	0	0	P-B	-	st-str	-	i	-	b	2.2	-	
Charophyta																		
Cosmarium baileyi Wolle	1	1	1	1	1	1	1	1	B	-	-	-	-	-	o	1.2	-	-
Cosmarium undulatum Corda ex Ralfs	1	0	0	0	0	0	0	0	P-B	-	-	acf	i	-	-	-	m	-

Table A3. Distribution of species richness in phyla, total species number in community, and ecological properties of bioindicators for sampling stations on Lake Borovoe, 2019. Abbreviations: B, benthic; P-B, plankto-benthic; P, planktonic; temp, inhabitants of temperate-temperature waters; eterm, eurythermic; str, inhabitants of streaming well-oxygenated waters; st-str, inhabitants of standing to streaming middle-oxygenated waters; st, inhabitants of standing low oxygenated waters; H2S, sulfides anoxia indicators; acf, acidophilic; ind, pH-indifferent; alf, alkaliphilic; alb, alkalibiontes; i, chlorides indifferents; hl, halophiles; mh, mesohalobes; sx, saproxenes; es, eurysaprobes; sp, saprophiles; Classes 1–4 of water-quality indicators according to Index S; o, oligotraphentes; o-m, oligo-mesotraphentes; m, mesotraphentes; me, meso-eutraphentes; e, eutraphentes; o-e, oligotraphentes to eutraphentes; ats, autotrophes-inhabited low-nitric-nitrogen waters; ate, autotrophes-inhabited high-nitric-nitrogen waters; hne, facultative-heterotrophes-inhabited low organically enriched waters.

Station	1	2	3	4	5	6	7	8
Species richness in phyla								
Bacillariophyta	7	11	8	10	9	11	9	7
Chlorophyta	7	6	7	2	8	8	8	10
Charophyta	2	1	1	1	1	1	1	1
Cyanobacteria	8	4	7	5	5	7	6	5
Euglenozoa	0	0	1	1	2	0	1	1
Miozoa	1	2	3	1	1	1	2	3
Ochrophyta	1	1	1	1	1	1	1	1
Total Species number	26	25	28	21	27	29	28	28
Substrate								
B	2	3	3	6	5	4	3	4
P-B	12	12	9	6	10	9	10	9
P	11	9	14	9	12	13	14	15
Temperature								
temp	3	4	3	5	4	2	4	3
eterm	0	0	1	1	2	1	1	2
Water moving and oxygenation								
str	0	1	0	0	0	1	0	0
st-str	10	9	11	11	11	13	9	11
st	2	5	5	3	5	2	4	6
H$_2$S	1	0	0	0	0	0	0	0
pH								
acf	1	0	0	1	0	0	0	0
ind	6	7	5	8	9	6	6	6
alf	3	7	4	5	6	6	6	3
alb	0	0	0	0	0	0	0	1
Salinity								
i	15	16	17	14	17	17	15	16
hl	8	6	5	4	5	4	5	5
mh	0	0	1	0	1	0	1	0
Watanabe								
sx	1	2	2	2	2	2	1	3
es	5	6	3	6	4	6	5	3
sp	0	1	1	0	1	0	1	0
Class of water quality based on species-specific index saprobity S								
Class 1	0	0	0	1	0	0	1	0
Class 2	6	10	7	8	7	9	8	9
Class 3	12	12	13	9	14	13	13	14
Class 4	3	2	2	3	3	3	3	2

Table A3. *Cont.*

Station	1	2	3	4	5	6	7	8
Trophic state								
o	1	1	1	1	0	1	0	0
o-m	6	4	3	4	4	6	3	3
m	2	2	0	0	0	0	0	0
me	4	4	7	6	5	4	5	6
e	3	4	3	4	4	3	4	2
o-e	0	0	0	0	0	1	1	0
Autotrophy–heterotrophy								
ats	1	1	0	2	2	0	1	1
ate	1	2	2	4	1	3	2	2
hne	2	2	3	2	3	2	3	2

References

1. Kurmanbaeva, A.S.; Khusainov, A.T.; Zhumai, E. Ecological condition of the Burabay lake of the Burabay State National Natural Park. *Sci. News Kazakhstan* **2019**, *3*, 202–209. (In Russian)
2. IUCN. Available online: https://www.iucn.org/about-iucn (accessed on 24 October 2022).
3. Goryunova, A.I.; Danko, E.K. (Eds.) *Lake Fund of Kazakhstan. Section 1. Lakes of the Kokchetav region (within the borders of 1964–1999)*; Bastau Publishing LLP: Almaty, Kazakhstan, 2009; 70p. (In Russian)
4. Skakun, V.A.; Kiseleva, V.A.; Goryunova, A.I. The ecosystem of Lake Borovoy and the possibilities of its transformation. *Selevinia* **2002**, *1–4*, 249–264. (In Russian)
5. Krupa, E.G. Zooplankton of the lakes of the Shchuchinsko-Borovsk system (northern Kazakhstan) as an indicator of their trophic status. *Terra* **2007**, *3*, 60–66. (In Russian)
6. Ismailova, A.A.; Zhamankara, A.K.; Akbaeva, L.K.; Adamov, A.A.; Abakumov, A.I.; Tulegenov, S.A.; Muratov, R.M. Hydro-chemical and hydrobiological indicators as characteristics of the ecological state of lakes (on the example of Burabay and Ulken Shabakty lakes). *Bull. Kazakh Natl. Univ. Biol. Ser.* **2013**, *3*, 503–507. (In Russian)
7. Voronikhin, N.N. To the algae flora of the Kazakh SSR. *Bull. Acad. Sci. Kazakh SSR Alma-Ata* **1947**, *30*, 70–73. (In Russian)
8. Voronikhin, N.N. About some algae of the Borovsky Reserve. *Proc. All-Union Hydrobiol. Soc.* **1951**, *3*, 5–52. (In Russian)
9. Voronikhin, N.N.; Krasnoperova, L.A. Zignemous algae of the Borovsky Reserve (Kokchetav region). *News Taxon. Low. Plants Leningr.* **1970**, *7*, 46–63. (In Russian)
10. Demchenko, L.A. Aquatic vegetation of the lake Borovoe. *Proc. State Nat. Reserve Borovoe* **1948**, *1*, 52–61. (In Russian)
11. Krasnoperova, L.A. Zignemous algae of Kazakhstan. (Chlorophyta: Zygnematales). Ph.D. Biol. Sciences, Leningrad State University, Leningrad, Russia, 1971. (In Russian).
12. Ismagulova, A.Z. Green algae of reservoirs of the Borovoe nature reserve (Kaz. SSRP). Proceedings of 1st International Conference "Actual Problems of Modern Algology", Cherkasy, Russia, 23–25 September 1987; Naukova Dumka: Kyiv, Ukraine, 1987; pp. 65–66. (In Russian).
13. Ismagulova, A.Z. Green algae of water bodies of the Kokchetav region. In *Botanicheskie Materialy Gerbariya Botanicheskogo Instituta Botaniki Akademii Nauk Kazakhskoi S S R. Alma Ata*; Institute of Botany of the Academy of Sciences of Kaz. SSR Press: Alma-Ata, Kazakhstan, 1989; pp. 119–133. (In Russian)
14. Sviridenko, B.F. *Flora and Vegetation of Water Bodies in Northern Kazakhstan*; Publishing house of the Omsk State Pedagogical University: Omsk, Russia, 2000; p. 196. (In Russian)
15. Abiev, S.A.; Nurashov, S.B.; Sametova, E.S.; Toleuzhanova, A.T. Algoflora of Kokshetau—Borovsky lake systems. In Proceedings of the International Scientific Conference "Modern Trends in the Study of the Flora of Kazakhstan and Its Protection" Almaty, Republic of Kazakhstan, 24–26 April 2014; Institute of Botany of the Academy of Sciences of Kaz. SSR Press: Almaty, Kazakhstan, 2014; pp. 206–209. (In Russian).
16. Dembowska, E.A.; Mieszczankin, T.; Napiórkowski, P. Changes of the phytoplankton community as symptoms of deterioration of water quality in a shallow lake. *Environ. Monit. Assess.* **2018**, *190*, 95. [CrossRef]
17. Aazami, J.; Sari, A.E.; Abdoli, A.; Sohrabi, H.; Van den Brink, P.J. Assessment of ecological quality of the Tajan river in Iran using a multimetric macroinvertebrate index and species traits. *Environ. Manag.* **2015**, *56*, 260–269. [CrossRef]
18. Svensson, O.; Bellamy, A.S.; Van den Brink, P.J.; Tedengren, M.; Gunnarsson, J.S. Assessing the ecological impact of banana farms on water quality using aquatic macroinvertebrate community composition. *Environ. Sci. Pollut. Res.* **2018**, *25*, 13373–13381. [CrossRef]
19. Qin, B.Q.; Gao, G.; Zhu, G.W.; Zhang, Y.L.; Song, Y.Z.; Tang, X.M.G.; Xu, H.; Deng, J.M. Lake eutrophication and its ecosystem response. *Chin. Sci. Bull. March* **2013**, *58*, 961–970. [CrossRef]

20. Krupa, E.G.; Barinova, S.S.; Romanova, S.M.; Khitrova, E.A. *Hydrochemical and Hydrobiological Characteristics of the Lakes of the Shchuchinsko-Borovsk Resort Zone (Northern Kazakhstan) and the Main Methodological Approaches to Assessing the Ecological State of Small Water Bodies*; Republican State Enterprise "Institute of Zooology" of The Science Committee of The Ministry of Education and Science of The Republic of Kazakhstan LLP "Kazakhstan Agency of Applied Ecology" Publishing house Etalon Print LLP: Almaty, Kazakhstan, 2021; p. 300. (In Russian)

21. Barinova, S.; Bilous, O.; Ivanova, N. New Statistical Approach to Spatial Analysis of Ecosystem of the Sasyk Reservoir, Ukraine. *Int. J. Ecotoxicol. Ecobiol.* **2016**, *1*, 118–126. [CrossRef]

22. Barinova, S. Ecological mapping in application to aquatic ecosystems bioindication: Problems and methods. *Int. J. Environ. Sci. Nat. Resour.* **2017**, *3*, 1–7. [CrossRef]

23. Krupa, E.G.; Barinova, S.S.; Ponamareva, L.; Tsoy, V.N. Statistical mapping and 3-D surface plots in phytoplankton analysis of the Balkhash Lake (Kazakhstan). *Transylv. Rev. Syst. Ecol. Res. Wetl. Divers.* **2018**, *20*, 1–16. [CrossRef]

24. Krupa, E.; Romanova, S.; Berkinbaev, G.; Yakovleva, N.; Sadvakasov, E. Zooplankton as Indicator of the Ecological State of Protected Aquatic Ecosystems (Lake Borovoe, Burabay National Nature Park, Northern Kazakhstan). *Water* **2020**, *12*, 2580. [CrossRef]

25. Romanov, R.E.; Zhamangara, A. Pre-Symposium Field Excursion Report: Results in context of regional charophyte knowledge. *IRGC News* **2017**, *28*, 11–13.

26. Uteshev, A.S. *Climate of Kazakhstan*; Hydrometeorological Publishing House: Leningrad, Russia, 1959; p. 371.

27. Abakumov, V.A. (Ed.) *Guidance on Methods of Hydrobiological Analysis of Surface Waters and Bottom Sediments*; Gidrometeoizdat: Leningrad, Russia, 1983; 240p. (In Russian)

28. Kiselev, I.A. Research methods of plankton. In *Life of the Fresh Water of the USSR*; Pavlovsky, E.N., Zhadin, V.I., Eds.; USSR Academy of Sciences: Moscow/Leningrad, Russia, 1956; pp. 188–253. (In Russian)

29. Kiselev, I.A. *Plankton of the Seas and Continental Water Bodies*; Nauka: Leningrad, Russia, 1969; p. 657. (In Russian)

30. Semenov, A.D. (Ed.) *Guidance on the Chemical Analysis of Land Surface Water*; Gidrometeoizdat: Leningrad, Russia, 1977; p. 541. (In Russian)

31. Komárek, J.; Anagnostidis, K. Cyanoprokaryota. Teil 1: Chroococcales. In *Süsswasser flora von Mitteleuropa 19/1*; Ettl, H., Gärtner, G., Heynig, H., Mollenhauer, D., Eds.; Gustav Fisher: Jena, Stuttgart, Lübeck, Ulm, Germany, 1998.

32. Komárek, J.; Anagnostidis, K. Cyanoprokaryota. Teil 2: Oscillatoriales. In *Süsswasserflora von Mitteleuropa 19/2*; In Büdel, B., Gärtner, G., Krienitz, L., Schagerl, M., Eds.; Elsevier: München, Germany, 2005; p. 759.

33. Krammer, K.; Lange-Bertalot, H. *Bacillariophyceae. 2. Bacillariaceae, Epithemiaceae, Surirellaceae. Süßwasserflora von Mitteleuropa, 2/2*; G. Fischer: Stuttgart/Jena, Germany, 1988; p. 596.

34. Krammer, K.; Lange-Bertalot, H. *Bacillariophyceae. 3. Centrales, Fragilariaceae, Eunotiaceae. Süsswasserflora von Mitteleuropa 2/3*; G. Fischer: Stuttgart/Jena, Germany, 1991; p. 576.

35. Krammer, K.; Lange-Bertalot, H. *Bacillariophyceae. 4. Achnanthaceae, Kritische Ergänzungen zu Navicula (Lineolatae) und Gomphonema Gesamtliteraturverzeichnis. Teil 1–4. Süßwasserflora von Mitteleuropa 2/4*; G. Fischer: Stuttgart/Jena, Germany, 1991; p. 437.

36. John, D.M.; Whitton, B.A.; Brook, A.J. (Eds.) *The Freshwater Algal Flora of the British Isles: An Identification Guide to Freshwater and terrestrial Algae*; Cambridge University Press: Cambridge, UK, 2011; p. 878.

37. Guiry, M.D.; Guiry, G.M.; AlgaeBase World-Wide Electronic Publication. National University of Ireland, Galway. Available online: http://www.algaebase.org (accessed on 24 June 2022).

38. Magurran, A.E. *Ecological Diversity and Its Measurements*; Princeton University Press: Princeton, NJ, USA, 1988; p. 192. [CrossRef]

39. Sládeček, V. Diatoms as indicators of organic pollution. *Acta Hydroch. Hydrobiol.* **1986**, *14*, 555–566. [CrossRef]

40. Barinova, S.S.; Medvedeva, L.A.; Anissimova, O.V. *Diversity of Algal Indicators in Environmental Assessment*; Pilies Studio: Tel Aviv, Israel, 2006; p. 495. Available online: https://drive.google.com/drive/folders/1FNW2oIMkvujsouuL9GF4c478k99qMPho (accessed on 30 August 2022). (In Russian)

41. Barinova, S.; Bilous, O.; Tsarenko, P.M. *Algae Indication of Waterbodies in Ukraine: Methods and Perspectives*; Haifa University Publishing House: Haifa, Kiev, 2019; pp. 297–367. Available online: https://drive.google.com/drive/folders/1K3U-eEo0aYRu0s84C9XNtKePY9jK32yC (accessed on 30 August 2022). (In Russian)

42. Barinova, S. Essential and practical bioindication methods and systems for the water quality assessment. *Int. J. Environ. Sci. Nat. Resour.* **2017**, *2*, 555588. [CrossRef]

43. Love, J.; Selker, R.; Marsman, M.; Jamil, T.; Dropmann, D.; Verhagen, A.J.; Ly, A.; Gronau, Q.F.; Smira, M.; Epskamp, S.; et al. JASP: Graphical statistical software for common statistical designs. *J. Stat. Softw.* **2019**, *88*, 1–17. [CrossRef]

44. Novakovsky, A.B. Abilities and base principles of program module "GRAPHS". *Sci. Rep. Komi Sci. Cent. Ural. Div. Russ. Acad. Sci.* **2004**, *27*, 28. (In Russian)

45. Ter Braak, C.J.F.; Šmilauer, P. *CANOCO Reference Manual and CanoDraw for Windows User's Guide: Software for Canonical Community Ordination (Version 4.5)*; Microcomputer Power Press: Ithaca, NY, USA, 2002; p. 500.

46. Bragin, E.A.; Bragina, T.M. (Eds.) Wetlands in system of nature complexes of Northern Kazakhstan. General characteristic and nature conservation importance. In *The Most Important Wetlands of Northern Kazakhstan (In Limits of Kostanay and West Part of North-Kazakhstan Regions)*; WWF series, issue 5; Russian University Press: Moscow, Russia, 2002; pp. 19–38. (In Russian)

47. Reynolds, C.S. *The Ecology of Phytoplankton*; Cambridge University Press: Cambridge, UK, 2006; p. 535.

48. Barinova, S.S.; Nevo, E.; Bragina, T.M. Ecological assessment of wetland ecosystems of northern Kazakhstan on the basis of hydrochemistry and algal biodiversity. *Acta Bot. Croat.* **2011**, *70*, 215–244. [CrossRef]
49. Barinova, S.S.; Bragina, T.M.; Nevo, E. Algal species diversity of arid region lakes in Kazakhstan and Israel. *Community Ecol.* **2009**, *10*, 7–16. [CrossRef]
50. Barinova, S.S.; Romanov, R.E. Towards an inventory of algal diversity of the Zerenda Lake, Northern Kazakhstan. In *Biological Diversity of Asian Steppe, Proceedings of the III International Scientific Conference, Kostanay, Kazakhstan, 24–27 April 2017*; Abil, E.A., Bragina, T.M., Eds.; KSPI: Kostanay, Kazakhstan, 2017; pp. 139–144.
51. Barinova, S.; Krupa, E.; Tsoy, V.; Ponamareva, L. The application of phytoplankton in ecological assessment of the Balkhash Lake (Kazakhstan). *Appl. Ecol. Environ. Res.* **2018**, *16*, 2089–2111. [CrossRef]
52. Krupa, E.G.; Barinova, S.M.; Romanova, S.M.; Malybekov, A.B. Hydrobiological assessment of the high mountain Kolsay Lakes (Kungey Alatau, Southeastern Kazakhstan) ecosystems in climatic gradient. *Br. J. Environ. Clim. Change* **2016**, *6*, 259–278. [CrossRef]
53. Krupa, E.G.; Barinova, S.S.; Amirgaliyev, N.A.; Issenova, G.; Kozhabayeva, G. Statistical approach to estimate the anthropogenic sources of potentially toxic elements on the Shardara Reservoir (Kazakhstan). *MOJ Eco. Environ. Sci.* **2017**, *2*, 8–14. [CrossRef]
54. Barinova, S.; Gabyshev, V.; Boboev, M.; Kukhaleishvili, L.; Bilous, O. Algal indication of climatic gradients. *Am. J. Environ. Protection. Spec. Issue: Appl. Ecol. Probl. Innov.* **2015**, *4*, 72–77. [CrossRef]
55. IUCN. *IUCN Red List Categories and Criteria*, Version 3.1, 2nd ed.; IUCN: Gland, Switzerland; Cambridge, UK, 2012; p. 32.
56. Muylaert, K.; Pérez-Martínez, C.; Sánchez-Castillo, P.; Lauridsen, T.L.; Vanderstukken, M.; Declerck, S.A.J.; Van der Gucht, K.; Conde-Porcuna, J.M.; Jeppesen, E.; De Meester, L.; et al. Influence of nutrients, submerged macrophytes and zooplankton grazing on phytoplankton biomass and diversity along a latitudinal gradient in Europe. *Hydrobiologia* **2010**, *653*, 79–90. [CrossRef]

Article

Beyond the Hydro-Regime: Differential Regulation of Plant Functional Groups in Seasonal Ponds

Jasmine Rios, Melanie Dibbell, Emely Flores and Jamie M. Kneitel *

Department of Biological Sciences, California State University—Sacramento, Sacramento, CA 95819-6077, USA; jasminerios2@csus.edu (J.R.); mjdibbell@csus.edu (M.D.); emelyflores@csus.edu (E.F.)
* Correspondence: kneitel@csus.edu

Abstract: Plant community assembly can be influenced by many environmental factors. In seasonal wetlands, most studies focus on the considerable variation that occurs from the hydro-regime (patterns of inundation and desiccation). Other factors and their interactions also influence plants but are seldom studied, including nutrient and thatch levels. Furthermore, the responses of aquatic and terrestrial functional groups can provide important insights into patterns of cover and richness. The aim of this study was to evaluate how algae and plant functional groups (aquatic and terrestrial plants) respond to variation in hydro-regime (stable and unstable), nutrient addition (none and added), and thatch (none, native plants, and exotic plants) addition. We measured algal cover, total species richness, and the cover of the functional groups over two years. Algal cover increased with unstable hydroperiods and the addition of nutrients. Algae were also negatively associated with aquatic plant cover and positively associated with terrestrial plant cover. Aquatic plant cover increased with a stable hydro-regime and decreased with increased thatch. Terrestrial plant cover increased with an unstable hydro-regime and decreased with thatch addition. Thatch accumulation and excess nutrients can be linked to human activities, which directly and indirectly alter plant community composition. The interactions of these factors with the hydro-regime should be considered when evaluating a plant community's response to changing environmental conditions. Overall, these results are necessary for the conservation and management of essential wetland functions and services.

Keywords: algae; California vernal pools; functional groups; macrophytes; mesocosms; seasonality; species richness

Citation: Rios, J.; Dibbell, M.; Flores, E.; Kneitel, J.M. Beyond the Hydro-Regime: Differential Regulation of Plant Functional Groups in Seasonal Ponds. *Diversity* 2023, *15*, 832. https://doi.org/10.3390/d15070832

Academic Editors: Mateja Germ, Igor Zelnik and Matthew Simpson

Received: 17 May 2023
Revised: 26 June 2023
Accepted: 27 June 2023
Published: 2 July 2023

1. Introduction

Plant recruitment and community assembly are commonly influenced by changes in environmental conditions [1–3]. These effects can include direct or indirect abiotic (e.g., soil moisture, light) and biotic (e.g., competition and herbivory) interactions that can vary in space and time [4]. Different environmental effects on individual species can also lead to changes in plant species composition, richness, and cover [5,6]. These differences in responses can stem from differences in life history traits (e.g., timing of germination and growth) that interact with the environment to facilitate species coexistence [7]. Therefore, plant functional groups may respond differently to the environment, which results in observed patterns of cover and richness [5–8]. The presence of extreme conditions (natural and human-mediated) provides an opportunity to examine questions related to species traits that can affect species distributions [4,8]. This situation is found in many habitats worldwide, including deserts, high-elevation locations, and wetlands.

In wetland habitats, the hydro-regime (inundation timing, duration, and frequency) is the principal factor determining plant community structure [9–12]. Species traits (e.g., life history, growth structure, and phenology of germination, growth, and flowering) all affect how plants respond to the hydro-regime [4,8,11–16]. For example, individual species from many taxonomic groups exhibited variation in germination and growth under different hydroperiod timings and lengths in greenhouse studies [8,17]. The variation in response to the environment, combined with species interactions, results in what is observed in the community composition [2,4,11,17]. These patterns also hold for seasonal wetland habitats, including vernal pools found in California.

Seasonal wetland habitats, including California vernal pools, consist of species adapted to wet and dry cycles [18–21]. Inundation timing and length affect plant germination and growth [8,17] and result in the distribution of plants within a vernal pool basin along an inundation gradient [4,18,22]. Further, long hydroperiods have been found to prevent the establishment of competitively dominant introduced species, which are invasive in California vernal pools [17]. Vernal pool plant communities exhibit patterns that include annual flood-tolerant species near flood zones and competitive species along edge zones of the vernal pool-grassland ecotone [20]. Changes in hydro-period can create a year-to-year variation that can carry across aquatic and terrestrial phases and determine plant community composition, richness, and cover [16,20,23,24].

Shifts in plant community composition can alter above-ground biomass and lead to thatch deposition [25–27]. Plant thatch affects the germination of native forbs and grasses by affecting light availability and soil temperature in vernal pools [25,27]. This results in overall declines in species richness and diversity. Changes in nutrient cycling (e.g., decomposition rates) and decreased inundation periods and pool depths have also been linked to increased thatch, making vernal pool habitats more susceptible to plant invasions, drastically affecting existing communities [27,28].

Nutrient levels in vernal pools can greatly vary and depend on different factors, including soil properties and human activities [24,29–31]. California vernal pools tend to be oligotrophic (i.e., have low nutrient levels) but are commonly surrounded by human-modified habitats, primarily urban and agricultural development. The modified habitats export excessive nutrients into surrounding ecosystems [31], which are known to be detrimental to vernal pools [24,29,30]. For example, it can lead to algal blooms and decreased richness of plants and invertebrates [24,29–31]

Many studies have investigated the individual effects of hydroperiod, thatch, and nutrient concentrations but have not looked at the potential for additive or interactive effects on plant community composition. Specifically, we conducted a two-year mesocosm experiment that measured the cover and richness of aquatic and terrestrial plants in response to different treatments of hydro-regime (stable and unstable), thatch (control, native, or exotic plant addition), and nutrient (control or nitrogen and phosphorus addition), for a total of 12 experimental treatments. We hypothesized that aquatic plants will respond more to hydroperiod and nutrients [27,28,32] because they directly interact with these treatments, while terrestrial plants will be more negatively affected by thatch and its shading effects during desiccation [25,27,33,34].

2. Materials and Methods

2.1. Experimental Design

Soil from vernal pool complexes in the Elder Creek Watershed and Gill Ranch in Sacramento County was used for the mesocosm experiment [20,24,29,30]. The top 6 cm of soil was collected to ensure the presence of a viable seed bank [34]. Dry soil was homogenized with a cement mixer to intersperse the seed bank.

The study occurred from December 2014 to May 2016. In December 2014, 48 mesocosms (plastic containers with volume = 151 L; mesocosm dimensions: diameter = 1.03 × 0.36 m, height = 0.72 m, area = 0.37 m^2) were established outdoors at the California State University, Sacramento Arboretum. Approximately 7 kg of the homogenized soil was added, leaving an approximately 2 cm depth of soil. Mesocosms were left uncovered over the course of the study. Well water was used to supplement natural rainfall to fill the entire mesocosm. This water was used because it is easily accessible at the site and lacked any chlorinated treatment. Nutrient levels were not measured but all treatments received approximately the same amount of supplemented water. Previous mesocosm studies using this water have found negligible nutrient increases in control treatments [30]. The timing of the treatments and sampling was the same in both years.

Twelve treatments were randomly assigned to four blocks of mesocosms in a randomized block design. The experimental design was a full factorial $2 \times 2 \times 3$ that consisted of hydroperiod (stable and unstable), nutrient addition (control and addition), and thatch (control, native plant thatch, and exotic plant thatch). A stable hydroperiod treatment was inundated for 20 weeks (December–May) and the unstable treatment was 2 short hydroperiods that consisted of a 9-week inundation (December–March), which was desiccated and kept dry for 2 weeks and then refilled for another 9 weeks (March–May). Desiccation at the end of the experiment occurred naturally, but approximately 20 L was removed twice weekly over the last 2 weeks. Removed water was poured through a net (0.2 mm mesh) and all filtered individuals (e.g., seed and eggs) were returned to the mesocosm because we only wanted to remove water to simulate the natural desiccation of vernal pools. To ensure the water removal did not create an added disturbance, the long hydroperiod treatments had the same procedure applied to them, but water was added to refill the mesocosm.

Nutrient addition treatments included nitrogen and phosphorus addition via an aqueous solution of $NaNO_3$ and KH_2PO_4. A 0.5 mg/L concentration was used for both N and P and added every two weeks. Plant thatch treatments included control, exotic plant thatch, and native plant thatch. Aboveground plant vegetation from vernal pools and adjacent upland habitat was collected from Mather Field (Sacramento County, CA, USA). Plants were haphazardly collected and included typical native and exotic plants from vernal pools. Native species included *Eleocharis macrostachya*, *Eryngium castrense*, and *Plagiobothrys stipitatus*, and exotics included *Erodium botys*, *Avena* spp., and *Hordeum* spp. [35]. Fifty grams of dried vegetation (thatch) were added to the appropriate treatment replicates prior to inundation.

2.2. Plant Sampling

Plant cover and richness were sampled at the end of the experiment. Percent cover was measured as the percent of mesocosm area covered. The algal cover was measured two weeks prior to the complete desiccation of mesocosms, while plant cover and richness were measured following desiccation. This allowed plants to complete growth, flower, and fruit. Nine macroalgae and two algal (multicellular) species were present (Table 1). Most of the macroalgae were vernal pool endemic species. Species were categorized into their hydro-phase affiliation (aquatic or terrestrial) based on when most of their growth and flowering occurred (Table 1).

Table 1. Species list with each characteristic of habitat affiliation, growth form, and phase affiliation.

Species	Habitat Distribution	Hydro-Phase Affiliation
Cladophora sp. [1]	Widespread	Aquatic
Nitella sp. [1]	Widespread	Aquatic
Eleocharis macrostachya	Endemic	Aquatic
Callitriche marginata	Endemic	Aquatic
Ranunculus aquatilis	Widespread	Aquatic
Marsilea vestita	Widespread	Aquatic
Plagiobothrys stipitatus	Endemic	Terrestrial
Downwingia bicornuta	Endemic	Terrestrial
Gratiola ebracteata	Endemic	Terrestrial
Navarretia leucocephala	Endemic	Terrestrial
Psilocarphus brevissimus	Endemic	Terrestrial

[1] Macroalgae.

2.3. Statistical Analysis

To assess the effects of treatments on the richness and cover (total) and functional group cover (aquatic and terrestrial affiliated plants) of each of the species, the General Linearized Model (GLMM) (gamma distribution with log link) was used with the year (random factor), hydro-regime, thatch addition, and nutrient addition as independent variables in a fully factorial model. All richness and densities were ln-transformed. Correlations among algal cover, total cover and richness, and functional group cover were assessed using partial correlations, which controlled for the year. These analyses were conducted using IBM SPSS, version 24 (IBM Corp., Armonk, NY, USA).

3. Results

3.1. Macroalgae

Cladophora sp. dominated the first year, but *Nitella* sp. invaded during the second year and dominated many of the treatments. Year, hydro-regime, and nutrients were all significant main effects on the algal cover (Figure 1a, Table 2). On average, algal cover in unstable hydroperiods increased by over 69% compared with stable hydroperiods, and nutrients increased algae by 38% on average. Several interactions were also found, including year–hydroperiod, year–thatch, and hydroperiod–nutrient. In the first year, unstable hydro-regime and nutrient addition increased algae, and their coupling resulted in the highest levels of algae. Significant year–hydroperiod–nutrient interactions resulted in nutrient addition increasing algae in unstable hydro-regime treatments in the first year.

Table 2. Generalized linear models for each of the dependent variables.

Dependent Variable	Independent Variable	Wald Chi-Square	df	*p*
Algae				
	Year	54.313	1	<0.001
	Hydroperiod	52.147	1	<0.001
	Thatch	3.170	2	0.205
	Nutrient	5.925	1	0.015
	Year * Hydroperiod	40.840	1	<0.001
	Year * Thatch	12.052	2	0.002
	Year * Nutrient	2.023	1	0.155
	Hydroperiod * Thatch	0.158	2	0.924
	Hydroperiod * Nutrient	5.163	1	0.023
	Thatch * Nutrient	1.985	2	0.371
	Year * Hydroperiod * Thatch	3.295	2	0.193
	Year * Hydroperiod * Nutrient	9.122	1	0.003
	Year * Thatch * Nutrient	6.624	2	0.036
	Hydroperiod * Thatch * Nutrient	3.089	2	0.213
	Year * Hydroperiod * Thatch * Nutrient	8.485	1	0.004
Species richness				
	Year	17.856	1	<0.001
	Hydroperiod	36.815	1	<0.001
	Thatch	39.877	2	<0.001
	Nutrient	13.598	1	<0.001
	Year * Hydroperiod	0.802	1	0.370
	Year * Thatch	2.906	2	0.234
	Year * Nutrient	0.359	1	0.549
	Hydroperiod * Thatch	3.998	2	0.135
	Hydroperiod * Nutrient	0.941	1	0.332
	Thatch * Nutrient	7.599	2	0.022
	Year * Hydroperiod * Thatch	1.999	2	0.368
	Year * Hydroperiod * Nutrient	3.287	1	0.070
	Year * Thatch * Nutrient	1.205	2	0.548
	Hydroperiod * Thatch * Nutrient	5.252	2	0.072
	Year * Hydroperiod * Thatch * Nutrient	1.230	2	0.541

Table 2. *Cont.*

Dependent Variable	Independent Variable	Wald Chi-Square	df	*p*
Total percent cover				
	Year	78.223	1	<0.001
	Hydroperiod	8.152	1	0.004
	Thatch	5.888	2	0.050
	Nutrient	0.010	1	0.922
	Year * Hydroperiod	1.893	1	0.169
	Year * Thatch	0.729	2	0.694
	Year * Nutrient	0.021	1	0.886
	Hydroperiod * Thatch	6.671	2	0.036
	Hydroperiod * Nutrient	2.629	1	0.105
	Thatch * Nutrient	5.555	2	0.062
	Year * Hydroperiod * Thatch	1.825	2	0.401
	Year * Hydroperiod * Nutrient	2.014	1	0.156
	Year * Thatch * Nutrient	3.353	2	0.187
	Hydroperiod * Thatch * Nutrient	1.800	2	0.407
	Year * Hydroperiod * Thatch * Nutrient	2.375	2	0.305
Aquatic plants				
	Year	1.300	1	0.254
	Hydroperiod	5.561	1	0.018
	Thatch	7.684	2	0.021
	Nutrient	0.412	1	0.521
	Year * Hydroperiod	0.192	1	0.661
	Year * Thatch	6.134	2	0.047
	Year * Nutrient	0.753	1	0.386
	Hydroperiod * Thatch	4.378	2	0.112
	Hydroperiod * Nutrient	2.008	1	0.156
	Thatch * Nutrient	1.583	2	0.453
	Year * Hydroperiod * Thatch	2.324	2	0.313
	Year * Hydroperiod * Nutrient	0.228	1	0.633
	Year * Thatch * Nutrient	1.004	2	0.605
	Hydroperiod * Thatch * Nutrient	1.916	2	0.384
	Year * Hydroperiod * Thatch * Nutrient	1.931	2	0.381
Terrestrial plants				
	Year	163.145	1	<0.001
	Hydroperiod	0.446	1	0.504
	Thatch	10.387	2	0.006
	Nutrient	0.087	1	0.768
	Year * Hydroperiod	0.016	1	0.898
	Year * Thatch	8.300	2	0.016
	Year * Nutrient	0.087	1	0.768
	Hydroperiod * Thatch	4.703	2	0.095
	Hydroperiod * Nutrient	0.042	1	0.837
	Thatch * Nutrient	0.887	2	0.642
	Year * Hydroperiod * Thatch	6.902	2	0.032
	Year * Hydroperiod * Nutrient	0.111	1	0.739
	Year * Thatch * Nutrient	1.930	2	0.381
	Hydroperiod * Thatch * Nutrient	1.715	2	0.424
	Year * Hydroperiod * Thatch * Nutrient	1.035	2	0.596

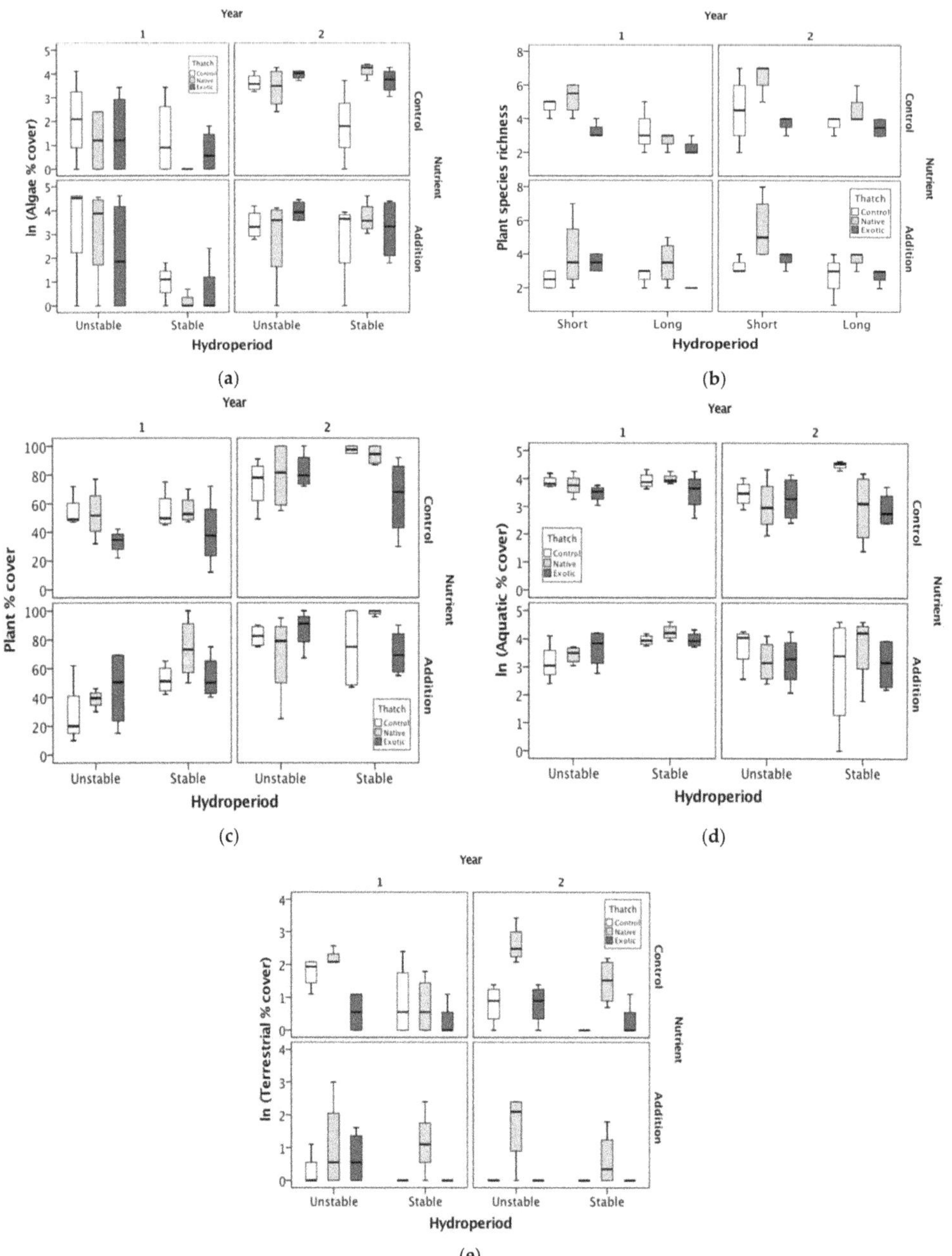

Figure 1. Box plots of dependent variables ((**a**) algal cover, (**b**) plant species richness, (**c**) total plant cover, (**d**) aquatic plant cover, and (**e**) terrestrial plant cover) in response to hydro-regime, nutrient addition, and thatch addition over the two years of the study.

3.2. Plants

All main effects of year, hydro-regime, thatch, and nutrients affected total species richness. Unstable hydro-regime increased, and nutrients decreased plant species richness, but thatch treatments had variable effects. Species richness increased with native thatch addition and decreased or was neutral with exotic thatch addition (Figure 1b, Table 2). Thatch–nutrient was the only significant interaction with thatch having variable effects depending on nutrient addition. In contrast, year and hydro-regime and a hydro-regime–thatch interaction affected total cover. Stable hydro-regimes alone increased total percent cover, and native thatch increased cover while exotic thatch decreased cover (Figure 1c, Table 2).

Aquatic plant cover was significantly affected by the main effects of hydroperiod and thatch. Thatch effects also changed over the two years (Figure 1d, Table 2). In general, aquatic plant cover increased with hydroperiod stability and decreased with thatch addition. Furthermore, aquatic plant cover was negatively associated with algal cover (r = −0.59, $p < 0.001$). Terrestrial plant cover changed across years and responded to the main effects of thatch and the year–hydroperiod–thatch interaction (Figure 1e, Table 2). Thatch effects varied based on type; native thatch increased cover while exotic thatch did not affect cover compared with the control treatment. The interaction resulted from increased terrestrial cover with an unstable hydroperiod. This effect occurred in both years but was more pronounced in the second year. Terrestrial plant cover and algal cover were positively correlated (r = 0.48, $p < 0.001$).

4. Discussion

The California vernal pool plant community is a seasonal habitat that experiences considerable environmental variation, including those influenced by human activities. Climate change affects both temperature and precipitation patterns [36], which determine the hydro-regime of seasonal wetlands. Nutrient inputs can indirectly increase following regional fertilizer application [30,31], and thatch can increase following the introduction of exotic plant species, all resulting in decreased plant productivity and diversity [37,38]. The present experiment examined how algae and plant species responded to manipulated hydro-regime, thatch, and nutrients. Each of the functional groups (algae, aquatic plants, and terrestrial plants) responded differently to the treatments and their interactions. Therefore, these are important regulation factors of richness and cover that create the patterns of species coexistence, species composition, and community structure.

As expected, algal cover increased with unstable hydroperiods and with nutrient addition, patterns commonly found in freshwater ecosystems [24,39–41]. Results were observed after the first year and carried over into the second year. Furthermore, the increase in algae was negatively associated with aquatic plant cover and positively with terrestrial plant cover. Filamentous algae, *Cladophora* spp. and *Nitella* spp., became dominant across many of the treatments after two years, likely reducing aquatic plant cover via competitive exclusion for resources (e.g., light and space) [24,27,42–44]. This is often a result of the formation of algal mats, which are associated with P retention and reduction in light attenuation [24,44–47]. During unstable hydroperiods, algae can increase in growth [40,48], which can also stabilize the spatial distribution of nutrients and water fluctuations, promoting plant establishment during the terrestrial hydro-phase [24,47]. Excess nutrients released from algal mats during the terrestrial hydro-phase are directly associated with an increase in exotic plants, facilitating increases in exotic thatch [24,25,27,29,47].

The aquatic plant functional group increased with increased hydroperiod and decreased with increased thatch. Hydro-regime is well-known for affecting plant germination and growth, as the native plant community is specially adapted to environmental cues from inundation and desiccation cycles [8–13]. The presence of a stable hydroperiod also ensures the completion of the plant life cycle. Decreased cover in response to thatch, like increased algae, was likely the result of light attenuation or creating a barrier to emerging seedlings [24,42]. Many wetland plant seeds rely on light availability for germination and are unable to germinate in unfavorable conditions (e.g., shaded, limited space), so they remain dormant in the soil until those conditions are met [34,49–51]. Increased exotic thatch in vernal pools has further been shown to inhibit native plant richness and cover in both hydro-phases by altering vernal pool inundation periods and depths (e.g., shortens hydroperiod and lowers depth of water in pools) [27]. This also creates divergent plant communities that favor invasive plant species [27]. The presence of exotic thatch, which decomposes more slowly than native thatch, and its accumulation can further promote invasive plants by adding to invasive seed banks, which readily germinate compared with native seed banks [27,34]. Our seed bank did not support invasive plants and therefore we could not assess a shift to an invasive-dominated community.

Terrestrial plants have a later phenology of germination, growth, and flowering during the hydroperiod. They exhibited complex responses to the hydro-regime and thatch that differed compared with the aquatic plants. Terrestrial plant cover increased with an unstable hydro-regime, because of a direct effect on their phenology (e.g., an environmental cue from desiccation) [8], or indirectly by decreasing the competitive advantage of aquatic plant cover. Thatch addition effects differed among the types (native or exotic) and these effects increased over the two years. Native plant thatch effects on plant cover were similar or greater than controls, which was likely due to the contributed seeds to the mesocosms. This has been shown to be effective for restoration in California vernal pools [2,51]. In contrast, exotic plants decreased the cover of the plant functional groups. Exotic plants in California are composed of mostly grasses and some forbs from the Mediterranean region and are known to have high phenotypic plasticity that aids in their ability to colonize various ecosystems [36,52,53]. Exotic plants have slower decomposition rates compared with natives [28] and, as a result, can have many influences on hydrology, soil chemistry, and plant species composition by altering inundation patterns, retaining more nutrients such as phosphorus, and further promoting plant invasions [27,28,54–57]. Our findings corroborate previous research that exotic plant thatch inhibits the growth of vernal pool natives by reducing light, creating a structural barrier, and overall, not meeting favorable environmental conditions needed for germination and growth [27,55].

Exotic thatch impacts the availability of nutrients and moisture in the soil, which impedes native plants and decreases species richness [25,27,34,56,58,59]. Environmental variation caused by the addition of exotic thatch combined with unstable hydro-regimes disrupts ecosystem functions, further preventing the establishment of native plants [27,59,60]. Our findings highlight the need for appropriate management of exotic species in vernal pool habits to mitigate the impacts of exotic thatch. The effects of exotic thatch during the terrestrial phase were shown to carry over into the aquatic phase and reduce cover and richness. Previous research has found that exotic plants can be managed using grazing and fire, which are commonly used in California vernal pool habitats [61,62]. Targeted, low-intensity grazing and fire have been found to reduce the above-ground biomass of exotic thatch and promote the recruitment of native forbs and grasses while facilitating stable hydroperiods [51,63–65]. Further investigations are needed to determine other appropriate management approaches in addition to grazing and fire to keep up with projected climate change.

The current study was conducted in mesocosms with homogenized soil, which has many advantages to understand environmental effects on biodiversity [66]. However, there are disadvantages to this approach, including simplified treatment variation, limited species pool, and smaller scales [66]. The treatment effects were largely consistent over the two years with a couple of exceptions, which may have been the result of temperature and precipitation differences or legacy effects of the first year [24]. Long-term effects of prolonged environmental variation, in conjunction with spatial dynamics (i.e., dispersal), should be further explored to better understand community changes in natural vernal pools over different spatiotemporal scales.

Another limitation of this study was that we could not distinguish the treatment effects on emergence, species interactions, and fitness. For example, we could not decipher whether algal cover and exotic thatch prevented species emergence from the seed bank or prevented growth once they germinated. All of the species present had an annual lifecycle with considerable seed banks that can remain dormant in the soil for decades [20,34,48–50]. Distinguishing how life history stages are affected by the relative effects of the treatments and species interactions (i.e., competition) is important for understanding community dynamics. Nevertheless, our results and those of previous studies [7,34] point to the storage effect being an important mechanism of species coexistence under environmental variation.

While our study showed that the hydro-regime played a dominant role, other factors (nutrients and thatch) and their interactions also affected species composition in significant ways. These results contribute to the knowledge gap regarding the high unexplained variation commonly found in wetland communities beyond the hydro-regime [67]. Climate change, habitat loss, invasive species, and pollution are the primary stressors threatening wetland function globally [36], including California vernal pools. Understanding how wetlands respond to interconnected environmental conditions [68] is critical for their conservation. The effects of abiotic and biotic factors on functional groups and overall biodiversity can provide insight for prioritization of management practices to mitigate impacts from human activity and climate change [36].

Author Contributions: Conceptualization, J.M.K. and J.R.; methodology, J.M.K.; software, J.M.K.; validation, J.M.K., J.R. and M.D.; formal analysis, J.M.K.; investigation, J.R., M.D. and E.F.; resources, J.M.K.; data curation, J.M.K.; writing—original draft preparation, J.R.; writing—review and editing, J.M.K. and E.F.; visualization, J.M.K., J.R. and M.D.; supervision, J.M.K., J.R. and M.D.; project administration, J.M.K.; funding acquisition, J.M.K. All authors have read and agreed to the published version of the manuscript.

Funding: This research was funded by a National Science Foundation grant (DEB 1354724).

Institutional Review Board Statement: Ethical review not applicable.

Data Availability Statement: The data presented in this study are available upon request from the corresponding authors.

Acknowledgments: We gratefully acknowledge those who assisted in the sampling and maintenance of this project: Nestor Samiylenko, Luis Rosas-Saenz, Alyssa Nerida, Trisha Velasquez, Nabilah Fareed, Adam Kneitel, and Nina Kneitel. We also appreciate Mike Baad's continued support at the CSUS Arboretum. Three anonymous reviewers provided useful comments that improved this paper. This research was sampled under USFWS Permit TE192702 to J.M.K.

Conflicts of Interest: The authors declare no conflict of interest.

References

1. McIntyre, S.; Díaz, S.; Lavorel, S.; Cramer, W. Plant functional types and disturbance dynamics–Introduction. *J. Veg. Sci.* **1999**, *10*, 603–608. [CrossRef]
2. Collinge, S.K.; Ray, C. Transient patterns in the assembly of vernal pool plant communities. *Ecology* **2009**, *90*, 3313–3323. [CrossRef] [PubMed]
3. Van der Knaap, Y.A.M.; Aerts, R.; Van Bodegom, P.M. Is the differential response of riparian plant performance to extreme drought and inundation events related to differences in intraspecific trait variation? *Funct. Plant Biol.* **2014**, *41*, 609–619. [CrossRef] [PubMed]
4. Kraft, N.J.B.; Crutsinger, G.M.; Forrestel, E.J.; Emery, N.C. Functional trait differences and the outcome of community assembly: An experimental test with vernal Pool annual plants. *Oikos* **2014**, *23*, 1391–1399. [CrossRef]
5. Komatsu, K.J.; Avolio, M.L.; Lemoine, N.P.; Isbell, F.; Grman, E.; Houseman, G.R.; Koerner, S.E.; Johnson, D.S.; Wilcox, K.R.; Alatalo, J.M.; et al. Global change effects on plant communities are magnified by time and the number of global change factors imposed. *Proc. Natl. Acad. Sci. USA* **2019**, *116*, 17867–17873. [CrossRef]
6. Schumacher, J.; Roscher, C. Differential effects of functional traits on aboveground biomass in semi-natural grasslands. *Oikos* **2009**, *118*, 1659–1668. [CrossRef]
7. Chesson, P. Mechanisms of Maintenance of Species Diversity. *Annu. Rev. Ecol. Syst.* **2000**, *31*, 343–366. [CrossRef]
8. Bliss, S.A.; Zedler, P.H. The Germination Process in Vernal Pools: Sensitivity to Environmental Conditions and Effects on Community Structure. *Oecologia* **1998**, *113*, 67–73. [CrossRef]
9. Tiner, R.W. Using Plants as Indicators of Wetland. *Proc. Acad. Nat. Sci. Phila.* **1993**, *144*, 240–253.
10. Euliss, N.H.; LaBaugh, J.W.; Fredrickson, L.H.; Mushet, D.M.; Laubhan, M.K.; Swanson, G.A.; Winter, T.C.; Rosenberry, D.O.; Nelson, R.D. The Wetland Continuum: A Conceptual Framework for Interpreting Biological Studies. *Wetlands* **2004**, *24*, 448–458. [CrossRef]
11. Van der Valk, A. The prairie potholes of North America. In *The World's Largest Wetlands: Ecology and Conservation*; Fraser, L., Keddy, P., Eds.; Cambridge University Pres: Cambridge, UK, 2005; pp. 393–423.

12. Toner, M.; Keddy, P. River hydrology and riparian wetlands: A predictive model for ecological assembly. *Ecol. Appl.* **1997**, *7*, 236–246. [CrossRef]

13. Kirkman, L.K.; Goebel, P.C.; West, L.; Drew, M.B.; Palik, B.J. Depressional wetland vegetation types: A question of plant community development. *Wetlands* **2000**, *20*, 373–385. [CrossRef]

14. Warwick, N.W.M.; Brock, M.A. Plant reproduction in temporary wetlands: The effects of seasonal timing, depth, and duration of flooding. *Aquat. Bot.* **2003**, *77*, 153–167. [CrossRef]

15. Capon, S.J.; Brock, M.A. Flooding, soil seed bank dynamics and vegetation resilience of a hydrologically variable desert floodplain. *Freshw. Biol.* **2006**, *51*, 206–223. [CrossRef]

16. Miao, S.; Zou, C.B.; Breshears, D.D. Vegetation responses to extreme hydrological events: Sequence matters. *Am. Nat.* **2009**, *173*, 113–118. [CrossRef]

17. Gerhardt, F.; Collinge, S.K. Abiotic constraints eclipse biotic resistance in determining invasibility along experimental vernal pool gradients. *Ecol. Appl. A Publ. Ecol. Soc. Am.* **2009**, *17*, 922–933. [CrossRef] [PubMed]

18. Bauder, E.T. Inundation effects on small-scale plant distributions in San Diego, California vernal pools. *Aquat. Ecol.* **2000**, *34*, 43–61. [CrossRef]

19. Platenkamp, G.A.J. Patterns of vernal pool biodiversity at Beale Air Force Base. In *Ecology, Conservation and Management of Vernal Pool Ecosystems*; Witham, C.W., Bauder, E.T., Belk, D., Ferren, W.R., Ornduff, R., Eds.; California Native Plant Society: Sacramento, CA, USA, 1998; pp. 151–160.

20. Capon, S. Plant community responses to wetting and drying in a large arid floodplain. *River Res. Appl.* **2003**, *19*, 509–520. [CrossRef]

21. Kneitel, J.M. Inundation timing, more than duration, affects the community structure of California vernal pool mesocosms. *Hydrobiologia* **2014**, *732*, 71–83. [CrossRef]

22. Gosejohan, M.C.; Weisberg, P.J.; Merriam, K.E. Hydrologic influences on plant community structure in vernal pools of northeastern California. *Wetlands* **2017**, *37*, 257–268. [CrossRef]

23. Pätzig, M.; Kalettka, T.; Glemnitz, M.; Berger, G. What governs macrophyte species richness in kettle hole types? A case study from Northeast Germany. *Limnol.-Ecol. Manag. Inland Waters* **2012**, *42*, 340–354. [CrossRef]

24. Kneitel, J.M.; Lessin, C.L. Ecosystem-phase interactions: Aquatic eutrophication decreases terrestrial plant diversity in California vernal pools. *Oecologia* **2010**, *163*, 461–469. [CrossRef]

25. Amatangelo, K.L.; Dukes, J.S.; Field, C.B. Responses of a California annual grassland to litter manipulation. *J. Veg. Sci.* **2008**, *19*, 605–612. [CrossRef]

26. Loydi, A.; Eckstein, L.; Otte, A.; Donath, T. Effects of litter on seedling establishment in natural and semi-natural grasslands: A meta-analysis. *J. Ecol.* **2013**, *101*, 454–464. [CrossRef]

27. Faist, A.M.; Beals, S.C. Invasive plant feedbacks promote alternative states in California vernal pools. *Restor. Ecol.* **2018**, *26*, 255–263. [CrossRef]

28. Churchill, A.C.; Faist, A.M. Consequences of above-ground invasion by non-native plants into restored vernal pools do not prompt same changes in below-ground processes. *AoB Plants* **2021**, *13*, plab042. [CrossRef] [PubMed]

29. Croel, R.C.; Kneitel, J.M. Cattle waste reduces plant diversity in vernal pool mesocosms. *Aquat. Bot.* **2011**, *95*, 140–145. [CrossRef]

30. Kido, R.R.; Kneitel, J.M. Eutrophication effects differ among functional groups in vernal pool invertebrate communities. *Hydrobiologia* **2021**, *848*, 1659–1673. [CrossRef]

31. Carpenter, S.R.; Caraco, N.F.; Correll, D.L.; Howarth, R.W.; Sharpley, A.N.; Smith, V.H. Nonpoint pollution of surface waters with phosphorus and nitrogen. *Ecol. Appl.* **1998**, *8*, 559–568. [CrossRef]

32. Avolio, M.L.; Koerner, S.E.; La Pierre, K.; Wilcox, K.; Wilson, G.; Smith, M.; Collins, S. Changes in plant community composition, not diversity, during a decade of nitrogen and phosphorus additions drive above-ground productivity in a tallgrass prairie. *J. Ecol.* **2014**, *102*, 1649–1660. [CrossRef]

33. Krause-Jensen, D.; McGlathery, K.; Rysgaard, S.; Christensen, P.B. Production within dense mats of the filamentous macroalga Chaetomorpha linum in relation to light and nutrient availability. *Mar. Ecol. Prog. Ser.* **1996**, *134*, 207–216. [CrossRef]

34. Faist, A.M.; Ferrenberg, S.; Collinge, S.K. Banking on the past: Seed banks as a reservoir for rare and native species in restored vernal pools. *AoB Plants* **2013**, *5*, plt043. [CrossRef]

35. Baldwin, B.G.; Goldman, D.H.; Keil, D.J.; Patterson, R.; Rosatti, T.J. (Eds.) *The Jepson Manual: Vascular Plants of California*; University of California Press: Berkeley, CA, USA, 2012.

36. Millennium Ecosystem Assessment. *Ecosystems and Human Well-Being: Wetlands and Water Synthesis*; World Resources Institute: Washington, DC, USA, 2005.

37. Bartolome, J.W.; Jackson, R.D.; Betts, A.D.K.; Connor, J.M.; Nader, G.A.; Tate, K.W. Effects of residual dry matter on net primary production and plant functional groups in Californian annual grasslands. *Grass Forage Sci.* **2007**, *62*, 445–452. [CrossRef]

38. Smith, M.J.; Kay, W.R.; Edward, D.H.D.; Papas, P.J.; Richardson, K.S.J.; Simpson, J.C.; Halse, S.A. AusRivAS: Using macroinvertebrates to assess ecological condition of rivers in Western Australia. *Freshw. Biol.* **1999**, *41*, 269–282. [CrossRef]

39. Boven, L.; Stoks, R.; Forró, L.; Brendonck, L. Seasonal dynamics in water quality and vegetation cover in temporary pools with variable hydroperiods in Kiskunság (Hungary). *Wetlands* **2008**, *28*, 401–410. [CrossRef]

40. Zhang, P.; Ayumi, K.; Van Leeuwen, C.H.A.; Velthuis, M.; Van Donk, E.; Xu, J.; Bakker, E.S. Interactive effects of rising temperature and nutrient enrichment on aquatic plant growth, stoichiometry, and palatability. *Front. Plant Sci.* **2020**, *11*, 58. [CrossRef]

41. Wijewardene, L.; Wu, N.; Fohrer, N.; Riis, T. Epiphytic biofilms in freshwater and interactions with macrophytes: Current understanding and future directions. *Aquat. Bot.* **2022**, *176*, 103–467. [CrossRef]
42. Hautier, Y.; Niklaus, P.A.; Hector, A. Competition for light causes plant biodiversity loss after eutrophication. *Science* **2009**, *324*, 636–638. [CrossRef]
43. Power, M.E. Benthic turfs vs floating mats of algae in river food webs. *Oikos* **1990**, *58*, 67–79. [CrossRef]
44. Bai, Y.; Han, X.; Wu, J.; Chen, Z.; Li, L. Ecosystem stability and compensatory effects in the inner Mongolia grassland. *Nature* **2004**, *431*, 181–184. [CrossRef]
45. Scheffer, M.; Rinaldi, S.; Gragnani, A.; Mur, L.R.; Van Nes, E.H. On the dominance of filamentous cyanobacteria in shallow, turbid lakes. *Ecology* **1997**, *78*, 272–282. [CrossRef]
46. Wu, X.; Mitsch, W.J. Spatial and temporal patterns of algae in newly constructed freshwater wetlands. *Wetlands* **1998**, *18*, 9–20. [CrossRef]
47. Marazzi, L.; Gaiser, E.E.; Jones, V.J.; Tobias, F.A.; Mackay, A.W. Algal richness and life-history strategies are influenced by hydrology and phosphorus in two major subtropical wetlands. *Freshw. Biol.* **2017**, *62*, 274–290. [CrossRef]
48. Grime, J.P. The role of seed dormancy in vegetation dynamics. *Ann. Appl. Biol.* **1981**, *98*, 555–558. [CrossRef]
49. Baskin, C.C.; Baskin, J.M. *Seeds: Ecology, Biogeography and Evolution of Dormancy and Germination*; Academic Press: San Diego, CA, USA, 1998.
50. Carta, A.; Bedini, G.; Mueller, J.V.; Probert, R.J. Comparative seed dormancy and germination of eight annual species of ephemeral wetland vegetation in a Mediterranean climate. *Plant Ecol.* **2013**, *214*, 339–349. [CrossRef]
51. Xiong, S.; Nilsson, C. The effects of plant litter on vegetation: A meta-analysis. *J. Ecol.* **1999**, *87*, 984–994. [CrossRef]
52. Collinge, S.K.; Ray, C.; Marty, J.T. A long-term comparison of hydrology and plant community composition in constructed versus naturally occurring vernal pools. *Restor. Ecol.* **2013**, *21*, 704–712. [CrossRef]
53. Heady, H.F. Vegetational changes in the California annual type. *Ecology* **1958**, *39*, 402–416. [CrossRef]
54. Pitt, M.D.; Heady, H.F. Responses of annual vegetation to temperature and rainfall patterns in northern California. *Ecology* **1978**, *59*, 336–350. [CrossRef]
55. Facelli, J.M.; Pickett, S.T. Plant litter: Its dynamics and effects on plant community structure. *Bot. Rev.* **1991**, *57*, 1–32. [CrossRef]
56. Molinari, N.A.; D'Antonio, C.M. Where have all the wildflowers gone? The role of exotic grass thatch. *Biol. Invasions* **2020**, *22*, 957–968. [CrossRef]
57. Hassan, N.; Sher, K.; Rab, A.; Abdullah, I.; Zeb, U.; Naeem, I.; Khan, A. Effects and mechanism of plant litter on grassland ecosystem: A review. *Acta Ecol. Sin.* **2021**, *41*, 341–345. [CrossRef]
58. Weltzin, J.F.; Keller, J.K.; Bridgham, S.D.; Pastor, J.; Allen, P.B.; Chen, J. Litter controls plant community composition in a northern fen. *Oikos* **2005**, *110*, 537–546. [CrossRef]
59. Craft, C.; Kull, K.; Graham, S. Ecological indicators of nutrient enrichment, freshwater wetlands, Midwestern United States. *Ecol. Indic.* **2007**, *7*, 733–750. [CrossRef]
60. Veen, G.F.; Fry, E.L.; Ten Hooven, F.C.; Kardol, P.; Morriën, E.; De Long, J.R. The role of plant litter in driving plant-soil feedbacks. *Front. Environ. Sci.* **2019**, *7*, 168. [CrossRef]
61. Marty, J. Fire effects on plant biodiversity across multiple sites in California vernal pool grasslands. *Ecol. Restor.* **2015**, *33*, 266–273. [CrossRef]
62. Marty, J.T. Loss of biodiversity and hydrologic function in seasonal wetlands persists over 10 years of livestock grazing removal. *Restor. Ecol.* **2015**, *23*, 548–554. [CrossRef]
63. Bailey, D.W.; Mosley, J.C.; Estell, R.E.; Cibils, A.F.; Horney, M.; Hendrickson, J.R.; Burritt, E.A. Synthesis paper: Targeted livestock grazing: Prescription for healthy rangelands. *Rangel. Ecol. Manag.* **2019**, *72*, 865–877. [CrossRef]
64. Michaels, J.; Batzer, E.; Harrison, S.; Eviner, V.T. Grazing affects vegetation diversity and heterogeneity in California vernal pools. *Ecology* **2021**, *102*, e03295. [CrossRef]
65. Wei, L.; Tan, R.; Yu-ming, Y.; Wang, J. Plant diversity as a good indicator of vegetation stability in a typical plateau wetland. *J. Mt. Sci.* **2014**, *11*, 464–474.
66. Spivak, A.C.; Vanni, M.J.; Mette, E.M. Moving on up: Can results from simple aquatic mesocosm experiments be applied across broad spatial scales? *Freshw. Biol.* **2011**, *56*, 279–291. [CrossRef]
67. Batzer, D.P. The seemingly intractable ecological responses of invertebrates in North American wetlands: A review. *Wetlands* **2013**, *13*, 1–15. [CrossRef]
68. Sardans, J.; Penuelas, J. The role of plants in the effects of global change on nutrient availability and stoichiometry in the plant soil system. *Plant Physiol.* **2012**, *160*, 1741–1761. [CrossRef] [PubMed]

Article

The Importance of Thermally Abnormal Waters for Bioinvasions—A Case Study of *Pistia stratiotes*

Nina Šajna [1,*], Tina Urek [1], Primož Kušar [2] and Mirjana Šipek [1]

[1] Faculty of Natural Sciences and Mathematics, University of Maribor, 2000 Maribor, Slovenia
[2] Meljski Hrib 30, 2000 Maribor, Slovenia
* Correspondence: nina.sajna@um.si; Tel.: +386-22-293-705

Abstract: Thermally abnormal waters represent safe sites for alien invasive plants requiring warmer conditions than provided by the ambient temperatures in the temperate zone. Therefore, such safe sites are frequently inhabited by tropical and sub-tropical plants. By performing a literature review we assessed that at least 55 alien aquatic plant taxa from 21 families were found in thermally abnormal waters in Europe. The majority of these taxa are submerged or rooted macrophytes. Six taxa are listed as quarantine pests according to EPPO. Among these, *Pistia stratiotes* is present in seven European countries, most of the records of this presence being recent. We studied *P. stratiotes* in a thermally abnormal stream where a persistent population was able to survive harsh winters. Models showed that the optimum temperature for *P. stratiotes* biomass was 28.8 ± 3.5 °C. Here, we show that air temperatures had a higher influence on the photosynthetic efficiency of *P. stratiotes*, estimated by chlorophyll fluorescence measurements, than did water temperatures. Generally, growth, and consequently surface cover for free-floating plants, cannot be explained solely by thermally abnormal water temperatures. We conclude that even though the majority of thermophile alien plant occurrences resulted from deliberate introductions, thermally abnormal waters pose an invasion risk for further deliberate, accidental, or spontaneous spread, which might be more likely for free-floating macrophytes.

Keywords: macrophytes; alien invasive plants; chlorophyll fluorescence; plant mass; temperature gradient

Citation: Šajna, N.; Urek, T.; Kušar, P.; Šipek, M. The Importance of Thermally Abnormal Waters for Bioinvasions—A Case Study of *Pistia stratiotes*. *Diversity* **2023**, *15*, 421. https://doi.org/10.3390/d15030421

Academic Editors: Mateja Germ, Igor Zelnik, Matthew Simpson and Michela Marignani

Received: 31 January 2023
Revised: 27 February 2023
Accepted: 9 March 2023
Published: 13 March 2023

1. Introduction

Freshwater ecosystems cover less than 1% of the Earth's surface but support around 10% of all known species, making them biodiversity hotspots [1]. However, they are one of the most threatened ecosystems on the global scale, mainly because of land-use change, hydrological system alterations, pollution, climate change, and invasive species establishment [2,3].

Biological invasions of European freshwater ecosystems by alien invasive species have a significant impact on biodiversity and ecosystem functions and cause economic damage [4,5]. However, a small share of freshwater alien plants has a higher economic and ecological impact in comparison to terrestrial plant species [4]. One of the most severe impacts of alien aquatic plants is reducing biodiversity because of rapid vegetative propagation and the ability to outcompete native flora [6,7].

The number of alien aquatic plants in Europe has doubled since 1980 and is still increasing [8]. Hussner [9] reports 96 alien freshwater plant species found in European countries. The majority of them originate from temperate regions similar to Europe (Northern America and Asia), while a smaller share is native to warmer parts of the world (e.g., sub-tropical and tropical regions: Africa, Australia, and Southern America; [9]). The latter group consists of frost-sensitive plants. Low temperatures during the coldest months prevent their establishment, and in general only ephemeral populations are formed during the warmer part of the year [10]. However, thermally abnormal waters represent an opportunity for frost-sensitive macrophytes to establish and survive cold winters [6,9,11]. Therefore, systems with heated waters, either natural (e.g., geothermal waters) or anthropogenic (e.g., thermal pollution), represent safe sites, which imitate tropical conditions [12]. Consequently, alien macrophyte communities in thermally abnormal waters differ from those in freshwaters with local temperature regimes [13]. However, in light of global warming, we can expect that at least some thermophilous alien macrophytes will increase their introduced distribution range beyond the present restriction to thermally abnormal waters [14]. In this regard, thermally abnormal waters might represent safe sites for potential sleeper invasive plants.

In the present study, we aim to assess a diversity of alien macrophytes recorded in European thermally abnormal waters by performing a literature overview. Later, we focus on a globally invasive pan-tropical macrophyte, *Pistia stratiotes* L. The species can survive winter conditions in thermally abnormal waters [11] and can occasionally occur outside safe sites with thermally abnormal waters during favorable climatic conditions [10]. What is more, a substantial increase in new occurrences in Europe reported recently suggests that deliberate releases might not be the only reason for the range expansion of *P. stratiotes*. Since climate warming can promote the expansion of aquatic plants [15], this might also be the case for *P. stratiotes* [11,16]. Therefore, *P. stratiotes* represents a suitable thermophile macrophyte for investigating the effects of temperature as the most challenging climate condition for tropical plants in the temperate zone. In the thermal backwater in Slovenia, where *P. stratiotes* successfully survived winter for several years, we had an opportunity to study the responses of plants to stress along a temperature gradient that included temperatures that were both too low and too high for *P. stratiotes* survival. We aimed to define the most suitable temperature range by studying the year-round surface cover, biomass, and quantum yield of photosynthesis of *P. stratiotes*, using chlorophyll fluorescence.

2. Materials and Methods

2.1. Literature Review

The list of alien macrophytes occurring in European thermally abnormal waters was made by reviewing the existing literature in Google Scholar and Web of Science. We used the following combinations of words for alien aquatic plants and thermal inland waters: "alien" OR "exotic" OR "invasive" OR "non-native" AND "thermal water" OR" thermally polluted water" OR "thermally abnormal water" OR "heated water" OR "warm water". We further searched specifically for all records relating to *P. stratiotes* and referring to Europe to obtain a list of countries where the species occurs.

2.2. Study Species

Pistia stratiotes L. (water lettuce) is a herbaceous perennial aquatic macrophyte. Its hairy leaves form free-floating rosettes. It is a clonal plant with stolons that form new rosettes, which eventually detach from the mother plant and start new colonies on their own. Clonal growth is extremely successful, and plants can very rapidly overgrow an entire water surface and form dense mats [17]. *Pistia stratiotes* is found in tropical areas around the world, and its origin is uncertain, though it is likely to be from South America [17]. The species has increased its distribution throughout the sub-tropical regions. It was recorded first in South Africa [18,19] and later also in North Africa [19]. It has a wide distribution in Asia, and it is recorded as invasive [19]. The same is true for the Northern Territory of Australia and for the USA, where it occurs in several states [19]. Its rapid growth alters aquatic ecosystems by decreasing oxygen and photosynthesis in the water underneath mats [20,21]. It clogs waterways, hinders navigation, and damages fisheries. It also enables the transmission in the tropics of certain human diseases spread by mosquitoes and snails. In many countries, it is known as one of the worst pantropical aquatic weeds [22].

2.3. Study Site

Our study site is a natural thermally abnormal backwater of the larger Sava River named Topla struga in Prilipe near the village Čatež (45°53′ N, 15°37′ E; Figure 1). A naturalized *P. stratiotes* population was established there in 2001 and persisted until its eradication by local authorities in the spring of 2022. The backwater is a 4 km long stream beginning at a geothermal spring and flowing toward the Sava River. An additional source of warm water is provided by a discharge of the thermal spa, which joins the stream after passing through a water treatment plant (Figure 1B). The temperature of the spring water is above 30 °C and cools downstream. The situation at the study site represents a permanent continuously decreasing temperature gradient all year round. We positioned 10 permanent sampling points along the stream for season-long continuous monitoring (Figure 1B). The first five sampling points were 100 m apart and the last five were 200 m apart. Data for the water temperatures of the Sava River at the Jesenice na Dolenjskem recording location (5 km downstream of the study site) and the air temperatures at the Novo Mesto meteorological station were obtained from the national Slovenian Environment Agency (ARSO) [23]. For comparisons, we used temperature data from known locations with *P. stratiotes* distribution in Corumba, Brazil (https://en.tutiempo.net/records/sbcr) (accessed on 7 November 2018) [24] and Eldoret, Kenya (https://en.tutiempo.net/records/hkel) (accessed on 7 November 2018) [25].

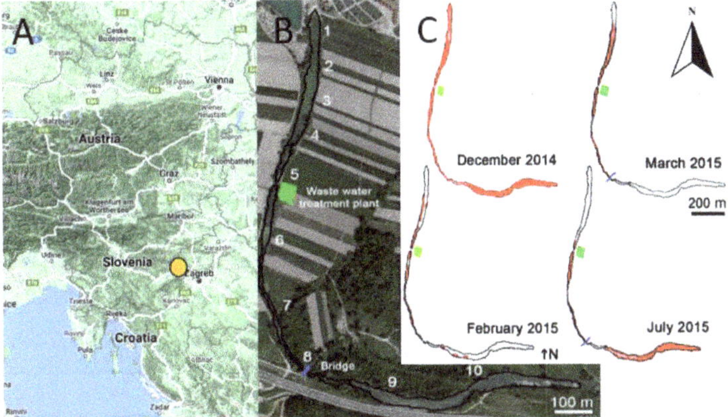

Figure 1. (**A**) Study site in Slovenia shown by the orange circle; (**B**) ten permanent sampling points along the thermal stream Topla (the green square indicates the location of the wastewater treatment plant); and (**C**) surface cover of *P. stratiotes* from December 2014 to July 2015 marked with color red.

2.4. Sampling Design

2.4.1. Abiotic Parameters

The following abiotic conditions were measured weekly between March and July 2015. Measurements took place at some point between 9 a.m. and 2 p.m. At each sampling point, the following were measured:

Air temperature and air humidity 1.5 m above the water surface, using an aspiration psychrometer (Ahlborn FNAD 46, Holzkirchen, Germany);

- Photosynthetically active radiation (PAR) on the water surface, using a quantum sensor (Quantum Meter; Apogee Instruments, Roseville, CA, USA);
- Water temperature at 10 cm depth, using a digital thermometer (HI 98501; Hanna Instruments, Woonsocket, RI, USA);
- Water temperature of the surface, using an infrared thermometer (572-2, Fluke, Washington, DC, USA).

Simultaneously, water samples were collected to obtain the pH value (HI 98103; Hanna Instruments, Woonsocket, RI, USA), the nitrate concentration (Visocolor®ECO Nitrate; Macherey-Nagel, Düren, Germany), and the phosphate concentration (Visocolor®ECO Phosphate; Macherey-Nagel, Düren, Germany). All data are included in the Supplementary Material (Table S1). Initial multiple regression analysis showed that, among measured temperatures, which significantly predicted fresh biomass and leaf length, air temperatures showed the highest correlation with beta (β = 0.53, $p < 0.001$) and were therefore included in models as well as water temperatures.

2.4.2. Seasonal Surface Cover

We monitored the surface cover of *P. stratiotes* in the stream Topla from December 2014 to July 2015. We marked distributional data for December, February, March, and July on the map and digitized it using ArcView GIS 3.2. [26].

2.4.3. Ecophysiological Measurements In Situ

Five plants were sampled randomly at each sampling point for the following ecophysiological measurements (Table S2). The photochemical activity of the plant was detected by measuring chlorophyll *a* (Chl-*a*) fluorescence of photosystem II (PS II) using a portable fluorescence spectrometer (Handy PEA; Hansatech Instruments, Norfolk, UK) following the manufacturer's instructions. We recorded the F_v/F_m value representing the maximum quantum yield of PS II, which is highly correlated with the quantum yield of net photosynthesis, whereby F_v is the variable fluorescence ($F_m - F_o$) and F_m is the maximal fluorescence value [27]. F_v/F_m in healthy plants is consistently around 0.83. The ratio F_v/F_m is a measure of stress, if smaller than 0.8, especially if the value falls after dark-adaptation measurement compared to ambient measurement. Chlorophyll fluorescence measurements were performed immediately and after a 10 min period of dark adaptation to obtain actual and dark-adapted fluorescence values, respectively. The dark-adaptation time of 10 min was determined experimentally according to the device manual. Fluorescence measurements were taken on healthy leaves without signs of damage to exclude the effect of stress.

2.4.4. Biomass Studies

After ecophysiological measurements in the field, plants were collected and stored in plastic bags in a refrigerator until measurements were taken of the rosette diameter and of the longest leaf length (Table S2). Because of the strong correlation between these two variables (Pearson's r = 0.97), only leaf length was included in the generalized linear models (GLMs). The plants' fresh and dry biomass (after drying at 60 °C for 72 h) were weighed.

2.5. Data Analyses

In the observed range, we had sites with no living specimens on both sides of the temperature gradient. To extract the water temperature range for *P. stratiotes* survival in our site, we used nonlinear regression models for modeling the temperature dependence of plant growth, which included the Gaussian functions of fresh plant mass and leaf length. Nonlinear parameter estimates were obtained using the nonlinear least squares method. We used zero values for cases where there was zero abundance. For the chlorophyll fluorescence data, we used GLMs to analyze the influence of the air temperature and sampling date (both as continuous variables) during the observation period. We applied GLMs to the actual and dark-adapted values of the fluorescence data.

3. Results

3.1. Aquatic Plants Alien to Europe Associated with Thermally Abnormal Waters

European thermally abnormal waters include natural thermal waters, discharges of thermal spas, industrially heated waters from power plants, and mining. Countries in which aquatic plants alien to Europe were recorded in thermally abnormal waters include

Austria, Germany, Hungary, Poland, Romania, Serbia, Slovakia, Slovenia, Ukraine, the European part of Russia, and Iceland, while no reports from other European countries were found (Appendix A, Table A1).

Altogether, 55 alien aquatic plant taxa from 21 families were found, each one having been observed in thermally abnormal waters in at least one location. Among them, six taxa are included in the EPPO A2 List (Appendix A, Table A1) [28]. The alien aquatic plants found comprised the following growth forms: 18% free-floating (10 taxa, two of them ferns), 13% leaf floating (7 taxa), 33% submersed (18 taxa), and 36% emersed (18 taxa, one of them a fern).

The most numerous were the Hydrocharitaceae family (8 taxa), followed by Nymphaeaceae (6), Pontederiaceae (5), Plantaginaceae (4), Onagraceae (4), Araceae (4), Acanthaceae (3), Haloragaceae (3) and Alismataceae (3). The remaining 12 families had one or two alien representatives (Appendix A, Table A1). The vast majority of alien aquatic plants associated with thermally abnormal waters originate from subtropical and tropical climates.

The most common alien aquatic plant in warm waters are *Pistia stratiotes* and *Vallisneria spiralis*, both found in seven European countries, followed by *Egeria densa*, which was recorded in six countries. *Cabomba caroliniana*, *Hygrophila polysperma*, *Monochoria korsakowii*, and *Shinnersia rivularis* were found in the warm waters of four European countries. Of the 55 alien aquatic plant taxa, 60% (33 taxa) were found in only one country, the majority of them in Hungarian thermal waters (Appendix A, Table A1).

3.2. The Current Distribution Range of Tropical Macrophyte P. stratiotes in Europe

The literature review assessed the occurrence of *P. stratiotes* in 21 European countries (Figure 2). The first record is dated 1966 [29]. In some locations, introductions happened several times. For example, in Germany, the latest introductions resulted in persistent populations (see discussion for the Erft River in Germany). In some countries *P. stratiotes* was recorded in the past but is no longer present. In Spain, it was immediately eradicated after it was found [30]. In Croatia, it was found once in the summer and completely died during the following winter. Occurrences in the 1980s and 1990s (Table 1; Figure 2) were recorded as ephemeral. Later records report increasing occurrences outside of thermally abnormal waters.

Figure 2. European countries where *Pistia stratiotes* has already been recorded are in grey. For details see Table 1.

Table 1. Distribution of *P. stratiotes* within European countries and year of the first report.

Country	The First Report (Location)	Reference
Austria	1980	[31]
Belgium	2000	[32]
Croatia	2017 (River Sava)	[10]
Czech Republic	1991	[33]
France	1992 (Auvergne)	[34]
Germany	1981 (thermally polluted River Erft)	[35]
Hungary	1966	[29]
Italy	1998 (Bodrio le Margherite)	[36]
Netherlands	1973	[37]
Norway	1989 (Lake Stilla, Ak Skedsmo)	[38]
Poland	2012 (Lake Stawiki in Sosnowiecu)	[39]
Portugal	1990	[40]
Romania	2005	[41]
Russia	1989 (River Volga)	[42]
Serbia	1994 (thermal spring Banjica)	[43]
Slovakia	2007	[44]
Slovenia	2001 (thermal stream Topla)	[11]
Spain	2001 (Guipúzcoa)	[30]
Sweden	2006	[45]
United Kingdom	1983 (pond in London)	[46]
Ukraine	2011 (River Seversky Donets)	[47]

3.3. A Case Study from Thermal Stream Topla

3.3.1. Physical and Chemical Parameters of the Environment

The air humidity during the sampling period was between 40% and 64%, with an average of 53%. Photosynthetically active radiation varied at between 365 and 1249 $\mu mol/m^2s$ (average 765 $\mu mol/m^2s$). The highest values were measured in June and July (Table S1).

The pH values varied at between 7.4 and 8.9 (average 8.1). The concentration of nitrates in the water was between 0.50 and 3.75 mg/L, with the highest values measured in April. The average nitrate content was 1.83 mg/L. The concentrations of phosphates in the water varied at between 0 and 0.43 mg/L (average value of 0.15 mg/L; Table S1).

3.3.2. Thermal Conditions

The study site is located in a typical temperate climate, characterized by a harsh winter lasting from 21st December until 21st March with temperatures well below 0 °C and snowfall. At the nearest meteorological station, the lowest average monthly air temperatures for 2015 were measured in January and February, −1.7 and −1.6 °C, respectively. In 2015, the lowest daily maximum of −15.4 °C was recorded in January. Summer lasts from June 21st until September 23rd. Temperatures can rise above 30 °C. In 2015 the highest monthly average temperatures were recorded in July and exceeded 29 °C, with a daily maximum of 36.2 °C [23]. Comparison with the temperatures in the tropics, where *P. stratiotes* is native (Corumba, Brazil), and in the subtropics, where *P. stratiotes* is invasive (Eldoret, Kenya), shows that from the beginning of April until the end of September the air temperatures in our study site were as high as in the tropics and subtropics (Figure 3).

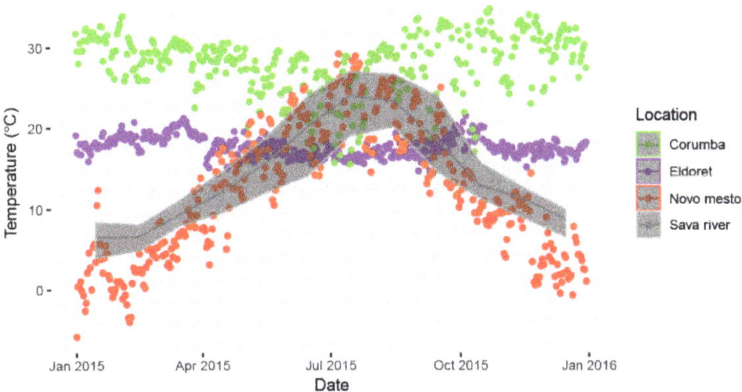

Figure 3. Comparisons of daily mean air temperatures in 2015 at the Novo Mesto meteorological station near Čatež (Slovenia), in Corumba (Brazil), and in Eldoret, (Kenya). Monthly mean water temperatures at the Jesenice na Dolenjskem recording site on the Sava River, 5 km downstream of the study site (Prilipe, Slovenia), obtained from the Slovenian Environment Agency are shown by the dark grey area: average temperature (dark grey line) and minimum and maximum temperatures (light grey lines).

The temperatures of the Sava River 5 km downstream of the study site (Figure 3) were strongly correlated with the air temperatures; however, they never fell below 6 °C. The highest average water temperatures were in July (24 °C) and August 2015 (23 °C), with the warmest day peaks reaching as high as 29 °C.

In our study site, the mean water temperature at the source was 27 °C in March and reached 33 °C in May, June, and July (Figure 4, Table S1). The last and also the coolest sampling point on our gradient had a mean temperature in March of about 15 °C and increased with the season to a mean of 27 °C in July. With the warming season, the water temperatures increased with time, resulting in the shift of temperature gradient downstream. In addition, the mean daytime air temperatures increased from 10 to 30 °C during our observation period (Table S1). Water temperatures varied less in time; however, they had a clear site dependency. Sites closer to the thermal source had 10 to 15 °C higher temperatures (Figure 4, Table S1), a temperature difference that persisted during our observations.

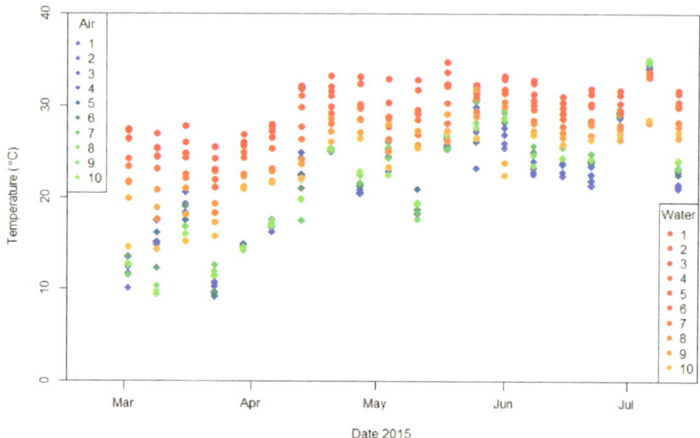

Figure 4. Air temperature 1.5 m above the water surface and water temperature at 10 cm depth at 10 permanent sampling points along thermal stream Topla measured weekly from March to July 2015.

3.3.3. Seasonal Surface Cover

The cover and spread of *P. stratiotes* were greatest at the end of the growing season in December, which was then followed by die-offs caused by winter (Figure 1C). Plants survived in the upper part of the stream and started to increase their cover again in March and April, occupying the stream at sampling points 1 to 8. In May, plants were distributed along all 10 sampling points on the gradient. In June and July, plants were missing at the warmest locations (sampling points 1 to 3). In the winter it was too cold at the end of the stream, while in the summer, with the increase in the air temperature, locations at the hot source were too hot for *P. stratiotes* (Figure 5B).

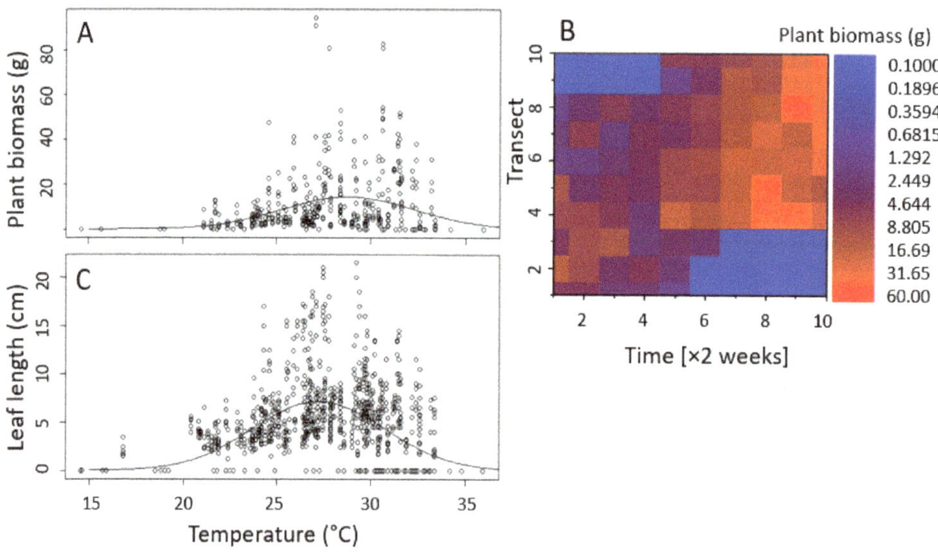

Figure 5. Effect of water temperature on fresh *Pistia stratiotes* biomass (**A**,**B**) and leaf length (**C**).

3.3.4. Effect of Temperature on Biomass and Chlorophyll Fluorescence

Water temperatures at which plant biomass reached the highest values were slightly higher than temperatures at which we observed plants with larger leaves (Figure 5A,C). The nonlinear least-squares predictions were centered at 28.8 °C, with 1 sigma of 3.5 °C for fresh plant mass and 27.2 °C with the same sigma for the leaf length (Table 2).

Table 2. Estimated regression parameters, standard errors, t-values, and p-values for the models explaining the effect of water temperature on fresh plant mass and leaf length of *Pistia stratiotes*. Model: $x \sim a \times \exp(-(T - T_0)^2/(2 \times sigma^2))$.

	Estimate	Std. Error	t-Value	p-Value
Fresh mass				
a	14.372	1.034	13.900	<0.0001
T_0	28.779	0.324	88.944	<0.0001
sigma	3.516	0.402	8.744	<0.0001
Leaf length				
a	7.131	0.211	33.80	<0.0001
T_0	27.231	0.133	204.16	<0.0001
sigma	3.516	0.158	22.26	<0.0001

The photochemical efficiency, estimated by measuring actual chlorophyll fluorescence, increased with the season. For the first few months (from March until May), the actual chlorophyll fluorescence ratio F_v/F_m was below 0.6 for the entire gradient (Figure 6A). Chlorophyll fluorescence measured after a 10-minute-long dark adaptation, during which reaction centers of photosystem II (PSII) were able to return to a relaxed state, showed some increase. However, the values never exceeded 0.75 in March, April, and May (Figure 6B). In June and July, the actual as well as the dark-adapted F_v/F_m increased (Figure 6), particularly at the last two sampling points of the gradient. The difference between actual and dark-adapted values was the smallest in July. At the same time, at more than half of the sampling points, plants had F_v/F_m above 0.8 on average. The outcome of the GLM shows that actual chlorophyll fluorescence was significantly dependent on the sampling date and air temperature (Table 3).

Table 3. Estimated regression parameters, standard errors, t-values, and p-values for the actual chlorophyll fluorescence and after-dark adaptation of *Pistia stratiotes* leaves.

	Estimate	Std. Error	t-Value	p-Value
Chl Fluorescence F_v/F_m—actual (GLM, Gaussian)				
Intercept	0.464	0.020	22.873	<0.0001
Sampling date	0.019	0.001	14.213	<0.0001
Temperature	−0.004	0.001	−3.509	<0.001
Chl Fluorescence F_v/F_m—dark adapted; F_v/F_m (GLM, Gaussian)				
Intercept	0.702	0.015	45.845	<0.0001
Sampling date	0.009	0.001	8.729	<0.0001
Temperature	−0.002	0.001	−2.198	0.028

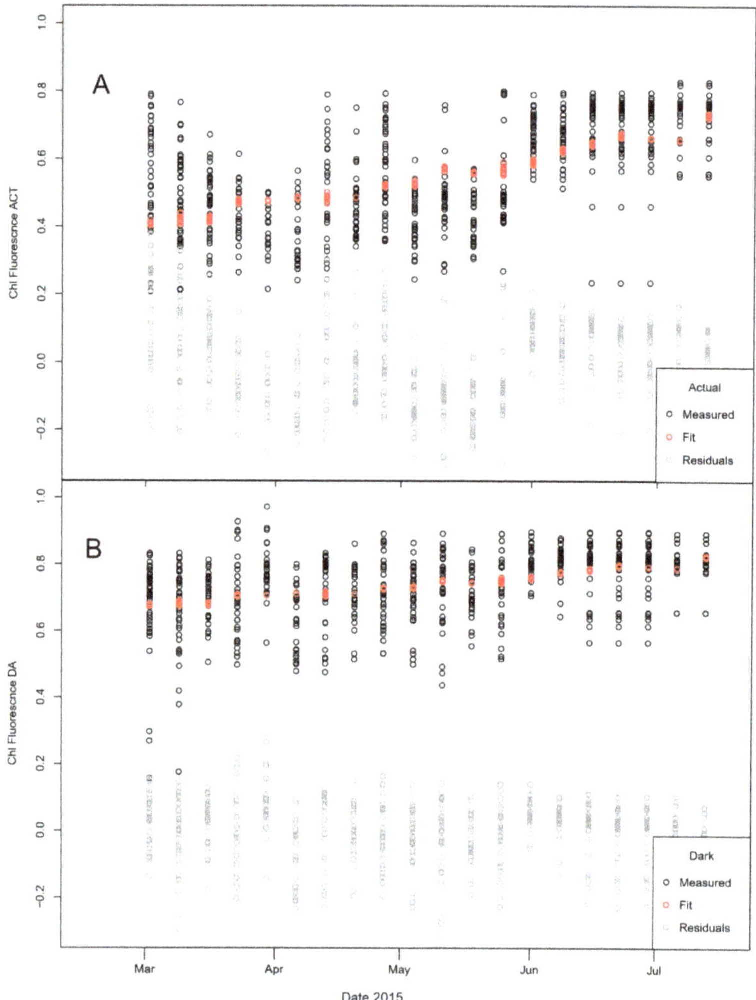

Figure 6. (**A**) Actual and (**B**) dark-adapted chlorophyll *a* fluorescence measured as F_v/F_m.

4. Discussion

4.1. Macrophytes in Thermally Abnormal European Waters

Fifty-five alien macrophytes were associated with thermally abnormal waters in Europe, which corresponds to about half of all alien macrophyte species recorded in Europe [9]. Lukács et al. [29] reported that, in Hungary, 80% of alien macrophytes were restricted to thermal waters. However, the frequency of these macrophytes was relatively low, pointing to their invasive potential being restricted by habitat availability, when compared to alien macrophytes that prefer cold waters.

The vast majority of alien macrophytes found in thermal waters originate from deliberate releases (93%), i.e., they are used as ornamental plants either in aquaria or outdoors (e.g., pond plants). Therefore, emersed or leaf-floating macrophytes with prominent flowers (e.g., *Nymphaea*) and/or free-floating macrophytes with large leaves (e.g., *Pistia stratiotes*) are overrepresented in the literature assessed. The significant role of aquarists and the trade in ornamental alien macrophytes in freshwater-ecosystem invasions is already well recognized and addressed in several studies [29,48,49]. The group of alien plants likely to have originated from aquaria was mainly represented by fast-growing submerged plants (e.g., *Vallisneria*). Alien plants introduced unintentionally were rare and included small free-floating plants (e.g., *Wollfia globosa*) or submerged plants (e.g., *Myriophyllum heterophyllum*). It is worth mentioning that, in European thermally abnormal waters, two of the most problematic aquatic plants worldwide were present: *Eichhornia crassipes* and *Pistia stratiotes*.

Only a smaller proportion of macrophytes found in thermally abnormal waters originate from climates similar to Europe, while approximately 80% of them have natural distribution in subtropical and tropical areas. For alien macrophytes originating from the temperate zone, European thermally abnormal waters are a habitat where the growing season is extended, while their survival is also possible at ambient conditions. For the invasive species that are native to warmer conditions, the current distribution range in Europe is likely to expand in the Mediterranean and Atlantic regions, and possibly also in continental parts, because of milder winters indicative of global warming. Further expansion in all regions can be expected because of increasing waterbodies changes: thermal pollution, channeling, urban spread incorporating waterbodies, gravel pits, shallow artificial lakes, or ponds [11,16,50]. Shallow low-velocity aquatic habitats might be particularly suitable for ephemeral colonization during summers. What is more, analysis of the field data for the Netherlands' ditches showed that milder winters enhanced eutrophication by increasing the total phosphorus concentrations, and favored evergreen submerged and, especially, free-floating macrophytes—the latter because of their ability to obtain nutrients from the water column [51,52]. This is of concern, given that some alien macrophytes recorded have a reputation as extremely problematic and costly weeds in subtropical regions (e.g., *Pistia stratiotes* and *Eichhornia crassipes*) [53]). According to our study, locations reported in the literature often contained several co-occurring invasive alien plants. One of the reasons—besides deliberate introduction—might be that the invaded habitat was altered by invasive macrophytes in a way that favors the establishment of other invasive species more than native species [54].

4.2. Pistia Stratiotes Is Expanding Introduced Distribution

We assessed *Pistia stratiotes* to be the most recorded alien invasive plant in thermally abnormal waters. Although the data could be biased, since the species is easily recognized, and smaller or submerged macrophytes can be easily overlooked, there is no doubt that *Pistia stratiotes* is a popular aquatic macrophyte for aquaria and outdoors. Its mass cultivation is increasing the probability of irresponsible releases into nature, which provide the main opportunity for ephemeral population establishment and naturalization possibilities across Europe. Especially worrying is the recent increase in records of established populations, and of populations consistently present for a longer time, in the southern part of Europe.

It is generally accepted that the first introduction to Europe happened in 1973 in the Netherlands [37]; however, there is also an earlier notation from Hungary in 1966 [29]. Early introductions are believed to have been deliberate releases of aquaria plants, especially since several of the first recordings were from ponds (e.g., 1983 in London, UK [46]); thermal springs (e.g., 1994 in Banjica, Serbia [43]), or thermally polluted waters (e.g., 1981 in Erft, Germany [6] and 1989 in Astrakhan, Russia [42]). Ephemeral occurrences sometimes resulted in expansions during summers in the 1980s and 1990s [55,56]. In 2001, the first overwintering of *Pistia stratiotes* in the temperate zone with winter temperatures well below 0 °C was recorded for a population in a thermal water location [11]. After that, the number of ephemeral observations in the temperate zone increased, including those outside thermally abnormal waters. In some locations, *P. stratiotes* was successfully eradicated (e.g., in Spain [30]), while in some countries the number of observations increased considerably after the year 2000 (Figure 7), e.g., from 49 to 200 in the Netherlands [57] and to 36 in the UK. In Somerset (UK), besides more frequent sightings, beginning with several records in 2004, the populations changed also in their abundance. In 2010, *P. stratiotes* was already well established in the Bridgewater and Taunton Canal [58]. In six European countries with the most records of *P. stratiotes*, 86% of all observations were recorded after 2012 (Figure 7) [45]. However, we have to consider that a high number of records does not necessarily reflect a wide occurrence of the species. Several records in the Global Biodiversity Information Facility (GBIF) [45] database include observations from wildlife recording applications (e.g., iNaturalist or Pl@ntNet), which allow multiple inputs of the same species at the same locality. Nevertheless, such data give information about the actual distribution of a species and the persistence of its populations.

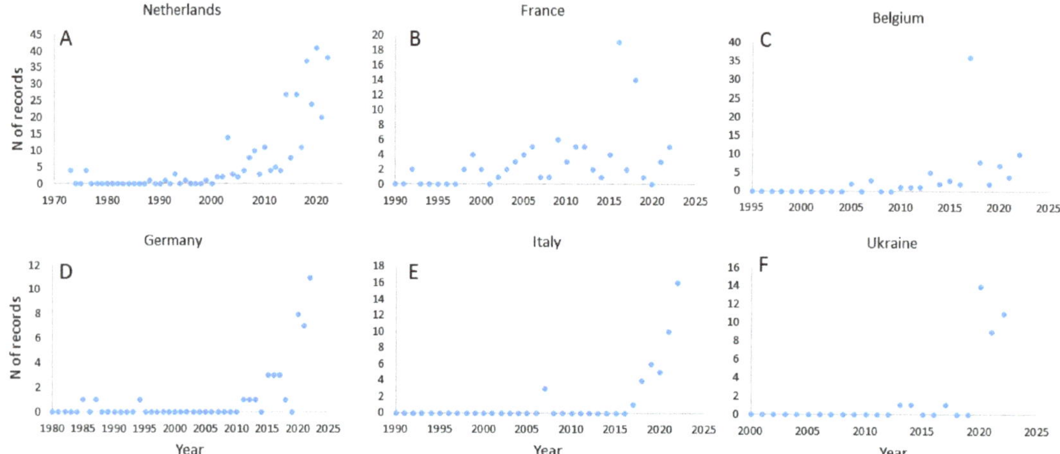

Figure 7. Increase in *Pistia stratiotes* records in the database GBIF since the year 1980 for selected European countries: (**A**) the Netherlands, (**B**) France, (**C**) Belgium, (**D**) Germany, (**E**) Italy, (**F**) Ukraine.

The majority of recent recordings remain ephemeral occurrences in normal waters during summers. However, in the last two decades, several established populations have been recorded in various habitats. Since its introduction, *Pistia stratiotes* has remained present in natural thermally abnormal waters in Hungary [29], and it was present until 2022 in Slovenia. Since 2008, it has been established in the thermally polluted River Erft in Germany [6], where it was previously introduced in 1981, but failed to survive [35]. Persisting populations are found in the Mediterranean area. According to the database SI Observation Flore [34], water lettuce was observed in France as early as 1992, in Auvergne. Two more locations were recorded in 1998 and 1999. After that, it occurred abundantly in the River Moselle between 2002 and 2015. In summer, it has been spotted in 31 locations throughout France, with dense, persisting populations in the canals along the River Rhone [59]. *Pistia stratiotes* was also recorded in Italy, where mass reproduction was reported in 1998 in Lombardy [36] and in 2007 in Toscana [60]. Recent distribution ranges include, additionally, the Campania, Emilia-Romagna, and Veneto regions [16,61]. In Spain, *P. stratiotes* was first observed in the province of Gipuzkoa in 2001. In 2004, a population was recorded in the vicinity of Doñana National Park in Andalusia [30]. The information was communicated rapidly to the environmental authorities, and the plant was removed from the continental part of Spain according to the EPPO [19], although it is invasive on the Canary Islands. In Portugal, the first reported naturalization of water lettuce was in 1990 [40].

The mild climate has also enabled persistently recurring populations for more than five years in the South of England and Belgium (since 2000) [32] and Slovakia (since 2007) [44,62]. Increases in population numbers have been reported from less favorable climatic regions as well, e.g., 15 populations in the Czech Republic since the first observation in 1999 [33], while mass reproduction was recorded in 2013 in Ukraine (near Ekshar) [47]. Persisting populations have been reported from the European part of Russia [63]. Habitats invaded by persistent *P. stratiotes* populations in Europe are very similar and are characterized by slow-flowing, shallow waters, often in canals, experiencing strong human influence. Plants in natural waters form smaller rosettes with smaller leaves than plants in thermally abnormal waters [62].

Additionally, new records became more frequent in regions with a more pronounced temperate climate. *P. stratiotes* was found in 2005 in Romania [41] and in 2012 in Poland [39], and in 2017 it was observed for the first time in a natural river in Vojvodina, Serbia, near the Romanian border [64]. It was also recorded as far north as Sweden (in 2006 and 2011) [45] and Norway, where it managed to form a dense population during the summer of 1989. However, it did not survive winters [19,38].

In several locations, thermally abnormal waters pose an invasion risk for further *P. stratiotes* spread. Spontaneous *P. stratiotes* spread from a thermal safe site was recorded for the first time in 2017, when plants were flushed from the thermal stream Topla in Slovenia into the Sava River and managed to colonize a natural reserve area in Croatia—the Sava-Strmec, 15 km downstream—during the summer [10]. A similar spread can be expected from the thermally polluted Erft River in Germany, because of the high number of plants and seeds drifting downstream into the River Rhine [65,66].

Similar to EU countries, the increase in *P. stratiotes* records in Japan was noted around the year 2000. Until then, *P. stratiotes* had been present, since the 1930s, only on the Okinawa Islands, while the expansion to many sites, including an overwintering population in thermally abnormal water in Japan's temperate zone, began after the late 1990s ([67] and references therein). This might be an indication that additional *P. stratiotes* expansion is occurring globally. Additionally, the spread of *P. stratiotes* in Europe recently might be attributed to global warming [16,17], which might also be the case globally because the spread of *P. stratiotes* into subtropical and tropical regions is enhanced [68].

4.3. Effect of Temperature on P. stratiotes Performance

Temperature and photoperiod are abiotic variables that directly influence *P. stratiotes* [69]. This is shown in our study. Even though in our study site winter air temperatures fell below 0 °C, plants survived in the warmer part of the gradient. The suitability of the thermal gradient for *P. stratiotes* growth changed during the season, since parts close to the thermal source became too warm in June and July, while the lower part warmed up as summer approached and turned out to be the most suitable for high photosynthetic yield from May onwards. The wider tolerance of free-floating freshwater macrophytes for low sub-optimal temperatures than for high sub-optimal temperatures [70] is also confirmed for *P. stratiotes* in our study. However, because one can measure more extreme temperature tolerance values if such temperature is applied for only a short time [71], we need to measure the response of plants which are under long-term exposure or have had the opportunity to adapt to an environment with periods of sub-optimal temperatures. In Brazil and in our study site, similar high water temperatures (24 to 28 °C) were most favorable for *P. stratiotes* growth. These water temperatures were present in Brazil in August [69] and in our study site in July. According to our study's regression models, the water temperatures at which maximum plant biomass was reached were 28.8 ± 3.5 °C.

Such mean temperatures were present in the lower part of the thermal gradient in July, which was also the part of the gradient that was closest and most similar to the natural Sava River. This poses the threat of a spillover of plants [10,65] or their seeds [7,11] from the thermal safe site into natural waters during the warm summer. If we take into consideration the sigma range of 3.5 °C, favorable temperatures above 25.3 °C were already present along the entire gradient in May. However, at that time, plants had a low F_v/F_m ratio, indicating a low quantum yield of PSII, which increased after dark adaptation to 0.72 on average. The dark-adapted fluorescence parameters reflect the degree of acclimation to local environmental conditions, and we can assume that in favorable environmental conditions the plants can express their best potential photosynthetic performance [72]. Therefore, the failure of the dark-adapted F_v/F_m to reach the critical value of 0.8 in May showed that plants were under stress. In June, the mean dark-adapted F_v/F_m reached 0.8

at the very end of our temperature gradient. In July, the mean dark-adapted F_v/F_m was higher than 0.8 for the last three sampling points.

In comparison, the water temperature gradient in May ranged from 33 °C to 25 °C, while the mean air temperature was between 22 and 23 °C. In June and July, the water temperature gradient (from 33 °C to 26 °C) did not change considerably compared to May, while the mean air temperatures rose along the entire gradient to 25–27 °C in July and to 26–28 °C in August. According to our results, we can conclude that increased air temperatures in June and July had an important role in the increased F_v/F_m ratio indicating that plants were not under stress. At the same time, dark-adapted F_v/F_m above 0.8 reflected a high quantum yield of PSII and therefore photosynthetic efficiency, which is shown also by increased plant biomass in June and particularly for the last sampling points in July. The importance of air temperature was also stressed in the tropics, where winter air temperatures (16 °C at rosette height) were unfavorable [69]. According to our results in the temperate zone with pronounced winters, the projected global warming will not enable *P. stratiotes* to overwinter in the form of rosettes outside thermally abnormal waters, because the average winter temperature would need to increase significantly, which is unlikely to happen.

5. Conclusions

According to the presented literature review, we can conclude that in thermally abnormal waters submerged thermophile macrophytes have a higher probability of survival. Submerged plants are surrounded by warm water with more or less stable temperatures. On the other hand, free-floating thermophile macrophytes depend also on air temperatures. This is also the reason why almost 50% of non-native macrophytes in European thermally abnormal waters are submerged or rooted, and free-floating macrophytes represent only 18% of species. Here we have shown that we should pay attention to free-floating macrophytes, which can survive in thermally abnormal waters and which, at the same time, can tolerate low air temperatures and even considerable frost damage to their leaves. Additionally, there might be an indication that free-floating plants such as *P. stratiotes* have a slightly higher possibility of dispersal compared to rooted macrophytes. Increased water flow or elevated water levels caused by rain or changed discharges can enable easier spillover of free-floating plants than of rooted, submerged plants. What is more, waterfowl, pet dogs, and water equipment constitute additional possible vectors of dispersal. Free-floating macrophytes originating from warmer regions, particularly those plants that do not have leaves close to the water surface, are very interesting subjects for investigation of their temperature tolerance when air temperatures are limited.

Supplementary Materials: The following supporting information can be downloaded at: https://www.mdpi.com/article/10.3390/d15030421/s1, Table S1: Physical and chemical parameters of the environment measured at 10 permanent sampling points along thermal stream Topla in 2015; Table S2: Morphological and ecophysiological characteristics of *P. stratiotes* measured at 10 permanent sampling plots along thermal stream Topla.

Author Contributions: Conceptualization, N.Š.; methodology, N.Š. and T.U.; formal analysis, M.Š. and P.K.; investigation, T.U.; data curation, N.Š., M.Š. and P.K.; writing—original draft preparation, N.Š. and M.Š.; writing—review and editing, N.Š., P.K., and M.Š.; visualization, N.Š., M.Š. and P.K.; supervision, N.Š.; project administration, N.Š.; funding acquisition, N.Š. All authors have read and agreed to the published version of the manuscript.

Funding: This research was partly performed within the project Creative Path to Practical Knowledge funded by the Slovenian Ministry of Education and Research and partly funded by the Slovenian Research Agency ARRS (program P1-0403 and project J1-2457).

Institutional Review Board Statement: Not applicable.

Data Availability Statement: The data supporting reported results are included as the supplementary material Tables S1 and S2.

Acknowledgments: The authors would like to acknowledge Petra Arh for her help in the field. The authors are grateful to three anonymous reviewers for their insightful comments that improved the final paper.

Conflicts of Interest: The authors declare no conflict of interest.

Appendix A

Table A1. List of alien aquatic plant taxa associated with thermally abnormal waters in Europe, including family, pathway, growth form, and EPPO A2 categorization. For each taxon, the year of the first record in a particular country and the native distribution are provided. Source of thermal abnormality (if the information was provided in the reference): G = natural geothermal spring, TP = thermal pollution, D = discharge of a thermal spa.

Taxon	Introduced Range (Year of the First Observation)	Native Distribution	Reference
Azolla filiculoides Lam. (Azollaceae); aquarium plant; free-floating fern	Germany (1991, River Erft—TP) Hungary (1940)	North, Central, and South America [9]	[6,29,73]
Bacopa caroliniana (Walt.) B.L. Robins (Plantaginaceae); aquarium plant; emersed	Hungary (2005)	Southern parts of North America [74]	[29]
Bacopa monnieri (L.) Wettst. (Plantaginaceae); aquarium plant; emersed	Hungary (2005)	Australia, Africa, Asia, South America, and southern parts of North America [74]	[29]
Cabomba caroliniana A. Gray (Cabombaceae); aquarium plant; submersed	Hungary (several localities; 1937, Héviz—G) European Russia (1990, backwater of the Moskva River—TP) Austria (2010 (20. cent.), thermal stream Warmbad Villach—D, G) Romania (2014 Peţa Lake—G)	South America [9]	[29,75–78]
Ceratopteris thalictroides (L.) Brongn. (Pteridaceae); aquarium plant; emersed fern	Hungary (1968) Austria (2013, thermal stream Warmbad Villach—D, G) Romania (2014, Peţa Lake—G)	Widespread in tropical regions [9]	[29,77,78]
Cryptocoryne crispatula subsp. *balansae* (Gagnep.) N.Jacobsen (Araceae); aquarium plant; submersed	Austria (2013, thermal stream Warmbad Villach—D, G)	N Vietnam, Thailand [77]	[77]
Cryptocoryne wendtii de Wit (Araceae); aquarium plant; submersed	Austria (2013, thermal stream Warmbad Villach—D, G)	Thailand [74]	[77]
Egeria densa Planch. (Hydrocharitaceae); aquarium plant; submersed	Germany (1980, River Erft—TP) Slovakia (1993) Hungary (1960) Russia (1983, Pekhorka River—TP) Iceland (2013, pond in Husavik—G; 2004, Opnur—G) Austria (1910, thermal stream Warmbad Villach—D, G)	South America [9]	[6,29,44,73, 76,77,79]

Table A1. *Cont.*

Taxon	Introduced Range (Year of the First Observation)	Native Distribution	Reference
Eichhornia crassipes (Mart.) Solms (Pontederiaceae); ornamental plant; free-floating; EPPO A2	Hungary (1950) Germany (2005, River Erft—TP)	South America [9]	[6,29]
Eichhornia diversifolia (Vahl) Urb. (Pontederiaceae); aquarium plant; free-floating	Hungary (2005)	South America [74]	[29]
Gymnocoronis spilanthoides DC. (Asteraceae); aquarium and aquatic ornamental plant; emersed; EPPO A2	Hungary (1988, Lake Héviz—G)	South America [74]	[29]
Heteranthera zosterifolia Mart. (Pontederiaceae); aquarium plant; submersed	Serbia (2016, Niška Banja—G) Austria (2013, thermal stream Warmbad Villach—D, G)	South America [74]	[77,80]
Houttuynia cordata Thumb. (Saururaceae); ornamental plant; emersed	Hungary (2005)	Southeast Asia [74]	[29]
Hydrilla verticillata (L. f.) Royle (Hydrocharitaceae); aquarium plant; submersed	Hungary (1980) Slovakia (1995) Austria (1907, thermal stream Warmbad Villach—D, G)	Asia [44]	[29,44,77]
Hydrocotyle ranunculoides Lf. (Apiaceae); ornamental pond plant; free-floating; EPPO A2	Germany (2003, River Erft—TP) Hungary (2005)	North, Central, and South America [9]	[6,29]
Hygrophila corymbosa Lindau. (Acanthaceae); aquarium plant; emersed	Hungary (2005)	Southeast Asia [9]	[29]
Hygrophila difformis Blume (Acanthaceae); aquarium plant; emersed	Hungary (2005, thermal pond and parts of the Danube near the Lukács Budapest—G)	S Asia and SE Asia [9]	[29,81]
Hygrophila polysperma (Roxb.) T. Anderson (Acanthaceae); aquarium plant; emersed	Hungary (1958, thermal pond and parts of the Danube near the Lukács Budapest—G) Germany (2005, River Erft—TP) Poland (2008, lakes of the Konin Valley area—TP) Austria (2005, thermal stream Warmbad Villach—D, G)	India and Malaysia [9]	[6,29,77,81,82]
Lagarosiphon major (Ridl.) Moss (Hydrocharitaceae); aquarium plant; submersed	Hungary (2005) Austria (1966, thermal stream Warmbad Villach—D, G)	S Africa [9]	[29,77]
Lemna minuta Kunth (Lemnaceae); unintentional introduction; free-floating	Germany (1981, River Erft—TP) Serbia (2016) Russia (2008, Pekhorka River—TP)	North, Central, and South America [9]	[6,73,76,80]
Lemna aequinoctialis Welw. (Lemnaceae); unintentional introduction; free-floating	Germany (1982, River Erft—TP)	E Asia, southern hemisphere [74]	[35]
Limnophila heterophylla (Roxb.) Benth. (Plantaginaceae); aquarium plant; submersed	Romania (2014, Peța Lake—G) Hungary (2018, thermal pond and parts of the Danube near the Lukács, Budapest—G)	India, SE China, Philippines [74]	[78,81]

Table A1. *Cont.*

Taxon	Introduced Range (Year of the First Observation)	Native Distribution	Reference
Limnophila sessiliflora Blume (Plantaginaceae); aquarium plant; submersed	Hungary (1940) Slovakia (1993, Bojnice)	Asia [44]	[29,44]
Ludwigia sp.—hybrid (Onagraceae); aquarium plant; emersed	Hungary (2018, thermal pond and parts of the Danube near the Lukács, Budapest—G)	/	[81]
Ludwigia alternifolia L. (Onagraceae); ornamental pond plant; emersed	Hungary (1940)	E North America [74]	[29]
Ludwigia grandiflora (Michx.) Greuter & Burdet (Onagraceae); ornamental pond plant; emersed; EPPO A2	Hungary (2005)	South America, parts of North America [74]	[29]
Ludwigia repens J.R. Forst. (Onagraceae); aquarium plant; emersed	Hungary (1924) Serbia (2011, Niška Banja—G) Slovakia (2017)	North and Central America [74]	[29,44,83]
Mimulus guttatus Fisch. ex DC. (Scrophulariaceae); ornamental pond plant; emersed	Hungary (1994)	North America [74]	[29]
Monochoria korsakowii Regel & Maack (Pontederiaceae); ornamental pond plant; emersed	Hungary (1988)	Ukraine, Caucasus, India, East Asia [74]	[29]
Myriophyllum aquaticum (Vell.) Verdc. (Haloragaceae); aquarium plant; submersed	Hungary (1968) Germany (2003, River Erft—TP) Romania (2014, Peţa Lake—G) Austria (1988, thermal stream Warmbad Villach—G, D)	South America [74]	[6,29,73,77, 78]
Myriophyllum heterophyllum Michx. (Haloragaceae); aquarium plant, ornamental pond plant; submersed; EPPO A2	Hungary (2006)	E-North America, Central America [74]	[29]
Myriophyllum tuberculatum Roxb. (Haloragaceae); aquarium plant; submersed	Hungary (2018, thermal pond and parts of the Danube near the Lukács, Budapest—G)	India to China, N Peninsula Malaysia [74]	[81]
Najas gracillima (A. Braun ex Engelm.) Magnus (Hydrocharitaceae); unintentional introduction; submersed	Hungary (2012)	Australasia, China, Eastern Asia, North America [74]	[29]
Najas guadalupensis (Spreng.) Magnus (Hydrocharitaceae); aquarium plant; submersed	Slovakia (1986) Hungary (2005) Romania (2014, Peţa Lake—G)	America [74]	[29,44,78]
Nelumbo nucifera Gaertn. (Nelumbonaceae); ornamental pond plant; leaf floating	Hungary (1955)	Ukraine to north Iran, Russian Far East to Tropical Asia, Australia [74]	[29]
Nuphar advena (Aiton) W.T. Aiton (Nymphaeaceae); ornamental pond plant; leaf floating	Hungary (1920)	North America [74]	[29]
Nymphaea "Blue Bird" (*N. micrantha* x *N. capensis*) (Nymphaeaceae); ornamental pond plant; leaf floating	Hungary (1900)	/	[29]
Nymphaea lotus var. *thermalis* L. (Nymphaeaceae); ornamental pond plant; leaf floating	Hungary (1842)	Endemic to the thermal water of the Peţa River in Romania [84]	[29]

Table A1. *Cont.*

Taxon	Introduced Range (Year of the First Observation)	Native Distribution	Reference
Nymphaea nouchali var. *caerulea* (Savigny) Verdc. (Nymphaeaceae); ornamental pond plant; leaf floating	Hungary (1891, thermal pond and parts of the Danube near the Lukács, Budapest—G)	E Africa [74]	[29,81]
Nymphaea rubra Roxb. ex Andrews (Nymphaeaceae); ornamental pond plant; leaf floating	Hungary (1891)	Tropical Asia [74]	[29]
Pistia stratiotes L. (Araceae); aquarium plant, ornamental pond plant; free-floating; EPPO A2	Hungary (1966) Germany (1981, River Erft—TP) Russia (1989, Volga—TP; 1998 Pekhorka River—TP) Serbia (1994, a thermal spring "Banjica"—G; 2005, Rgoška Banja spa—G) Slovenia (2001, stream Topla—G, D) Slovakia (2007) Ukraine (2011, Seversky Donets—TP)	Pantropical distribution, probably originating from South America [17]	[6,11,29,35, 42,44,47,76, 80]
Pontederia cordata L. (Pontederiaceae); ornamental pond plant; emersed	Hungary (2005)	North America, South America [74]	[29]
Rotala rotundifolia (Buch.-Ham. ex Roxb.) Koehne (Lythraceae); ornamental pond plant; emersed	Hungary (1998) Serbia (2016, Niška Banja—G)	Tropical Asia [74]	[29,80]
Sagittaria latifolia Willd. (Alismataceae); aquarium plant; emersed	Austria (1951, thermal stream Warmbad Villach—D, G	North America, Central America, N South America [74]	[77]
Sagittaria platyphylla (Engelm.) J.G. Sm. (Alismataceae); aquarium plant, ornamental pond plant; emersed	Russia (2002, Pekhorka River—TP)	S North America, Central America [74]	[76]
Sagittaria subulata (L.) Buchenau (Alismataceae); aquarium plant; submersed	Hungary (1965) Slovakia (1995) Germany (1984, Warme Wuhle—TP)	SE North America, N South America [74]	[29,44,85]
Salvinia auriculata Aubl. (Salviniaceae); aquarium plant; free-floating fern	Hungary (1964)	Central and South America [74]	[29]
Saururus cernuus L. (Saururaceae); ornamental plant; emersed	Hungary (2005)	E North America [74]	[29]
Shinnersia rivularis (A.Gray) R.M.King & H.Rob. (Asteraceae); aquarium plant; emersed	Germany (1992, River Erft—TP$_m$) Slovakia (1998, 2002, a waste canal discharging thermal water from the bath house "Kalinka", Bojnice—D, G) Hungary (1998, a thermal lake Hévíz—G) Austria (2000, thermal streams in Warmbad Villach—G)	Central America [9]	[29,44,86]
Utricularia gibba L. (Lentibulariaceae); aquarium plant; free-floating	Hungary (1936) Slovakia (1993)	America, Africa, Asia [74]	[29,44]
Vallisneria americana Michx. (Hydrocharitaceae); aquarium plant; submersed	Russia (2010, Pekhorka River—TP)	North America, Central America, N South America [74]	[76]

Table A1. *Cont.*

Taxon	Introduced Range (Year of the First Observation)	Native Distribution	Reference
Vallisneria gigantea Graebn. (Hydrocharitaceae); aquarium plant; submersed	Hungary (1891)	SE Asia, Australia [74]	[29]
Vallisneria spiralis L. (Hydrocharitaceae); aquarium plant; submersed	Hungary (1808) Russia (1999, Desnogorsk Reservoir—TP; 1972, Belovskoe Reservoir—TP) Germany (2003, River Erft—TP; 2017, Reden—TP) Serbia (2011, Niška Banja—G) Poland (1993, Konin Lakes—TP) Iceland (2013, pond Husavik—G) Austria (1880, thermal streams in Warmbad Villach—D, G)	N Africa, Asia, S Europe [9]	[6,29,73,77, 79,83,87– 91]
Victoria amazonica Sowerby (Nymphaeaceae); ornamental pond plant; leaf floating	Slovakia (1998)	South America [74]	[44]
Wolffia globosa (Roxb.) Hartog & Plas (Araceae); unintentional introduction; free-floating	Russia (2002, Pekhorka River—TP)	Pakistan to Japan, Malaysia [74]	[76]

References

1. Strayer, D.L.; Dudgeon, D. Freshwater biodiversity conservation: Recent progress and future challenges. *J. N. Am. Benthol. Soc.* **2010**, *29*, 344–358. [CrossRef]
2. Grzybowski, M.; Glińska-Lewczuk, K. Principal threats to the conservation of freshwater habitats in the continental biogeographical region of Central Europe. *Biodivers. Conserv.* **2019**, *28*, 4065–4097. [CrossRef]
3. Díaz, S.; Settele, J.; Brondízio, E.; Ngo, H.; Guèze, M.; Agard, J.; Arneth, A.; Balvanera, P.; Brauman, K.; Butchart, S.; et al. Summary for Policymakers of the Global Assessment Report on Biodiversity and Ecosystem Services of the Intergovernmental Science-Policy Platform on Biodiversity and Ecosystem Services. Project Report. Available online: https://uwe-repository.worktribe.com/output/1493508 (accessed on 5 January 2023).
4. Vilà, M.; Basnou, C.; Pyšek, P.; Josefsson, M.; Genovesi, P.; Gollasch, S.; Nentwig, W.; Olenin, S.; Roques, A.; Roy, D.; et al. How well do we understand the impacts of alien species on ecosystem services? A pan-European, cross-taxa assessment. *Front. Ecol. Environ.* **2010**, *8*, 135–144. [CrossRef]
5. Hulme, P.E. Biological invasions in Europe: Drivers, pressures, states, impacts and responses. In *Biodiversity under Threat*; Hester, R., Harrison, R.M., Eds.; Cambridge University Press: Cambridge, UK, 2007; ISBN 978-0-85404-251-7.
6. Hussner, A.; Heidbuechel, P.; Heiligtag, S. Vegetative overwintering and viable seed production explain the establishment of invasive *Pistia stratiotes* in the thermally abnormal Erft River (North Rhine-Westphalia, Germany). *Aquat. Bot.* **2014**, *119*, 28–32. [CrossRef]
7. Jaklič, M.; Koren, Š.; Jogan, N. Alien water lettuce (*Pistia stratiotes* L.) outcompeted native macrophytes and altered the ecological conditions of a Sava oxbow lake (SE Slovenia). *Acta Bot. Croat.* **2020**, *79*, 35–42. [CrossRef]
8. Hussner, A.; Nehring, S.; Hilt, S. From first reports to successful control: A plea for improved management of alien aquatic plant species in Germany. *Hydrobiologia* **2013**, *737*, 321–331. [CrossRef]
9. Hussner, A. Alien aquatic plant species in European countries. *Weed Res.* **2012**, *52*, 297–306. [CrossRef]
10. Boršić, I.; Rubinić, T. First record of *Pistia stratiotes* L. (Araceae) in Croatia, with the consideration of possible introduction pathways. *Period. Biol.* **2021**, *123*, 35–39. [CrossRef]
11. Šajna, N.; Haler, M.; Škornik, S.; Kaligarič, M. Survival and expansion of *Pistia stratiotes* L. in a thermal stream in Slovenia. *Aquat. Bot.* **2007**, *87*, 75–79. [CrossRef]
12. Wilk-Woźniak, E.; Najberek, K. Towards clarifying the presence of alien algae in inland waters: Can we predict places of their occurrence? *Biologia* **2013**, *68*, 838–844. [CrossRef]
13. Hussner, A.; van Dam, H.; Vermaat, J.E.; Hilt, S. Comparison of native and neophytic aquatic macrophyte developments in a geothermally warmed river and thermally normal channels. *Fundam. Appl. Limnol.* **2014**, *185*, 155–166. [CrossRef]

14. Piria, M.; Radočaj, T.; Vilizzi, L.; Britvec, M. Climate change may exacerbate the risk of invasiveness of non-native aquatic plants: The case of the Pannonian and Mediterranean regions of Croatia. In *Recent advancements in the Risk Screening of Freshwater and Terrestrial Non-Native Species*; Giannetto, D., Piria, M., Tarkan, A.S., Zięba, G., Eds.; NeoBiota: Sofia, Bulgaria, 2022; pp. 25–52. [CrossRef]
15. Ormerod, S.J.; Dobson, M.; Hildrew, A.G.; Townsend, C.R. Multiple stressors in freshwater ecosystems. *Freshw. Biol.* **2010**, *55*, 1–4. [CrossRef]
16. Brundu, G.; Stinca, A.; Angius, L.; Bonanomi, G.; Celesti-Grapow, L.; D'Auria, G.; Griffo, R.; Migliozzi, A.; Motti, R.; Spigno, P. *Pistia stratiotes* L. and *Eichhornia crassipes* (Mart.) Solms.: Emerging invasive alien hydrophytes in Campania and Sardinia (Italy). *Bull. OEPP* **2012**, *42*, 568–579. [CrossRef]
17. Neuenschwander, P.; Julien, M.H.; Center, T.D.; Hill, M.P. Pistia stratiotes L. (Araceae). In *Biological Control of Tropical Weeds Using Arthropods*; Muniappan, R., Reddy, G.V.P., Raman, A., Eds.; Cambridge University Press: London, UK, 2009; pp. 332–352. ISBN 978-0-521-87791-6.
18. Cilliers, C.J. Biological control of water lettuce, *Pistia stratiotes* (Araceae), in South Africa. *Agric. Ecosyst. Environ.* **1991**, *37*, 225–229. [CrossRef]
19. EPPO. *Pistia stratiotes* L. Datasheets on pests recommended for regulation. EPPO Bull. **2017**, *47*, 537–543. [CrossRef]
20. Van Dijk, G. *Vallisneria* and its interactions with other species. *Aquatic* **1985**, *7*, 6–10.
21. Simberloff, D.; Schmitz, D.C.; Brown, T.C. *Strangers in Paradise: Impact and Management of Nonindigenous Species in Florida*; Island Press: Washington, DC, USA, 1997.
22. Labrada, R.; Fornasari, L. *Management of the Worst Aquatic Weeds in Africa. FAO Efforts and Achievements during the Period 1991–2001*; FAO: Rome, Italy, 2002.
23. ARSO 2023. ARHIV—Opazovani in Merjeni Meteorološki Podatki po Sloveniji. Available online: https://meteo.arso.gov.si/met/sl/archive/ (accessed on 11 January 2023).
24. Records ICAO/OACI: SBCR. Available online: https://en.tutiempo.net/records/sbcr (accessed on 7 November 2018).
25. Records ICAO/OACI: HKEL. Available online: https://en.tutiempo.net/records/hkel (accessed on 7 November 2018).
26. ESRI. *ArcView GIS 3.2.*; Environmental Systems Research Institute, Inc.: Redlands, CA, USA, 1999.
27. Roháček, K.; Soukupová, J.; Barták, M. Chlorophyll fluorescence: A wonderful tool to study plant physiology and plant stress. *Res. Signpost.* **2008**, *37*, 41–104.
28. EPPO A2 List of Pests Recommended for Regulation as Quarantine Pests. Version 2022-09. Available online: https://www.eppo.int/ACTIVITIES/plant_quarantine/A2_list (accessed on 11 January 2023).
29. Lukács, B.A.; Mesterházy, A.; Vidéki, R.; Király, G. Alien aquatic vascular plants in Hungary (Pannonian ecoregion): Historical aspects, data set and trends. *Plant Biosyst.* **2016**, *150*, 388–395. [CrossRef]
30. García Murillo, P.; Dana Sanchez, E.D.; Rodrigez Hiraldo, C. *Pistia stratiotes* L. (Araceae) una planta acuatica en las proximidades del parquet nacional de Doñana (SW Espana). *Acta Bot. Malacit.* **2005**, *30*, 235–236. [CrossRef]
31. Pall, K.; Mayerhofer, V.; Mayerhofer, S. *Aquatische Neobiota in Österreich-Stand 2013*; BMLFUW: Wien, Austria, 2013. Available online: https://info.bml.gv.at/service/publikationen/wasser/Aquatische-Neobiota-in--sterreich---Stand-2013.html (accessed on 11 January 2023).
32. Verloove, F. *Catalogue of Neophytes in Belgium (1800–2005)*; Scripta Botanica Belgica; National Botanic Garden of Belgium: Meise, Belgium, 2006; Volume 39.
33. Kaplan, Z.; Danihelka, J.; Lepsi, M.; Lepší, P.; Ekrt, L.; Chrtek, J.; Kocián, J.; Prančl, J.; Kobrlová, L.; Hroneš, M.; et al. Distributions of vascular plants in the Czech Republic. Part 3. *Preslia* **2016**, *88*, 459–544.
34. SI Observation Flore. *Pistia stratiotes* L. Available online: https://siflore.fcbn.fr/?cd_ref=447733&r=metro (accessed on 11 January 2023).
35. Diekjobst, H. *Pistia stratiotes* L. und *Lemna aequinoctialis* Welwitsch vorübergehend im Gebiet der unteren Erft. *Gött. Flor. Rundbr.* **1984**, *18*, 90–95.
36. D'Auria, G.; Zavagno, F. *Indagine sui "Bodri" della Provincia di Cremona. Monografie di «Pianura» N. 3 -1999*; Provincia di Cremona: Cremona, Italy, 1999.
37. Mennema, J. Is water lettuce (*Pistia stratiotes* L.) becoming a new aquatic weed in the Netherlands? *Natura* **1977**, *74*, 187–190.
38. Elven, R.; Hegre, H.; Solstad, H.; Pedersen, O.; Pedersen, P.A.; Åsen, P.A.; Bjureke, K.; Vandvik, V. *Pistia stratiotes*, Vurdering av Økologisk Risiko. Artsdatabanken. Available online: https://artsdatabanken.no/Fab2018/N/1794 (accessed on 15 November 2018).
39. Krajewski, Ł. *Pistia* rozetkowa—Efemerofit flory regionu. *Przyr. Górnego Śląska* **2013**, *72*, 13.
40. Almeida, J.D.; Freitas, H. Exotic naturalized flora of continental Portugal—A reassessment. *Bot. Complut.* **2006**, *30*, 117–130.
41. Lansdown, R.V.; Anastasiu, P.; Barina, Z.; Bazos, I.; Çakan, H.; Caković, D.; Delipetrou, P.; Matevski, V.; Mitić, B.; Ruprecht, E.; et al. Review of Alien Freshwater Vascular Plants in South-east Europe. In *ESENIAS Scientific Reports 1. State of the Art of Alien Species in South-Eastern Europe*; Rat, M., Trichkova, T., Scalera, R., Tomov, R., Uludag, A., Eds.; University of Novi Sad Faculty of Sciences: Novi Sad, Serbia, 2016; pp. 137–154. ISBN 978-86-7031-3316.
42. Pilipenko, V.N. The tropical species *Pistia stratiotes* (Araceae) in the delta of Volga river. *Bot. Zhurnal* **1993**, *78*, 119–120.
43. Ranđelović, V.; Zlatković, B.; Jović, D. Tropska biljna vrsta *Pistia stratiotes* L. (Araceae) u flori Srbije. In *II Simpozijum o flori Srbije, IV Simpozijum o flori Jugoistočne Srbije*; Zbornik rezimea: Vranje, Srbia, 1995; pp. 33–34.
44. Hrivnák, R.; Medvecká, J.; Baláži, P.; Bubíková, K.; Otahelová, H.; Svitok, M. Alien aquatic plants in Slovakia over 130 years: Historical overview, current distribution and future perspectives. *NeoBiota* **2019**, *49*, 37–56. [CrossRef]

45. Oliati, M.; Coulson, S. Artportalen (Swedish Species Observation System). Version 92.144. ArtDatabanken. Occurrence Dataset. 2019. Available online: https://www.gbif.org/occurrence/1835762190 (accessed on 12 January 2023).
46. Newman, J. Rapid Risk Assessment Summary Sheet: Water Lettuce (*Pistia stratiotes*). Available online: https://www.nonnativespecies.org/assets/Uploads/RSS_RA_Pistia_stratiotes.pdf (accessed on 12 January 2023).
47. Dikiy, N.P.; Dovbnya, A.N.; Lyashko Yu, V.; Medvedev, D.V.; Medvedeva, E.P.; Botova, M.A.; Khlapova, N.P.; Fedorets, I.D. Radionuclide Biosorption by the Aquatic Plants of *Pistia stratiotes*. *Probl. Act. Sci. Technol.* **2014**, *5*, 50–53.
48. Hussner, A.; van de Weyer, K.; Gross, E.M.; Hilt, S. Comments on increasing number and abundance of non indigenous aquatic macrophyte species in Germany. *Weed Res.* **2010**, *50*, 519–526. [CrossRef]
49. Peres, C.K.; Lambrecht, R.W.; Tavares, D.A.; Chiba de Castro, W.A. Alien Express: The threat of aquarium e-commerce introducing invasive aquatic plants in Brazil. *Perspect. Ecol. Conserv.* **2018**, *16*, 221–227. [CrossRef]
50. Zelnik, I.; Haler, M.; Gaberščik, A. Vulnerability of a riparian zone towards invasion by alien plants depends on its structure. *Biologia* **2015**, *70*, 869–878. [CrossRef]
51. Feuchtmayr, H.; Moran, R.; Hatton, K.; Connor, L.; Heyes, T.; Moss, B.; Harvey, I.; Atkinson, D. Global warming and eutrophication: Effects on water chemistry and autotrophic communities in experimental hypertrophic shallow lake mesocosms. *J. Appl. Ecol.* **2009**, *46*, 713–723. [CrossRef]
52. Netten, J.C.; Arts, G.H.P.; Gylstra, R.; Van Nes, E.H.; Scheffer, M.; Roijackers, R.M.M. Effect of temperature and nutrients on the competition between free-floating *Salvinia natans* and submerged *Elodea nuttallii* in mesocosms. *Fund. Appl. Limnol.* **2010**, *177*, 125–132. [CrossRef]
53. Adebayo, A.A.; Briski, E.; Briski, E.; Kalaci, O.; Hernandez, M.; Ghabooli, S.; Beric, B.; Chan, F.T.; Zhan, A.; Fifield, E.; et al. Water hyacinth (*Eichhornia crassipes*) and water lettuce (*Pistia stratiotes*) in the Great Lakes: Playing with fire? *Aquat. Invasions* **2011**, *6*, 91–96. [CrossRef]
54. Simberloff, D.; Von Holle, B. Positive interactions of nonindigenous species: Invasional meltdown? *Biol. Invasions* **1999**, *1*, 21–32. [CrossRef]
55. Pieterse, A.H.; Lange, L.D.; Verhagen, L. A study on certain aspects of seed germination and growth of *Pistia stratiotes* L. *Acta Bot. Neerl.* **1981**, *30*, 47–57. [CrossRef]
56. Venema, P. Fast spread of water lettuce (*Pistia stratiotes* L.) around Meppe. *Gorteria* **2001**, *27*, 133–135.
57. FLORON Verspreidingsatlas Vaatplanten: *Pistia stratiotes* L. Available online: https://www.verspreidingsatlas.nl/5269# (accessed on 15 January 2023).
58. Parker, S. Reports of Meetings. Somerset Rare Plant Group Newsletter 2010, 11. Available online: http://www.somersetrareplantsgroup.org.uk/wp-content/uploads/2014/11/2010-Newsletter-11.pdf (accessed on 12 January 2023).
59. Thiébaut, G.; Dutartre, A. Management of invasive aquatic plants in France. In *Aquatic Ecosystem Research Trends*; Nairne, G.H., Ed.; Nova Science Publishers: Hauppauge, NY, USA, 2009.
60. Ercolini, P. *Pistia stratiotes* L. (Alismatales: Araceae) in Versilia (Toscana nordoccidantale). *Biol. Ambient.* **2008**, *22*, 45–49.
61. Stinca, A.; D'Auria, G.; Motti, R. Sullo status invasivo di *Bidens bipinnata*, *Phoenix canariensis*, *Pistia stratiotes* e *Tradescantia fluminensis* in Campania (Sud Italia). *Inf. Bot. Ital.* **2012**, *44*, 295–299.
62. Ružičková, J.; Lehotská, B.; Takáčová, A.; Semerád, M. Morphometry of alien species *Pistia stratiotes* L. in natural conditions of the Slovak Republic. *Biologia* **2020**, *75*, 1–10. [CrossRef]
63. Shapovalov, M.I.; Saprykin, M.A. Alien Species *Pistia Stratiotes* L. (Araceae) in Water Bodies of Urbanized Territories of Southern Russia. *Russ. J. Biol. Invasions* **2016**, *7*, 195–199. [CrossRef]
64. Živković, M.M.; Anđelković, A.A.; Cvijanović, D.L.; Novković, M.Z.; Vukov, D.M.; Šipoš, Š.Š.; Ilić, M.M.; Pankov, N.P.; Miljanović, B.M.; Marisavljević, D.P.; et al. The beginnings of *Pistia stratiotes* L. invasion in the lower Danube delta: The first record for the Province of Vojvodina (Serbia). *BioInvasions Rec.* **2019**, *8*, 218–229. [CrossRef]
65. Hussner, A.; Heiligtag, S. *Pistia stratiotes* L. (Araceae), die Muschelblume, im Gebiet der unteren Erft (Nordrhein-Westfalen): Ausbreitungstendenz und Problempotenzial. *Veröffentlichungen Des. Boch. Bot. Ver.* **2013**, *5*, 1–6.
66. Hussner, A.; Heidbüchel, P.; Coetzee, J.; Gross, E.M. From introduction to nuisance growth: A review of traits of alien aquatic plants which contribute to their invasiveness. *Hydrobiologia* **2021**, *848*, 2119–2151. [CrossRef]
67. Tamada, K.; Itoh, K.; Uchida, Y.; Higuchi, S.; Sasayama, D.; Azuma, T. Water lettuce overwinter with heat waste. *Weed Biol. Manag.* **2012**, *15*, 20–26. [CrossRef]
68. Hoveka, L.N.; Bezeng, B.S.; Yessoufou, K.; Boatwright, J.S.; Van der Bank, M. Effects of climate change on the future distributions of the top five freshwater invasive plants in South Africa. *S. Afr. J. Bot.* **2016**, *102*, 33–38. [CrossRef]
69. Cancian, L.F.; Camargo, A.F.M.; Silva, G.H.G. Crescimento de *Pistia stratiotes* em diferentes condiçõesde temperatura e fotoperíodo. *Acta Bot. Bras.* **2009**, *23*, 552–557. [CrossRef]
70. van der Heide, T.; Roijackers, R.M.; van Nes, E.H.; Peeters, E.T.H.M. A simple equation for describing the temperature dependent growth of free-floating macrophytes. *Aquat. Bot.* **2006**, *84*, 171–175. [CrossRef]
71. Whiteman, J.B.; Room, P.M. Temperatures lethal to *Salvinia molesta* Mitchell. *Aquat. Bot.* **1991**, *40*, 27–35. [CrossRef]
72. Desotgiu, R.; Pollastrini, M.; Cascio, C.; Gerosa, G.; Marzuoli, R.; Bussotti, F. Chlorophyll *a* fluorescence analysis along a vertical gradient of the crown in a poplar (Oxford clone) subjected to ozone and water stress. *Tree Physiol.* **2012**, *32*, 976–986. [CrossRef]
73. Hussner, A.; Lösch, R. Alien aquatic plants in a thermally abnormal river and their assembly to neophyte-dominated macrophyte stands (River Erft, Northrhine-Westphalia). *Limnologica* **2005**, *35*, 18–30. [CrossRef]

74. Plants of the World Online. Facilitated by the Royal Botanic Gardens, Kew. Available online: http://www.plantsoftheworldonline. org/ (accessed on 9 January 2023).
75. Király, G.; Steták, D.; Bányász, A. Spread of invasive macrophytes in Hungary. In *Biological Invasions—From Ecology to Conservation*; Rabitsch, W., Essl, F., Klingenstein, F., Eds.; NeoBiota: Sofia, Bulgaria, 2008; pp. 123–130.
76. Shcherbakov, A.V.; Ershkova, E.V.; Khapugin, A.A.; Lyubeznova, N.V. Alien Aquatic Plant Species in European Russia. *Bot. Zh.* **2022**, *107*, 70–80. [CrossRef]
77. Ofenböck, G. *Aquatische Neobiota in Österreich*; Bundesministerium für Land—Und Forstwirtschaft, Umwelt und Wasserwirtschaft: Wien, Österreich, 2018; ISBN 978-3-89432-287-8.
78. Bunea, G.; Oroian, I.G.; Todea, A. Law helps, but shows up rather late: The case of Biharean thermal endemic species. *Poeciliid Res.* **2014**, *4*, 13–18.
79. Wasowicz, P.; Przedpelska-Wasowicz, E.M.; Guðmundsdóttir, L.; Tamayo, M. *Vallisneria spiralis* and *Egeria densa* (Hydrocharitaceae) in arctic and subarctic Iceland. *New J. Botany* **2014**, *4*, 85–89. [CrossRef]
80. Zlatković, B.K.; Bogosavljević, S.S. Risk analysis of alien plants recorded in thermal waters of Serbia. *Weed Res.* **2020**, *60*, 85–95. [CrossRef]
81. Kempkes, M.; Lukas, J.; Bierbach, D. *Tropische Neozoen in heimischen Fließgewässern Guppys und andere Exoten in Gillbach und Erft—Ursachen, Folgen, Perspektiven*; Wolf, VerlagsKG: Magdeburg, Germany, 2018; ISBN 978-3-89432-287-8.
82. Gabka, M.; Owsianny, P. First records of the *Hygrophila polysperma* Roxb T. Anderson (Acanthaceae) in Poland. *Bot. Steciana* **2009**, *13*, 9–14.
83. Milenković, M.; Žikić, V.; Stanković, S.S.; Marić, S. First study of the guppy fish (*Poecilia reticulate* Peters,) occurring in natural thermal waters of Serbia. *J. Appl. Ichthyol.* **2014**, *30*, 160–163. [CrossRef]
84. Veler, A. *Nymphaea lotus* up North, Naturally. Water Gardeners International 2008, 3. Available online: http://www. watergardenersinternational.org/journal/3-4/ana/page1.html (accessed on 4 January 2023).
85. Mühlberg, H. *Sagittaria subulata* im Stadtgebiet von Berlin. *Schlechtendalia* **2000**, *5*, 27–30.
86. Eliáš, P.; Hájek, M.; Hájková, P. A European warm waters neophyte *Shinnersia rivularis*—New alien species to the Slovak flora. *Biologia* **2009**, *64*, 684–686. [CrossRef]
87. Katsman, E.A.; Kuchkina, M.A. Introduction of *Vallisneria spiralis* into the Desnogorsk Reservoir. *Russ. J. Biol. Invasions* **2010**, *1*, 159–161. [CrossRef]
88. Babko, R.; Fyda, J.; Kuzmina, T.; Hutorowicz, A. Ciliates on the Macrophytes in Industrially Heated Lakes (Kujawy Lakeland, Poland). *Vestn. Zool.* **2010**, *44*, 1–11. [CrossRef]
89. Zdanowski, B.; Napiorkowska-Krzebietke, A.; Stawecki, K.; Świątecki, A.; Babko, R.; Bogacka-Kapusta, E.; Czarnecki, B.; Kapusta, A. Heated Konin Lakes: Structure, Functioning, and Succession. In *Polish River Basins and Lakes—PART I: Hydrology and Hydrochemistry*; Korzeniewska, E., Harnisz, M., Eds.; Springer International Publishing: Cham, Switzerland, 2020; pp. 321–349.
90. Yanygina, L.; Kirillov, V.; Zarubina, E.Y. Invasive species in the biocenosis of the cooling reservoir of Belovskaya power plant (Southwest Siberia). *Russ. J. Biol. Invasions* **2010**, *1*, 50–54. [CrossRef]
91. Lukas, J.; Kalinkat, G.; Kempkes, M.; Rose, U.; Bierbach, D. Feral guppies in Germany-a critical evaluation of a citizen science approach as biomonitoring tool. *Bull. Fish Biol.* **2017**, *17*, 13–27.

Article

Classification of Wetland Forests and Scrub in the Western Balkans

Dragan Koljanin [1], Jugoslav Brujić [1], Andraž Čarni [2,3], Đorđije Milanović [1], Željko Škvorc [4] and Vladimir Stupar [1,*]

[1] Department of Forest Ecology, Faculty of Forestry, University of Banja Luka, S. Stepanovića 75A, 78000 Banja Luka, Bosnia and Herzegovina

[2] Jovan Hadži Institute of Biology, Research Centre of the Slovenian Academy of Sciences and Arts, Novi trg 2, 1000 Ljubljana, Slovenia

[3] School for Viticulture and Enology, University of Nova Gorica, Vipavska 13, 5000 Nova Gorica, Slovenia

[4] Faculty of Forestry and Wood Technology, University of Zagreb, Svetošimunska 23, 10000 Zagreb, Croatia

* Correspondence: vladimir.stupar@sf.unibl.org

Abstract: Wetland forests and scrub (WFS) are conditioned by the strong impact of water. They consist of various vegetation types, depending on many factors such as type and duration of flooding, water table level and its fluctuation, river current strength, substrate ability to retain water, etc. WFS vegetation has been insufficiently studied in the Balkan Peninsula, especially in Bosnia and Herzegovina. By means of numerical classification, we aimed to classify Western Balkans WFS at the alliance level, and to identify the main underlying ecological gradients driving the variation in species composition. The dataset containing all published and available unpublished relevés from Slovenia, Croatia and Bosnia and Herzegovina was first classified using the EuroVegChecklist Expert System in Juice software in order to assign the corresponding class to each of the relevés. Relevés were subsequently analyzed within each of the four WFS classes (*Alno glutinosae-Populetea albae*, *Salicetea purpureae*, *Alnetea glutinosae* and *Franguletea*). Cluster analysis resulted in eight alliances, *Salicion albae*, *Salicion triandrae*, *Salicion eleagno-daphnoidis*, *Alno-Quercion*, *Alnion incanae*, *Alnion glutinosae*, *Betulion pubescentis* and *Salicion cinereae*, while one cluster could not be assigned with certainty. Edafic factors were found to be the most important factors determining the floristic composition and syntaxa differentiation of WFS in the study area.

Keywords: *Alnetea glutinosae*; *Alno-Populetea*; ecological factors; ecological gradient; floodplain; *Franguletea*; riparian forests; *Salicetea purpureae*; swamp forests; vegetation

Citation: Koljanin, D.; Brujić, J.; Čarni, A.; Milanović, Đ.; Škvorc, Ž.; Stupar, V. Classification of Wetland Forests and Scrub in the Western Balkans. *Diversity* **2023**, *15*, 370. https://doi.org/10.3390/d15030370

Academic Editors: Mateja Germ, Igor Zelnik, Matthew Simpson and Adriano Stinca

Received: 31 January 2023
Revised: 27 February 2023
Accepted: 2 March 2023
Published: 4 March 2023

1. Introduction

There are three types of natural vegetation, i.e., zonal, extrazonal and azonal, which generally develop in accordance with the biotic, climatic and soil conditions [1–3]. While zonal vegetation is the large-scale expression of climate dominating a particular area (extrazonal vegetation is found in microclimatically suitable habitats outside of its climatic zone), azonal communities exist in different macroclimatic belts due to the strong influence of specific ecological factors that do not allow zonal vegetation to prevail. Examples of such azonal vegetation are forest and scrub communities developed on sites with periodic/regular flooding and/or high groundwater level. In such sites, zonal vegetation is replaced by azonal, i.e., hygrophilous and mesohygrophilous forest and scrub vegetation. Many terms encompass such plant communities (floodplain, alluvial, riparian, swamp forests); however, none of the terms embrace the entirety of this type of vegetation. Junk and Piedade [4] used the term wetland forests to refer to all types of forests that are subject to irregular, seasonal or long-term flooding, but this definition overlooked scrub vegetation. In this paper, therefore, we refer to forest and scrub communities subjected to irregular, seasonal or long-term flooding as "wetland forests and scrub" (WFS). Based on the type of flooding and spatial position of the community in relation to the river stream, WFS are divided into alluvial and swamp WFS [5].

Alluvial WFS are mostly confined to rivers and other smaller streams. There, plant communities are often under the impact of flooding by flowing water. The floristic composition of these stands reflects the specific habitat conditions, such as flood duration, soil relocation, accumulation of nutrients, physical damage to and uprooting of plants, seed transportation and sometimes intensive changes in soil moisture [6–9]. Such events favor plant species that are able to utilize accessible resources, and tolerate disturbance events as well as competitive relationships [10].

On the other hand, swamp WFS are conditioned by micro-relief depressions that can occur near a river but are often not related to flowing water [11]. Important factors for forming and maintaining swamp habitats are a combination of relief depressions, a substrate capability to retain the water on the surface, and the presence of a water source to fill and maintain a high groundwater table. Such ecosystems often lack oxygen and are thus composed of species tolerant to oxidative stress [11]. It should be pointed out that it is sometimes hard to draw a line between a swamp and alluvial WFS, since both types of flooding might be present at the same sites. Stagnant water may also be present only for a part of the year but still have a huge impact on the floristic composition.

WFS ecosystems encompass the physical environment and biological communities of the inland–freshwater interface and are recognized as highly diverse compared to surrounding areas [12]. Their conservation is crucial for preserving biological diversity, since they contain specialist ecological communities and provide crucial ecosystem services, such as species and gene pool diversity conservation, prevention of riverbank erosion, prevention of floods and water retention, nutrients and contaminant retention, carbon fixation and storage, cultural heritage and ecotourism and many others, while occupying a relatively small landscape area [13,14]. However, they are globally suffering intense anthropogenic pressures (e.g., altered natural water regime, habitat fragmentation, invasive species, to name just a few), which puts them among the most endangered ecosystems of all [7,15,16]. This has led to some of the WFS types (i.e., temperate and boreal hardwood riparian woodland, Mediterranean and Macaronesian riparian woodland, and broadleaved swamp woodland) being listed as endangered habitats in the Red List of European Habitats [17]. They are also listed in Annex I of the European Union Habitats Directive as habitats of community interest for conservation [18].

We consider that the basic prerequisite for successful legal protection, preservation, monitoring and restoration of habitats is a good understanding of ecological factors and drivers that make and maintain those habitats. This can hardly be achieved without adequate classification. In Europe, WFS communities have to some extent been relatively well studied on the European [11,19] and the national levels [20–25]. However, different communities of WFS are differentiated by subtle differences in the water table, which often leads to mosaics and transitional communities, making them difficult to classify as a complex. Douda et al. [11] thus did not analyze *Salix* dominated communities, while Kalníková et al. [19] treated only gravel-bar scrub vegetation, which is represented by only one alliance in this part of Europe (i.e., *Salicion eleagno-daphnoidis*). Furthermore, in the study of Douda et al. [11], the Balkan Peninsula was heavily underrepresented, which is especially true for Bosnia and Herzegovina (B&H), where there were virtually no relevés included in the study.

Although there have been some reviews of WFS in the Western Balkans [9,26–29], they were not made based on numerical analyses, while syntaxonomic frameworks and concepts are often outdated and non-congruent among themselves. One of the major reasons for the lack of comprehensive analyses has been the insufficient number of published relevés of these communities in part of the research area. For instance, only a few dozen relevés have been published in B&H [30–33].

Another issue is the ambiguous treatment of some alliances listed in Mucina et al. [5]. For example, the *Alno-Quercion* alliance has been a matter of contention from the beginning [9,11,34,35], whereby it has sometimes been considered part of *Alnion incanae*, and sometimes as an alliance of its own. Another example is the alliance *Fraxino-Quercion*, which includes elm–ash and oak riparian forests and was previously considered to be part of *Alnion incanae*, under the suballiance name *Ulmenion* Oberd, 1953. Albeit *Ulmenion* is known to occur in the study area [9], *Fraxino-Quercion* is by definition geographically limited to Central Europe and, consequently, is, a priori, not to be found in the areas south of Austria, Hungary and Romania (i.e., southern Pannonia and the Balkans) [5,36]. Additionally, poplar (*Populus nigra* and *P. alba*)-dominated forests from the region of the Western Balkans do not have a proper syntaxonomic status, with *Populion albae* being by definition a Mediterranean alliance.

This study aimed to provide a comprehensive overview of Western Balkans WFS based on the numerical classification of all available published and unpublished phytosociological relevés from this region. Specifically, we aimed to (a) provide a consistent classification of Western Balkans WFS at the alliance level; (b) characterize the identified vegetation types by their species composition, ecology and distribution; and (c) identify the main underlying gradients driving the variation in species composition of WFS in the Western Balkans.

2. Materials and Methods

2.1. Study Area

The study area encompasses the Western Balkans region, i.e., the southwestern margin of the Pannonian Basin and Dinaric and Julian Alps in Slovenia, Croatia and B&H. Biogeographically, it includes Continental and Alpine biogeographic regions [37]. For the purpose of this study, the Mediterranean biogeographic region was not considered (Figure 1). The area covers approximately 103,000 km^2 (13.3468° E to 19.6534° E and 43.2207° N to 46.8758° N). The northern part of the study area is mostly represented by lowlands with floodplains of large and slow flowing rivers (Sava, Mura, Drava, Krka, Kupa, Danube, Una, Vrbas, Bosna and Drina), depositing finer sediment. The area has been subjected to intensive anthropogenic pressure for a long period of time, with one of the most affected areas being north B&H (Sava River floodplain with its tributaries), which has undergone significant water regime changes, deforestation and conversion of land into intensive agricultural fields in the past few centuries, with the majority of natural forests destroyed [32,38,39]. Lowland forests in Croatia and Slovenia are better preserved but still under great pressure [9,40,41]. The southern and western parts of the research area are represented by the hills and mountains of the Dinaric and Julian Alps. Most of the rivers and streams in this area are smaller and faster, depositing alluvial material of coarser structure. Forest vegetation is usually developed in narrow strips along the streams with the exception of karst poljes, where there is flat terrain and soil conditions similar to those in lowland floodplains.

2.2. Data Collection and Preparation

Data (phytosociological relevés; vegetation plots) were obtained from three vegetation databases: Slovenia (EU-00-021), Croatia (EU-HR-002) and B&H (EU-BA-001). Codes refer to the Global Index of Vegetation-Plot Databases (www.givd.info (accessed on 30 January 2023)). Relevés were selected from databases based on their original author assignment to WFS or, in the absence of assignment, indicator species based on various sources [5,11,22,24] were used (i.e., *Acer negundo, Alnus glutinosa* agg., *Alnus incana, Betula pendula, Betula pubescens, Frangula alnus, Fraxinus angustifolia, Fraxinus excelsior, Myricaria germanica, Populus alba, Populus canadensis, Populus canesens, Populus nigra, Quercus robur, Salix alba, Salix cinerea, Salix eleagnos, Salix euxina, Salix myrsinifolia, Salix purpurea, Salix triandra, Salix viminalis, Salix x rubens, Ulmus laevis* and *Ulmus minor*). Indicator species were also used in order the clear the dataset of non-forest/scrub relevés, in which the criterion for keeping a relevé in the dataset was the presence of at least one of the indicator species with a combined cover value greater than 25% in upper layers (shrub layer for scrub and tree layer for forest). All relevés without coordinates or sharing the same coordinates, as well as relevés from studies related to forest dieback, were excluded. After a closer inspection of the dataset, we also omitted several relevés originally assigned to *Salicetum albae* and *Lamio orvalae-Salicetum albae* by Dakskobler [42] and Dakskobler et al. [43] due to their transitional and mixed character. Additionally, around 230 relevés from WFS were recorded in B&H in the last several years and added to the dataset. A total of 1994 relevés was compiled in a Turboveg database [44], and exported to JUICE 7 software [45] for further analysis.

Figure 1. Location of the study area (in saturated tints). Inset map shows the position of the study area in Europe.

Taxa recorded for more than one layer were merged into one layer because of inconsistent sampling. Records of species determined to the genus level were deleted. Plant nomenclature followed Euro+Med [46]. Species from taxonomically critical groups that were not always identified by the relevé authors were combined into aggregates (agg.), and species that included several subspecies that were not always recorded or recognized by authors were combined and marked with the abbreviation "s.l." (sensu lato) and are also listed in Appendix A. The newly described taxon *Alnus rohlenae* [47] was treated as part of *Alnus glutinosa* agg. We also merged all subspecies of *Fraxinus angustifolia*, since those were not consistently recorded. Although many authors did not record mosses, we kept them in the dataset for the purpose of expert classification.

Although WFS belong to four vegetation classes, they usually share a number of species, and often only the cover ratio of these species makes a difference even between different classes. Furthermore, one class (*Frunguletea*) is differentiated more physiognomically (scrub) than floristically, which made it hard to delineate the class from the remaining swamp communities within the original dataset. We thus used the EuroVegChecklist (EVC) expert system of classification of vegetation plots to classes [5] to divide the initial dataset into four vegetation classes of WFS that would be subsequently analyzed separately. It was performed in JUICE software using a sum of powered species cover with no transformation.

Since the original EVC expert system species list was missing some important characteristic species required for the proper assignment of relevés to some of the classes (especially *Salicetea purpureae*), we modified it based on species diagnostic for classes according to various authors [11,22,24]. Specifically, *Urtica dioica*, *Echinocystis lobata*, *Humulus lupulus*, *Phalaroides arundinacea*, *Poa trivialis*, *Galium aparine*, *Solidago gigantea* and *Acer negundo* were assigned to *Salicetea purpureae*; *Carex riparia* to *Alno-Populetea*; *Carex riparia* and *Caltha palustris* to *Alnetea glutinosae*; and *Salix pentandra* to *Franguletea*. Since the EVC expert system outputs a list of classes ordered by the decreasing value of a relevé's affiliation to the given class for each and every relevé, all relevés with initial best scores for non-forest/scrub vegetation class were reassigned to the first scrub or forest class with the next best value. After this, we only kept relevés assigned to *Alnetea glutinosae*, *Franguletea*, *Salicetea purpureae* and *Alno glutinosae-Populetea albae*, while other classes were omitted from the further analyses (mainly *Quercus robur*-dominated plots belonging to *Carpino-Fagetea*). At this point, we removed all mosses from the dataset because of inconsistent sampling. The dataset was then divided into four subdatasets belonging to four WFS classes. Within each subdataset, we performed outlier analysis using PC-ORD 5.0 [48], and relevés whose species composition deviated more than ±2SD from the mean calculated Euclidean distance of all plots within the subdataset were omitted. This resulted in a total of 1086 relevés in all four subdatasets combined (*Salicetea purpureae* subdataset—210 relevés; *Alno-Populetea* subdataset—685 relevés; *Alnetea glutinosae* subdataset—135 relevés; and *Franguletea* subdataset—56 relevés). Prior to the numerical analysis of subdatasets, we deleted species occurring in up to two relevés in a subdataset since the removal of rare species has proven to be useful in minimizing noise of classification.

2.3. Data Analysis

Hierarchical classification was performed on three subdatasets (*Salicetea purpureae*, *Alno-Populetea* and *Alnetea glutinosae*). We did not divide the *Franguletea* subdataset further, since this class is only represented by one alliance in the studied area. Classification was carried out using cluster analysis in PC-ORD 5.0. Data were transformed with an ordinal scale with cut levels: 0 3 5 15 25, as proposed by Tichý et al. [49]. The relative Sørensen index, as the distance measure, and beta flexible set to −0.25 for group linkage, were used. Three clusters were accepted as the optimal level of division for *Salicetea purpureae* and *Alno-Populetea* subdatasets, while two clusters were chosen for the *Alnetea glutinosae* subdataset. As well as being best ecologically and floristically interpretable, cluster numbers for *Alno-Populetea* and *Alnetea glutinosae* were also confirmed by the Crispness of Classification method for identifying the optimum number of clusters [50], which was also performed in JUICE software. The optimum number of clusters for *Salicetea purpureae* was two; however, since the next division resulted in differentiating *Salicion triandrae* from *Salicion albae*, we settled for three clusters in this subdataset, too. After classification at the level of subdatasets (classes), all four subdatasets, with previously removed rare species restored, were unified into the final WFS dataset for further analysis.

To showcase the differences and similarities in species composition among alliances across all four classes of WFS, as well as within each of the classes (except *Franguletea*), an overall synoptic table of alliances within the unified WFS dataset, as well as synoptic tables of alliances within each of the classes *Salicetea purpureae*, *Alno glutinosae-Populetea albae* and *Alnetea glutinosae*, were generated in the JUICE program and the phi coefficient was used as the measure of fidelity. For each combination of clusters, each of the nine, three, three and two groups was virtually adjusted to 1/9, 1/3, 1/3 and 1/2 of the size of the entire dataset, while holding the percentage occurrences of species within and outside the target group the same as in the original dataset [51]. The threshold of the phi value was set at 0.30 for a species to be considered diagnostic. Fisher's exact test was calculated and gave a zero fidelity value to species whose phi values were not statistically significant ($p > 0.001$).

All 1086 relevés of the unified final WFS dataset, together with the selected ecological variables, were projected onto a DCA plot. Non-transformed percentage covers of species were used, with rare species downweighted. Species ecological indicator values (EIVs) for temperature, light, moisture, soil reaction and nutrients according to Pignatti et al. [52] were used as explanatory ecological variables. Unweighted average EIVs were calculated in JUICE. The significance of their correlation with the DCA relevé scores was tested using the modified permutation test [53]. Other explanatory variables (bioclimatic, elevation, chorotypes, lifeforms, CSR ecological strategies, urbanity type, type of reproduction and origin of taxa) were tested for the strength of correlation with the first and second DCA axis. Bioclimatic variables were obtained from the WorldClim 2 database [54], chorotypes were determined following Pignatti et al. [52] and Gajić [55], life forms according to Raunkiaer [56], while CSR strategies, urbanity type, type of reproduction and origin of taxa were obtained from the BIOLFLOR database [57]. The significances of correlations between these explanatory variables and DCA relevé scores were calculated using the Kendall tau coefficient in Statistica v. 14.0 software (TIBCO Software Inc.). Only three of those variables with the highest explanatory value were selected for further analysis and projected onto a DCA plot.

Syntaxonomical concepts and nomenclature of higher syntaxa followed Mucina et al. [5]. Complete names of associations and subasociations used in text (with author citation) are listed in Appendix B.

3. Results

3.1. Classification and Ordination

Nine ecologically and floristically distinct clusters of relevés of WFS were obtained after expert classification of the initial dataset and numerical classification of the sub-datasets (Tables 1 and S1–S4, Appendix C, Figures 2–4). Three out of four subdatasets, each representing an individual class gained during expert classification, were subjected to numerical classification, which resulted in: (a) three clusters within *Salicetea purpureae*; (b) three clusters within *Alno glutinosae-Populetea albae*; and (c) two clusters within *Alnetea glutinosae*. We tried to classify the fourth subdataset, i.e., the class *Franguletea*, but it turned out to be very homogeneous group. Bearing in mind that only one alliance (*Salicion cinereae*) from this class is recognized to be present in the study area [36], we have decided not to further divide it.

Table 1. Synoptic table of WFS types in the Western Balkans. Frequencies of species are presented as percentages with phi values multiplied by 100 shown in superscript. Diagnostic species (phi values higher than 0.30) are shaded. Diagnostic species are sorted by decreasing fidelity. Species with a frequency lower than 30% in a cluster for which they are diagnostic are not shown. Only up to 12 species with the highest phi value are presented. Cluster numbers: 1—*Salicion albae*, 2—*Salicion triandrae*, 3—*Salicion eleagno-daphnoidis*, 4—*Alnion incanae*, 5—*not assigned*, 6—*Alno-Quercion*, 7—*Alnion glutinosae*, 8—*Betulion pubescentis*, 9—*Salicion cinereae*. Cluster numbers correspond to Figures 2–4, Tables S1–S4, Appendix C and to those used in the text. The full version of this table is available in Table S1.

Cluster Number	1	2	3	4	5	6	7	8	9
No. of Relevés	90	37	83	281	178	226	121	14	56
Salicion albae									
Salix alba	91 [58.1]	8	23	21	47 [20.4]	1	2	.	9
Salix euxina	36 [33.3]	.	13	14	10	.	5	.	2
Salicion triandrae									
Salix triandra	19 [4]	97 [82]	11	2	1	.	.	.	4
Rorippa sylvestris	6	49 [56.5]	8	1	2	1	.	.	.
Echinocystis lobata	39 [24.2]	59 [44.7]	5	2	11	5	3	.	7
Phalaroides arundinacea	56 [25.6]	78 [44.3]	34 [7.6]	7	15	4	19	.	7
Agrostis stolonifera agg.	34 [10.9]	70 [41.7]	18	3	19	8	2	29	12
Calystegia sepium	46 [25.1]	62 [40.3]	14	5	11	7	4	.	14

Table 1. *Cont.*

Cluster Number	1	2	3	4	5	6	7	8	9
No. of Relevés	90	37	83	281	178	226	121	14	56
Persicaria dubia	12	43 [36.7]	29 [20.4]	6	2	1	2	.	2
Rorippa amphibia	10	30 [35.3]	2	.	1	6	3	.	2
Bidens tripartitus	14	43 [30.9]	4	4	4	26 [12.7]	12	.	14
Salicion eleagno-daphnoidis									
Salix eleagnos	4	.	70 [70.4]	14	2	.	.	.	.
Salix purpurea	18	16	84 [62.9]	9	8	.	5	.	14
Petasites hybridus	1	.	53 [50.8]	31 [24.4]	2	.	1	.	2
Saponaria officinalis	3	.	31 [46.4]	2	3	.	.	.	.
Clematis vitalba	3	.	43 [37.8]	27 [19.4]	17	1	2	.	.
Taraxacum sect. *Taraxacum*	8	16	42 [37.7]	6	8	4	1	.	5
Galium mollugo	8	11	37 [35.5]	6	7	1	4	.	5
Chaerophyllum hirsutum	3	.	35 [30.2]	30 [23.8]	2	.	12	.	5
Alnion incanae									
Corylus avellana	2	.	27	63 [44.8]	20	19	15	.	.
Acer pseudoplatanus	.	.	35 [26.1]	50 [42.6]	3	1	13	.	2
Lamium galeobdolon agg.	3	.	11	49 [42.4]	24	8	6	.	.
Sambucus nigra	30	3	12	74 [38.9]	61 [28.4]	21	21	.	12
Acer campestre	2	.	12	57 [38.2]	21	48 [30]	9	.	.
Fraxinus excelsior	2	.	39 [25.9]	47 [35]	8	2	20	.	4
Brachypodium sylvaticum	13	.	53 [25.7]	64 [34.8]	46 [19.8]	8	16	.	4
Symphytum tuberosum agg.	3	.	4	35 [34.2]	27 [24.1]	1	4	.	.
Geum urbanum	2	.	7	47 [33.6]	30	29	10	.	.
Lamium orvala	.	.	23	33 [33]	13	.	2	.	.
Carex sylvatica	2	.	5	35 [31.8]	17	15	8	.	.
Aegopodium podagraria	26	11	35	70 [31.1]	57 [20.9]	14	26	21	11
Cluster 5									
Populus alba	7	.	.	1	38 [47.5]	8	1	.	.
Ulmus laevis	13	.	1	4	54 [43.1]	38 [26]	7	.	.
Populus nigra	22	.	33 [20.2]	7	54 [42.6]	2	1	.	.
Prunus padus	6	.	4	5	40 [37.9]	4	21	.	4
Arum maculatum	2	.	1	17 [14.7]	30 [34.2]	6	.	.	2
Solidago gigantea	16	.	18	14	44 [33.9]	7	11	.	2
Acer negundo	23 [22.4]	.	2	3	30 [32.1]	4	1	.	.
Galium aparine	56 [26.4]	14	16	27	62 [32]	23	13	.	4
Pulmonaria officinalis	1	.	10	34 [27.5]	37 [30.9]	8	2	.	2
Alno-Quercion									
Fraxinus angustifolia s.lat.	7	.	1	9	43 [19.9]	84 [55.3]	37	.	4
Ulmus minor	4	.	1	10	14	52 [47.3]	15	.	.
Quercus robur	1	.	4	11	37	79 [46.5]	33	36	9
Crataegus laevigata	1	.	1	9	3	39 [42.6]	14	.	.
Acer tataricum	2	.	.	8	1	31 [40.4]	5	.	.
Rumex sanguineus	10	.	2	12	12	42 [37.2]	8	.	5
Carex remota	14	.	.	33	17	61 [35.6]	26	29	2
Glechoma hederacea	44	11	7	31	37	65 [34.5]	11	.	9
Stachys palustris	11	5	1	2	10	46 [34.3]	34 [22]	.	7
Lysimachia nummularia	24	5	5	23	24	62 [30.7]	27	21	27

Table 1. *Cont.*

Cluster Number	1	2	3	4	5	6	7	8	9
No. of Relevés	90	37	83	281	178	226	121	14	56
Alnion glutinosae									
Carex elongata	.	.	.	1	.	20 [12]	55 [53.1]	.	12
Carex vesicaria	7	.	.	2	3	11	38 [39.3]	.	11
Peucedanum palustre	.	.	.	6	2	23 [8.1]	52 [36.2]	36	18
Carex riparia	3	.	.	2	3	24 [20.8]	36 [36]	.	4
Valeriana dioica s.lat.	1	.	4	10	5	16	50 [35.4]	21	25
Lycopus europaeus	14	8	6	24	8	46 [16.3]	65 [32.3]	.	57 [25.7]
Betulion pubescentis									
Betula pubescens	.	.	.	.	.	1	.	86 [91.5]	.
Molinia caerulea agg.	.	5	11	2	.	1	7	93 [81.8]	4
Pinus sylvestris	.	.	4	.	.	.	.	71 [80.7]	.
Sorbus aucuparia	.	.	.	3	.	.	2	57 [70.2]	.
Betula pendula	.	.	.	1	.	1	1	50 [66.5]	.
Salix aurita	.	.	.	1	.	1	2	43 [61.2]	.
Knautia sarajevensis	.	.	.	.	.	.	.	36 [57.5]	.
Calamagrostis villosa	.	.	.	.	.	.	.	36 [57.5]	.
Carex rostrata	.	.	.	.	.	.	.	36 [57.5]	.
Frangula alnus	1	.	20	21	7	41 [7.6]	66 [26.7]	100 [52.5]	23
Epilobium palustre	.	5	.	1	1	.	2	36 [50.3]	.
Rubus hirtus s.lat.	.	.	1	14 [8.8]	.	4	1	43 [49.1]	2
Salicion cinereae									
Salix cinerea	3	8	4	2	.	8	28 [6.3]	36	100 [68.8]
Filipendula ulmaria	14	.	8	26	12	8	45 [23.9]	.	55 [32.8]
Diagnostic species for more than one cluster									
Alnus glutinosa agg.	12	.	11	81 [37.6]	31	47	94 [47.8]	.	7
Other species with high frequency									
Urtica dioica	82 [21]	86 [24]	48	59	66	61	43	.	27
Rubus caesius	73	19	58	70	90 [29.5]	69	46	.	12
Solanum dulcamara	38	81 [29.6]	27	22	13	38	54 [9.9]	29	59 [13.7]
Galium palustre agg.	29	49	6	8	10	58 [16.1]	60 [17.3]	36	70 [24.7]
Ranunculus repens	23	35	16	37	16	48 [10.6]	49 [11.3]	14	64 [23]
Cornus sanguinea	43	.	43	65 [24.7]	64 [24.2]	37	22	.	14
Lysimachia vulgaris	28	19	8	16	10	33	67 [28]	43	50 [15]
Angelica sylvestris	24	11	46 [14.8]	37 [8.1]	12	10	31	29	45
Lythrum salicaria	27	57 [23.9]	18	6	7	31 [3.6]	44 [13.5]	.	52 [19.9]
Iris pseudacorus	43	14	4	6	29	54 [22.3]	52 [20.4]	.	38
Viburnum opulus	10	.	13	43 [13.1]	35	25	53 [21]	36	25
Euonymus europaeus	32	3	12	55 [25.2]	47 [18.9]	31	26	.	12
Persicaria hydropiper	28	57 [28.9]	5	15	10	50 [22.8]	31	.	9
Caltha palustris	7	.	11	30	8	31	48 [21.5]	36	34
Humulus lupulus	43 [17.4]	27	6	32	39 [13.6]	10	26	.	21
Poa trivialis	43 [19.8]	14	22	20	38 [14.8]	21	11	.	18
Deschampsia cespitosa	4	.	40 [17.1]	29 [7.5]	13	22	29	36	11
Crataegus monogyna	7	.	22	48 [26.1]	34	46 [24.4]	14	.	2
Myosotis palustris agg.	12	30	4	10	2	35 [15.8]	40 [20.4]	7	23
Circaea lutetiana	9	.	.	44 [25.7]	35	38 [20.3]	22	. 2	

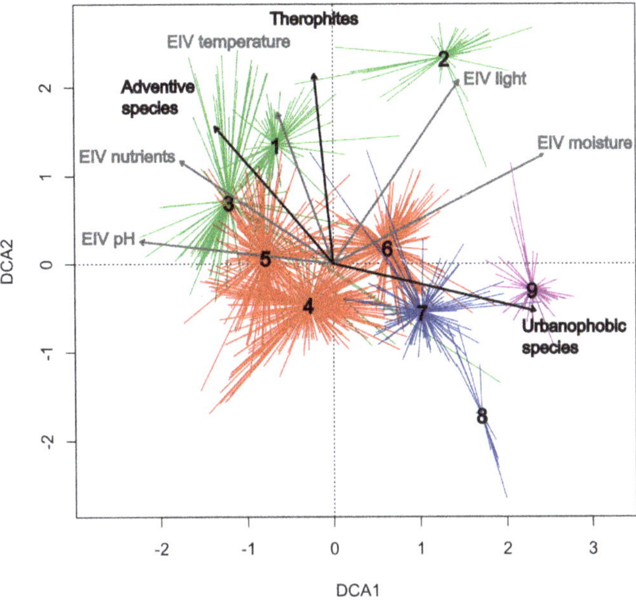

Figure 2. DCA spider plot of the final dataset (1086 relevés). Centroids of clusters are indicated by numbers corresponding to Table 1, Figures 3 and 4, Tables S1–S4, Appendix C and to those used in the text: 1—*Salicion albae*, 2—*Salicion triandrae*, 3—*Salicion eleagno-daphnoidis*, 4—*Alnion incanae*, 5—*not assigned*, 6—*Alno-Quercion*, 7—*Alnion glutinosae*, 8—*Betulion pubescentis*, 9—*Salicion cinereae*. The colors represent groups of clusters (classes): *Salicetea purpureae* (green), *Alno-Populetea* (red), *Alnetea* (blue) and *Franguletea* (purple).

Comparing these results with the accepted definitions of the syntaxa at the alliance level, we found for the most part a good correspondence. The syntaxonomic interpretation of the clusters of the presented classification was as follows: cluster 1—*Salicion albae*; cluster 2—*Salicion triandrae*; cluster 3—*Salicion eleagno-daphnoidis*; cluster 4—*Alnion incanae*; cluster 6—*Alno-Quercion roboris*; cluster 7—*Alnion glutinosae*; cluster 8—*Betulion pubescentis*; cluster 9—*Salicion cinereae*. Cluster 5 is mainly made of relevés traditionally classified as *Ulmenion* (without *Alno-Quercion*), but it does not fit into the current concept of geographical differentiation of European hardwood riparian forests, whereas *Fraxino-Quercion roboris* is limited to Central Europe, and *Alno-Quercion* is the only alliance to appear in the Balkans. Nevertheless, we decided to keep this cluster as it is, because it is ecologically and floristically very well differentiated from the rest of the dataset.

Classification is backed by the DCA ordination plot (Figure 2), in which EIVs for moisture, soil reaction, nutrients, temperature, and light are significantly related to the first two DCA axes ($p < 0.05$). The percentages of therophytes, urbanophobic, and adventive species were selected among the other explanatory variables having the highest score of statistically significant correlation with the first two DCA axes ($p < 0.001$). The main ecological factors influencing the variation in the floristic composition along the first axis are EIVs for moisture (positively correlated with the first axis), soil reaction and nutrients (negatively correlated with the first axis), suggesting that the main gradient in species composition is the gradient of site moisture, productivity and soil reaction, running from the driest, nutrient-rich and basophilous *Salicion eleagno-daphnoidis*, *Alnion incanae* and cluster 5 (left side of the diagram), to the wettest, nutrient-poor, and acidophilous *Betulion pubescentis* and *Salicion cinereae*. The first axis is also correlated with the type of urbanity, where the most urbanophobic species are found on the right side of the DCA plot. The second DCA axis is most strongly correlated with EIVs for light and temperature (both positively), differentiating the coldest and darkest communities (*Betulion pubescentis* and *Alnion incanue*) from the most temperature and light-demanding *Salicion triandrae*, *Salicion albae* and *Salicion eleagno-daphnoidis*. The second axis is also positively correlated with the percentage of annual and adventive species, indicating that *Salicetea purpureae* communities are the most disturbed and most endangered by invasive species.

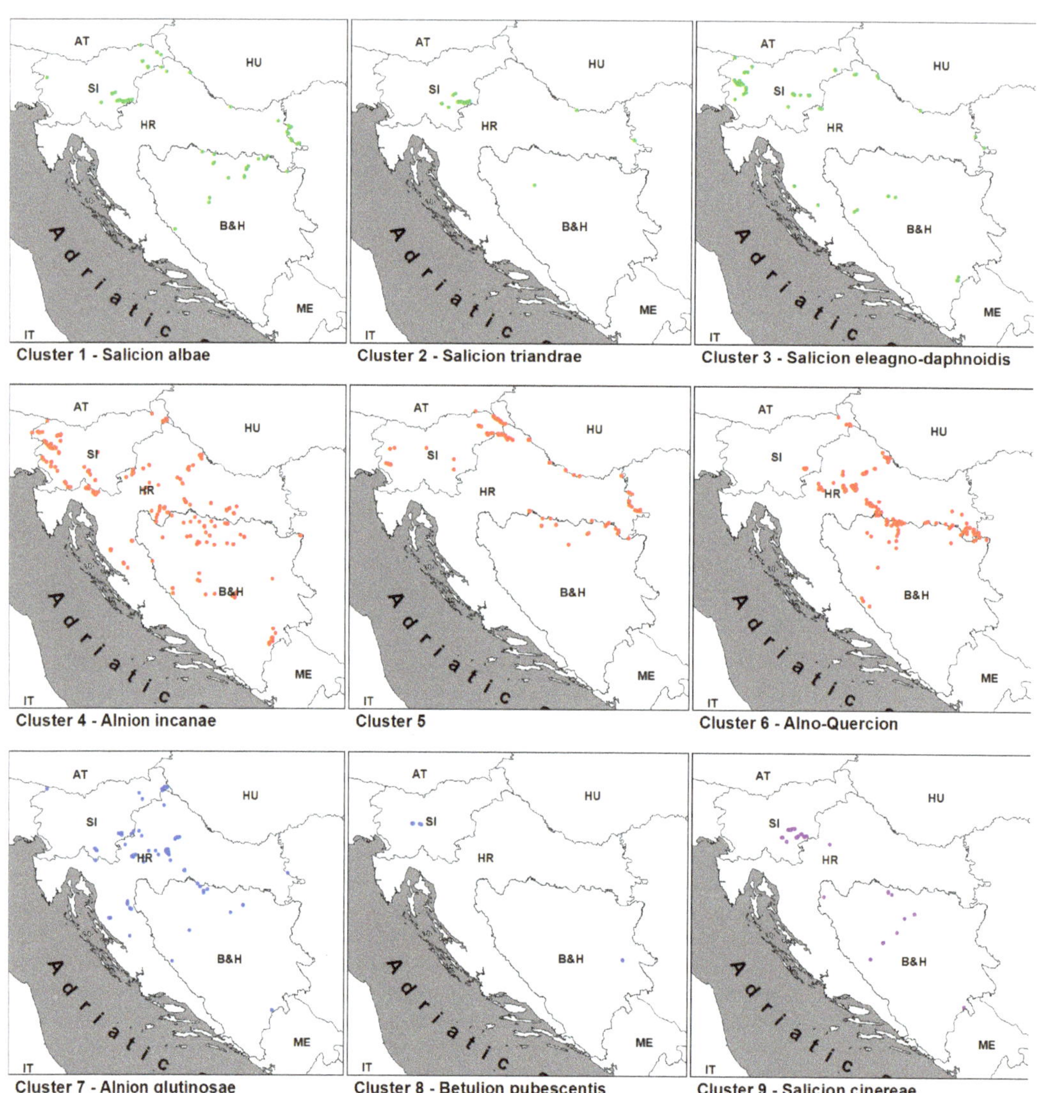

Figure 3. Distribution of relevés classified into the particular cluster. Cluster numbers correspond to Table 1, Figures 2 and 4, Tables S1–S4, Appendix C and to those used in the text. The colors represent groups of clusters (classes): *Salicetea purpureae* (green), *Alno-Populetea* (red), *Alnetea* (blue) and *Franguletea* (purple).

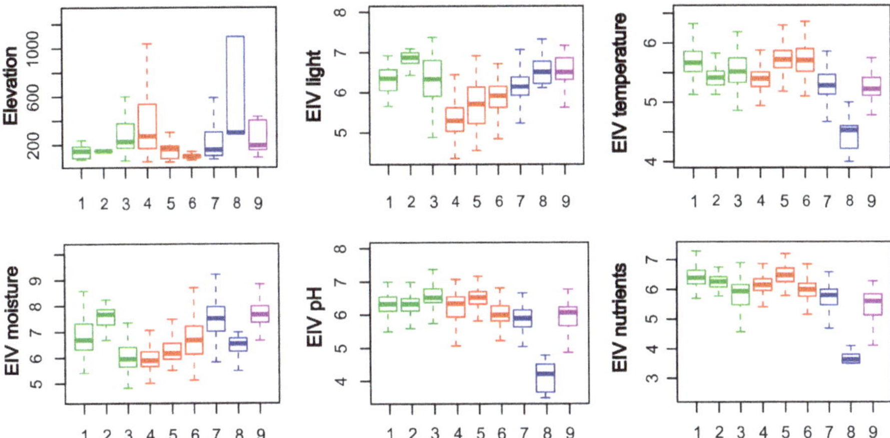

Figure 4. Comparison of the selected EIVs and elevation among clusters. Boxes indicate the 25–75% interquartile range with their median (bold line). Cluster numbers correspond to Table 1, Figures 2 and 3, Tables S1–S4, Appendix C and to those used in the text. The colors represent groups of clusters (classes): *Salicetea purpureae* (green), *Alno-Populetea* (red), *Alnetea* (blue) and *Franguletea* (purple).

3.2. Overview of the Classified Communities

3.2.1. *Salicetea purpureae* Group of Clusters (Clusters 1–3; Table 1, Columns 1–3; Table A1)

This group of clusters consists of willow scrub and woodland communities that are found near stream banks or on regularly flooded floodplain sites. This class is represented by three alliances in the researched area, which was confirmed by the results of unsupervised classification of the first subdataset.

Cluster 1 (Table 1, column 1; Table A1, column 1)

Syntaxonomy: ***Salicion albae***

This cluster is mostly comprised of relevés of tall *Salix alba*-dominated communities. *Salix euxina* and *Populus nigra* are also sometimes present in the tree layer. Invasive species such as *Acer negundo* and *Amorpha fruticose* can often be important.

Diagnostic (bold) and constant species within the WFS: ***Salix alba, Salix euxina***, *Rubus caesius, Galium aparine, Phalaroides arundinacea, Urtica dioica*.

Diagnostic (bold) and constant species within the *Salicetea purpureae* group of clusters: ***Acer negundo, Amorpha fruticosa, Euonymus europaeus, Rubus caesius, Salix alba, Salix euxina, Carex remota, Galium aparine, Glechoma hederacea, Iris pseudacorus***, *Rubus caesius, Phalaroides arundinacea, Urtica dioica*.

Ecology and distribution: These communities are usually located on the lower part of river terraces or in regularly flooded micro-depressions formed outside of the main riverbanks. In both cases, floodings with flowing water are regular and relatively long-lasting events. Soils are nutrient-rich fluvisols with a fine granulometric composition capable of retaining water for a long period of the year, although topsoil layers can dry out during summer. They are found in the floodplains of large lowland rivers throughout the whole area of research: Drava, Sava, Danube, Mura, Krka, Una, Vrbas, Bosna and Drina.

Published relevés from this cluster were mainly referred to as *Salicetum albae, Galio-Salicetum albae* and *Salici-Populetum*. *Populus nigra* dominated or co-dominated communities were not classified within this cluster or even this group of clusters. Additionally, 24 new and unpublished relevés from B&H were classified within this cluster.

Cluster 2 (Table 1, column 2; Table A1, column 2)

Syntaxonomy: ***Salicion triandrae***

This cluster consists of *Salix triandra*-dominated scrub, with *Salix viminalis* sometimes present. The tree layer is absent, while the height of stands is up to 5 m.

Diagnostic (bold) and constant species within the WFS: ***Salix triandra, Agrostis stolonifera* agg., *Bidens tripartitus, Calystegia sepium, Echinocystis lobata, Persicaria dubia, Phalaroides arundinacea, Rorippa amphibia, Rorippa sylvestris, Rumex crispus,*** *Solanum dulcamara, Lythrum salicaria, Persicaria hydropiper, Urtica dioica.*

Diagnostic (bold) and constant species within the *Salicetea purpureae* group of clusters: ***Salix triandra, Solanum dulcamara; Agrostis stolonifera* agg., *Alisma plantago-aquatica, Bidens tripartitus, Calystegia sepium, Echinocystis lobata, Galium palustre* agg., *Lythrum salicaria, Persicaria hydropiper, Phalaroides arundinacea, Rorippa amphibia, Rorippa sylvestris,*** *Urtica dioica.*

Ecology and distribution: These communities are usually located on the lowest part of river terraces along the slower downstream of large rivers. They form narrow vegetation strips along riverbanks and on sandbars, where they are under constant accumulation of new sandy and loamy sediment brought by the river current for as many as 100 days a year. With new material accumulating the ground gets higher, and the flood dynamics changes towards fewer days under flood. Hence, these short-lived pioneer communities, after no more than ten years, give away to the next stages in the succession. Although there is a lack of relevés from these communities, they are present in all three countries: Drava, Danube, Vrbas and Drina.

All published relevés from this cluster were originally assigned the name *Salicetum triandrae.* Only one new and unpublished relevé from B&H was classified within this cluster.

Cluster 3 (Table 1, column 3; Table A1, column 3)

Syntaxonomy: ***Salicion eleagno-daphnoidis***

This cluster comprises relevés of *Salix eleagnos* and/or *Salix purpurea*-dominated scrub. Diagnostic species are light-demanding species with moisture requirements varying from moisture-demanding to mesophilic species.

Diagnostic (bold) and constant species within the WFS: ***Clematis vitalba, Salix eleagnos, Salix purpurea, Centaurea nigrescens* ssp. *vochinensis, Chaerophyllum hirsutum, Galium mollugo, Helianthus tuberosus, Knautia drymeia* s.lat., *Lathyrus sylvestris, Melilotus albus, Mentha longifolia, Pastinaca sativa, Petasites hybridus, Petasites paradoxus, Peucedanum altissimum, Pimpinella major, Plantago lanceolata, Saponaria officinalis, Silene vulgaris, Taraxacum* sect. *Taraxacum, Tussilago farfara, Vicia cracca* s.lat.,** *Rubus caesius, Brachypodium sylvaticum.*

Diagnostic (bold) and constant species within the *Salicetea purpureae* group of clusters: ***Acer pseudoplatanus, Alnus incana, Carpinus betulus, Clematis vitalba, Corylus avellana, Frangula alnus, Fraxinus excelsior, Hedera helix, Salix eleagnos, Salix purpurea, Salvia glutinosa, Ulmus glabra, Brachypodium sylvaticum, Chaerophyllum hirsutum, Cirsium oleraceum, Deschampsia cespitosa, Equisetum arvense, Erigeron annuus, Eupatorium cannabinum, Festuca gigantea, Galium mollugo, Geranium robertianum, Helianthus tuberosus, Heracleum sphondylium, Knautia drymeia* s.lat., *Lamium orvala, Lunaria rediviva, Melilotus albus, Mentha longifolia, Mycelis muralis, Pastinaca sativa, Petasites hybridus, Petasites paradoxus, Peucedanum altissimum, Pimpinella major, Ranunculus lanuginosus, Saponaria officinalis, Silene vulgaris, Stachys sylvatica, Taraxacum* sect. *Taraxacum, Tussilago farfara, Vicia cracca* s.lat.,** *Rubus caesius.*

Ecology and distribution: These communities are usually developed on gravel or sandy beds of small and medium rivers with fast-flowing water and with regular and intense short floods. Fluctuations are intensified by pronounced drought periods that occur in summer caused by a significant drop in the water table, which is intensified by the inability of gravel and sand to retain water. Relevés are primarily concentrated in Slovenia, while scattered over only a couple of localities in Croatia and Bosnia and Herzegovina at different altitudes.

Published relevés from this cluster were originally assigned the following names: *Salici-Myricarietum, Salicetum incano-purpureae, Lamio orvalae-Salicetum eleagni, Lamio orvalae-Salicetum purpureae, Carici-Salicetum myrsinifoliae, Salicetum purpureae, Salicetum cinereo-purpureae, Saponario-Salicetum*. Additionally, nine new and unpublished relevés from B&H were classified within this cluster.

3.2.2. *Alno glutinosae-Populetea albae* Group of Clusters (Clusters 4–6; Table 1, Columns 4–6; Table A2)

The *Alno glutinosae-Populetea albae* group of clusters contains floodplain riparian alder–ash, elm–ash and oak forests on nutrient-rich soils and characterized by inter- and intra-annual fluctuations in the water level. This class is represented by three alliances in the researched area, which was confirmed by the results of unsupervised classification of the second subdataset.

Cluster 4 (Table 1, column 4; Table A2, column 1)

Syntaxonomy: ***Alnion incanae* s. str.**

This cluster consists of forests dominated by *Alnus incana* and/or *A. glutinosa*, as well as *Salix eleagnos*, and sometimes also *S. alba* and/or *S. euxina*. Trees related to mesophilous and ravine forests, such as *Acer pseudoplatanus, Fagus sylvatica, Fraxinus excelsior* and *Ulmus glabra*, are also frequent. The understory is also a mixture of hygrophilous, mesophilous and nitrophilous species.

Diagnostic (bold) and constant species within the WFS: ***Acer campestre, Acer pseudoplatanus, Alnus glutinosa* agg., *Corylus avellana, Fraxinus excelsior, Sambucus nigra, Aegopodium podagraria, Brachypodium sylvaticum, Cardamine bulbifera, Carex pendula, Carex sylvatica, Geum urbanum, Lamium galeobdolon* agg., *Lamium orvala, Lunaria rediviva, Mercurialis perennis, Oxalis acetosella, Primula acaulis, Symphytum tuberosum* agg.**, *Cornus sanguinea, Euonymus europaeus, Rubus caesius, Urtica dioica*.

Diagnostic (bold) and constant species within the *Alno glutinosae-Populetea albae* group of clusters: ***Acer pseudoplatanus, Alnus glutinosa* agg., *Corylus avellana, Fagus sylvatica, Fraxinus excelsior, Salvia glutinosa, Sambucus nigra, Ulmus glabra, Aegopodium podagraria, Angelica sylvestris, Brachypodium sylvaticum, Cardamine bulbifera, Chaerophyllum hirsutum, Cirsium oleraceum, Equisetum arvense, Knautia drymeia s.lat., Lamium galeobdolon* agg., *Lamium orvala, Lunaria rediviva, Mercurialis perennis, Petasites hybridus, Primula acaulis, Ranunculus lanuginosus**, *Acer campestre, Cornus sanguinea, Euonymus europaeus, Rubus caesius, Urtica dioica*.

Ecology and distribution: Stands classified in this cluster occur on stream banks and at headwater seepages, which are usually flooded in spring for several days or weeks and usually dry out during the summer. Stands dominated by *Alnus incana* and/or *A. glutinosa*, and sometimes *Salix eleagnos, S. alba* and *S. euxina*, together with *Acer pseudoplatanus, Fagus sylvatica, Fraxinus excelsior* and *Ulmus glabra*, usually occupy banks of small to medium-sized streams of the colline to montane belt, on stony to sandy, nutrient rich colluvial soil. On the other hand, stands dominated by *Alnus glutinosa* are mainly confined to lower and mid-elevations along smaller streams or at headwater seepages with sandy to loamy, slightly acidic and moderately rich soil. They are common in suitable habitats throughout the study area.

Published relevés from this cluster were referred to as: *Alnetum incanae, Lamio orvalae-Alnetum incanae, Carici acutiformis-Alnetum glutinosae, Carici brizoidis-Alnetum glutinosae* p.p., *Carici elongatae-Alnetum* p.p., *Frangulo-Alnetum glutinosae, Lamio orvalae-Alnetum glutinosae, Pruno padi-Fraxinetum, Stellario-Alnetum glutinosae, Lamio orvalae-Salicetum eleagni, Lamio orvalae-Salicetum albae ranunculetosum lanuginosae*. Additionally, 87 new and unpublished relevés from B&H were classified within this cluster.

Cluster 5 (Table 1, column 5; Table A2, column 2)

Syntaxonomy: *not assigned*

This cluster contains floodplain hardwood (*Ulmus laevis, Fraxinus angustifolia* and sometimes *Quercus robur*) and/or poplar (*Populus alba* and *P. nigra*) forests. The shrub layer is well developed, with *Cornus sanguinea, Sambucus nigra, Prunus padus, Euonymus europaeus* and *Prunus padus*, among others, while the herb layer is typically made of nemoral mesophilous and hygromesophilous species. Invasive alien species such as *Acer negundo, Solidago gigantea, Impatiens glandulifera* and *Robinia pseudoacacia* can be common.

Diagnostic (bold) and constant species within the WFS: ***Acer negundo, Populus alba, Populus nigra, Prunus padus, Ulmus laevis, Anemone ranunculoides, Arum maculatum, Galium aparine, Leucojum vernum, Pulmonaria officinalis* agg., *Solidago gigantea, Veronica hederifolia,*** *Cornus sanguinea, Rubus caesius, Sambucus nigra, Aegopodium podagraria, Urtica dioica.*

Diagnostic (bold) and constant species within the *Alno glutinosae-Populetea albae* group of clusters: ***Acer negundo, Populus alba, Populus nigra, Prunus padus, Robinia pseudoacacia, Salix alba, Ulmus laevis, Galium aparine, Impatiens glandulifera, Solidago gigantea, Veronica hederifolia,*** *Cornus sanguinea, Rubus caesius, Sambucus nigra, Aegopodium podagraria, Urtica dioica.*

Ecology and distribution: These forests are developed on floodplains of the middle and lower reaches of the largest rivers in the study area (Sava, Drava, Danube, Una, Vrbas, Bosna and Drina). They form on alluvial deposits on the highest terraces within the floodplain, which are only under water during the highest, mainly spring floods. The soil is mainly sandy and, due to the pronounced water regime dynamics, it can become very dry during the summer months.

Published relevés from this cluster were referred to as: *Equiseto-Alnetum incanae, Fraxino-Ulmetum effusae, Salicetum albae* p.p. (polidominant communities), *Lamio orvalae-Salicetum albae caricetosum pendulae, Populetum nigro-albae, Salici-Populetum* and *Carduo crispi-Populetum nigrae*. Additionally, 22 new and unpublished relevés from B&H were classified within this cluster.

Cluster 6 (Table 1, column 6; Table A2, column 3)
Syntaxonomy: ***Alno-Quercion roboris***

The cluster encompasses floodplain hardwood forests dominated by *Quercus robur* and/or *Fraxinus angustifolia* s.lat. with *Ulmus minor, Alnus glutinosa* and *Acer campestre*, frequently admixed. In some cases, *Alnus glutinosa* has the role of edifier (probably in secondary succession stages). The shrub layer is not as developed as in Cluster 5, with *Crataegus* sp., *Frangula alnus* and *Cornus sanguinea* being the most important, with a frequency of around 40%. The herb layer is represented mainly by hygrophilous and hygromesophilous forest species.

Diagnostic (bold) and constant species within the WFS: ***Acer tataricum, Crataegus laevigata, Fraxinus angustifolia* s.lat., *Quercus robur, Ulmus minor, Carex remota, Carex strigosa, Glechoma hederacea, Lysimachia nummularia, Rumex sanguineus, Stachys palustris,*** *Rubus caesius, Galium palustre* agg., *Iris pseudacorus, Urtica dioica.*

Diagnostic (bold) and constant species within the *Alno glutinosae-Populetea albae* group of clusters: ***Acer tataricum, Crataegus laevigata, Fraxinus angustifolia* s.lat., *Quercus robur, Ulmus minor, Bidens tripartitus, Carex elongata, Carex remota, Carex riparia, Galium palustre* agg., *Glechoma hederacea, Iris pseudacorus, Leucojum aestivum, Lycopus europaeus, Lysimachia nummularia, Lythrum salicaria, Myosotis palustris* agg., *Persicaria hydropiper, Rumex sanguineus, Stachys palustris,*** *Rubus caesius, Urtica dioica.*

Ecology and distribution: These communities are mostly distributed in lowlands but are not confined to floodplains, since they can be quite distant from a river. The commonality of these forests is the presence of stagnant water at the surface during a longer or shorter time during the year (mostly in spring and autumn), which is influenced by the flat relief and clayey soil. When within a floodplain, they develop in a transitional zone between the highest river terraces (Cluster 5) and depressions with stagnant water (Cluster 7). The fluctuation of water level in the soil can vary greatly and is often a key factor determining the type of community to develop. In the period between floods, the soil may be dry or wet, depending on the flood duration and groundwater table. In the research area, these communities are widespread within alluvia of large rivers, but also on flat, periodically waterlogged, terrains outside the alluvium, such as karst poljes (e.g., Livanjsko polje in B&H).

Published relevés from this cluster were referred to as: *Genisto elatae-Quercetum, Leucojo-Fraxinetum, Frangulo-Alnetum glutinosae, Carici elongatae-Alnetum* p.p. (less swampy relevés), *Carici brizoidis-Alnetum glutinosae* p.p. (only two relevés from original description [34]). Additionally, 53 new and unpublished relevés from B&H were classified within this cluster.

3.2.3. *Alnetea glutinosae* Group of Clusters (Clusters 7–8; Table 1, Columns 7–8; Table A3)

The *Alnetea glutinosae* group of clusters consists of swamp alder forests and birch wooded mires on gleic soils of permanently waterlogged sites. Swamp species tolerant of oxidative stress at permanently waterlogged sites dominate in the herb layer. This class is represented in the researched area by two alliances, which was confirmed by the results of unsupervised classification of the third subdataset.

Cluster 7 (Table 1, column 7; Table A3, column 1)

Syntaxonomy: ***Alnion glutinosae***

Cluster 7 encompasses relevés of mesotrophic regularly flooded alder carr dominated by *Alnus glutinosa* and sometimes accompanied by *Quercus robur* and/or *Fraxinus angustifolia*. The understory is mainly represented by tall sedges (*Carex elongate, C. acutiformis* and *C. riparia*) and other wetland plant species.

Diagnostic (bold) and constant species within the WFS: ***Alnus glutinosa* agg., *Carex elongata, Carex riparia, Carex vesicaria, Lycopus europaeus, Peucedanum palustre, Valeriana dioica* s.lat.**, *Frangula alnus, Solanum dulcamara, Viburnum opulus, Dryopteris carthusiana, Galium palustre* agg., *Iris pseudacorus, Lysimachia vulgaris*.

Diagnostic (bold) and constant species within the *Alnetea glutinosae* group of clusters: ***Alnus glutinosa* agg., *Rubus caesius, Carex elongata, Filipendula ulmaria, Iris pseudacorus, Lycopus europaeus, Lythrum salicaria, Urtica dioica***, *Frangula alnus, Solanum dulcamara, Viburnum opulus, Dryopteris carthusiana, Galium palustre* agg., *Lysimachia vulgaris, Peucedanum palustre*.

Ecology and distribution: In the study area, these forests develop in shallow waterlogged depressions usually inundated by groundwater for considerable parts of the growing season. Soils lack well-aerated horizons and are often characterized by a significant accumulation of undecomposed organic matter. Although this habitat often occurs on sites not related to rivers, they can also be found along oxbows of large rivers (Sava, Vrbas, Bosna, Drina).

Published relevés from this cluster were referred to as: *Carici elongatae-Alnetum, Carici acutiformis-Alnetum glutinosae, Carici brizoidis-Alnetum glutinosae* p.p., *Leucojo-Fraxinetum* p.p., *Genisto elatae-Quercetum roboris* p.p. and *Pseudostellario-Quercetum roboris* p.p. (the last four names are related only to several relevés with a pronounced swamp character and dominated by *Alnus glutinosa* (besides *Q. robur* and *F. angustifolia*)). Additionally, nine new and unpublished relevés from B&H were classified within this cluster.

Cluster 8 (Table 1, column 8; Table A3, column 2)

Syntaxonomy: ***Betulion pubescentis***

Cluster 8 contains acidophilous and poor in nutrients forests on bog, dominated by *Betula pubescens* and sometimes accompanied by *Pinus sylvestris* or *Betula pendula*. The herb layer is represented by acidophilous species and species of nutrient-poor soils. The moss layer is well developed and with a significant participation of various *Sphagnum* species.

Diagnostic (bold) and constant species within the WFS: ***Betula pendula, Betula pubescens, Frangula alnus, Picea abies, Pinus sylvestris, Populus tremula, Rubus hirtus* s.lat., *Salix caprea, Salix pentandra, Sorbus aucuparia, Vaccinium myrtillus, Salix aurita, Lonicera nigra, Vaccinium vitis-idaea, Agrostis canina, Aruncus dioicus, Calamagrostis villosa, Calluna vulgaris, Carex echinata, Carex pallescens, Carex paniculata, Carex*** *rostrata, Carex spicata, Cirsium palustre, Danthonia decumbens, Dryopteris carthu-siana, Eleocharis palustris, Epilobium palustre, Equisetum palustre, Equisetum syl-vaticum, Gentiana pneumonanthe, Knautia sarajevensis, Molinia caerulea* agg., *Orthilia secunda, Parnassia palustris, Persicaria bistorta, Pyrola media, Viola canina.*

Diagnostic (bold) and constant species within the *Alnetea glutinosae* group of clusters: **Betula pendula, Betula pubescens, Pinus sylvestris, Populus tremula, Rubus hirtus s.lat., Salix pentandra, Sorbus aucuparia, Salix aurita, Lonicera nigra, Agrostis canina, Aruncus dioicus, Calamagrostis villosa, Calluna vulgaris, Carex echinata, Carex ros-trata, Carex spicata, Cirsium palustre, Danthonia decumbens, Epilobium palustre, Eq-uisetum palustre, Equisetum sylvaticum, Knautia sarajevensis, Molinia caerulea** agg., **Parnassia palustris, Pyrola media, Viola canina,** *Frangula alnus, Dryopteris carthusiana.*

Ecology and distribution: These communities are far to the south of the center of their distribution, and there are only a few relict sites with this habitat type in the research area (Slovenia and Bosnia and Herzegovina). The stands occur on acidic and nutrient poor waterlogged habitats with *Sphagnum* peat.

Published relevés from this cluster were referred to as: *Pineto-Betuletum pubescentis, Sphagno nemorei-Betuletum pubescentis* and *Betulo-Quercetum roboris.*

3.2.4. *Franguletea* Group of Clusters (Cluster 9; Table 1, Column 9)

There is only one cluster in this group of willow swamp scrub.

Cluster 9 (Table 1, column 9)

Syntaxonomy: *Salicion cinereae*

Cluster 9 encompasses willow carr dominated by *Salix cinerea*, sometimes accompanied by *S. pentandra*. The herb layer is heterogeneous, represented by hygrophilous species of wet meadows and swamps.

Diagnostic (bold) and constant species within the WFS: **Salix cinerea, Carex nigra, Carex panicea, Filipendula ulmaria, Rhinanthus rumelicus, Succisella inflexa,** *Solanum dulca-mara, Galium palustre* agg., *Lycopus europaeus, Lythrum salicaria, Ranunculus repens.*

Ecology and distribution: This scrub can be found in river alluviums, wet meadows, fens and lake shores throughout the research area. They are a stage in the natural succession of lakes and fens, as well as the secondary succession following the abandonment of wet meadows or the removal of alder carrs.

Published relevés from this cluster were referred to as *Salicetum cinereae*. Additionally, 23 new and unpublished relevés from B&H were classified within this cluster.

4. Discussion

The application of different methods of unsupervised classification often results in different, incompatible classification results, often calling for compromise when choosing the final classification system [58]. Moreover, when dealing with broad-scale datasets of different but similar vegetation types, the practice of manual re-arrangement of the numerical classification results indicates that formalization of the traditional expert-based classification by cluster analysis is difficult to achieve on large datasets [59,60]. The expert classification of our initial dataset into classes enabled the preservation of the previously defined syntaxonomic system of alliances at the class level, while at the same time we were able to analyze species variation within each of the classes of WFS. We were thus able to recognize all of the eight already considered alliances for the study area [7,28,36], while another one which is ecologically and floristically well distinguished from the remainder of the dataset (cluster 5) emerged. Apart from the cluster 5, the obtained syntaxonomic scheme of alliances within WFS is in accordance with the broadly accepted syntaxonomic scheme [5].

The alliance *Salicion albae* encompasses only forest communities and is distributed in the majority of Europe [36]. This alliance is relatively well documented in Slove-nia [43,61–64] and Croatia [65–67]. In Bosnia and Herzegovina, only a couple of relevés transitional between *Salicion albae* and *Alnion incanae* have been published as a part of a synoptic table [68]. Here, *Salix alba* and/or *S. fragilis* stands along faster streams are subjected to shorter, stronger and irregular flooding, making them dryer for longer periods in the season, allowing forest mesophytes to prevail. Although some authors have considered poplar-dominated communities to be a part of *Salicion albae* [9,21,65,67], our results suggest that dryer sites on higher parts of alluvial plains with a shorter period of flooding and lower groundwater level, that are occupied by poplar-dominated communities, should be classified within *Alno-Populetea*. The soil is more stable and shows the first signs of pedogenetic evolution. As a result, the shrub layer is relatively abundant with mesophilous and hygro-mesophyllous species, while the herb layer comes with more forest mesophytes, which makes these communities similar to elm–ash communities of what was formerly known as the *Ulmenion* suballiance of *Alnion incanae*. This is in line with the fact that *Fraxino excelsioris-Populetum albae* Jurko 1958, i.e., poplar floodplain forests dominated by *Populus alba* and *P. nigra* distributed along large rivers in lowland areas of the nemoral zone of Europe, was classified within *Alnion incanae* [11]. Since poplars (especially *P. nigra*) require flooding when young but dryer conditions afterwards, natural communities with *P. nigra* are becoming scarcer, since the river dynamics are not as pronounced as before due to flow regulations [69].

Salicion triandrae encompasses scrub communities and forest mantle of *Salix alba* communities if the natural vegetation is preserved, and hence has a similar distribution to *Salicion albae*. However, some authors do not consider *Salicion triandrae* to be a separate alliance but rather a part of *Salicion albae* [64,67]. It is well documented only in Slovenia [64], while it potentially has a much bigger distribution area along large rivers in Croatia and B&H, although there are only a couple of relevés published from Croatia [9,65,67], and no published relevés from B&H. Large areas of its potential habitat along the Sava river in B&H are mine contaminated and impossible to sample.

Salicion eleagno-daphnoidis is a scrub occupying gravelly stream beds or bars in submontane to subalpine belts in different parts of Europe. Since these are under the strong impact of flowing water, they are also often in close contact with early-successional vegetation from the class *Epilobietalia fleischeri* [19]. Dominant willows in these alliances are *Salix eleagnos* on coarser gravelly sediment in upper elevations, while *Salix purpureae* occupies finer gravelly sediments of the lower parts of the river course. It should be noted that *Salix eleagnos* can also be an important species in forest communities of *Alnion incanae*, in which it is found in the tree layer. This difference in physiognomy between scrub and forest communities dominated by *Salix eleagnos* is the determining factor for their syntaxonomical differentiation (*Salicetea purpureae* or *Alno-Populetea*). Furthermore, dynamic and fast successional changes also make these communities challenging to classify. Thus, although the association *Lamio orvalae-Salicetum eleagni* is considered to be a part of *Alnion incanae* [43,62,63], our results suggest that some relevés from the Idrijca Valley belong to *Salicion eleagno-daphnoidis*. Although these relevés of *Salix eleagnos* are tall communities, they have high percentages of stoniness [62], which probably favors species that are characteristic of the alliance *Salicion eleagno-daphnoidis* and eliminates species of more developed soils that are characteristic of *Alnion incanae*. High *Salix eleagnos* communities have been recorded and analyzed in Bosnia and Herzegovina, from the Sutjeska river [68]. Those relevés were not included in further analyses because they were classified as *Quercetea pubescentis* because of the thermophilous character they displayed. The association *Salici eleagni-Juniperetum communis* has been described from Italy and placed into the alliance *Berberidion vulgaris* [70], indicating a much wider ecological amplitude of *Salix eleagnos* in terms of soil humidity, which can be especially pronounced in later life stages. The association *Petterio-Salicetum eleagni* has been described from the southern part of Bosnia and Herzegovina, from the Neretva river catchment [71], but it was not analyzed due to its geographical position in the Mediterranean biogeographic region. Comprehensive analyses of *Salix eleagnos* tall communities with special attention given to Mediterranean communities should be performed to determine their syntaxonomical position and the number of alliances to which they are related.

The alliance *Alnion incanae* is present in almost all European countries [36]. In Slovenia and Croatia, sites in the upper parts of stream catchments are often dominated by *Alnus incana* while in the lower parts *Alnus glutinosa* is adjoined and often prevails [9,42,43,63,72–75]. On the other hand, in Bosnia and Herzegovina, *Alnus glutinosa* is much more often in the upper courses, while *Alnus incana* is rarely found at lower and middle altitudes. In a recently published paper from Sutjeska National Park [68], all records of *Alnus glutinosa* from this habitat type refer to newly described species from the *Alnus glutinosa* complex, i.e., *Alnus rohlenae*, and sometimes *Salix eleagnos*, *S. alba* and *S. fragilis* also form the tree layer. Our results indicate that most of the lowland meso-hygrophilous *Alnus glutinosa*-dominated forests along small streams classified in *Alno-Populetea* belong to this alliance, which is not in line with several authors who have placed it in *Alno-Quercion* [27,72,76,77] or even *Carpinion betuli*/*Erythronio-Carpinion* [35,61]. It should also be noted that some authors put *Lamio orvalae-Alnetum glutinosae* (originally described from south-western Slovenia [42]) from north-eastern Italy into *Ligustro vulgaris-Alnion glutinosae*, the alliance encompassing riparian forests of the sub-Mediterranean regions of the northern and central Apennine Peninsula [78]. This, however, was not supported by Poldini and Sburlino [79] who placed it within the *Alnion incanae*.

Lowland hardwood riparian forests dominated by *Quercus robur* and *Fraxinus angustifolia* in the study area have usually been assigned to the alliance *Alno-Quercion roboris* [34,35,72,80–84]. However, it has sometimes been considered to be part of *Ulmenion* [9,15], or more often to be part of *Alnion incanae* [11,85]. In general, this type of forest is well-documented for most of the study area. While in Croatia these forests have a long history of research and are very well documented [34,65,80–82,84,86], since they are among the economically most important forests in the country, in Slovenia they have only been reported in a few papers [41,61]. In contrast, in Bosnia and Herzegovina, where these forests have largely been removed in the last 200 years, there is a significant lack of published relevés. Fukarek [32] published only six relevés of degraded *Q. robur* stands from northern B&H, which are to date the only published relevés of these forests in B&H. Although these forests are mostly distributed in lowlands, they can be quite distant from a river. On the other hand, they are rarely present in the Dinaric mountains or Alps. These communities require stagnant water on the surface during a longer or shorter part of the year [9,61,77], which is influenced by the flat relief. The fluctuation of the water level throughout the year can be very great and is often the key factor that determines which type of vegetation will develop [87]. *Fraxinus angustifolia* is present in southern Europe and parts of Central Europe [88]. In Austria, the Czech Republic and Slovakia, this species is usually confined to meso-hygrophillous forests and is absent from swamp microdepressions, in which *Alnus glutinosa* prevails [22,24,25,89]. However, in the southern part of its distribution, this species is often present in swamp habitats [9,21,35,90,91] where it replaces/adjoins *Alnus glutinosa*. *Fraxinus angustifolia* swamp communities are prone to drying out in the summer [21,92] and therefore can contain significantly less peat than *Alnus glutinosa* swamps. Nevertheless, Douda et al. [11] classified *F. angustifolia* dominated communities (*Leucojo-Fraxinetum*) within *Alnetea glutinosae*. However, since *Fraxinus angustifolia* was listed in the expert system as a characteristic species of *Alno-Populetea*, most of our *F. angustifolia* transitional relevés between *Alno-Populetea* and *Alnetea glutinosae* were classified into the class *Alno-Populetea* and, consequently, into *Alno-Quercion*.

Cluster 5 is close to *Alnion incanae* and *Alno-Quercion* but ecologically and floristically well distinguished (Tables 1 and A2; Figure 2). A large number of the relevés from this cluster were originally assigned to the association *Fraxino-Ulmetum effusae*. In general, this elm–ash dominated community is developed on well-drained sandy alluvial soils on elevated river terraces [41,84]. The poplar-dominated communities *Populetum nigro-albae* and *Salici-Populetum nigrae*, which are developed on sandy fluvisols on middle and high-positioned terraces along rivers banks of large rivers, show a similar ecological pattern [65]. A pioneer community of *Alnus incana* is known from the lowland floodplain of the Drava river in Croatia under the name *Equiseto hyemali-Alnetum incanae*, where it occupies gravelly and sandy fluvisols with a developed humus layer [9]. Although these are *Alnus incana*-dominated communities, Vukelić et al. [75] noted that the tree layer is rich in species typical of lowland hardwood riparian forests (*Ulmus laevis, Ulmus minor, Quercus robur, Fraxinus angustifolia* and *Populus alba*). The syntaxon *Lamio orvalae-Salicetum albae caricetosum pendulae* was described from the Vipava Valley in Slovenia and encompasses mostly polydominant communities of *Populus nigra* and *Salix alba*, also growing on elevated fluvisols [42]. These communities are well differentiated from other analyzed alliances since they develop on well-aerated water-permeable soils in lowlands. Although this cluster is a combination of relevés originally classified into different alliances (*Alnion incanae, Salicion albae* and *Alno-Quercion*), it always occupies similar positions and soil types along a river. This cluster best corresponds to the description and floristic composition of the suballiance *Ulmenion* provided by Petrášová-Šibíková et al. [93]. For a long time, and by many authors, *Ulmenion* was considered to be part of *Alnion incanae* [9,21,22,25,94]. However, Mucina et al. [5] considered *Ulmenion* to be a corresponding name of the alliance *Fraxino-Quercion roboris*, i.e., elm–ash and oak riparian floodplain forests on nutrient-rich brown soils in the nemoral zone of Europe, which is geographically constrained to Central Europe [36]. On the other hand, *Populion albae* which was formerly known as an alliance that encompassed poplar-dominated communities of the nemoral zone [29], is now biogeographically constrained to the Mediterranean region [5,36]. The ambiguous position of this vegetation type is supported by the recently described alliance *Dioscoreo communis-Populion nigrae* from Italy, which shows similar ecological and floristic traits to our cluster 5 [95]. This implies that the syntaxonomical position of cluster 5, although closest to the concept of *Fraxino-Quercion*, is still unclear and should be further investigated on a larger geographical scale.

The alliance *Alnion gutinosae* has a wide distribution and is present in almost the whole of Europe [36]. These communities are well documented in Croatia [73,80,96,97] and Slovenia [61,72], but there have been no published relevés from B&H. Although this alliance mainly encompasses communities dominated by *Alnus glutinosa*, some authors also include monodominant wet forests of other species such as *Fraxinus angustifolia* (*Leucojo-Fraxinetum*) [11,83], *Salix alba* (*Galio-Salicetum albae*) [70] and *Quercus robur* (*Cardamini parviflorae-Quercetum roboris* and *Carici elongatae-Quercetum roboris*) [21,98]. However, our results suggest that hardwood forests with a pronounced swamp character, even though ecologically transitional, are dry for a significant time during the vegetation season, and thus harbor a significant number of mesophitic species, which makes them closer to *Alno-Quercion*. Moreover, the whole original table with the type relevé [34] was classified into *Alno-Quercion*. Another alliance from *Alnetea* glutinosae that could enter into the consideration is *Frangulo alni-Fraxinion oxycarpae*, which encompasses interdunal or karstic swamps developed on hydromorphic soils with large amounts of slightly decomposed organic matter that are dominated by *Fraxinus angustifolia* [90]. This alliance has a narrow distribution (Italy, Croatia and Albania) and, although it is reported for Croatia [36], we did not recognize it in our dataset. The reason may be that it is confined to the Mediterranean region, which we omitted in our analysis. Furthermore, since it is a relatively recently described alliance, its exact distribution is not known and, moreover, it is noted that the distinction between the alliances *Alnion glutinosae* and *Frangulo-Fraxinion* is sometimes not clear [91].

Forests from the alliance *Betulion pubescentis* are widespread in Europe, except for the southern parts of the continent. In the study area, they have been recorded only at two localities, i.e., Ljubljansko Barje in Slovenia [99] and Han Kram in B&H [33]. Bearing in mind their marginal position in the distribution area of the alliance [36] and small number of recorded relevés, these might show some peculiarities in comparison with typical stands from Central and Northern Europe.

The alliance *Salicion cinereae* is distributed in most European countries, except for the southernmost parts [36]. Many authors have considered these communities to be part of the class *Alnetea glutinosae* [9,21,22]. However, Mucina et al. [5] proposed relocating this alliance into the scrub class *Franguletea*, based on the principle of distinction between forest and scrub communities in different classes. Nevertheless, it should be noted that, without expert classification, relevés from *Salicion cinereae* could not be distinguished from *Alnion glutinosae*. This alliance has been well documented and analyzed in Slovenia [100]. On the other hand, only one relevé of this alliance has been published in Croatia [9], while in Bosnia and Herzegovina no relevés have been published to date, although Milanović and Stupar [101] reported the class *Franguletea* in the checklist of vegetation classes of B&H.

Soil moisture, soil reaction and nutrient availability were found to be the most important factors determining the floristic composition and, consequently, the alliance differentiation of WFS in the study area (Figure 2). On the other hand, the correlation with climatic variables was not found to be statistically significant. Few smaller-scale studies within WFS consider soil moisture and nutrient-related variables to be dominant drivers of variation in species composition [41,102], while Douda et al. [11] reported only site moisture as an important driver. On the other hand, the differentiation of the Iberian floodplain forest at the alliance level was mainly influenced by climatic drivers (i.e., continentality and precipitation) [20]. Our ordination results suggest that Western Balkans WFS are azonal vegetation influenced by edaphic factors and physiological stresses that floodplain plants share, regardless of climatic differences.

5. Conclusions

Our study supported the division of WFS of the Western Balkans into nine clusters that corresponded well with accepted syntaxa at the alliance level. Additionally, the classification of around 230 new and unpublished relevés from B&H contributed to the knowledge of WFS in this part of Europe, since only a few dozen relevés had been published from B&H to date.

The main ecological factors influencing the variation of the floristic composition are soil moisture, soil reaction, and nutrients, while there is a weak correlation with macroclimatic variables, implying that WFS represent azonal plant communities, with no significant geographical patterns.

However, further research is needed to determine the syntaxonomic position of cluster 5. Although this cluster is floristically and ecologically similar to *Fraxino-Quercion*, it does not fit into its current geographical concept, which is limited to Central Europe.

Supplementary Materials: The following supporting information can be downloaded at: https://www.mdpi.com/article/10.3390/d15030370/s1, Table S1: Full synoptic table of the of WFS types in the Western Balkans; Table S2: Full synoptic table for the cluster group Salicetea purpureae; Table S3: Full synoptic table for the cluster group Alno-Populetea; Table S4: Full synoptic table for the cluster group Alnetea glutinosae.

Author Contributions: Conceptualization, D.K., J.B., A.Č., Đ.M., Ž.Š. and V.S.; methodology, D.K., J.B., A.Č., Đ.M., Ž.Š. and V.S.; formal analysis, D.K. and V.S.; investigation, D.K., J.B., A.Č., Đ.M., Ž.Š. and V.S.; writing—original draft preparation, D.K. and V.S.; writing—review and editing, D.K., J.B., A.Č., Đ.M., Ž.Š. and V.S.; visualization, D.K. and V.S.; supervision, V.S. All authors have read and agreed to the published version of the manuscript.

Funding: This study was partly funded by the Slovenian Research Agency (ARRS P1-0236).

Institutional Review Board Statement: Not applicable.

Informed Consent Statement: Not applicable.

Data Availability Statement: Not applicable.

Acknowledgments: The study was supported by COST Action (CA16208)–CONVERGES: Knowledge Conversion for Enhancing Management of European Riparian Ecosystems and Services.

Conflicts of Interest: The authors declare no conflict of interest.

Appendix A

List of species merged to aggregates (agg.) and broadly defined taxa (s.l.).

Alnus glutinosa agg. (*A. glutinosa* and *A. rohlenae*)
Aquilegia vulgaris agg. (*A. nigricans* and *A. vulgaris*)
Aconitum variegatum agg. (*A. variegatum* and *A. degenii*)
Agrostis stolonifera agg. (*A. gigantea* and *A. stolonifera*)
Galium palustre agg. (*G. palustre* and *G. elongatum*)
Lamium galeobdolon agg. (*L. galeobdolon* ssp. *argentatum*, *L. galeobdolon* ssp. *flavidum*, *L. galeobdolon* ssp. *galeobdolon* and *L. galeobdolon* ssp. *montanum*)
Molinia caerulea agg. (*M. caerulea* and *M. arundinacea*)
Myosotis palustris agg. (*M. scorpioides* and *Myosotis palustris*)
Rubus fruticosus agg. (*R. plicatus*, *R. silvaticus* and *Rubus fruticosus*)
Crocus vernus agg. (*Crocus vernus* and *C. vernus* ssp. *albiflorus*)
Malus sylvestris agg. (*M. sylvestris* and *M. pumila*)
Ranunculus auricomus agg. (*R. cassubicus.* and *R. auricomus*)
Stellaria media agg. (*S. media* and *S. neglecta*)
Rosa canina (all species from *Rosa canina* group sensu Tutin et al. [103])
Aconitum lycoctonum s.lat. (*A. lycoctonum* ssp. *lycoctonum* and *A. lycoctonum* ssp. vulparia)
Asarum europaeum s.lat. (*A. europaeum* ssp. *caucasicum* and *A. europaeum* ssp. *europaeum*)
Centaurea scabiosa s.lat (*C. scabiosa* ssp. *scabiosa* and *Centaurea scabiosa* ssp. *fritschii*)
Dactylis glomerata s.lat. (*D. glomerata* ssp. *glomerata* and *D. glomerata* ssp. *lobata*)
Fraxinus angustifolia s.lat. (*F. angustifolia* ssp. *angustifolia* and *Fraxinus angustifolia* ssp. *oxycarpa*)
Hesperis matronalis s.lat. (*H. matronalis* ssp. *matronalis* and *H. matronalis* ssp. *candida*)
Knautia drymeia s.lat. (*K. drymeia* ssp. *drymeia* and *K. drymeia* ssp. *intermedia*)
Phyteuma spicatum s.lat. (*P. spicatum* ssp. *spicatum* and *P. spicatum* ssp. *coeruleum*)
Plantago major s.lat (*P. major* ssp. *major* and *P. major* ssp. *intermedia*)
Prunus domestica s.lat. (*P. domestica* ssp. *domestica* and *P. domestica* ssp. *insititia*)
Solanum nigrum s.lat. (*S. nigrum* ssp. *nigrum* and *S. nigrum* ssp. *schultesii*)
Arabidopsis halleri s.lat. (*A. halleri* ssp. *halleri* and *A. halleri* ssp. *ovirensis*)
Carex divulsa s.lat. (*C. divulsa* ssp. *divulsa* and *C. divulsa* ssp. *leersii*)
Euphorbia esula s.lat (*E. esula* ssp. *esula* and *E. esula* ssp. *tommasiniana*)
Leucanthemum ircutianum s.lat. (*L. ircutianum* ssp. *ircutianum* and *L. ircutianum* ssp. *leucolepis*)
Pyrus communis s.lat. (*P. communis* ssp. *communis* and *P. communis* ssp. *pyraster*)
Rhamnus alpina s.lat. (*R. alpina* ssp. *alpina* and *Rhamnus alpina* ssp. *fallax*)
Helleborus dumetorum s.lat. (*H. dumetorum* ssp. *dumetorum* and *H. dumetorum* ssp. *atrorubens*)

Appendix B

List of associations used throughout the text with authorship indicated

Alnetum incanae Lüdi 1921
Betulo-Quercetum roboris Martinčič 1987
Cardamini parviflorae-Quercetum roboris Molnár Zs. 2010
Carduo crispi–Populetum nigrae Kevey in Borhidi and Kevey 1996
Carici acutiformis-Alnetum glutinosae Scamoni 1935
Carici brizoides-Alnetum glutinosae Horvat 1938
Carici elongatae-Alnetum glutinosae Koch 1926
Carici elongatae-Quercetum Sokołowski 1972
Carici paniculatae-Salicetum myrsinifoliae Dakskobler in Vreš, Seliškar et Dakskobler 2012
Equiseto hyemali-Alnetum incanae Moor 1958
Frangulo-Alnetum glutinosae Rauš 1971 (1973)
Fraxino excelsioris-Populetum albae Jurko 1958
Fraxino angustifoliae-Ulmetum effusae Slavnić 1952
Galio palustri–Salicetum albae Rauš 1973
Genisto elatae-Quercetum roboris Horvat 1938
Lamio orvalae-Alnetum glutinosae Dakskobler 2016
Lamio orvalae-Alnetum incanae Dakskobler 2010
Lamio orvalae-Salicetum albae Dakskobler 2016
Lamio orvalae-Salicetum albae caricetosum pendulae Dakskobler 2016

Lamio orvalae-Salicetum albae ranunculetosum lanuginosae Dakskobler 2016
Lamio orvalae-Salicetum eleagni Dakskobler, Šilc and Čušin ex Dakskobler 2007
Lamio orvalae-Salicetum purpureae nom. prov. (Dakskobler, 2016)
Leucojo aestivi-Fraxinetum angustifoliae Glavač 1959
Petterio-Salicetum eleagni Redžić, Muratspahić and Lakušić 1992
Pino-Betuletum pubescentis Stefanović 1961
Populetum nigro-albae Slavnić 1952
Pruno padi-Fraxinetum angustifoliae Čarni et al. 2008
Pseudostellario europaeae-Quercetum roboris Accetto 1974
Salicetum albae Issler 1926
Salicetum cinereae Zólyomi 1931
Salicetum cinereo-purpureae Pelcer 1975 prov.
Salicetum purpureae Wendelberger-Zelinka 1952
Salicetum incano-purpureae Sillinger 1933
Salicetum triandrae Malcuit ex Noirfalise in Lebrun et al. 1955
Salici eleagni-Juniperetum communis Poldini, Francescato, Vidali and Castello 2020
Salici purpureae-Myricarietum germanicae Moor 1958
Salici-Populetum nigrae (R. Tx. 1931) Meyer Drees 1936
Saponario officinalis-Salicetum purpureae Tchou 1948
Sphagno nemorei-Betuletum pubescentis (Libbert 1933) Passarge 1968
Stellario nemorum-Alnetum glutinosae Lohmeyer 1957

Appendix C

Table A1. Synoptic table for the cluster group *Salicetea purpureae*. Frequencies of species are presented as percentages with phi values multiplied by 100 shown in superscript. Diagnostic species (phi values higher than 0.30) are shaded. Species with frequency lower than 30% in a cluster for which they are diagnostic are not shown. Only up to 15 species with the highest phi value are presented. Cluster numbers: 1—*Salicion albae*, 2—*Salicion triandrae*, 3—*Salicion eleagno-daphnoidis*. Cluster numbers correspond to Table 1, Figures 2–4 and to those used in the text. The full version of this table is available in Table S2.

Cluster Number	1	2	3
Number of Relevés	90	37	83
Salicion albae			
Salix alba	91[72.6]	8	23
Galium aparine	56[42.9]	14	16
Glechoma hederacea	44[41.1]	11	7
Iris pseudacorus	43[40.9]	14	4
Salix euxina	36[36.9]	.	13
Rubus caesius	73[33]	19	58
Euonymus europaeus	32[32.2]	3	12
Salicion triandrae			
Salix triandra	19	97[78.6]	11
Alisma plantago-aquatica	1	38[50.8]	1
Rorippa sylvestris	6	49[48.3]	8
Solanum dulcamara	38	81[46.2]	27
Agrostis stolonifera agg.	34	70[42.2]	18
Persicaria hydropiper	28	57[41.7]	5
Bidens tripartitus	14	43[40]	4
Echinocystis lobata	39	59[37.3]	5
Lythrum salicaria	27	57[34.3]	18
Galium palustre agg.	29	49[32.8]	6
Phalaroides arundinacea	56	78[32]	34

Table A1. *Cont.*

Cluster Number	1	2	3
Number of Relevés	**90**	**37**	**83**
Rorippa amphibia	10	$30^{31.9}$	2
Calystegia sepium	46	$62^{30.9}$	14
Salicion eleagno-daphnoidis			
Salix eleagnos	4	.	$70^{73.9}$
Salix purpurea	18	16	84^{65}
Petasites hybridus	1	.	$53^{64.3}$
Clematis vitalba	3	.	$43^{54.2}$
Brachypodium sylvaticum	13	.	$53^{52.6}$
Fraxinus excelsior	2	.	$39^{51.5}$
Acer pseudoplatanus	.	.	$35^{51.3}$
Deschampsia cespitosa	4	.	$40^{49.9}$
Eupatorium cannabinum	7	.	$42^{49.6}$
Salvia glutinosa	.	.	$31^{48.3}$
Chaerophyllum hirsutum	3	.	35^{47}
Equisetum arvense	6	.	$37^{46.6}$
Ranunculus lanuginosus	1	.	$30^{45.6}$
Saponaria officinalis	3	.	$31^{43.7}$
Cirsium oleraceum	11	3	$39^{39.3}$
Other species with high frequency			
Urtica dioica	$82^{15.7}$	86	48
Cornus sanguinea	43	.	43
Persicaria dubia	12	43	29
Angelica sylvestris	24	11	$46^{29.9}$
Poa trivialis	$43^{27.6}$	14	22
Lamium maculatum	29	24	24
Humulus lupulus	43^{29}	27	6
Alliaria petiolata	34	14	27
Ranunculus repens	23	35	16
Aegopodium podagraria	26	11	35

Table A2. Synoptic table for the cluster group *Alno-Populetea*. Frequencies of species are presented as percentages, with phi values multiplied by 100 shown in superscript. Diagnostic species (phi values higher than 0.30) are shaded. Species with frequency lower than 30% in a cluster for which they are diagnostic are not shown. Only up to 15 species with the highest phi value are presented. Cluster numbers: 4—*Alnion incanae*, 5—*not assigned*, 6—*Alno-Quercion*. Cluster numbers correspond to Table 1, Figures 2–4 and to those used in the text. The full version of this table is available in Table S3.

Cluster Number	4	5	6
Number of Relevés	**281**	**178**	**226**
Alnion incanae			
Acer pseudoplatanus	$50^{59.5}$	3	1
Fraxinus excelsior	$47^{50.7}$	8	2
Ranunculus lanuginosus	$32^{45.7}$	2	1
Petasites hybridus	$31^{44.5}$	2	.
Chaerophyllum hirsutum	$30^{43.5}$	2	.
Equisetum arvense	$39^{43.2}$	9	1
Corylus avellana	$63^{42.9}$	20	19
Alnus glutinosa agg.	$81^{39.6}$	31	47

Table A2. *Cont.*

Cluster Number	4	5	6
Number of Relevés	**281**	**178**	**226**
Brachypodium sylvaticum	64$^{35.4}$	46	8
Lamium galeobdolon agg.	49$^{35.1}$	24	8
Lamium orvala	33$^{34.3}$	13	.
Aegopodium podagraria	70^{33}	57	14
Sambucus nigra	74$^{31.5}$	61	21
Angelica sylvestris	37$^{31.1}$	12	10
Cluster 5			
Populus nigra	7	54^{57}	2
Prunus padus	5	40$^{45.7}$	4
Populus alba	1	38$^{43.9}$	8
Salix alba	21	47$^{39.8}$	1
Acer negundo	3	30$^{38.5}$	4
Solidago gigantea	14	44$^{38.3}$	7
Impatiens glandulifera	8	30^{37}	.
Galium aparine	27	62$^{36.2}$	23
Ulmus laevis	4	54$^{33.6}$	38$^{8.7}$
Alno-Quercion			
Fraxinus angustifolia s.lat.	9	43	84$^{54.5}$
Galium palustre agg.	8	10	58$^{53.1}$
Quercus robur	11	37	79$^{52.5}$
Stachys palustris	2	10	46$^{46.8}$
Ulmus minor	10	14	52$^{43.5}$
Crataegus laevigata	9	3	39^{42}
Persicaria hydropiper	15	10	50$^{40.6}$
Iris pseudacorus	6	29	54$^{38.5}$
Myosotis palustris agg.	10	2	35$^{38.1}$
Lysimachia nummularia	23	24	62$^{37.2}$
Leucojum aestivum	2	9	33$^{36.3}$
Acer tataricum	8	1	31$^{36.3}$
Rumex sanguineus	12	12	42$^{34.5}$
Carex remota	33	17	61$^{34.5}$
Lythrum salicaria	6	7	31$^{32.9}$
Other species with high frequency			
Rubus caesius	70	90$^{23.4}$	69
Urtica dioica	59	66	61
Cornus sanguinea	65$^{13.6}$	64	37
Euonymus europaeus	55$^{14.7}$	47	31
Crataegus monogyna	48	34	46
Acer campestre	57$^{21.1}$	21	48
Geum urbanum	47$^{16.9}$	30	29
Ranunculus repens	37	16	48$^{21.1}$
Hedera helix	41$^{19.5}$	22	23
Ligustrum vulgare	34$^{13.9}$	30	13

Table A3. Synoptic table for the cluster group *Alnetea glutinosae*. Frequencies of species are presented as percentages with phi values multiplied by 100 shown in superscript. Diagnostic species (phi values higher than 0.30) are shaded. Species with frequency lower than 30% in a cluster for which they are diagnostic are not shown. Only up to 15 species with the highest phi value are presented. Cluster numbers: 7—*Alnion glutinosae*, 8—*Betulion pubescentis*. Cluster numbers correspond to Table 1, Figures 2–4 and to those used in the text. The full version of this table is available in Table S4.

Cluster Number	7	8
Number of Relevés	121	14
Alnion glutinosae		
Alnus glutinosa agg.	$94^{94.4}$	.
Lycopus europaeus	$65^{69.6}$	.
Carex elongata	$55^{61.2}$	.
Iris pseudacorus	$52^{59.3}$	.
Rubus caesius	$46^{54.9}$	.
Filipendula ulmaria	$45^{54.2}$	.
Lythrum salicaria	44^{53}	.
Urtica dioica	$43^{52.3}$	.
Betulion pubsecentis		
Betula pubescens	.	$86^{86.6}$
Molinia caerulea agg.	7	$93^{86.2}$
Pinus sylvestris	.	$71^{74.5}$
Sorbus aucuparia	2	$57^{60.9}$
Betula pendula	1	$50^{56.5}$
Rubus hirtus s.lat.	1	$43^{50.9}$
Salix aurita	2	$43^{49.5}$
Calamagrostis villosa	.	$36^{46.6}$
Lonicera nigra	.	$36^{46.6}$
Cirsium palustre	.	$36^{46.6}$
Parnassia palustris	.	$36^{46.6}$
Knautia sarajevensis	.	$36^{46.6}$
Carex rostrata	.	$36^{46.6}$
Salix pentandra	1	$36^{45.1}$
Equisetum sylvaticum	1	$36^{45.1}$
Other species with high frequency		
Frangula alnus	66	100
Dryopteris carthusiana	53	71
Lysimachia vulgaris	67	43
Galium palustre agg.	60	36
Viburnum opulus	53	36
Peucedanum palustre	52	36
Caltha palustris	48	36
Solanum dulcamara	54	29
Valeriana dioica s.lat.	50	21
Quercus robur	33	36

References

1. Dierschke, H. *Pflanzensoziologie*; Eugen Ulmer Verlag: Stuttgart, DE, USA, 1994.
2. Ellenberg, H. *Vegetation Ecology of Central Europe*, 4th ed.; Cambridge University Press: Cambridge, UK, 2009.
3. Stupar, V.; Čarni, A. Ecological, Floristic and Functional Analysis of Zonal Forest Vegetation in Bosnia and Herzegovina. *Acta Bot. Croat.* **2017**, *76*, 15–26. [CrossRef]
4. Junk, W.J.; Piedade, M.T.F. An Introduction to South American Wetland Forests: Distribution, Definitions and General Characterization. In *Amazonian Floodplain Forests*; Junk, W.J., Piedade, M.T.F., Wittmann, F., Schöngart, J., Parolin, P., Eds.; Springer Science & Business Media: Berlin/Heidelberg, Germany, 2010; pp. 3–25.
5. Mucina, L.; Bültmann, H.; Dierßen, K.; Theurillat, J.-P.; Raus, T.; Čarni, A.; Šumberová, K.; Willner, W.; Dengler, J.; García, R.G.; et al. Vegetation of Europe: Hierarchical Floristic Classification System of Vascular Plant, Bryophyte, Lichen, and Algal Communities. *Appl. Veg. Sci.* **2016**, *19*, 3–264. [CrossRef]
6. Eremiášová, R.; Skokanová, H. Response of Vegetation on Gravel Bars to Management Measures and Floods: Case Study from the Czech Republic. *Ekologia* **2014**, *33*, 274–285. [CrossRef]
7. Mandžukovski, D.; Čarni, A.; Biurrun, I.; Douda, J.; Škvorc, Ž.; Stupar, V.; Slezák, M.; Ćušterevska, R.; Rodríguez-González, P.; Mendías, C.; et al. *Interpretative Manual of European Riparian Forests and Shrublands*; Ss Cyril and Methodius University in Skopje, Hans Em Faculty of Forest Sciences, Landscape Architecture and Environmental Engineering: Skopje, Republic of Macedonia, 2021; ISBN 978-9989-132-22-3.
8. Vogt, K.; Rasran, L.; Jensen, K. Water-Borne Seed Transport and Seed Deposition during Flooding in a Small River-Valley in Northern Germany. *Flora–Morphol. Distrib. Funct. Ecol. Plants* **2004**, *199*, 377–388. [CrossRef]
9. Vukelić, J. *Šumska Vegetacija Hrvatske*; Šumarski fakultet, Sveučilište u Zagrebu, DZZP: Zagreb, Croatia, 2012; ISBN 978-953-292-024-6.
10. Slezák, M.; Hrivnák, R.; Petrášová, A. Numerical Classification of Alder Carr and Riparian Alder Forests in Slovakia. *Phytocoenologia* **2014**, *44*, 283–308. [CrossRef]
11. Douda, J.; Boublík, K.; Slezák, M.; Biurrun, I.; Nociar, J.; Havrdová, A.; Doudová, J.; Aćić, S.; Brisse, H.; Brunet, J.; et al. Vegetation Classification and Biogeography of European Floodplain Forests and Alder Carrs. *Appl. Veg. Sci.* **2016**, *19*, 147–163. [CrossRef]
12. Rodríguez-González, P.M.; Abraham, E.; Aguiar, F.; Andreoli, A.; Baležentienė, L.; Berisha, N.; Bernez, I.; Bruen, M.; Bruno, D.; Camporeale, C.; et al. Bringing the Margin to the Focus: 10 Challenges for Riparian Vegetation Science and Management. *WIREs Water* **2022**, *9*, e1604. [CrossRef]
13. Riis, T.; Kelly-Quinn, M.; Aguiar, F.C.; Manolaki, P.; Bruno, D.; Bejarano, M.D.; Clerici, N.; Fernandes, M.R.; Franco, J.C.; Pettit, N.; et al. Global Overview of Ecosystem Services Provided by Riparian Vegetation. *BioScience* **2020**, *70*, 501–514. [CrossRef]
14. Verhoeven, J.T.A. Wetlands in Europe: Perspectives for Restoration of a Lost Paradise. *Ecol. Eng.* **2014**, *66*, 6–9. [CrossRef]
15. Cestarić, D.; Škvorc, Ž.; Franjić, J.; Sever, K.; Krstonošić, D. Forest Plant Community Changes in the Spačva Lowland Area. *Plant Biosyst.* **2017**, *151*, 584–597. [CrossRef]
16. Havrdová, A.; Douda, J.; Doudová, J. Threats, Biodiversity Drivers and Restoration in Temperate Floodplain Forests Related to Spatial Scales. *Sci. Total Environ.* **2023**, *854*, 158743. [CrossRef]
17. Janssen, J.A.M.; Rodwell, J.S.; Garcia Criado, M.; Gubbay, S.; Haynes, T.; Nieto, A.; Sanders, N.; Landucci, F.; Loidi, J.; Ssymank, A.; et al. *European Red List of Habitats—Part 2. Terrestrial and Freshwater Habitats*; European Commission: Brussels, Belgium, 2016; ISBN 978-92-79-61588-7.
18. Council of the European Communities. Council Directive 92/43/EEC of 21 May 1992 on the Conservation of Natural Habitats and of Wild Fauna and Flora. *Off. J. Eur. Communities* **1992**, *L206*, 7–50.
19. Kalníková, V.; Chytrý, K.; Biţa-Nicolae, C.; Bracco, F.; Font, X.; Lakushenko, D.; Kącki, Z.; Kudrnovsky, H.; Landucci, F.; Lustyk, P.; et al. Vegetation of the European Mountain River Gravel Bars: A Formalized Classification. *Appl. Veg. Sci.* **2021**, *24*, e12542. [CrossRef]
20. Biurrun, I.; Campos, J.A.; García-Mijangos, I.; Herrera, M.; Loidi, J. Floodplain Forests of the Iberian Peninsula: Vegetation Classification and Climatic Features. *Appl. Veg. Sci.* **2016**, *19*, 336–354. [CrossRef]
21. Borhidi, A.; Kevey, B.; Lendvai, G. *Plant Communities of Hungary*; Akadémiai Kiadó: Budapest, Hungary, 2012.
22. Chytry, M.; Douda, J.; Roleček, J.; Sádlo, J.; Boublík, K.; Hédl, R.; Vítková, M.; Zelený, D.; Navrátilová, J.; Neuhäuslová, Z.; et al. *Vegetace České Republiky 4. Lesní a Křovinná Vegetace (Vegetation of the Czech Republic 4. Forest and Scrub Vegetation)*; Chytrý, M., Ed.; Academia: Praha, CZ, USA, 2013.
23. Poldini, L.; Vidali, M.; Ganis, P. Riparian *Salix Alba*: Scrubs of the Po Lowland (N-Italy) from an European Perspective. *Plant Biosyst.* **2011**, *145*, 132–147. [CrossRef]
24. Valachovič, M.; Kliment, J.; Hegedüšová Vantarová, K. *Rastlinné Spoločenstvá Slovenska 6. Vegetácia Lesov a Krovín*; Veda vydavateľstvo Slovenskej akadémie vied: Bratislava, Slovakia, 2021.
25. Willner, W.; Grabherr, G.; Drescher, A.; Eichberger, C.; Exner, A.; Franz, W.R.; Grabner, S.; Heiselmayer, P.; Karner, P.; Steiner, G.M.; et al. *Die Wälder und Gebüsche Österreichs: Ein Bestimmungswerk mit Tabellen (in zwei Bänden)*; Willner, W., Grabherr, G., Eds.; Spektrum Akademischer Verlag: Berlin/Heidelberg, Germany, 2007.
26. Barudanović, S.; Macanović, A.; Topalić-Trivunović, L.; Cero, M. *Ekosistemi Bosne i Hercegovine u Funkciji Održivog Razvoja*; Univerzitet u Sarajevu, Prirodno-matematički fakultet: Sarajevo, BA, USA, 2015; ISBN 978-9958-592-60-7.
27. Šilc, U.; Čarni, A. Conspectus of Vegetation Syntaxa in Slovenia. *Hacquetia* **2012**, *11*, 113–164. [CrossRef]

28. Škvorc, Ž.; Jasprica, N.; Alegro, A.; Kovačić, S.; Franjić, J.; Krstonošić, D.; Vraneša, A.; Čarni, A. Vegetation of Croatia: Phytosociological Classification of the High-Rank Syntaxa. *Acta Bot. Croat.* **2017**, *76*, 200–224. [CrossRef]

29. Trinajstić, I. *Biljne Zajednice Republike Hrvatske*; Akademija šumarskih znanosti: Zagreb, Croatia, 2008.

30. Fabijanić, B.; Fukarek, P.; Stefanović, V. Lepenica: Pregled osnovnih tipova šumske vegetacije. *Naučn. Druš. BH Poseb. Izd.* **1963**, *3*, 85–129.

31. Fukarek, P. Šumske zajednice prašumskog rezervata Perućice u Bosni. *Akad. Nauk. Umjet. BH Od. Prir. Druš. Nauk. Poseb. Izd.* **1970**, *15*, 157–262.

32. Fukarek, P. Hrastove šume bosanskog posavlja u prošlosti i sadašnjosti. *Jugosl. Akad. Znan. Umjet. Poseb. Izd.* **1975**, *2*, 371–379.

33. Stefanović, V.; Sokač, A. Fitocenoza bijelog bora i maljave breze na rubu tresetišta kod Han-krama (*Pineto-Betuletum pubescentis*, Stef.). *Rad. Naučn. Druš. BH* **1962**, *5*, 97–126.

34. Horvat, I. Biljnosociološka Istraživanja Šuma u Hrvatskoj. *Glas. Šumske Pokuse* **1938**, *6*, 127–279.

35. Horvat, I.; Glavač, V.; Ellenberg, H. *Vegetation Südosteuropas*; Geobotanica selecta; Gustav Fischer Verlag: Stuttgart, DE, USA, 1974.

36. Preislerová, Z.; Jiménez-Alfaro, B.; Mucina, L.; Berg, C.; Bonari, G.; Kuzemko, A.; Landucci, F.; Marcenò, C.; Monteiro-Henriques, T.; Novák, P.; et al. Distribution Maps of Vegetation Alliances in Europe. *Appl. Veg. Sci.* **2022**, *25*, e12642. [CrossRef]

37. European Environment Agency (EEA). *Biogeographic Regions in Europe*; European Environment Agency (EEA): Copenhagen, Denmark, 2011.

38. Mrgić, J. *Severna Bosna: 13–16. Vek*; Istorijski Institut: Beograd, RS, USA, 2008; ISBN 1019-469X.

39. Mrgić, J. Lijevče polje–beleške o naseljima i prirodi 15-19. vek. *Hist. Rev.* **2007**, *5*, 171–199.

40. Dakskobler, I.; Kutnar, L.; Šilc, U. *Poplavni, Močvirni in Obrežni Gozdovi v Sloveniji. Gozdovi vrb, jelš, Dolgopecljatega Bresta, Velikega in Ozkolistnega Jesena, doba in Rdečega bora ob Rekah in Potokih*; Gozdarski inštitut Slovenije: Ljubljana, SI, USA, 2013.

41. Košir, P.; Čarni, A.; Marinšek, A.; Šilc, U. Floodplain Forest Communities along the Mura River (NE Slovenia). *Acta Bot. Croat.* **2013**, *72*, 71–95. [CrossRef]

42. Dakskobler, I. Phytosociological Analysis of Riverine Forests in the Vipava and Reka Valleys (Southwestern Slovenia). *Folia Biol. Geol.* **2016**, *57*, 5–61. [CrossRef]

43. Dakskobler, I.; Šilc, U.; Čušin, B. Riverine Forests in the Upper Soča Valley (the Julian Alps, Western Slovenia). *Hacquetia* **2004**, *3*, 51–80.

44. Hennekens, S.M.; Schaminée, J.H.J. TURBOVEG, a Comprehensive Data Base Management System for Vegetation Data. *J. Veg. Sci.* **2001**, *12*, 589–591. [CrossRef]

45. Tichý, L. JUICE, Software for Vegetation Classification. *J. Veg. Sci.* **2002**, *13*, 451–453. [CrossRef]

46. Euro+Med Euro+Med PlantBase–the Information Resource for Euro-Mediterranean Plant Diversity. Available online: http://ww2.bgbm.org/EuroPlusMed/query.asp (accessed on 12 October 2015).

47. Vít, P.; Douda, J.; Krak, K.; Havrdová, A.; Mandák, B. Two New Polyploid Species Closely Related to *Alnus Glutinosa* in Europe and North Africa–An Analysis Based on Morphometry, Karyology, Flow Cytometry and Microsatellites. *Taxon* **2017**, *66*, 567–583. [CrossRef]

48. McCune, B.; Mefford, M.J. *PC-ORD. Multivariate Analysis of Ecological Data. Version 5.0 1999*; Fundacion BBVA: Bilbao, Spain, 2014.

49. Tichý, L.; Hennekens, S.M.; Novák, P.; Rodwell, J.S.; Schaminée, J.H.; Chytrý, M. Optimal Transformation of Species Cover for Vegetation Classification. *Appl. Veg. Sci.* **2020**, *23*, 710–717. [CrossRef]

50. Botta-Dukát, Z.; Chytrý, K.; Hájková, P.; Havlová, M. Vegetation of Lowland Wet Meadows along a Climatic Continentality Gradient in Central Europe. *Preslia* **2005**, *77*, 89–111.

51. Tichý, L.; Chytrý, M. Statistical Determination of Diagnostic Species for Site Groups of Unequal Size. *J. Veg. Sci.* **2006**, *17*, 809–818. [CrossRef]

52. Pignatti, S.; Menegoni, P.; Pietrosanti, S. Valori di bioindicazione delle piante vascolari della flora d'Italia. *Braun-Blanquetia* **2005**, *39*, 1–97.

53. Zelený, D.; Schaffers, A.P. Too Good to Be True: Pitfalls of Using Mean Ellenberg Indicator Values in Vegetation Analyses. *J. Veg. Sci.* **2012**, *23*, 419–431. [CrossRef]

54. Fick, S.E.; Hijmans, R.J. WorldClim 2: New 1-Km Spatial Resolution Climate Surfaces for Global Land Areas. *Int. J. Climatol.* **2017**, *37*, 4302–4315. [CrossRef]

55. Gajić, M. Pregled vrsta flore SR Srbije sa biljnogeografskim oznakama. *Glas. Šumar. Fak.* **1980**, *54*, 111–141.

56. Raunkiaer, C. *The Life Forms of Plants and Statistical Plant Geography*; Oxford University Press: Oxford, UK, 1934.

57. Klotz, S.; Kühn, I.; Durka, W. (Eds.) *BIOLFLOR–Eine Datenbank Zu Biologisch-Ökologischen Merkmalen Der Gefäßpflanzen in Deutschland*; Bundesamt für Naturschutz: Bonn, DE, USA, 2002.

58. Novák, P.; Willner, W.; Zukal, D.; Kollár, J.; Roleček, J.; Świerkosz, K.; Ewald, J.; Wohlgemuth, T.; Csiky, J.; Onyshchenko, V.; et al. Oak-Hornbeam Forests of Central Europe: A Formalized Classification and Syntaxonomic Revision. *Preslia* **2020**, *92*, 1–34. [CrossRef]

59. Kočí, M.; Chytrý, M.; Tichý, L. Formalized Reproduction of an Expert-Based Phytosociological Classification: A Case Study of Subalpine Tall-Forb Vegetation. *J. Veg. Sci.* **2003**, *14*, 601–610. [CrossRef]

60. Willner, W.; Jiménez-Alfaro, B.; Agrillo, E.; Biurrun, I.; Campos, J.A.; Čarni, A.; Casella, L.; Csiky, J.; Ćušterevska, R.; Didukh, Y.P.; et al. Classification of European Beech Forests: A Gordian Knot? *Appl. Veg. Sci.* **2017**, *20*, 494–512. [CrossRef]

61. Čarni, A.; Košir, P.; Marinček, L.; Marinšek, A.; Šilc, U.; Zelnik, I. *Komentar k Vegetacijski Karti Gozdnih Združb Slovenije v Merilu 1: 50.000-List Murska Sobota*; Pomurska Akademsko Znanstvena Unija-PAZU: Murska Sobota, Slovenia, 2008; ISBN 961-91503-6-8.
62. Dakskobler, I. Development of Vegetation on Gravel Sites of the Idrijca River in Western Slovenia. *Folia Biol. Geol.* **2010**, *51*, 5–90.
63. Dakskobler, I.; Rozman, A. Phytosociological Analysis of Riverine Forests along the Sava Bohinjka, Radovna, Učja and Slatenik Rivers in Northwestern Slovenia. *Folia Biol. Geol.* **2013**, *54*, 37–105.
64. Šilc, U. Vegetation of the Class *Salicetea Purpureae* in Dolenjska (SE Slovenia). *Fitosociologia* **2003**, *40*, 3–27.
65. Rauš, Đ. Vegetacija Ritskih Šuma Dijela Podunavlja Od Aljmaša Do Iloka. *Glas. Šum. Pokuse* **1976**, *19*, 5–75.
66. Rauš, Đ. *Vegetacija Podravskih Ritskih Šuma u Okolici Legrada Na Ušću Mure u Dravu*; Trinajstić, I., Ed.; Šumarski fakultet i Hrvatske šume po Zagreb: Zagreb, Croatia, 1994; pp. 87–100.
67. Vukelić, J.; Baričević, D.; Perković, Z. Vegetacijske i Druge Značajke Zaštičenog Dijela "Slatinskih Podravskih Šuma". *Šumar. List* **1999**, *123*, 287–299.
68. Milanović, Đ.; Stupar, V. Riparian forest communities along watercourses in the Sutjeska National Park (SE Bosnia and Herzegovina). *Glas. Šumar. Fak. Univ. Banjal.* **2017**, *26*, 95–111. [CrossRef]
69. Vilhar, U.; Čarni, A.; Božič, G. Growth and vegetation characteristics of European black poplar (*Populus nigra* L.) in a floodplain forest along the river Sava and temperature differencies among selected sites. *Folia Biol. Geol.* **2013**, *54*, 193–214.
70. Poldini, L.; Vidali, M.; Castello, M.; Sburlino, G. A Novel Insight into the Remnants of Hygrophilous Forests and Scrubs of the Po Plain Biogeographical Transition Area (Northern Italy). *Plant Soc.* **2020**, *57*, 17–69. [CrossRef]
71. Redžić, S.; Muratspahić, D.; Lakušić, R. Neke fitocenoze šuma i šikara iz doline Neretve. *Poljopr. Šumar. Podgor.* **1992**, *38*, 95–101.
72. Accetto, M. *Močvirni in Poplavni Gozdovi. Zasnova Rajonizacije Ekosistemov Slovenije*; Oddelek za Biologijo, Biotehniška Fakulteta: Ljubljana, Slovenia, 1994.
73. Plišo Vusić, I.; Šapić, I.; Vukelić, J. Prepoznavanje i Kartiranje Šumskih Staništa Natura 2000 u Hrvatskoj (I)–91E0*, Aluvijalne Šume s Crnom Johom *Alnus Glutinosa* i Običnim Jasenom *Fraxinus Excelsior* (*Alno-Padion, Alnion Incanae, Salicion Albae*). *Šumar. List* **2019**, *5–6*, 255–263. [CrossRef]
74. Vukelić, J.; Šapić, I.; Alegro, A.; Šegota, V.; Stankić, I.; Baričević, D. Phytocoenological Analysis of Grey Alder (*Alnus incana* L.) Forests in the Dinarides of Croatia and Their Relationship with Affiliated Communities. *Tuexenia* **2017**, *37*, 65–78. [CrossRef]
75. Vukelić, J.; Baričević, D.; Poljak, I.; Vrček, M.; Šapić, I. Fitocenološka analiza šuma bijele johe (*Alnus incana*/L./Moench subsp. *incana*) u Hrvatskoj. *Šumar. List* **2018**, *142*, 123–135. [CrossRef]
76. Brullo, S.; Spampinato, G. Syntaxonomy of Hygrophilous Woods of the *Alno-Quercion Roboris. Ann. Bot.* **1999**, *57*, 133–146.
77. Rauš, Đ. Šuma crne johe (*Frangulo-Alnetum glutinosae* Rauš 68) u bazenu Spačva. *Šumar. List* **1975**, *99*, 431–446.
78. Sciandrello, S.; Angiolini, C.; Bacchetta, G.; Cutini, M.; Dumoulin, J.; Fois, M.; Gabellini, A.; Gennai, M.; Gianguzzi, L.; Landi, M.; et al. *Alnus glutinosa* Riparian Woodlands of Italy and Corsica: Phytosociological Classification and Floristic Diversity. *Land* **2023**, *12*, 88. [CrossRef]
79. Poldini, L.; Sburlino, G. *Lamio orvalae-Alnetum glutinosae* Dakskobler 2016, nuova associazione ripariale per l'Italia settentrionale (Friuli Venezia Giulia, Veneto e Lombardia) con note sulle cenosi corrispondenti poste a sud del Po. *Gortania* **2020**, *42*, 5–21.
80. Baričević, D. Ecological-Vegetational Properties of Forest "Žutica". *Glas. Šum. Pokuse* **1998**, *35*, 1–91.
81. Glavač, V. O Vlažnom Tipu Hrasta Lužnjaka i Običnog Graba. *Šumar. List* **1961**, *85*, 342–347.
82. Rauš, Đ. Fitocenološke Značajke i Vegetacijska Karta Fakultetskih Šuma Lubardenik i Opeke. *Šumar. List* **1973**, *97*, 190–221.
83. Stefanović, V. *Fitocenologija Sa Pregledom Šumskih Fitocenoza Jugoslavije*, 2nd ed.; Svjetolost, Zavod za udžbenike i nastavna sredstva: Sarajevo, Bosnia and Herzegovina, 1986.
84. Vukelić, J.; Baričević, D. The Association of Spreading Elm and Narrow-Leaved Ash (*Fraxino-Ulmetum Laevis* Slav. 1952) in Floodplain Forests of the Podravina and Podunavlje. *Hacquetia* **2004**, *3*, 49–60.
85. Plišo Vusić, I.; Šapić, I.; Vukelić, J. Prepoznavanje i kartiranje šumskih staništa Natura 2000 u Hrvatskoj (II)–91F0, poplavne šume s vrstama *Quercus robur, Ulmus laevis, Ulmus minor, Fraxinus angustifolia*; 91L0, hrastovo-grabove šume ilirskoga područja. *Šumar. List.* **2019**, *143*, 461–467. [CrossRef]
86. Horvat, I. Pregled Šumske Vegetacije u Hrvatskoj. *Šumar. List* **1937**, *61*, 337–344.
87. Dekanić, I. Utjecaj podzemne vode na pridolazak i uspijevanje šumskog drveća u posavskim šumama kod Lipovljana. *Glas. Šum. Pokuse* **1962**, *15*, 5–117.
88. Caudullo, G.; Houston Durrant, T. *Fraxinus Angustifolia* in Europe: Distribution, Habitat, Usage and Threats. In *European atlas of Forest Tree Species*; San-Miguel-Ayanz, J., de Rigo, D., Caudullo, G., Houston Durrant, T., Mauri, A., Eds.; Publ. Off. EU: Luxembourg, UK, 2016; Volume 97.
89. Douda, J. Formalized Classification of the Vegetation of Alder Carr and Floodplain Forests in the Czech Republic. *Preslia* **2008**, *80*, 199–224.
90. Biondi, E.; Allegrezza, M.; Casavecchia, S.; Galdenzi, D.; Gasparri, R.; Pesaresi, S.; Poldini, L.; Sburlino, G.; Vagge, I.; Venanzoni, R. New Syntaxonomic Contribution to the Vegetation Prodrome of Italy. *Plant Biosyst.* **2015**, *149*, 603–615. [CrossRef]
91. Gennai, M.; Gabellini, A.; Viciani, D.; Venanzoni, R.; Dell'Olmo, L.; Giunti, M.; Lucchesi, F.; Monacci, F.; Mugnai, M.; Foggi, B. The Floodplain Woods of Tuscany: Towards a Phytosociological Synthesis. *Plant Soc.* **2021**, *58*, 1–28. [CrossRef]
92. Glavač, V. O šumi poljskog jasena sa kasnim drijemovcem (*Leucoieto-Fraxinetum angustifoliae* ass. nov.). *Šumar. List* **1959**, *1–3*, 39–45.

93. Petrášová-Šibíková, M.; Bacigál, T.; Jarolímek, I. Fragmentation of Hardwood Floodplain Forests—How Does It Affect Species Composition? *Community Ecol.* **2017**, *18*, 97–108. [CrossRef]
94. Petrášová, M.; Jarolímek, I. Hardwood Floodplain Forests in Slovakia: Syntaxonomical Revision. *Biologia* **2012**, *67*, 889–908. [CrossRef]
95. Poldini, L.; Sburlino, G.; Vidali, M. New Syntaxonomic Contribution to the Vegetation Prodrome of Italy. *Plant Biosyst.* **2017**, *151*, 1111–1119. [CrossRef]
96. Rauš, D. Fitocenoloska Osnova i Vegetacijska Karta Nizinskih Suma Srednje Hrvatske. *Glas. Šum. Pokuse* **1993**, *29*, 335–364.
97. Rauš, Đ. Nizinske Šume Pokupskog Bazena. *Šumar. Inst. Jastrebar. Rad.* **1996**, *31*, 17–37.
98. Sokołowski, W.A. Zespół *Carici Elongatae-Quercetum*–Dębniak Turzycowy. *Acta Soc. Bot. Pol.* **1972**, *41*, 113–120. [CrossRef]
99. Martinčič, A. Fragmenti Visokega Barja Na Ljubljanskem Barju. *Scopolia* **1987**, *14*, 1–53.
100. Šilc, U. Asociacija *Salicetum Cinereae* Zölyomi 1931 v JV Sloveniji. *Hacquetia* **2002**, *1*, 165–184.
101. Milanović, Đ.; Stupar, V. Checklist of Vegetation Classes of Bosnia and Herzegovina: How Much Do We Know? *Rad. Šumar. Fak. Univ. Sarajev.* **2019**, *49*, 9–20. [CrossRef]
102. Slezák, M.; Hrivnák, R.; Petrášová, A.; Dítě, D. Variability of Alder-Dominated Forest Vegetation along a Latitudinal Gradient in Slovakia. *Acta Soc. Bot. Pol.* **2013**, *81*, 25–35. [CrossRef]
103. Tutin, T.G.; Heywood, V.H.; Burges, N.A.; Moore, D.M.; Valentine, D.H.; Walters, S.M.; Webb, D.A. (Eds.) *Flora Europaea*; Cambridge University Press: Cambridge, UK, 1968.

Article

Environmental Drivers of Functional Structure and Diversity of Vascular Macrophyte Assemblages in Altered Waterbodies in Serbia

Dragana Vukov *, Miloš Ilić, Mirjana Ćuk and Ružica Igić

Department of Biology and Ecology, Faculty of Sciences, University of Novi Sad, 21000 Novi Sad, Serbia
* Correspondence: dragana.vukov@dbe.uns.ac.rs

Abstract: There is a gap in the knowledge about how environmental factors affect functional diversity and trait structures of macrophyte communities in altered waterbodies. We used macrophyte and environmental data collected from 46 waterbodies; we extracted data on 14 traits with 43 attributes for 59 species and calculated seven functional diversity indices. We used redundancy analysis (RDA) to investigate the response of functional diversity indices to the environmental variables. To relate traits to environment we performed the analysis on three data matrices: site by environmental variables (R), site by species (L), and species by traits (Q)—the RLQ analysis, and the 4th corner analyses. The RDA showed that the environmental variables explained 47.43% of the variability in the functional diversity indices. Elevation, hemeroby (integrative measure of the impact of all human intervention) of the land cover classes on the banks, and water conductivity were correlated with all diversity indices. We found that the traits characteristic of floating and emergent plants represents a strategy to increase efficiency in light interception under high nutrient concentrations in lowland waterbodies, while submerged plants dominate nutrient-poorer waterbodies at higher altitudes. Future investigations should be focused on the role of functional diversity and the structure of macrophyte communities in the indication of tradeoffs and/or facilitation between ecosystem services that altered waterbodies provide, in order to guide their adequate management.

Keywords: heavily modified waterbodies; artificial waterbodies; macrophytes; ecological processes

Citation: Vukov, D.; Ilić, M.; Ćuk, M.; Igić, R. Environmental Drivers of Functional Structure and Diversity of Vascular Macrophyte Assemblages in Altered Waterbodies in Serbia. *Diversity* **2023**, *15*, 231. https://doi.org/10.3390/d15020231

Academic Editors: Matthew Simpson and Michael Wink

Received: 30 October 2022
Revised: 27 January 2023
Accepted: 3 February 2023
Published: 6 February 2023

1. Introduction

Three types of ecological processes, or filters (i.e., dispersal, abiotic environment, and biotic interactions), shape local species assemblages by progressively filtering species from the regional species pool to local communities, producing non-random patterns in community structure, by acting on species traits rather than on species themselves [1–5]. Species traits are morphological, physiological, and phenological characteristics, measurable at the species or individual level [2]. Dispersal influences the community structure by selecting species according to their ability to disperse to a site [6]. The abiotic environment acts as filter that selects species with suitable traits that can persist in the given habitat, leading to convergent traits and reducing functional diversity [7,8]. Biotic interactions (niche differentiation) are ecological processes prevailing on a finer scale where coexisting species have small overlaps in their functional niches due to competitive exclusion, leading to divergent traits and increased functional diversity [9]. The interactions of these three filters are complex and can lead to shifts in the community structure, changing the abundance and presence of species [10]. It is suggested that environmental filtering is important under disturbed conditions, while biotic interactions prevail in less-disturbed environments [5,11,12].

Macrophytes are key elements of freshwater ecosystems, playing a central role in biogeochemical processes, and as primary producers, they constitute an important food resource [13,14]. Due to their ecological importance, macrophyte species are formally recognized as one out of four biological quality elements used in the monitoring of surface water ecosystems [15]. Macrophyte communities are strongly influenced by hydrology, reflecting both anthropogenic and natural disturbances, with hydrology having a stronger effect on the trait composition than on the species composition of the community [16–19].

There is a growing body of literature dealing with the relationships between environmental factors and functional diversity and trait distributions of macrophyte communities in altered waterbodies (e.g., [4–6,13–18,20–23]). The methods applied to investigate these relationships include the use of multivariate characterization of functional assemblage structures, functional diversity measures, and/or specific trait–environment correlations. In general, all these studies supported the conclusion first drawn by Baattrup-Pedersen et al. [4] that trait-based rather than species-specific approaches might provide better insight into the biological mechanisms underlying the changes in macrophyte communities induced by habitat disturbances.

When analyzing trait–environment relationships in European lowland streams, Baattrup-Pedersen et al. [4] found that eutrophication affected macrophyte community trait characteristics by filtering species with efficient light utilization, indicating that light is a limiting factor for growth in nutrient-enriched environments. Lukács et al. [5] also found that the mechanisms underlying the changes in stream plant communities are mostly related to light capture and utilization, although their results indicated that the plant traits showed stronger associations with carbon gradient than with nutrient gradients.

It was suggested that the functional trait composition of aquatic plants can distinguish hydromorphological degradation from eutrophication in streams [20]. Mouton et al. [22] demonstrated that native and non-native stream macrophyte assemblages responded differently to habitat disturbances, with riparian shading and hydromorphological conditions being the strongest variables shaping the macrophyte functional structure by selecting the species with suitable traits related to colonization and competition strategies. They found that disturbances such as a lack of riparian shading and eutrophication increase functional diversity. A study carried out in permanent and intermittent streams in Cyprus [14] revealed that hydromorphological factors, primarily flow duration patterns, shape macrophyte communities by filtering species with traits related to population resistance and resilience. Comparing channelized and un-channelized streams, Paz et al. [6] showed that macrophyte communities in channelized streams were dominated by emergent species and had lower functional diversity.

Previous studies conducted in the Danube River section in Serbia showed that both species and functional diversity increased along the anthropogenic pressure gradient (i.e., damming and loss of forest cover in the riparian zone) [24,25]. Since similar pressures operate in the altered waterbodies analyzed here, we hypothesized that a similar functional diversity response of macrophyte communities will be found. We expected a significant influence of the water chemistry, particularly nutrient content, on the functional diversity indices and on the individual traits. A higher nutrient content should favor a higher number of strategies (i.e., traits) to exploit these resources [5,26], leading to higher functional diversity. Similarly, species in nutrient-poor environments are constrained to a smaller range of these strategies [5], which will lead to a decrease in functional diversity.

Therefore, here, we evaluated the functional response of macrophyte communities to the environmental drivers in the waterbodies with altered hydromorphological features, using both functional diversity indices and trait–environment correlations. Our main objectives were: 1. to investigate the effects of water chemistry and river habitat features on macrophyte functional diversity indices; and 2. to identify macrophyte traits that respond to the analyzed environmental factors.

2. Materials and Methods

Original data on the abundance and distribution of macrophytes were collected during the surveys conducted between 2017 and 2019, under the WFD scheme in order to achieve the ecological classification of waterbodies in Serbia. We sampled 72 altered bodies of running water (Figure 1). The macrophytes were surveyed according to EN 14184:2014 Standard [27], using a five-level scale metric for the estimation of their abundance (1—very rare, only single plants, up to 5 specimens; 2—rare, approximately 6 to 10 single plants, loosely scattered over sampling unit, or up to 5 single plant stocks; 3—frequent, cannot be overlooked, but not frequent; 4—abundant, occurring frequently, but not in masses, with incomplete cover, exhibiting large gaps; 5—very abundant, dominant, found more or less everywhere, cover markedly more than 50%) in relation to the volume and length of the sampling unit. The survey of macrophytes was carried out over the whole length of the sampling unit visually and/or by raking, using a small boat. The sampling units were 100 m, 500 m, and 1000 m long, depending on the size and characteristics of the waterbody [27]. In each waterbody, a minimum of three sampling units were surveyed on the left and on the right riverside, and the collected data were averaged for later use. The macrophytes were not recorded in 26 out of a total of 72 surveyed waterbodies. Further analyses included data from the remaining 46 waterbodies.

Figure 1. Map of surveyed altered waterbodies in Serbia.

The environmental data analyzed here were divided into two groups: water chemistry and river habitat features (Table 1). We used time- and space-specific data on water chemistry and water temperature published in yearly reports available on the Serbian Environmental Agency website (www.sepa.gov.rs (accessed on 27 April 2022)) as well as in the European Environmental Agency Central Data Repository—EIONET (https://cdr.eionet.europa.eu/rs/eea (accessed on 13 June 2022)). For the purposes of this study, the annual average values were calculated. Data on the bank slope (1—flat, gentle slope; 2—intermediate; 3—very steep slope) and CORINE Land Cover classes were estimated in the field, according to the guidelines provided by the Copernicus Land Monitoring Service (https://land.copernicus.eu/pan-european/corine-land-cove (accessed on 12 September 2022)). The mean width of the riparian zone and the mean width of the waterbodies (i.e., channel width) were measured using landscape images from Google Earth. Data on the CORINE Land Cover classes were used to extract the number of land cover classes (No_LUt) and average degree of hemeroby (avg.hemeroby) of land cover classes along the banks of the waterbody. To measure hemeroby, we applied the approach used by Walz and Stein [28], where CORINE Land Cover classes were assigned to seven degrees of hemeroby, between degree 1—land cover with almost no human impact (e.g., CORINE class 332—bare rocks), and degree 7, assigned for land cover with excessively strong human impact (e.g., 111—continuous urban fabric, 121—industrial or commercial units, etc.). We treated the degrees of hemeroby as numerical values, and for each waterbody we calculated the average degree of hemeroby.

Table 1. Environmental variables used in data analyses, with the grouping applied in the Variation Partitioning (VarPart) procedure: Wchem—water chemistry; habitat—river habitat parameters.

VarPart	Environmental Variable	Unit	Code	Mean	Std.Dev	Min	Median	Max
Wchem	dissolved oxygen	mg/L	dis_O	8.55	1.55	5.21	8.93	11.29
	pH	-	pH	8.03	0.18	7.56	8.01	8.39
	conductivity	µS/cm	cond	404	164	126	395	1234
	ammonium	mg/L	NH4_N	0.15	0.18	0.03	0.11	1.14
	nitrite	mg/L	NO2_N	0.02	0.01	0.01	0.01	0.09
	nitrate	mg/L	NO3_N	0.70	0.34	0.23	0.66	1.53
	total nitrogen	mg/L	total_N	1.48	0.72	0.64	1.45	4.57
	orthophosphate	mg/L	PO4_P	0.07	0.08	0.01	0.05	0.37
	total phosphorus	mg/L	total_P	0.14	0.09	0.04	0.12	0.46
habitat	water temperature	°C	w_t	15.08	2.16	7.95	15.35	18.79
	channel width	m	chn_w	357	364	9	220	1622
	elevation	m a.s.l.	elev	180	185	32	80	832
	bank slope degree	0–3	bnk_slp	1.98	0.69	1.00	2.00	3.00
	riparian width	m	rip_w	116	253	0.00	5	1140
	average degree of hemeroby	1–7	avg_hmrb	3.63	1.10	1.50	3.84	6.00
	number of land cover classes	-	No_LUt	2.80	1.44	1.00	3.00	7.00

For 59 recorded vascular macrophyte species, data on 14 traits with 43 trait attributes were extracted from the literature [29–31] (Table 2). These encompassed morphological traits including growth forms and traits important for species dispersal, reproduction, and survival. The trait attributes had values of 0 for absence, 1 for occasional, but not general presence, and 2 for general presence of the attribute [20,31]. We also included ecological preference traits in form of Ellenberg's indicator values on a nine-level scale for temperature, reaction, nitrogen, and light; and a twelve-level scale for water [29,30].

Table 2. List of 14 plant species traits and their attributes included in the analyses.

Trait	Attribute	Code
Ellenberg Temperature	1–9	TB
Ellenberg Water	1–12	WB
Ellenberg Reaction	1–9	RB
Ellenberg Nitrogen	1–9	NB
Ellenberg Light	1–9	LB
Growth form:	free-floating, surface	Ffl_srfc
	free-floating, submerged	Ffl_sbm
	floating leaves, anchored	Ac_fllv
	submerged leaves, anchored	Ac_sbmlv
	emergent leaves, anchored	Ac_emglv
	heterophylly, anchored	Ac_htrlv
Vertical shoot architecture:	single apical growth point	snglapgr
	single basal growth point	snglbsgr
	multiple apical growth point	mltpapgr
Leaf type:	tubular	tblr
	capillary	cplr
	entire	entr
Leaf area:	small (<1 cm^2)	LA1
	medium (1–20 cm^2)	LA2
	large (20–100 cm^2)	LA3
	extra-large (>100 cm^2)	LA4

Table 2. *Cont.*

Trait	Attribute	Code
Morphology index (score):	<1	MI1
	1–10	MI2
	10–40	MI3
	40–100	MI4
	>100	MI5
Mode of reproduction	rhizome	rhzm
	fragmentation	frgm
	budding	bdng
	turions	trns
	stolon	stln
	tubers	tbrs
	seeds	sds
Perennation:	annual	annl
	biennial	bnnl
	perennial	prnnl
Body flexibility:	low (<45°)	BF1
	intermediate (45–300°)	BF2
	high (>300°)	BF3
Leaf texture:	soft	lfsft
	rigid	lfrgd
	waxy	lfwx
	non-waxy	lfnwx

Based on the plant abundance dataset and trait dataset, we computed seven distance-based multi-trait alpha diversity indices: FRic—functional richness, representing the amount of functional space filled by a community; FEve—functional evenness, measuring the regularity of the species abundance distribution along the minimum spanning tree that links the species points in multidimensional functional space; FDiv—functional divergence, relating to how species abundances are distributed within the functional trait space; FDis—functional dispersion, representing the mean distance of individual species to the centroid of all species in the multidimensional trait space, taking into account the species relative abundances; RaoQ—the sum of the dissimilarities between all possible pairs of species, weighted by the product of species proportions; FD—functional diversity, an index based on the Rao quadratic entropy, representing the sum of dissimilarities in functional traits between all possible pairs of species weighted by the product of species proportions; and Fred—functional redundancy, index based on the Rao quadratic entropy, representing the difference between maximum functional diversity—the inverse Simpson taxonomic diversity—and FD [32–36]. Since the trait attribute values were not quantitative and continuous, Principal Coordinate Analysis (PCoA) of the Gower dissimilarity matrix computed on trait attributes matrix was used to obtain the functional trait space [35]. To explore the correlations between functional diversity indices, linear correlation coefficients were calculated and the correlations between paired samples were tested. Analyses were conducted using R package "FD" [37].

To analyze the relations between the functional diversity indices and environmental variables, we applied redundancy analysis (RDA). We chose RDA because: (a) linear relationships between functional diversity indices and environmental variables were indicated in a number of previous studies, e.g., [6,22,23,38,39]; (b) during the data exploration procedure (not shown here), detrended correspondence analysis (DCA) of the functional diversity indices matrix gave gradient lengths <3 standard deviations for the longest axis, indicating that a linear response would adequately fit the data [40]; (c) this method allows the extraction and summarization of variation in the whole set of components of functional diversity, in form of functional diversity indices, explained by the environmental variables. Prior to the analysis, both the functional diversity indices and the environmental variables were standardized. The statistical significance of the final RDA model and of the individual canonical axes was tested using permutation tests. An unbiased amount of variation in the response data (functional diversity indices) explained by environmental variables was measured as an adjusted R^2 and was used later in the Variation Partitioning (VarPart). To explore the correlations between the functional diversity indices and environmental variables, linear correlation coefficients were calculated and the correlations between paired samples were tested. VarPart was applied to distinguish between the amount of variation in the response data explained by water chemistry, by river habitat features, and by their joint effect (Table 1). The analyses were carried out using the R package "vegan" [41].

To evaluate the relationships between the species traits and environmental data, mediated by the species abundance data, two complementary methods were applied: RLQ and fourth-corner analysis [21,42–44]. RLQ analysis was performed on three data matrices: site by species (L), site by environmental variables (R), and species by traits (Q). RLQ analysis is an ordination method allowing the visualization of the joint structure resulting from the three data tables, based on the coefficients for the environmental variables and species traits. It is followed by single global test [42]. Fourth-corner analysis is a series of statistical tests of individual trait–environment relationships. Since the fourth-corner method involves multiple testing, the overall rate of type I error is increased, and therefore correction for multiple testing should be performed [21,40,43,44]. Here, the false discovery rate (FDR) method with $\alpha = 0.05$ was applied [41,43–47]. RLQ and fourth-corner analysis were carried out using the "ade4" R package [48,49].

3. Results

We recorded 59 plant species (Supplementary Material Table S1) in 46 out of the 72 surveyed waterbodies. Based on the species traits (Table 2), seven distance-based functional indices were calculated (Table 3).

Table 3. Functional diversity indices used in the analyses.

Index	Abbreviation	Mean	Std.Dev	Min	Median	Max
Functional Richness	FRic	0.38	0.28	0	0.37	0.92
Functional Evenness	FEve	0.73	0.3	0	0.83	0.99
Functional Divergence	FDiv	0.69	0.28	0	0.79	0.88
Functional Dispersion	FDis	5.63	1.12	0	5.88	6.85
Rao Quadratic Entropy	RaoQ	33.8	9.68	0	35.91	47.76
Functional Diversity	FD	1.37	0.12	1	1.38	1.56
Functional Redundancy	FRed	6.36	5.14	0	4.86	19.06

All linear correlation coefficients between functional diversity indices were statistically significant (Table S2). The highest correlation coefficient was found between Rao quadratic entropy and Rao-based diversity, while the lowest value was found between evenness and redundancy.

The RDA model explained 47.43% of the variation in the response data (Figure 2). According to the permutation tests, both the whole RDA model and the first canonical axes were highly statistically significant ($p < 0.001$). The first two RDA axes explained 45.42%, and the first RDA axis alone explained 40.16% of the unbiased variation in the response data. Therefore, the major trends were well modeled in the RDA. Furthermore, the first unconstrained eigenvalue (PC1) is relatively small (1.207) compared to the first constrained eigenvalue (3.919), indicating that any dominant residual structure is not displayed in the RDA model. Variables: the elevation ($r = 0.822$), average hemeroby ($r = -0.665$), and conductivity ($r = -0.531$) were strongly correlated with the first RDA axis, whereas the strongest correlation with the second RDA axis was detected for the width of the riparian zone ($r = 0.551$).

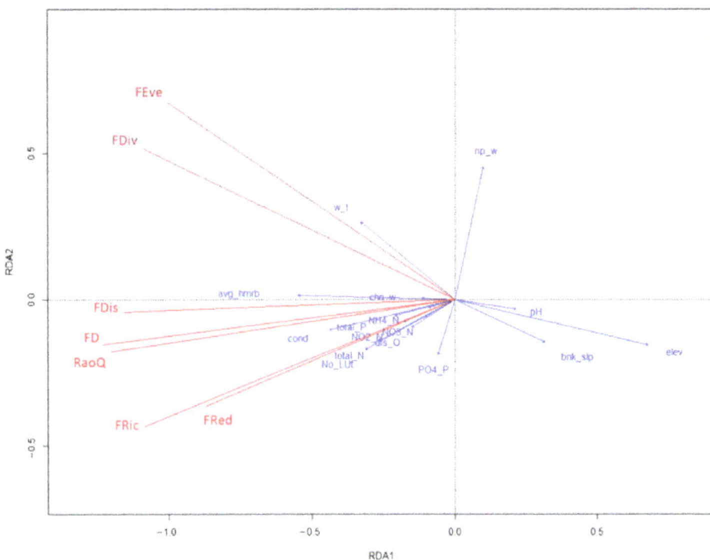

Figure 2. RDA plot of the environmental predictors and functional diversity indices as response variables, with scaling 2 applied. The abbreviated names of the environmental variables are given in Table 1, and those of functional diversity indices in Table 3.

Conductivity and average degree of hemeroby had significant positive correlations and elevation a negative correlation with all functional diversity indices (Figure 3A). a negative correlation was found between bank slope and functional evenness and divergence. These two indices, as well as functional diversity, had a positive correlation with water temperature. The number of CORINE land cover classes had a positive correlation with functional richness, diversity, and redundancy. The coefficients of correlation between the other environmental variables and functional diversity indices were not statistically significant.

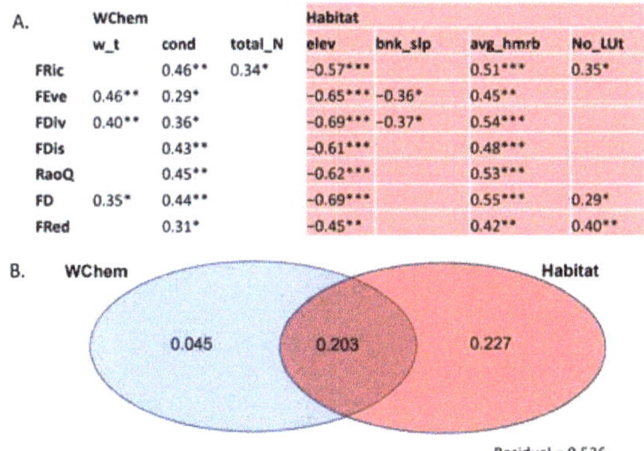

A.	WChem			Habitat			
	w_t	cond	total_N	elev	bnk_slp	avg_hmrb	No_LUt
FRic		0.46**	0.34*	−0.57***		0.51***	0.35*
FEve	0.46**	0.29*		−0.65***	−0.36*	0.45**	
FDiv	0.40**	0.36*		−0.69***	−0.37*	0.54***	
FDis		0.43**		−0.61***		0.48***	
RaoQ		0.45**		−0.62***		0.53***	
FD	0.35*	0.44**		−0.69***		0.55***	0.29*
FRed		0.31*		−0.45**		0.42**	0.40**

Figure 3. (**A**) Statistically significant coefficients of linear correlation between functional diversity indices and environmental variables ($p < 0.001$—***; $p < 0.01$—**; $p < 0.05$—*). (**B**) Venn diagrams of variation partitioning. The abbreviated names of the environmental variables are given in Table 1, and those of the functional diversity indices in Table 3.

The fraction of the variation in the functional diversity indices explained only by river habitat features was the highest at 22.7% (Figure 3B). The joint effect of both subsets of environmental variables on the variability in functional diversity indices was 20.3%, while water chemistry alone had the smallest effect on the response variables, explaining only 4.5% of their variability.

The RLQ analysis (Figure S1) yielded a combined *p*-value < 0.001, which means that the links between the matrices sites by species (L) and species by traits (Q), and between site by environmental variables (R) and site by species (L) were significant. The fourth-corner analysis revealed eleven positive and seven negative associations (Figure 4). Positive associations were found between the concentration of dissolved oxygen and ecological indicator value for light; water conductivity and extra-large leaf area; nitrite concentration and "surface free-floating" growth form, and low body flexibility; nitrate concentration and low body flexibility, and waxy leaf texture; and altitude and "anchored with submerged leaves" growth form, soft leaf texture, and non-waxy leaf texture. Negative associations were found between the concentration of dissolved oxygen and non-waxy leaf texture; nitrite concentration and "anchored with submerged leaves" growth form, and soft leaf texture; nitrate concentration with ecological indicator value for water, multiple apical growth, and non-waxy leaf texture; and average hemeroby of the land use types on the banks and morphology index between 40 and 100.

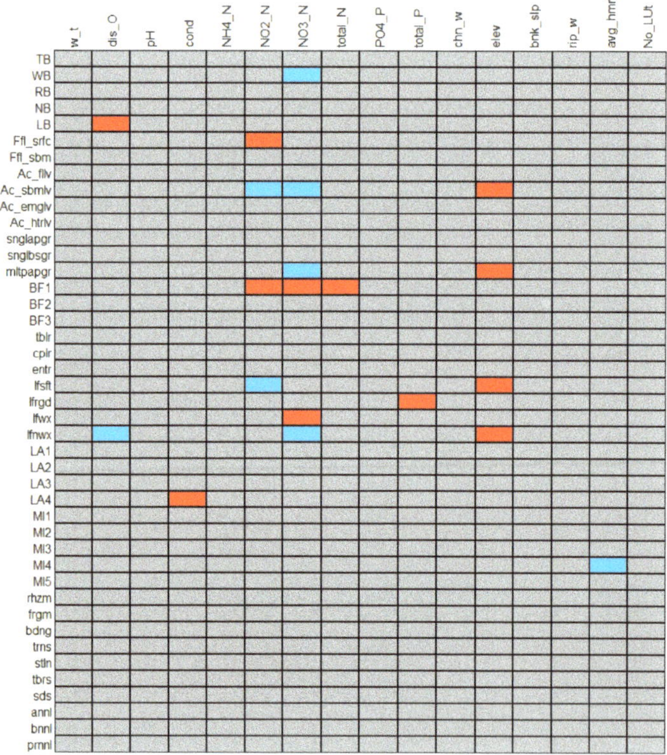

Figure 4. Results of the fourth-corner tests, corrected for multiple testing using FDR procedure. At α = 0.05 level, significant positive associations are represented by red cells and negative ones by blue cells. The abbreviated names of environmental variables are given in Table 1, and those of functional traits in Table 2.

4. Discussion

We investigated the effects of environmental factors—water chemistry and river habitat features—on the functional diversity and trait structure of vascular macrophyte assemblages in canals and heavily modified rivers in Serbia.

4.1. Correlations between Macrophyte Functional Diversity Indices

We found significant positive correlations between all functional diversity indices, since they measure different aspects of the same facet of diversity. Notably, a strong positive correlation was observed between RaoQ and FDis. That was expected, since both indices, although different, estimate the dispersion of species in trait space, weighted by their relative abundances [34]. Similarly, a strong positive correlation was found between RaoQ and Rao-based FD. We also found a strong positive correlation between FDiv and FEve, indices that measure two components of functional diversity—divergence and evenness. A high FDiv indicates a high degree of niche differentiation and low resource competition, while a high FEve indicates a high degree to which the biomass of a community is distributed in niche space to allow the effective utilization of the entire range of available resources [50]. Therefore, in the altered waterbodies analyzed here, as the degree of niche differentiation increases, the abundance distribution degree of the macrophyte communities in niche space tends to be even, allowing the effective utilization of resources, with low resource competition. This tends to increase the productivity and reliability of the ecosystem, decreasing the opportunity for invaders [50].

4.2. Effects of Water Chemistry and River Habitat Features on Macrophyte Functional Diversity Indices

The environmental constraints analyzed here were able to explain nearly half of the variability in functional diversity indices. River habitat features, compared to water chemistry, were responsible for the largest part of that variability. Our results showed similar responses of all functional diversity indices to the three main gradients: positive correlations with the essentially related factors—average degree of hemeroby and conductivity; and negative correlations with the elevational gradient, indicating an increase in all components of functional diversity towards the lowland waterbodies that are under stronger anthropogenic influence.

Notably strong negative correlations between elevation and all the analyzed components of functional diversity suggested that species with similar traits were filtered along the elevational gradient. Elevational gradient is a gradient of natural variability [22] and was found to be an important predictor of macrophyte vegetation in lakes and rivers [15,22,51–55]. It reflects changes in slope, type of substrate, water temperature, nutrient load, and the intensity of agricultural pressures typical for lowland reaches. Along the elevational gradient, all factors contributing to the increase in the diversity of vascular macrophyte vegetation are decreasing. Several studies have shown an opposite pattern, suggesting that the observed response of macrophyte communities to altitude might depend on the analyzed range of the elevational gradient [52,53,56,57].

In the studied waterbodies, the water temperature was related to the elevational gradient, with higher values at lower altitudes. Our results showed that a higher water temperature leads to an increase in functional evenness, functional divergence, and Rao-based FD, which measures both components [50]. Although all components of functional diversity shift along the elevational gradient, changes in temperature along this gradient particularly influence niche differentiation and the abundance distribution degree of macrophyte communities in niche space (i.e., functional evenness). At lower altitudes, the abundance distribution in niche space tends to be even, allowing the effective utilization of resources with low resource competition [50].

Functional evenness and divergence were also related to the littoral slope degree. They were found to be lower in waterbodies with a steep littoral slope, indicating the filtering of species with similar traits allowing them to occupy deeper water, with a low degree of niche differentiation, high resource competition, uneven abundance distribution in niche space, and the underutilization of resources [50]. In this case, the average degree of hemeroby was significantly positively correlated to all aspects of functional diversity. Hemeroby represents an integrative measure of the anthropogenic impact on ecosystems [58,59]. Different human activities have shaped the area surrounding the studied waterbodies, decreasing their naturalness [28] and directly and/or indirectly influencing other environmental factors. Similarly, a positive relationship between anthropogenic disturbances and diversity, both taxonomical and functional, was found for macrophytes in the Danube River [24,25,60] and in Slovakian ponds [54]. The intensity of human impact on the waterbodies studied here, reflected mainly in the deforestation of regions in the riparian zone, the reduction in flow velocity, the increase in transparency, sediment accumulation, as well as the subsequent higher nutrient supply, can promote an increased abundance of functionally distinct species, providing niche diversification for the establishment of diverse assemblages of macrophytes [23]. It was suggested that the functional response of the communities depends on the intensity of stress (both natural and anthropogenic), and that ecosystems that are under intensified stress will exhibit functionally homogeneous communities [61]. Our results indicate that the anthropogenic stressors are not intense enough to lead to the functional homogenization of macrophyte communities, while the intensity of natural stressors operating at higher altitudes (shading, bedrock in littoral areas, oligotrophic environment, lower temperatures, etc.) leads to trait convergence and functionally more homogeneous communities.

An additional possible explanation for the observed positive correlation between functional diversity and the intensity of human impact (i.e., hemeroby) might be (often anthropogenically induced) habitat heterogeneity, expressed here as the number of land cover classes along the banks of the waterbodies. The number of land cover classes was found to be significantly positively correlated to functional richness, the Rao-based diversity index, and functional redundancy. Larger number of land cover classes on banks provide habitat heterogeneity offering a more heterogeneous range of niches (open, shaded, different littoral slope, nutrient supply, etc.) and favoring plants with different resource utilization strategies, increasing productivity and stability, since the number of species with the same traits (i.e., functional redundancy) is also increasing [5,50].

Like the average hemeroby, water conductivity was also significantly and positively correlated to all functional diversity indices. Conductivity has frequently been reported as a key driver of richness and diversity of macrophyte communities [52,53,55,61–65], where high conductivity was indicative of waters disturbed by anthropogenic processes [52,53,55,63].

We found that the amount of niche space filled by species in the community (i.e., functional richness) increased in waterbodies with higher nitrogen concentrations. Previous studies reported nitrogen having a positive effect on species richness and diversity [55]. However, it was also documented that some species, particularly submerged macrophytes, may be lost when nitrogen concentration exceeds a certain threshold, and when the phosphorus concentration is high enough [66,67]. At high or very low nutrient levels, species and functional richness is expected to be lower, since low nutrient concentrations filter species with similar traits adapted to it, while competitive interactions usually increase when nutrient levels increase, and the highest diversity is expected at intermediate nutrient levels [68]. Therefore, we can assume that the nitrogen concentrations in the waterbodies with high functional richness are in the intermediate range.

4.3. Effect of Water Chemistry and River Habitat Features on Macrophyte Functional Trait Structure

Our results confirmed the conclusions of previous studies [4,6] that floating and emergent plants represent strategies to increase the efficiency of light interception under high nutrient concentrations, while submerged plants dominate in nutrient-poorer waterbodies at higher altitudes. A similar shift from more productive and eutrophic waterbodies dominated by emergent and floating plants to nutrient-poorer waterbodies dominated by submerged macrophytes was found in a cascade series of reservoirs at the medium and low Tietê River in Brazil [69]. This shift between two dominating growth forms is the consequence of competition for resources, primarily nitrogen and light. It was reported that floating plants have primacy for light, while submerged plants can grow at lower nutrient concentrations in water and can reduce nutrient concentrations in water to lower levels [70,71].

5. Conclusions

The simultaneous use of functional diversity indices and trait–environment correlations applied here allow us to draw several conclusions regarding the community changes along the environmental gradients, and the traits that are related to those changes. The shift from low to high functional diversity along the gradient of anthropogenic stressors, and the opposite trend of functional diversity along the elevational gradient, might indicate the prevailing processes that structure macrophyte communities. Our results confirm the previous findings that biotic filters, competition, and niche differentiation play a role in aquatic ecosystems under high anthropogenically induced productivity [72,73]. In contrast, the functional diversity indices were found to decrease with elevation, indicating the shift to predominating abiotic environmental filtering and more stressful conditions for vascular macrophytes along the elevational gradient. Trait–environment analyses indicated the change in community structure, from emergent and floating macrophytes dominant in more productive and eutrophic lowland waterbodies to submerged macrophyte-dominated waterbodies at higher altitudes with lower nutrient load and conductivity.

Our results indicate that both functional diversity indices and functional traits might prove to be useful diagnostic tools to guide the choices of relevant management measures in altered waterbodies in Serbia. We found that the current intensity of anthropogenic stressors is low enough to prevent functional homogenization and the decline of macrophyte communities. However, the high productivity and dominance of emergent and floating plants in lowland waterbodies might indicate the eutrophication-related process [4] that will, in the future, impair the ecosystem services that these waterbodies were designed to provide. This is particularly important for waterbodies designed to store water for the water supply, where in many cases, intensive land use close to the shoreline was observed (pastures, farming, summer cottages, villages, agriculture and plantations, beaches, camping areas). Therefore, in these waterbodies, management efforts should be directed toward the reevaluation of land use decisions and the reduction of nutrient inputs. Although altered waterbodies are designed to provide different services for human wellbeing, they might be harboring biota of conservational value [74]. Future investigations of altered waterbodies must include more detailed biodiversity studies and the selection of sites of conservational importance, as well as the application of adequate measures for their protection. Moreover, it is important to focus further investigation on the role of functional diversity and the structure of macrophyte communities in the indication of possible tradeoffs and/or facilitation between the ecosystem services that those altered waterbodies provide in order to direct their adequate management.

Supplementary Materials: The following supporting information can be downloaded at: https://www.mdpi.com/article/10.3390/d15020231/s1, Table S1: Species list; Table S2: Linear correlation coefficients between functional diversity indices; Figure S1: Results of the RLQ analysis.

Author Contributions: Conceptualization and methodology, D.V.; data collection and preparation, D.V., M.I., M.Ć. and R.I., analyses and visualization, D.V., M.I. and M.Ć.; resources and supervision, R.I.; writing, reviewing, and editing, D.V., M.I., M.Ć. and R.I. All authors have read and agreed to the published version of the manuscript.

Funding: This research was funded by the Provincial Secretariat for Higher Education and Scientific Research, through the grant number 142-451-2213/2022-01/1, and the Ministry for Education, Science and Technological Development of the Republic of Serbia, through the Scientific Research Program, in 2022, grant number 451-03-68/2022-14/200125.

Institutional Review Board Statement: Not applicable.

Data Availability Statement: Data may be provided upon reasonable request.

Acknowledgments: The authors thank Vanja Vukov for the help during the field surveys. We would like to thank the reviewers for taking the necessary time and effort to review the manuscript. We sincerely appreciate all their valuable comments and suggestions, which helped us in improving the quality of the manuscript.

References

1. McGill, B.J.; Enquist, B.; Weiher, E.; Westoby, M. Rebuilding community ecology from functional traits. *Trends Ecol. Evol.* **2006**, *21*, 178–185. [CrossRef] [PubMed]
2. Violle, C.; Navas, M.-L.; Vile, D.; Kazakou, E.; Fortunel, C.; Hummel, I.; Garnier, E. Let the concept of trait be functional! *Oikos* **2007**, *116*, 882–892. [CrossRef]
3. Villéger, S.; Miranda, J.R.; Hernández, D.F.; Mouillot, D. Contrasting changes in taxonomic vs. functional diversity of tropical fish communities after habitat degradation. *Ecol. Appl.* **2010**, *20*, 1512–1522. [CrossRef] [PubMed]
4. Baattrup-Pedersen, A.; Göthe, E.; Larsen, S.E.; O'Hare, M.; Birk, S.; Riis, T.; Friberg, N. Plant trait characteristics vary with size and eutrophication in European lowland streams. *J. Appl. Ecol.* **2015**, *52*, 1617–1628. [CrossRef]
5. Lukács, B.A.; E-Vojtkó, A.; Erős, T.; Molnár, V.A.; Szabó, S.; Götzenberger, L. Carbon forms, nutrients and water velocity filter hydrophyte and riverbank species differently: A trait-based study. *J. Veg. Sci.* **2019**, *30*, 471–484. [CrossRef]
6. Paz, L.; Altieri, P.; Ferreira, A.; Ocon, C.; Capítulo, A.R.; Cortelezzi, A. Macrophyte functional traits in channelized streams. *Aquat. Bot.* **2021**, *175*, 103434. [CrossRef]
7. MacArthur, R.; Levins, R. The Limiting Similarity, Convergence, and Divergence of Coexisting Species. *Am. Nat.* **1967**, *101*, 377–385. [CrossRef]
8. Kraft, N.J.B.; Valencia, R.; Ackerly, D.D. Functional Traits and Niche-Based Tree Community Assembly in an Amazonian Forest. *Science* **2008**, *322*, 580–582. [CrossRef] [PubMed]
9. Mason, N.W.H.; Lanoiselée, C.; Mouillot, D.; Irz, P.; Argillier, C. Functional characters combined with null models reveal inconsistency in mechanisms of species turnover in lacustrine fish communities. *Oecologia* **2007**, *153*, 441–452. [CrossRef]
10. Cadotte, M.W.; Tucker, C.M. Should Environmental Filtering be Abandoned? *Trends Ecol. Evol.* **2017**, *32*, 429–437. [CrossRef]
11. Weiher, E.; Keddy, P.A. Assembly Rules, Null Models, and Trait Dispersion: New Questions from Old Patterns. *Oikos* **1995**, *74*, 159–164. [CrossRef]
12. Swenson, N.G.; Enquist, B.J. Ecological and evolutionary determinants of a key plant functional trait: Wood density and its community-wide variation across latitude and elevation. *Am. J. Bot.* **2007**, *94*, 451–459. [CrossRef]
13. Bornette, G.; Puijalon, S. Macrophytes: Ecology of aquatic plants. In *Encyclopedia of Life Sciences*; John Wiley & Sons, Ltd.: Chichester, UK, 2009. [CrossRef]
14. Manolaki, P.; Guo, K.; Vieira, C.; Papastergiadou, E.; Riis, T. Hydromorphology as a controlling factor of macrophytes assemblage structure and functional traits in the semi-arid European Mediterranean streams. *Sci. Total. Environ.* **2020**, *703*, 134658. [CrossRef]
15. Papastergiadou, E.; Stefanidis, K.; Dörflinger, G.; Giannouris, E.; Kostara, K.; Manolaki, P. Exploring biodiversity in riparian corridors of a Mediterranean island: Plant communities and environmental parameters in Cyprus rivers. *Plant Biosyst. Int. J. Deal. All Asp. Plant Biol.* **2014**, *150*, 91–103. [CrossRef]
16. Göthe, E.; Baattrup-Pedersen, A.; Wiberg-Larsen, P.; Graeber, D.; Kristensen, E.A.; Friberg, N. Environmental and spatial controls of taxonomic versus trait composition of stream biota. *Freshw. Biol.* **2016**, *62*, 397–413. [CrossRef]
17. Baattrup-Pedersen, A.; Garssen, A.; Göthe, E.; Hoffmann, C.C.; Oddershede, A.; Riis, T.; van Bodegom, P.; Larsen, S.E.; Soons, M. Structural and functional responses of plant communities to climate change-mediated alterations in the hydrology of riparian areas in temperate Europe. *Ecol. Evol.* **2018**, *8*, 4120–4135. [CrossRef] [PubMed]
18. Bejarano, M.D.; Jansson, R.; Nilsson, C. The effects of hydropeaking on riverine plants: A review. *Biol. Rev.* **2018**, *93*, 658–673. [CrossRef] [PubMed]
19. Directive 2000/60/EC of October 23, 2000. Of the European Parliament and of the council establishing a framework for community action in the field of water policy. *Off. J. Eur. Communities* **2000**, *L327*, 1–72.
20. Baattrup-Pedersen, A.; Göthe, E.; Riis, T.; O'Hare, M.T. Functional trait composition of aquatic plants can serve to disentangle multiple interacting stressors in lowland streams. *Sci. Total. Environ.* **2016**, *543*, 230–238. [CrossRef]
21. Stefanidis, K.; Oikonomou, A.; Papastergiadou, E. Responses of different facets of aquatic plant diversity along environmental gradients in Mediterranean streams: Results from rivers of Greece. *J. Environ. Manag.* **2021**, *296*, 113307. [CrossRef] [PubMed]
22. Mouton, T.L.; Matheson, F.E.; Stephenson, F.; Champion, P.D.; Wadhwa, S.; Hamer, M.P.; Catlin, A.; Riis, T. Environmental filtering of native and non-native stream macrophyte assemblages by habitat disturbances in an agricultural landscape. *Sci. Total. Environ.* **2018**, *659*, 1370–1381. [CrossRef]
23. Vukov, D.; Ilić, M.; Ćuk, M.; Radulović, S.; Igić, R.; Janauer, G.A. Combined effects of physical environmental conditions and anthropogenic alterations are associated with macrophyte habitat fragmentation in rivers—Study of the Danube in Serbia. *Sci. Total. Environ.* **2018**, *634*, 780–790. [CrossRef]

24. Dray, S.; Choler, P.; Dolédec, S.; Peres-Neto, P.R.; Thuiller, W.; Pavoine, S.; ter Braak, C.J.F. Combining the fourth-corner and the RLQ methods for assessing trait responses to environmental variation. *Ecology* **2014**, *95*, 14–21. [CrossRef]
25. Vukov, D.; Ilić, M.; Ćuk, M.; Igić, R. The Effect of Hydro-Morphology and Habitat Alterations on the Functional Diversity and Composition of Macrophyte Communities in the Large River. *Front. Environ. Sci.* **2022**, *10*, 863508. [CrossRef]
26. Reich, P.B.; Wright, I.J.; Cavender-Bares, J.; Craine, J.M.; Oleksyn, J.; Westoby, M.; Walters, M.B. The Evolution of Plant Functional Variation: Traits, Spectra, and Strategies. *Int. J. Plant Sci.* **2003**, *164*, S143–S164. [CrossRef]
27. *European Standard EN 14184:2014*; Water Quality—Guidance Standard for the Surveying of Macrophytes in Running Waters. Comité Européen de Normalisation: Bruxelles, Belgium, 2014.
28. Walz, U.; Stein, C. Indicators of hemeroby for the monitoring of landscapes in Germany. *J. Nat. Conserv.* **2014**, *22*, 279–289. [CrossRef]
29. Ellenberg, H. *Vegetation Ecology of Central Europe*, 4th ed.; Cambridge University Press: Cambridge, UK, 1988.
30. Borhidi, A. *Social Behaviour Types of the Hungarian flora, its Naturalness and Relative Ecological Indicator Values*, 1st ed.; Janus Pannonius Tudományegyetem Kiadványa: Pécs, Hungary, 1993.
31. Willby, N.J.; Abernethy, V.J.; Demars, B.O.L. Attribute-based classification of European hydrophytes and its relationship to habitat utilization. *Freshw. Biol.* **2000**, *43*, 43–74. [CrossRef] [PubMed]
32. Botta-Dukát, Z. Rao's Quadratic Entropy as a Measure of Functional Diversity Based on Multiple Traits. *J. Veg. Sci.* **2005**, *16*, 533–540. [CrossRef]
33. Villéger, S.; Mason, N.W.H.; Mouillot, D. New multidimensional functional diversity indices for a multifaceted framework in functional ecology. *Ecology* **2008**, *89*, 2290–2301. [CrossRef]
34. Laliberté, E.; Legendre, P. A distance-based framework for measuring functional diversity from multiple traits. *Ecology* **2010**, *91*, 299–305. [CrossRef]
35. Borcard, D.; Gillet, F.; Legendre, P. *Numerical Ecology with R*, 2nd ed.; Springer: New York, NY, USA, 2018; pp. 369–412.
36. De Bello, F.; Lavergne, S.; Meynard, C.N.; Lepš, J.; Thuiller, W. The partitioning of diversity: Showing Theseus a way out of the labyrinth. *J. Veg. Sci.* **2010**, *21*, 992–1000. [CrossRef]
37. Laliberté, E.; Legendre, P.; Shipley, B. FD: Measuring Functional Diversity from Multiple Traits, and Other Tools for Functional Ecology. 2022, R package version 1.0-12.1. Available online: https://CRAN.R-project.org/package=FD (accessed on 5 October 2022).
38. Ricotta, C.; De Bello, F.; Moretti, M.; Caccianiga, M.S.; Cerabolini, B.E.L.; Pavoine, S. Measuring the functional redundancy of biological communities: A quantitative guide. *Methods Ecol. Evol.* **2016**, *7*, 1386–1395. [CrossRef]
39. Heino, J.; Girón, J.G.; Hämäläinen, H.; Hellsten, S.; Ilmonen, J.; Karjalainen, J.; Mäkinen, T.; Nyholm, K.; Ropponen, J.; Takolander, A.; et al. Assessing the conservation priority of freshwater lake sites based on taxonomic, functional and environmental uniqueness. *Divers. Distrib.* **2022**, *28*, 1966–1978. [CrossRef]
40. ter Braak, C.J.F. Canonical correspondence analysis: A new eigenvector technique for multivariate direct gradient analysis. *Ecology* **1986**, *64*, 454–462. [CrossRef]
41. Oksanen, J.; Simpson, G.; Blanchet, F.; Kindt, R.; Legendre, P.; Minchin, P.; O'Hara, R.; Solymos, P.; Stevens, M.; Szoecs, E.; et al. Vegan: Community Ecology Package. 2022, R Package Version 2.6-5. Available online: https://github.com/vegandevs/vegan (accessed on 10 October 2022).
42. Dolédec, S.; Chessel, D.; ter Braak, C.J.F.; Champely, S. Matching species traits to environmental variables: A new three-table ordination method. *Environ. Ecol. Stat.* **1996**, *3*, 143–166. [CrossRef]
43. Legendre, P.; Galzin, R.; Harmelin-Vivien, M.L. Relating Behavior to Habitat: Solutions to the Fourth-corner Problem. *Ecology* **1997**, *78*, 547–562. [CrossRef]
44. Dray, S.; Legendre, P. Testing the species traits–environment relationships: The fourth-corner problem revisited. *Ecology* **2008**, *89*, 3400–3412. [CrossRef] [PubMed]
45. Legendre, P.; Legendre, L. *Numerical Ecology*, 3rd ed.; Elsevier: Amsterdam, The Netherlands, 2012; pp. 613–622.
46. ter Braak, C.J.F.; Cormont, A.; Dray, S. Improved testing of species traits–environment relationships in the fourth-corner problem. *Ecology* **2012**, *93*, 1525–1526. [CrossRef]
47. Benjamini, Y.; Hochberg, Y. Controlling the False Discovery Rate: A Practical and Powerful Approach to Multiple Testing. *J. R. Stat. Soc. Ser. B* **1995**, *57*, 289–300. [CrossRef]
48. Pebesma, E.J.; Dufour, A.; Thioulouse, J. The ade4 Package—I: One-Table Methods. *R News* **2004**, *4*, 5–10. Available online: https://cran.r-project.org/doc/Rnews/ (accessed on 14 October 2022).
49. Pebesma, E.J.; Dufour, A.; Chessel, D. The ade4 Package—II: Two-Table and K-Table Methods. *R News* **2007**, *7*, 47–52. Available online: https://cran.r-project.org/doc/Rnews/ (accessed on 14 October 2022).
50. Mason, N.W.H.; Mouillot, D.; Lee, W.G.; Wilson, J.B. Functional richness, functional evenness and functional divergence: The primary components of functional diversity. *Oikos* **2005**, *111*, 112–118. [CrossRef]
51. Lacoul, P.; Freedman, B. Relationships between aquatic plants and environmental factors along a steep Himalayan altitudinal gradient. *Aquat. Bot.* **2006**, *84*, 3–16. [CrossRef]
52. Chappuis, E.; Gacia, E.; Ballesteros, E. Environmental factors explaining the distribution and diversity of vascular aquatic macrophytes in a highly heterogeneous Mediterranean region. *Aquat. Bot.* **2014**, *113*, 72–82. [CrossRef]

53. Fernández-Aláez, C.; Fernández-Aláez, M.; García-Criado, F.; García-Girón, J. Environmental drivers of aquatic macrophyte assemblages in ponds along an altitudinal gradient. *Hydrobiologia* **2016**, *812*, 79–98. [CrossRef]
54. Kochjarová, J.; Novikmec, M.; Oťaheľová, H.; Hamerlík, L.; Svitok, M.; Hrivnák, M.; Senko, D.; Bubíková, K.; Matúšová, Z.; Paľove-Balang, P.; et al. Vegetation-Environmental Variable Relationships in Ponds of Various Origins along an Altitudinal Gradient. *Pol. J. Environ. Stud.* **2017**, *26*, 1575–1583. [CrossRef]
55. Stefanidis, K.; Sarika, M.; Papastegiadou, E. Exploring environmental predictors of aquatic macrophytes in water-dependent Natura 2000 sites of high conservation value: Results from a long-term study of macrophytes in Greek lakes. *Aquat. Conserv. Mar. Freshw. Ecosyst.* **2018**, *29*, 1133–1148. [CrossRef]
56. Alahuhta, J.; Kosten, S.; Akasaka, M.; Auderset, D.; Azzella, M.M.; Bolpagni, R.; Bove, C.P.; Chambers, P.A.; Chappuis, E.; Clayton, J.; et al. Global variation in the beta diversity of lake macrophytes is driven by environmental heterogeneity rather than latitude. *J. Biogeogr.* **2017**, *44*, 1758–1769. [CrossRef]
57. Oikonomou, A.; Stefanidis, K. α- and β-Diversity Patterns of Macrophytes and Freshwater Fishes are Driven by Different Factors and Processes in Lakes of the Unexplored Southern Balkan Biodiversity Hotspot. *Water* **2020**, *12*, 1984. [CrossRef]
58. Kowarik, I. Zum menschlichen Einfluss auf Flora und Vegetation: Theoretische Konzepte und ein Quantifizierungsansatz am Beispiel von Berlin (West). *Landsch. Umweltforsch.* **1988**, *56*, 1–280.
59. Sukopp, H. Dynamik und Konstanz in der Flora der Bundesrepublik Deutschland. *Schr. Veg.* **1976**, *10*, 9–26.
60. Janauer, G.; Exler, N.; Anačkov, G.; Barta, V.; Berczik, Á.; Boža, P.; Dinka, M.; Georgiev, V.; Germ, M.; Holcar, M.; et al. Distribution of the Macrophyte Communities in the Danube Reflects River Serial Discontinuity. *Water* **2021**, *13*, 918. [CrossRef]
61. Gutiérrez-Cánovas, C.; Sánchez-Fernández, D.; Velasco, J.; Millán, A.; Bonada, N. Similarity in the difference: Changes in community functional features along natural and anthropogenic stress gradients. *Ecology* **2015**, *96*, 2458–2466. [CrossRef] [PubMed]
62. Sleith, R.S.; Wehr, J.D.; Karol, K.G. Untangling climate and water chemistry to predict changes in freshwater macrophyte distributions. *Ecol. Evol.* **2018**, *8*, 2802–2811. [CrossRef]
63. Lukács, B.A.; Tóthmérész, B.; Borics, G.; Várbíró, G.; Juhász, P.; Kiss, B.; Müller, Z.; G-Tóth, L.; Erős, T. Macrophyte diversity of lakes in the Pannon Ecoregion (Hungary). *Limnologica* **2015**, *53*, 74–83. [CrossRef]
64. Bini, L.M.; Thomaz, S.M.; Murphy, K.J.; Camargo, A.F.M. Aquatic macrophyte distribution in relation to water and sediment conditions in the Itaipu Reservoir, Brazil. *Hydrobiologia* **1999**, *415*, 147–154. [CrossRef]
65. Rolon, A.S.; Lacerda, T.; Maltchik, L.; Guadagnin, D.L. Influence of area, habitat and water chemistry on richness and composition of macrophyte assemblages in southern Brazilian wetlands. *J. Veg. Sci.* **2008**, *19*, 221–228. [CrossRef]
66. Barker, T.; Hatton, K.O.; O'Connor, M.; Connor, L.; Moss, B. Effects of nitrate load on submerged plant biomass and species richness: Results of a mesocosm experiment. *Fundam. Appl. Limnol.* **2008**, *173*, 89–100. [CrossRef]
67. Moss, B.; Jeppesen, E.; Søndergaard, M.; Lauridsen, T.L.; Liu, Z.W. Nitrogen, macrophytes, shallow lakes and nutrient limita-tion: Resolution of a current controversy? *Hydrobiologia* **2013**, *710*, 3–21. [CrossRef]
68. Bornette, G.; Puijalon, S. Response of aquatic plants to abiotic factors: A review. *Aquat. Sci.* **2011**, *73*, 1–14. [CrossRef]
69. Smith, W.S.; Espíndola, E.L.G.; Rocha, O. Environmental gradient in reservoirs of the medium and low Tietê River: Limnological differences through the habitat sequence. *Acta Limnol. Bras.* **2014**, *26*, 73–88. [CrossRef]
70. Scheffer, M.; Szabó, S.; Gragnani, A.; van Nes, E.H.; Rinaldi, S.; Kautsky, N.; Norberg, J.; Roijackers, R.M.M.; Franken, R.J.M. Floating plant dominance as a stable state. *Proc. Natl. Acad. Sci. USA* **2003**, *100*, 4040–4045. [CrossRef] [PubMed]
71. Szabó, S.; Koleszár, G.; Braun, M.; Nagy, Z.; Vicei, T.T.; Peeters, E.T.H.M. Submerged Rootless Macrophytes Sustain a Stable State Against Free-Floating Plants. *Ecosystems* **2021**, *25*, 17–29. [CrossRef]
72. Engelhardt, K.A.M.; Ritchie, M.E. Effects of macrophyte species richness on wetland ecosystem functioning and services. *Nature* **2001**, *411*, 687–689. [CrossRef] [PubMed]
73. Fu, H.; Zhong, J.; Yuan, G.; Ni, L.; Xie, P.; Cao, T. Functional traits composition predict macrophytes community productivity along a water depth gradient in a freshwater lake. *Ecol. Evol.* **2014**, *4*, 1516–1523. [CrossRef]
74. Hill, M.J.; White, J.C.; Biggs, J.; Briers, R.A.; Gledhill, D.; Ledger, M.E.; Thornhill, I.; Wood, P.J.; Hassall, C. Local contributions to beta diversity in urban pond networks: Implications for biodiversity conservation and management. *Divers. Distrib.* **2021**, *27*, 887–900. [CrossRef]

 *diversity*

Article

Current Distribution and Conservation Issues of Aquatic Plant Species Protected under Habitats Directive in Lithuania

Zofija Sinkevičienė *, Liucija Kamaitytė-Bukelskienė, Lukas Petrulaitis and Zigmantas Gudžinskas *

Nature Research Centre, Institute of Botany, Žaliųjų Ežerų Str. 49, 12200 Vilnius, Lithuania
* Correspondence: zofija.sinkeviciene@gamtc.lt (Z.S.); zigmantas.gudzinskas@gamtc.lt (Z.G.)

Abstract: The European Habitats Directive was adopted to halt the rapid loss of biodiversity and has become an important instrument for protecting biodiversity in the European Union. Three aquatic plant species protected under the European Habitats Directive have so far been found in Lithuania: *Aldrovanda vesiculosa*, *Caldesia parnassifolia*, and *Najas flexilis*. Our aim in this study was to evaluate the former and current distribution and the status of conservation of the target species. Screening for the above-mentioned protected aquatic plant species was performed in 73 natural lakes throughout Lithuania in 2019–2021. We confirmed extant populations of *Aldrovanda vesiculosa* in four lakes, *Caldesia parnassifolia* in two lakes, and *Najas flexilis* in four lakes in the northeastern part of the country. We studied *Aldrovanda vesiculosa* populations three times (2015, 2019, and 2022) in Lake Rūžas and once each in Lake Apvardai and Lake Dysnai (2020). The population density of *Aldrovanda vesiculosa* ranged from 193.4 ± 159.7 to 224.0 ± 211.0 individuals/m^2, the mean length of plants ranged from 12.5 ± 2.1 to 14.3 ± 2.7 cm, and the mean number of apices ranged from 2.0 ± 0.7 to 2.2 ± 0.9 per individual. The habitat of *Aldrovanda vesiculosa* in Lake Rūžas covered an area of about 3 ha. The number of generative individuals of *Caldesia parnassifolia* widely varied between years in Lake Rūžas. All populations of *Najas flexilis* were small, although the potential habitats in the studied lakes cover relatively large areas. We propose designating all lakes with populations of *Aldrovanda vesiculosa*, *Caldesia parnassifolia*, and *Najas flexilis* as special areas of conservation, as well as developing and implementing action plans for the conservation of these species and their habitats.

Keywords: *Aldrovanda vesiculosa*; *Caldesia parnassifolia*; communities; distribution; *Najas flexilis*; population size; special areas of conservation; turions; vegetative propagation

Citation: Sinkevičienė, Z.; Kamaitytė-Bukelskienė, L.; Petrulaitis, L.; Gudžinskas, Z. Current Distribution and Conservation Issues of Aquatic Plant Species Protected under Habitats Directive in Lithuania. *Diversity* **2023**, *15*, 185. https://doi.org/10.3390/d15020185

Academic Editors: Mateja Germ, Igor Zelnik, Matthew Simpson and Michael Wink

Received: 21 December 2022
Revised: 25 January 2023
Accepted: 25 January 2023
Published: 28 January 2023

1. Introduction

Inland freshwater bodies and wetlands are crucial ecosystems for maintaining ecological stability and preserving biodiversity [1–4]. Ecologically, economically, and socially important freshwater ecosystems are vulnerable, especially in agrarian and urbanised areas [5]. Freshwater ecosystems are changing because of both natural causes and anthropogenic pressure, particularly as a result of pollution and the resulting eutrophication of water bodies [3,6,7]. Climate change and the resulting alterations in water regimes also seriously impact the state of water bodies and wetlands [1,8]. The effects of these factors are altered habitats, reduced biodiversity, and threat of extinction in certain species [9–12].

The European Habitats Directive (Council Directive 92/43/EEC) was adopted to halt the rapid loss of biodiversity and has become an important instrument for protecting biodiversity in the European Union. The preamble to the Habitats Directive specifies that its primary objective is to ensure the restoration or maintenance of natural habitats and species of community interest at a favourable status. The Habitats Directive now aims to protect the 233 habitat types and 1389 characteristic, rare, or endangered species of flora and fauna in the European Union [13]. Twelve vascular plant species protected under the Habitats Directive are found in Lithuania, of which three (*Aldrovanda vesiculosa* L., *Caldesia parnassifolia* (L.) Parl., and *Najas flexilis* (Willd.) Rostk. & Schmidt) are aquatic plant species.

Aldrovanda vesiculosa is among the most renowned and studied aquatic plant species in the world because of its distinctive appearance, carnivorous habits, and worldwide rarity [14]. Globally, *A. vesiculosa* is considered an endangered (EN) species [15]. In North America, populations of this species are classified as non-native but not invasive [16,17]. New occurrences of the species have recently been discovered in India [18] and Mongolia [19], and several new localities have been reported in Ukraine [20]. In Europe, extant populations of *A. vesiculosa* have been recorded in 10 countries, where it occurs in natural habitats [15,21]. The species has been successfully reintroduced in the Czech Republic, and has been introduced in Germany [22], Switzerland, and the Netherlands [14]. Although the species has been recorded in 14 European countries, the continent-wide population is considered to be declining. Recently, 184 populations were confirmed as extinct, while only 49 populations were extant [15]. Relatively large and stable populations of *A. vesiculosa* persist in Poland, Lithuania, Russia, and Ukraine. More than half of the extant populations are concentrated in Ukraine, where they are currently most threatened [15,20].

Caldesia parnassifolia is widespread in the tropical and subtropical regions of Africa, Asia, Australia, and the temperate zone of Europe. Globally, the population status of the species is of least concern (LC) [23]. *C. parnassifolia* is a critically endangered (CR) species in Poland [24,25], Hungary [26], and Lithuania [27], whereas in Italy, it is declared extinct [10,28]. In Belarus, it is listed as a protected species [29].

Najas flexilis, in contrast to *Aldrovanda vesiculosa* and *Caldesia parnassifolia*, is mainly distributed in the boreal and temperate regions of Europe and North America, with isolated occurrences in Asia [19,30–32]. The status of the population of *Najas flexilis* is of least concern (LC) at the global level [31], but more than a decade ago, it was recognised as a vulnerable (VU) species [33]. *N. flexilis* is considered extinct in Germany, Switzerland, and Poland [34–36]. Most extant localities in Europe are concentrated in the British Isles [37], and a considerable number of occurrences have been recorded in Latvia [38]. The species has also been confirmed to occur in Norway, Finland, Sweden, Estonia, Lithuania, Austria, Belarus, and Russia [39–46].

The degradation and loss of suitable habitats because of natural and anthropogenic causes are among the most important drivers of species decline [7,47]. Aquatic, riparian, and wetland plants are particularly sensitive to changes in habitat conditions; therefore, their conservation and the protection of their habitats are challenging [7,47–50]. As such, precise knowledge of the distribution, habitat ecology, and population size of an endangered species is essential to ensure its favourable conservation [28].

At the initial stages of the implementation of the Habitats Directive in Lithuania, when the special areas of conservation (SACs) were first established, the focus was on the already known habitats of protected aquatic plants. Initially, experts assessed the status of the populations, and the Conservation and Action Plan for *Aldrovanda vesiculosa* [51] was prepared. Following the implementation of the nature management measures, in 2015, more extensive studies on *A. vesiculosa* were initiated. In 2019–2021, a new round of search and assessment for *A. vesiculosa*, *Caldesia parnassifolia*, and *Najas flexilis* was undertaken throughout the territory of Lithuania. The aim of this study was to assess the status of three aquatic plant species, *Aldrovanda vesiculosa*, *Caldesia parnassifolia*, and *Najas flexilis*, protected under the European Habitats Directive in Lithuania. With this study, we attempted to answer the following questions: (a) What is the current distribution of the three protected aquatic plant species in Lithuania? (b) What is the current state of the habitats of these species? (c) In which plant communities are *Aldrovanda vesiculosa* and *Caldesia parnassifolia* occurring? (d) How do the population density and morphological parameters of *Aldrovanda vesiculosa* depend on habitat conditions? (e) What are the conservation status and requirements of the target species?

2. Materials and Methods

2.1. Study Species

Aldrovanda vesiculosa L. (Droseraceae) is a perennial, rootless, free-floating carnivorous aquatic plant with a stem 6–20 cm long, and occasionally longer. The stem of a mature plant has 15–20 whorls of leaves, with 1–8 branches, or is unbranched. The leaf whorl consists of 6–9 leaves with snapping traps for small aquatic animals. In temperate regions, it can flower and produce viable seeds, but usually reproduces vegetatively. Branches detached from the parent individual in summer form a new individual, the apices of which turn into turions (wintering buds) in autumn [14,24].

This species has a wide range, extending from the tropical regions of Australia, Asia, and Africa, to the temperate regions of Europe. The northernmost occurrences of the species were recorded in northwestern Russia [15].

Caldesia parnassifolia (L.) Parl. (Alismataceae) is a perennial rhizomatous aquatic plant that grows up to 80 cm tall. Its leaves are floating or emerged, with petioles 5–100 cm long and ovate or elliptic leaf blades with a cordate base and obtuse apex. Its inflorescence is emergent and paniculate, with whorled branches. Its flowers are bisexual, with persistent sepals and ovate petals. Its fruitlets are obovoid, with 3–5 longitudinal ribs on each side. It reproduces vegetatively via turions, which form at the tips of rhizomes and sometimes in inflorescences [52–54].

This species is distributed mainly in the tropical and subtropical regions of Africa, Asia, and Australia, and occurs in the temperate regions of Europe [23]. The newly discovered locality on the border between Lithuania and Latvia [55] is the northernmost record for this species in Europe.

Najas flexilis (Willd.) Rostk. & Schmidt (Hydrocharitaceae) is an annual aquatic plant that grows completely submerged and rooted on the bottom. Its stem is 2.5–50 cm long. Its leaves are sessile, its lamina is minutely serrulate with unicellular teeth, and its apex is acute. The leaf base is slightly wider than the lamina and minutely serrulate with teeth like those of the lamina. The plants are monoecious. Female and male flowers are solitary, sessile, and enveloped in membranous involucre. Its fruits consist of one narrowly-to-broadly obovate seed. Its seed coat has 5–6 angled areolae [34,36].

This species is distributed in the boreal and temperate regions of North America and Europe, with a few isolated parts of its range known in Asia [30–32]. Many localities of *Najas flexilis* are recorded in Latvia [38]. The species also occurs in Belarus [44], whereas it is considered extinct in Poland [56].

2.2. Study Area and Sites

The study area was in the northeastern part of Lithuania, including parts of the Zarasai, Ignalina, and Visaginas administrative districts (Figure 1). The territory is in the Aukštaičiai Upland, which is a part of the Baltic Highlands. The relief of the region was shaped by two glacial flows of the last glaciation [57]. The Aukštaičiai Upland is characterised by an abundance of lakes, which occupy approximately 6% of its area [57]. Because the region has a more continental climate than other parts of Lithuania, it is characterised by a wide range of temperatures, colder winters, longer snow cover, and a shorter plant-growing season [58]. The mean annual precipitation in this area is 650 mm, approximately 65% of which falls during the warm season. Evaporation from the water surface is as high as 538 mm in May–October [58].

Most of the studied lakes belong to the Daugava River basin, except for Lake Dūkštas, which belongs to the Nemunas River basin [59]. All the studied lakes, except Lakes Avilys and Ažvintis, are in protected areas (including SACs of the *NATURA 2000* network).

The studied lakes have at least one inflow and outflow, meandering shorelines, and sheltered shallow bays. Alksnas, Apvardai, Dysnai, and Rūžas Lakes are natural, shallow, polymictic, unstratified, and mostly surrounded by mires (Table 1, Figure 1). Because the areas around the lakes experienced intensive agriculture in the past, much of the waterlogged land in the lake basins has been drained. Deep, dimictic, stratified, mesotrophic lakes, such as Sągardas, Ažvintis, and Dūkštas Lakes are natural, and are surrounded by forests, fields, and sparsely populated settlements (Table 1, Figure 1). The bottom substrate of the lakes with *Najas flexilis* populations consist of sand and gravel with a thin layer of silt, whereas the bottom substrate of the lakes with *Aldrovanda vesiculosa* and *Caldesia parnassifolia* consist of organic sediments, silt, and clay.

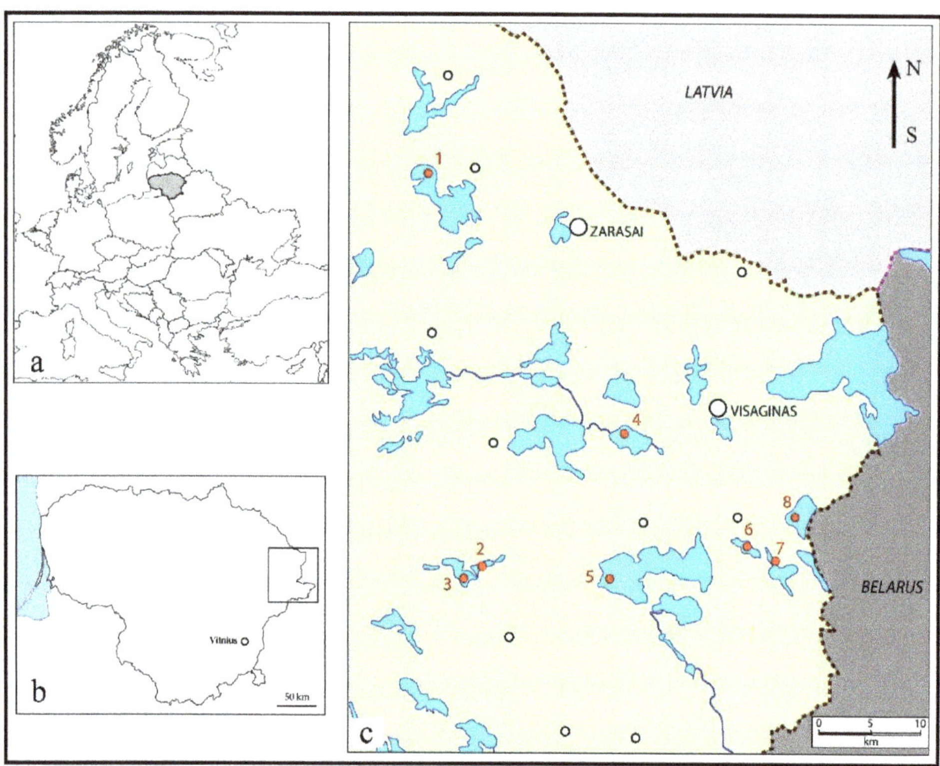

Figure 1. Position of Lithuania in Europe (**a**). A rectangle delineates the northeastern region of Lithuania (**b**). Lakes where the target species were found are marked by red dots (**c**): *Najas flexilis* in lakes Avilys (1), Ažvintis (2), Sągardas (3), and Dūkštas (4); *Aldrovanda vesiculosa* in lakes Dysnai (5) and Alksnas (6); *Aldrovanda vesiculosa* and *Caldesia parnassifolia* in lakes Rūžas (7) and Apvardai (8).

Table 1. Hydromorphological characteristics of studied lakes and physicochemical parameters of their water [60]. Physicochemical properties of water were determined in 2007 (Lake Apvardai), 2017 (Lakes Avilys and Rūžas), 2019 (Lakes Dysnai and Alksnas), 2020 (Lake Dūkštas), and 2021 (Lakes Sągardas and Ažvintis). Abbreviations: DO, dissolved oxygen; TDS, total dissolved substances; EC, electric conductivity; TN, total nitrogen; TP, total phosphorus.

No.	Lake	Area (ha)	Mean Depth (m)	Maximum Depth (m)	Secchi Depth (m)	DO (mg/L)	pH	TDS (mg/L)	EC (μS/cm)	TN (mg/L)	TP (mg/L)
1	Avilys	1224	3.0	13.5	3.7	8.8	8.6	1.7	245	0.54	0.01
2	Sągardas	114	7.6	26.5	5.2	9.9	8.4	1.0	165	0.48	0.01
3	Ažvintis	264	5.7	23.0	4.0	9.9	8.4	1.5	127	0.46	0.01
4	Dūkštas	520	5.4	10.5	4.8	9.4	8.6	1.3	207	0.53	0.01
5	Dysnai	2401	3.0	6.0	1.1	10.6	8.7	–	321	0.56	0.03
6	Alksnas	176	2.6	4.6	4.2	9.8	8.5	1.5	343	0.46	0.01
7	Rūžas	219	2.5	4.3	2.5	8.0	8.4	–	–	0.54	0.02
8	Apvardai	425	2.6	4.9	–	7.6	8.2	–	310	–	–

2.3. Historical and Current Distribution of Species

We critically reassessed the historical and current distributions of *Aldrovanda vesiculosa*, *Caldesia parnassifolia*, and *Najas flexilis* based on studies of the herbarium collections and analyses of the literature and other information sources. Because the literature in the first half of the 20th century contained inaccurate references to the locations of the studied species, we relied on the information provided by primary sources or on the labels of the herbarium specimens to assess the historical records of species. We used the plant specimens from the herbaria of the Institute of Botany of the Nature Research Centre (BILAS) and Vilnius University (WI) in this study.

We performed targeted searches for protected aquatic plant species of European importance in Lithuania from mid-July to mid-September 2019–2021. For the search for *Aldrovanda vesiculosa* populations, we selected 48 shallow lakes in the whole territory of Lithuania that were predominantly vegetated by floating-leaved plants and with banks fully or partially surrounded by mires. We selected the lakes according to orthophotographic images and previously collected information on plant species diversity and plant community composition. We also screened lakes where *A. vesiculosa* had been previously recorded. We did not perform targeted screening for *Caldesia parnassifolia*, but lakes where this species had been recorded previously were inspected.

To screen for *Najas flexilis*, we selected 25 mostly large and deep lakes with no or a weakly developed belt of floating-leaved plants and predominantly submerged vegetation consisting mainly of charophyte (or bryophyte) species in the whole territory of Lithuania. We paid particular attention to lakes where plants of the genus *Najas* had previously been found or were found during the survey. We surveyed the submerged vegetation in the lakes, searching for *N. flexilis* in transects perpendicular to the shoreline. We surveyed the shallows for *N. flexilis* using an aquascope, whereas in the deep areas, we obtained plant samples using a Bernatowicz's grab (0.4 × 0.4 m) or a grapnel for detailed identification and analyses. We recorded data on water and bottom sediment characteristics in each transect, and species diversity and abundance in different depth zones.

2.4. Assessment of Aldrovanda vesiculosa Populations

We started regular assessments of the population status and habitat condition of *A. vesiculosa* in Lake Rūžas in 2015, following the implementation of the Species Conservation Action Plan in 2013–2014. We conducted studies in Lake Rūžas in 2015, 2019, and 2022; we assessed Lake Apvardai and Lake Dysnai *A. vesiculosa* populations in August–September 2020. In Lake Rūžas, we performed studies at 20 sampling plots during each year of the study; in the small populations in Lakes Apvardai and Dysnai, we performed studies at 10 and 15 sampling plots, respectively. We analysed a total of 85 sampling plots.

At each survey point, we performed a phytosociological relevé using the Braun-Blanquet [61] approach. We assessed the plant community on minimum and maximum areas of 1 and 4 m^2, respectively, with uniform plant cover. After the community assessment, we placed a 0.25 m^2 floating square frame (all sides 0.5 m long) with a grid of 25 cells (10 × 10 cm) to delimit the sampling plot (Figure 2). We assessed the percentage cover of each plant species present in the sampling plot (with a precision of 0.1%). Then, we randomly selected ten individuals of *A. vesiculosa* from the sampling plot for length measurements and apex counts. We counted all individuals of *A. vesiculosa* in the sampling plot to determine their density. Then, we measured the water depth at the survey point. *Caldesia parnassifolia* occurred in communities with *Aldrovanda vesiculosa* in Lake Rūžas and Lake Apvardai; therefore, we did not separately assess the abundance of this species.

Figure 2. Sampling plot was delimited by a graded wooden frame to study *Aldrovanda vesiculosa* populations (photo by Z. Sinkevičienė).

The nomenclature of phytocenoses followed Šumberová [62]. The names of vascular plants were determined according to Euro+Med PlantBase [63].

2.5. Statistical Analyses

The results of the Shapiro–Wilk test showed that parts of the dataset (plant length, number of apices, number of individuals per sampling plot, cover of species or their groups, and depth) did not meet the criteria of normal distribution. Therefore, we compared the datasets using nonparametric statistical analysis methods. We used the Kruskal–Wallis H-test to detect differences between plant length, number of apices, number of individuals per sampling plot, and cover of *Aldrovanda vesiculosa* from different study sites and years, and we applied the Mann–Whitney U-test for post hoc pairwise comparison. When presenting descriptive statistics, we report the mean and the standard deviation (mean ± SD), and the minimum, maximum, and median values. Relationships between the number of individuals and water depth, and between the coverage of individuals and water depth, were estimated using Spearman's rank-order correlation. Our results and data on the length of *A. vesiculosa* individuals obtained from references were compared using a *t*-test. Because we counted individuals of *A. vesiculosa* in 0.25 m^2 sampling plots during the study, we multiplied the number of individuals in each plot by four, and we calculated the mean density of individuals per 1 m^2 from the result. We performed principal component analysis (PCA) using the number of individuals of *A. vesiculosa*, the cover of submerged, floating, and emergent plants, and the water depth in the sample plot. We used a correlation matrix between groups in the analysis. We considered each study in different years in the same lake or in different lakes as a separate group. We performed a cluster analysis of the communities with *A. vesiculosa* and *Caldesia parnassifolia*, and we created a dendrogram using the paired group (UPGMA) algorithm and Euclidean distance. We performed all calculations using PAST 4.10 software [64].

3. Results

3.1. Historical and Current Distribution of Species

Our comprehensive analysis of the literature and herbarium data showed that *Aldrovanda vesiculosa* was first recorded in Lithuania in 1955 in Lake Dysnai (WI) [65]. An earlier reference [66] to the occurrence of *A. vesiculosa* in the vicinity of Vilnius in the 19th century is a misinterpretation, and this error has been repeated by several recent authors [15,67]. Diels [66] cited Gorski [68] as a source of information, but he was referring to the locality of *A. vesiculosa* in the Pinsk region (present-day southern Belarus), not in the vicinity of Vilnius. The discovery of this species in 1821 in the Pinsk region was reported by Wolfgang [69], who was probably the primary source of information.

The search for *A. vesiculosa* in Lake Dysnai, which lasted almost half a century, was unsuccessful. A relatively large population of the species was discovered only in 2001 in Lake Rūžas, a few kilometres northeast of Lake Dysnai [70]. During subsequent investigations of lakes in the northeastern part of Lithuania, in 2005, *A. vesiculosa* was rediscovered in Lake Dysnai [51], but in a different part of the lake to which it had been found in 1955 [65]. Furthermore, new localities of the species were found in Lake Alksnas [51] and Lake Apvardai [71] in 2005 and 2006, respectively. The record of *A. vesiculosa* in the Ignalina district in Lake Daržinėlė [71] was not confirmed by herbarium specimens and has not been confirmed in any subsequent study.

Our extensive searches for *A. vesiculosa* in 48 selected potential lakes between 2019 and 2021 did not reveal any new localities of this species. Considering the similarity of the vegetation and ecological conditions to the already known localities, we found ten lakes to be suitable for *A. vesiculosa*, but we did not record the species. Despite diligent searching, we did not find the species in any of the suitable bays of Lake Dysnai, except for the northwestern bay. The surveys confirmed that *A. vesiculosa* is currently growing in previously recorded localities in Lakes Rūžas, Apvardai, and Alksnas.

Caldesia parnassifolia was first found in present-day Lithuania in the early 19th century [72]. We confirmed the record of the species in the vicinity of the city of Vilnius through herbarium specimens (WI). In the first half of the 19th century, another locality of this species was registered in western Lithuania, in Kretinga, in a pond near a monastery (BILAS), but whether the plant grew naturally or was planted remains unclear [73,74]. In the middle of the 20th century, two localities of *C. parnassifolia* were recorded in southern Lithuania, in Lakes Daugai [75] and Ilgis [65]. The species was subsequently not detected at these sites and was considered extinct in Lithuania [27,76,77]. More than half a century later, *C. parnassifolia* was rediscovered at a new site in northeastern Lithuania, in Lake Rūžas [78]. Another locality of this species was found in the nearby Lake Apvardai during a survey of an *Aldrovanda vesiculosa* habitat. Both populations of *Caldesia parnassifolia* that are currently known in Lithuania are more than 100 km to the north of the previously recorded localities. In 2021, *C. parnassifolia* was found in Lake Kampiniškiai in the territorial waters of Latvia and Lithuania [55]. The lake is divided into two parts by the state border between Latvia and Lithuania; in Latvia, it is named Lake Lielais Kumpaniški (Medumi municipality).

Najas flexilis was first found in Lithuania in the second half of the 20th century. Fragments of plant parts of this species were intermixed in the herbarium specimen of *Chara aspera* Wild. (BILAS) collected in 1966 in Lake Germantas (western Lithuania) [40,76]. During subsequent studies, *Najas flexilis* was not rediscovered in this lake. A new locality of the species was discovered in 1998 in Lake Sagardas (eastern Lithuania) [40,76]. During an extended search for the species in 2019–2021, we identified three new localities in Lakes Avilys, Ažvintis, and Dūkštas, and we confirmed that *N. flexilis* still occurs in Lake Sagardas.

3.2. Assessment of Aldrovanda vesiculosa and Caldesia parnassifolia Populations

3.2.1. Habitats and Communities

Aldrovanda vesiculosa in Lake Rūžas (Figure 3) occurred in an approximately 730 m long northern inlet, with a width ranging from 15 to 80 m, and a total area of about 33,000 m^2 (Table 2). The inlet was connected by a stream with Lake Žilmas to the north. *A. vesiculosa* was distributed throughout the length of the inlet, but mainly concentrated along its shores and amongst floating-leaved plants (ca. 11,000 m^2). The water depth in the inlet varied from (0.1) 0.5 m at the coast of the mire to 2 m at its deepest points. The water was clear, slightly brownish, and transparent to the bottom.

Table 2. Area of *Aldrovanda vesiculosa* habitats, area of occurrence, and calculated number of individuals (mean $\pm$ SD) in the studied lakes.

Lake	Year	Area of Habitat (m^2)	Area of Occupancy (m^2)	Number of Individuals
Rūžas	2015	33,000	11,000	2,127,400 $\pm$ 1,756,700
Rūžas	2019	33,000	11,000	2,464,000 $\pm$ 2,321,000
Rūžas	2022	33,000	11,000	2,248,400 $\pm$ 1,005,400
Apvardai	2020	6400	2400	219,840 $\pm$ 81,840
Dysnai	2020	10,000	3300	306,240 $\pm$ 145,200
Alksnas	2021	<10	1	5

In Lake Dysnai, the species occurred in the southwestern bay at the mouth of the Svetyčia stream. The plants were scattered in a shallow part of the bay, up to 1 m deep, and the habitat covered approximately 10,000 m^2 (Table 2). This shallow area was protected from direct wave action by a wide belt of *Nuphar lutea* (L.) Sm. stands. Loose stands of *Typha angustifolia* L., floating-leaved plants, and swampy shores were the main refuge for *Aldrovanda vesiculosa*. The water was turbid and greenish or yellowish.

In Lake Apvardai, *A. vesiculosa* was fragmentarily distributed along the southwestern and western shoreline and occupied a total area of approximately 6400 m^2 (Table 2). We found it in open areas between floating swampy islets and sparse stands of helophytes (*Typha angustifolia* and *Schoenoplectus lacustris* (L.) Palla). These areas were also separated from the main part of the lake by a belt of *Nuphar lutea* stands. The water was slightly greenish or brownish and moderately turbid.

We confirmed only a few individuals of *Aldrovanda vesiculosa* occurring in Lake Alksnas at the mouth of the stream that connected it to Lake Liūneliai in a small area (Table 2). The dominant vegetation in Lake Alksnas was charophytes, and it was therefore unsuitable for *A. vesiculosa*; it can only grow in shallow water in a mire. In all the other lakes, the localities of *A. vesiculosa* were surrounded by transitional mires or quaking bogs and at the mouths of inflowing or outflowing streams. These parts of mainly eutrophic lakes showed signs of dystrophy. The bottoms of the lakes were composed of thick blackish organic sediments.

We recorded 37 plant species in the 85 phytosociological relevés with *A. vesiculosa* from Lakes Rūžas, Dysnai, and Apvardai (Appendix A; Supplementary Material, Table S1). The communities were dominated by vascular plants, and only one species, *Nitellopsis obtusa* (Desv.) J. Groves belonged to the charophytes. We did not identify the species of filamentous algae, so they were not included in the number of species, but they are an important indicator of the ecological conditions of habitats. We combined *Utricularia australis* R. Br. and *Utricularia vulgaris* s.str., which are rarely found with the flowers necessary for accurate identification, into a single taxon: *Utricularia vulgaris* s.l. We recorded the highest number of species, 27, in Lakes Apvardai and Rūžas in 2020 and 2022, respectively. We recorded floating-leaved and free-floating plants (*Nuphar lutea, Potamogeton natans* L., *Nymphaea candida* J. Presl & C. Presl., *Stratiotes aloides* L., *Hydrocharis morsus-ranae* L., and *Utricularia vulgaris* s.l.), as well as submerged plants (*Potamogeton* $\times$ *bambergesnis* Fisch. and *Myriophyllum* cf. *verticillatum* L.), in all the studied communities (Figure 3). We assumed that *Myriophyllum* cf. *verticillatum* included not only the species itself but also its hybrids.

Figure 3. Plant communities of floating-leaved and floating plants with dominant *Aldrovanda vesiculosa* (**a**,**b**) and communities with *Aldrovanda vesiculosa* and *Caldesia parnassifolia* (**c**,**d**) in Lake Rūžas. Photos by Z. Sinkevičienė (**a**,**b**) and L. Petrulaitis (**c**,**d**).

We found *Aldrovanda vesiculosa* in all 85 phytosociological relevés, which was usually abundant. Most of the phytocenoses with dominant *A. vesiculosa* were grouped in one cluster (Figure 4, cluster C). We could hardly attribute these relevés to the association *Spirodelo-Aldrovandetum vesiculosae* Borhidi et Járai-Komlódi 1959, as the presence of the characteristic species of the *Lemnetea* class (*Utricularion vulgaris* alliance) was negligible compared with that of the species of the *Potametea* class (*Nymphaeion alliance*). Some of the relevés from this cluster, especially those recorded in the shallows of Lakes Apvardai and Dysnai, were similar to the relevés from cluster J (Figure 4) recorded in the same lakes, where none of the species present were dominant.

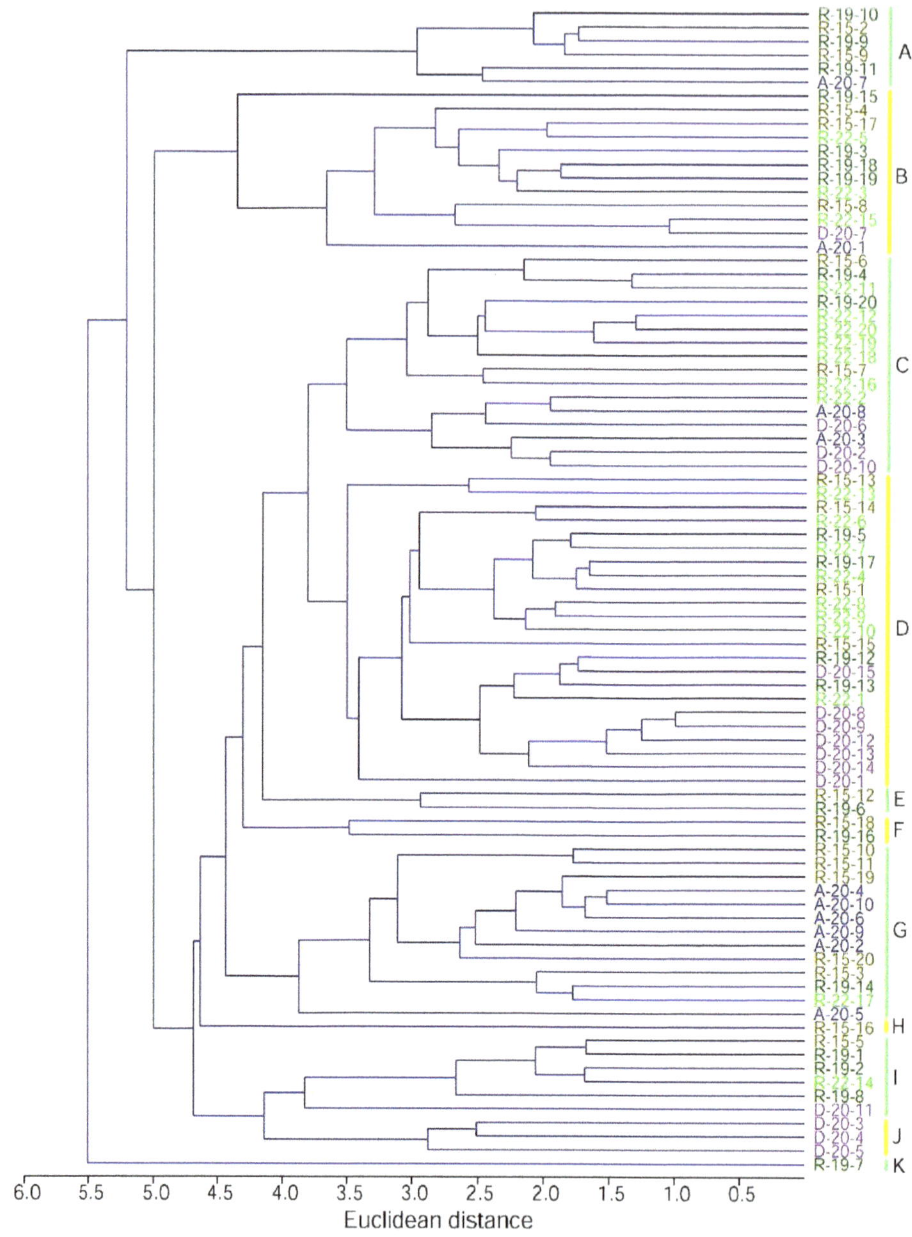

Figure 4. Dendrogram of communities with *Aldrovanda vesiculosa*, created using paired-group (UP-GMA) algorithm and Euclidean distance. Different colours indicate different study years and lakes: Lake Rūžas (olive, green, and light green in 2015 (R-15), 2019 (R-19), and 2022 (R-22), respectively), Lake Apvardai (blue in 2020 (A-20)), and Lake Dysnai (dark purple in 2020 (D-20)). In relevé numbers, last two digits indicate sequence number of relevé (Supplementary Material, Table S1). Capital letters on the right refer to clusters explained in the text.

We commonly found *A. vesiculosa* in stands dominated by *Nuphar lutea* (association *Nymphaeo albae-Nupharetum luteae* Nowiński 1927), especially in Lakes Rūžas and Dysnai (Figure 4, cluster D). Less frequently, this species occurred in phytocenoses of the associations *Potametum natantis* Hild 1959 (Figure 4, cluster G) and *Stratiotetum aloidis* Miljan 1933 (Figure 4, cluster B,), mainly recorded in Lakes Rūžas and Apvardai. We also found *Aldrovanda vesiculosa* in the association *Myriophylletum verticillati* Gaudet ex Šumberová in Chytrý 2011 (Figure 4, cluster I) in Lakes Rūžas and Dysnai. Only occasionally was the study species recorded in stands of *Sparganium emersum* L. (cluster E), *Ceratophyllum demersum* L. (cluster H), or *Sparganium natans* L. (cluster K) in Lake Rūžas (Figure 4).

Caldesia parnassifolia was a constant and sometimes abundant (Figure 3) co-occurring species with *Aldrovanda vesiculosa* in Lakes Rūžas and Apvardai. Phytosociological relevés with dominant *Caldesia parnassifolia* were clustered together (Figure 4, cluster A). We inventoried communities dominated by flowering individuals of *C. parnassifolia* in Lake Rūžas in 2015 and 2019, but we recorded no such stands in 2022. Flowering individuals were mainly located along the swampy shores up to a depth of 0.5 m in the lake and were most abundant in 2019. Nonflowering individuals of *C. parnassifolia* were widespread throughout the inlet and were a quite constant component of the communities surveyed in all years of the study.

3.2.2. Morphology of *Aldrovanda vesiculosa*

We recorded the longest individual (23 cm) of *A. vesiculosa* in Lake Rūžas in 2015. The mean length of *Aldrovanda vesiculosa* individuals only slightly varied between study years (Table 3). We found the highest mean length of individuals in Lake Rūžas in 2015 (14.3 ± 2.7 cm), and their mean length was significantly longer ($p < 0.01$) than that in other years in Lake Rūžas and in Lakes Apvardai and Dysnai. We found no significant differences ($p > 0.05$) between the lengths of *A. vesiculosa* individuals in Lake Rūžas in 2019 and 2022 and the lengths of individuals in Lakes Apvardai and Dysnai in 2020 (Table 3).

Table 3. Descriptive statistics of *Aldrovanda vesiculosa* individuals in different years and lakes (mean $\pm$ SD, minimum and maximum values, median). Different lower-case letters in superscript indicate significant differences ($p < 0.05$) across rows according to Mann–Whitney pairwise comparison.

Lake	Rūžas	Rūžas	Rūžas	Apvardai	Dysnai
Study year	2015	2019	2022	2020	2020
Number of studied individuals	197	196	200	100	150
Plant length (cm)	14.3 ± 2.7 [a]	12.5 ± 2.5 [b]	12.5 ± 2.1 [b]	13.2 ± 3.2 [b]	12.6 ± 2.6 [b]
Minimum–maximum	8–23	7–20	6–19	7–21	8–22
Median	14.5	12.0	12.5	13.0	12.0
Number of apices	2.2 ± 0.9 [a]	2.0 ± 0.9 [b]	2.0 ± 0.7 [b]	3.4 ± 1.2 [c]	3.6 ± 1.2 [c]
Minimum–maximum	1–4	1–5	1–4	1–8	1–7
Median	2	2	2	3	3

The number of shoots, and thus, the number of apices of *A. vesiculosa* indicates their reproductive capacity because each apex forms a turion from which a new plant may grow the following year. During the study, we found that individuals had a maximum of eight apices. *A. vesiculosa* had significantly fewer apices (Table 3) in Lake Rūžas than in Lakes Apvardai and Dysnai ($p < 0.001$), and we found no difference between the populations of the latter two lakes in the number of apices ($p = 0.939$). In Lake Rūžas, the number of apices of *A. vesiculosa* individuals was significantly higher (Table 3) in 2015 than in 2019 ($p = 0.023$) and 2022 ($p = 0.044$), whereas the number of apices did not significantly differ between 2019 and 2022 ($p = 0.570$).

3.2.3. Density and Cover of *Aldrovanda vesiculosa* Individuals

The density of *A. vesiculosa* individuals varied more between the lakes studied than between years in Lake Rūžas (Table 4). In this lake, the density ranged from 193.4 ± 159.7

to 224.0 ± 211.0 individuals/m^2, but the density did not significantly differ between years ($p > 0.05$). The density of *A. vesiculosa* in Lakes Apvardai and Dysnai was significantly ($p < 0.01$) lower than that in Lake Rūžas in all years of the study (Table 4). We found the same patterns when assessing the coverage of *A. vesiculosa* individuals in the study plots (Table 4). In Lake Rūžas, we found no significant differences between the coverage of individuals in the study plots in any survey year ($p > 0.05$). In the individual study plots, *A. vesiculosa* covered between 3.2% and 75.5% of the surface area (Table 4). We found a weak significant correlation between the number of individuals in the study plot and water depth ($r_s = 0.32$; $p = 0.003$) and between the coverage of individuals and water depth ($r_s = 0.35$; $p = 0.001$).

Table 4. Characteristics of sampling plots and descriptive statistics of *Aldrovanda vesiculosa* density and cover in different years and lakes (mean $\pm$ SD, minimum and maximum values, median). Different lower-case letters in superscript indicate significant differences across rows according to Mann–Whitney pairwise comparison.

Lake	Rūžas	Rūžas	Rūžas	Apvardai	Dysnai
Study year	2015	2019	2022	2020	2020
Number of sampling plots	20	20	20	10	15
Mean depth (m)	1.0 ± 0.2 [a]	0.9 ± 0.2 [a]	1.0 ± 0.2 [a]	0.6 ± 0.2 [b]	0.4 ± 0.1 [c]
Density of individuals per m^2	193.4 ± 159.7 [a]	224.0 ± 211.0 [ab]	204.4 ± 91.4 [a]	91.6 ± 34.1 [b]	92.8 ± 44.0 [b]
Minimum–maximum	32–564	24–744	79–400	40–140	44–212
Median	146	158	165	90	84
Mean cover (%)	28.8 ± 19.5 [a]	25.4 ± 18.7 [ab]	29.7 ± 13.2 [a]	14.3 ± 5.9 [b]	14.0 ± 6.5 [b]
Minimum–maximum	7.5–75.5	3.2–65.0	11.5–58.0	5.1–26.9	7.3–28.9
Median	24.0	18.1	24.0	12.9	11.7

The results of the PCA performed using data from 85 sampling plots showed that the first component (PC1) explained 70.1% (eigenvalue: 3.506) and the second component (PC2) explained 15.2% (eigenvalue: 0.761) of the variance (Figure 5). The loadings for each factor in PC1 and PC2 are presented in Table 5.

Table 5. Loadings of the analysed factors in first two principal components.

Component	PC1	PC2
Number of *Aldrovanda vesiculosa* individuals per sampling plot	0.459	0.437
Cover of submerged plants (%)	−0.319	0.716
Cover of floating plants (%)	0.436	−0.372
Cover of emergent plants (%)	−0.506	0.009
Water depth in the sampling plot (m)	0.491	0.397

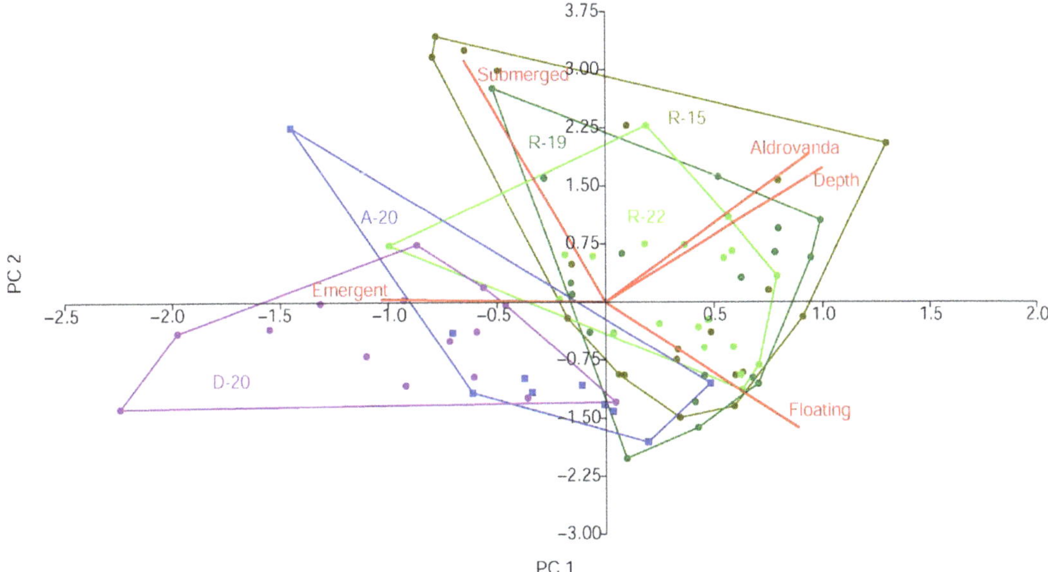

Figure 5. Principal component analysis of studied *Aldrovanda vesiculosa* populations based on density of individuals; cover of submerged, floating-leaved (excluding *Aldrovanda vesiculosa*), and emergent plants; and water depth in sampling plots. Different-coloured symbols and convex hulls indicate different study years and lakes: Lake Rūžas (olive, green, and light green in 2015 (R-15), 2019 (R-19), and 2022 (R-22), respectively), Lake Apvardai (blue in 2020 (A-20)), and Lake Dysnai (dark purple in 2020 (D-20)).

3.3. Assessment of Najas Flexilis Habitats

At all the studied sites, the abundance of *N. flexilis* was very low. At two sites in Lake Ažvintis, we found only solitary individuals of *N. flexilis* growing at a depth of 1.5 m on a sparsely vegetated bottom, in the transition zone between helophytes and submerged plants. We recorded a slightly more abundant population of this species at two sites in Lake Avilys. We found sparse *N. flexilis* at a depth of 1.2 m among sparse swards of *Fontinalis antipyretica* Hedw., *Hydrilla verticillata* (L. f.) Royle, *Potamogeton rutilus* Wolfg., *Potamogeton perfoliatus* L., and *Schoenoplectus lacustris*. *Najas flexilis* was somewhat more frequent in Lake Dūkštas, but, as in other areas, it was not abundant. We usually observed it at depths of 1–3 m in stands of charophytes (*Chara filiformis* A. Braun in Hertzsch, *Chara globularis* Thuill., *Chara virgata* Kütz., and *Chara tomentosa* L.) and bryophytes (*Fontinalis antipyretica*, *Drepanocladus aduncus* (Hedw.) Warnst., and *Rhynchostegium riparioides* Hedw.), often together with *Najas minor* All.

The population of *Najas flexilis* in Lake Sągardas has been known for more than two decades. Our current studies showed that the species grows in this lake at a depth of 1.5–3.5 m, together with *Nitella flexilis* (L.) C. Agardh, *Chara strigosa* A. Braun, and *Elodea canadensis* Michx., although it is also not abundant.

Most of the assessed populations of *Najas flexilis* occurred in lakes with bottom sediments consisting mainly of gravel or sand and a thin layer of silt. The bottom vegetation in these lakes was relatively sparse, formed mainly by mosses and macroalgae.

3.4. Conservation Status and Prospects

3.4.1. Conservation of Aldrovanda vesiculosa

Two special areas of conservation (SACs) of the *NATURA 2000* network were established in Lithuania within the conservation objectives of protecting *A. vesiculosa* populations (the *Lake Rūžas* SAC and the *Environs of Ažušilė–Didžiagiris* SAC). However, this species has not been confirmed in Lake Daržinėlė, situated in the *Environs of Ažušilė–Didžiagiris* SAC. Despite the proposals, no SACs have yet been established for the protection of the species in Lakes Apvardai and Dysnai. Lake Alksnas is in the *Pušnis Bog* SAC; however, objectives for the conservation of *A. vesiculosa* populations have not been defined in this SAC.

A conservation plan for *A. vesiculosa* has been developed to restore and maintain the favourable conservation status of its habitats in Lithuania [51]. Based on this document, a management plan for the habitat of the species in Lake Rūžas was developed in 2012 and implemented in 2013–2014. Based on the results of an expert assessment of the habitat, the management plan for *A. vesiculosa* included the following measures: (a) the removal of *Nuphar lutea* rhizomes forming floating islands on the surface of the water; (b) the removal of floating islets formed on rhizome confluences or detached from the shore from the lake (Figure 6); (c) the removal of tree and shrub thickets on the shores and on large islets, which provide shade in the lake; and (d) the removal of obstructions on the bed of the stream linking Lakes Rūžas and Žilmas to re-establish the natural circulation of water.

Figure 6. Aggregations of uprooted *Nuphar lutea* rhizomes (**a**), and uprooted individuals of *Caldesia parnassifolia* (**b**). Photos by Z. Sinkevičienė.

In late autumn, after the growing season had ended and *Aldrovanda vesiculosa* turions had sunk to the bottom, floating aggregations of *Nuphar lutea* rhizomes were removed from the lake and the stream bed using a special floating technique. Small floating islets and aggregations of floating islets and dams were also removed from the stream bed. Trees growing on the banks of the lake and larger islets were felled and all their biomass removed from the habitats. This improved the light regime of the habitat, as the trees growing on the shores of the small bays, where most of the *Aldrovanda vesiculosa* individuals were concentrated, provided strong shade.

Following the implementation of the measures outlined in the action plan for the conservation of *A. vesiculosa* in Lake Rūžas, we assessed the species population in 2015, 2019, and 2020 (Table 2). The results of the studies showed that the number of individuals in the population had remained relatively stable. The state of this population could be assessed as favourable; however, further regular observation and assessment of the state is necessary to analyse the effect of filamentous algae and assemblages of uprooted floating *Nuphar lutea* rhizomes.

The populations of *Aldrovanda vesiculosa* in Lakes Apvardai and Dysnai were much smaller and contained fewer individuals than those in Lake Rūžas (Table 2). We constantly observed the presence of filamentous algae in both lakes. They had been severely affected in the past by agricultural activities, such as the draining of the surrounding wetlands and the presence of livestock farms in the catchment area of the lakes. The negative effects of agricultural activities are continuing, as drainage water from cultivated fields is still flowing into the western part of Lake Apvardai. The state of both populations could be assessed as unfavourable.

The true state of the entire population in Lake Alksnas remains unknown due to its inaccessibility. By 2021, *A. vesiculosa* was almost extinct in the accessible part of the site at the mouth of the stream flowing from Lake Liūneliai.

3.4.2. Conservation of *Caldesia parnassifolia*

In Lithuania, *C. parnassifolia* occurs in one of the previously designated SACs, the *Lake Rūžas SAC*, but the species has not yet been included in the conservation targets. We propose amending the conservation targets of the *Lake Rūžas* SAC to include *C. parnassifolia*. We also propose designating Lake Apvardai as an SAC for the conservation of the habitats and populations of *C. parnassifolia* and *Aldrovanda vesiculosa*.

The habitat of *Caldesia parnassifolia* in Lake Rūžas partially coincided with the habitat of *Aldrovanda vesiculosa*, but it covered a larger area outside the narrow inlet. We recorded the largest population conditions in 2015 and 2019. In those years, both vegetative and generative individuals were abundant in the lake. We observed the highest number of flowering plants and individuals forming turions in the inflorescence on 1 September 2019. A few plants had fully developed but their seeds were immature. During the study in 2022, only one generative individual was present in the whole population of *Caldesia parnassifolia*, but it had not produced mature seeds or formed turions in the inflorescence. Many vegetative individuals had been uprooted and floated on the surface of the water (Figure 6).

We observed the negative impacts of the increasing eutrophication of water and of the abundance of filamentous algae in Lake Apvardai, where *C. parnassifolia* together with *Aldrovanda vesiculosa* occurred in only two locations.

3.4.3. Conservation of *Najas flexilis*

The *Lake Sągardas* SAC is the only area that has been purposefully designated for the conservation of *N. flexilis*. The population of this species in Lake Dūkštas belongs to a previously designated *Gražutė Regional Park* SAC, and additional objectives for the conservation of *N. flexilis* have been established. The historical locality of this species in Lake Germantas (Telšiai district, western Lithuania) is in the *Lake Germantas* SAC, but *N. flexilis* has not been detected in recent decades and is therefore not listed in the objectives for conservation. No SACs have been designated for the conservation of populations of *N. flexilis* in Lake Avilys or Ažvintis.

After screening for *N. flexilis* in the selected lakes and assessing the status of the previously recorded populations, we found that the population of the species in Lithuania is scarce. In most cases, we found a few solitary individuals at the observed sited. We observed the most favourable habitat and the largest currently known population in Lake Dūkštas, where we found *N. flexilis* at several sites. The area of habitat suitable for *N. flexilis* in the lakes was several times larger than the current area of its occurrence. More lakes in Lithuania may be suitable for *N. flexilis*, but the species has not yet been surveyed in these lakes.

4. Discussion

4.1. Historical and Current Distribution of Species

The aquatic plant species protected under the European Habitats Directive, *Aldrovanda vesiculosa*, *Caldesia parnassifolia*, and *Najas flexilis*, currently occur only in the northeastern part of Lithuania, in a relatively small area. These species are also unevenly distributed throughout the world, and this distribution pattern is probably mainly determined by their specific requirements for habitat conditions and sensitivity to environmental changes [14,79]. The northeastern part of Lithuania, as well as the adjacent regions in Latvia and Belarus, which are part of the Baltic Uplands, has a high concentration of lakes of different sizes and trophic levels [80,81]. Therefore, we assumed that the abundance of diverse lakes was the main factor determining the high concentration of rare aquatic plant species both in this region of Lithuania and in adjacent areas in Latvia [38,55] and Belarus [44,82].

In their historical localities, *Caldesia parnassifolia* [78] and *Najas flexilis* [40] in other parts of Lithuania are probably extinct because of changes in these habitats caused by natural and anthropogenic factors. The populations of *Caldesia parnassifolia* in the southern part of Lithuania were small [65,75] and may have disappeared because of natural succession. Small populations may have become extinct because of several consecutive unfavourable growing seasons, or the activities of aquatic plant-feeding birds, fish, or other animals [83–85], which may have consumed most or all of the formed turions. The current known populations of *C. parnassifolia* from northeastern Lithuania and southern Latvia are located more than 100 km north of the historic sites [78]. However, whether the species was previously unnoticed in the current sites or whether it has naturally spread to these sites remains unclear. We think that small populations of the species may have existed for a long time in these locations but were overlooked. With climate change, the warmer and longer growing season [86] may have led to an increase in its abundance and a spread to larger parts of the lakes. Additionally, *C. parnassifolia* was only discovered in Lake Rūžas after management measures had been implemented [78]. This suggests that the improved habitat conditions have led to a significant increase in the plant population and increased the ease of detecting the species.

The new records of *Najas flexilis* from three localities in this region suggest that the species may be more widespread in northeastern Lithuania, as quite abundant populations exist in the adjacent territories of Latvia and Belarus [38,44]. The species may also occur in other parts of the country, but detailed studies on potential habitats are required. The current surveys identified several water bodies with a combination of ecological and geomorphological characteristics suitable for *N. flexilis*, but we found no individuals in these lakes. Among the reasons why the species has not been detected may be the annual variation in population size due to the life history of *N. flexilis*, as well as the small size of the individuals and the difficulty in detecting them, especially at greater depths. In Lithuania, most surveys have been conducted from a boat using a grapnel, preventing sufficiently accurate assessments of species diversity and abundance [79].

4.2. Assessment of Aldrovanda vesiculosa and Caldesia parnassifolia Populations

4.2.1. Habitats and Communities

Throughout its wide range, *Aldrovanda vesiculosa* grows in a variety of habitats, ranging from shallow and stagnant to slow-flowing natural and artificial water bodies [14]. In Europe, *A. vesiculosa* is commonly found in nutrient-poor oligo-mesotrophic and dystrophic (humic) wetland systems, such as small marshes, peat bog pools, and dystrophic or peaty lakes, as well as lagoons and river deltas [14]. Sometimes, it occurs in highly eutrophic habitats such as fishponds and rice paddies [15]. The main habitats of *A. vesiculosa* at the northern edge of its range (e.g., Poland) are eutrophic–dystrophic lakes [24]. This type of habitat is also typical for *A. vesiculosa* in Lithuania. Notably, all the lakes in Lithuania where this species has been found are a part of wetland systems with various-sized mires. The mires around the lakes are crucial for the maintenance of habitats and for the protection of plants from the direct negative effects of agricultural pollution in the lake basins. Streams flowing into lakes or from lakes, which ensure water turnover, are also vital for the maintenance of favourable conditions in the habitats of *A. vesiculosa*.

The habitat of the largest population of *A. vesiculosa* in Lake Rūžas is remarkable for its species richness, low eutrophication, and relatively deep water. The relatively deep water protects the plant from the negative effects of fluctuations in water levels, provides favourable wintering conditions for turions during the cold season, and ensures the longevity of the population [53]. The relatively deep parts of a lake overgrow much slower than the shallows. Adamec [14] stated that *A. vesiculosa* grows faster in shallow areas, but shallow areas are at increased risk of drying out, faster eutrophication, the faster development of filamentous algae mats, and faster growth of helophytes. We observed such natural vegetation succession in the shallowest part of Lake Alksnas, where *A. vesiculosa* had almost vanished within a decade.

Water depth also determines the dominant vegetation. At three studied sites in Lithuania, *A. vesiculosa* usually occurred in communities of floating-leaved (alliances *Nymphaeion* and *Potamion*) or free-floating (association *Stratiotetum aloidis*) plants. This species occupies similar communities in Poland [24]. The communities of tall sedges (alliance *Magnocaricion*) were indicated as being particularly important habitats for the species in Poland [24], but such communities are not common in Lithuania. In Lake Rūžas, communities of tall helophytes (*Typha angustifolia* and *Phragmites australis*) were restricted to marshy shores (such as *Thelypterido palustris-Phragmitetum australis*), and *Aldrovanda vesiculosa* was absent in those areas. Adamec [14] assumed that *A. vesiculosa* habitats at all European sites may be stable, only occurring among or near tall helophytes such as *Phragmites australis* and *Typha angustifolia*, and tall species of *Carex*. The habitat in Lake Rūžas was protected by the surrounding marsh, whereas in Lakes Apvardai and Dysnai, tall helophytes (*Typha angustifolia*, *Schoenoplectus lacustris*) were the most notable shelter formation, providing a wind- and wave-protected microenvironment. In addition, these microenvironments were protected from the direct impact of waves by a wide belt of *Nuphar lutea* stands. Typical communities in which *Aldrovanda vesiculosa* occurs belong to the association *Spirodelo-Aldrovandetum vesiculosae*, with the presence of characteristic species of the class *Lemnetea* and the alliance *Utricullarion vulgaris* (*Spirodela polyrhiza*, *Riccia fluitans*, *Salvinia natans*, *Utricularia vulgaris*, and *Utricularia australis*). However, in such communities, the species is likely to occur only in the southern regions of its range, e.g., in Ukraine [87], but has not been recorded in the northern parts of its range, such as in Poland [24], the Czech Republic [62], or Lithuania.

We investigated communities of *Caldesia parnassifolia* in parallel with those of *Aldrovanda vesiculosa*, as the two species co-existed in the two lakes. We found no records to confirm that the two species occur in the same communities in other regions of Europe. Lake Rūžas supports the largest population of *Caldesia parnassifolia* compared with other historical and current localities. In this lake, we found only vegetating plants at depths greater than 0.5 m, which we mainly found in stands of floating-leaved plants (*Nuphar lutea* and *Potamogeton natans*). We observed similar species composition in the phytocenoses of Lake Apvardai and in the historical locality of Lake Daugai [75], as well as in the recently recorded locality of Lake Kampiniškiai on both sides of the Latvian–Lithuanian border [55]. We observed stands with dominant and flowering *Caldesia parnassifolia* in shallow (up to 0.5 m deep) areas of Lake Rūžas. Under similar conditions among helophytes (*Carex* spp., *Phragmites australis*, *Menyanthes trifoliata*, and *Comarum palustre*), it occurred at the historical site in Lake Ilgis [65]. Among helophytes (*Phragmites australis* and *Schoenoplectus lacustris*), the species has also been observed in Lakes Apvardai and Kampiniškiai [55]. In general, *Caldesia parnassifolia* is mainly distributed in communities of the class *Potametea* (alliance *Nymphaeion*) at the northern edge of the range; in shallow parts of the lakes, it may occur in helophyte communities (class *Phragmito-Magno-Caricetea*). During surveys in fishponds in France [52], *C. parnassifolia* was recorded in communities of five vegetation classes, but mainly in communities of the classes *Littorelletea* and *Potametea*, and less frequently in communities of the *Nymphaeion* alliance.

The flowering of *C. parnassifolia* is limited by water depths greater than 0.5 m [88]. This phenomenon has been confirmed by studies in northern populations in Lithuania [78] and Latvia [55], as well as in much more southerly populations in France [52]. Throughout the temperate zone of Europe, reproduction via turions is the main mode of reproduction, even though plants may sometimes produce mature seeds [52]. We observed stands of flowering and turion-forming *C. parnassifolia* plants in Lake Rūžas in 2015 and particularly in 2019, but in 2022, we noted few flowering individuals. This finding may be associated with the relatively low water levels in the lake in the first two years and the much higher water levels in the latter years. The ability of plants to grow at water depths greater than 0.5 m and to form turions has not been assessed and requires further investigation.

4.2.2. Morphology of *Aldrovanda vesiculosa*

The mean lengths of *A. vesiculosa* individuals were similar in the lakes studied, but only in Lake Rūžas in 2015 were the plants significantly longer than in other years in the same and different lakes. This suggests that more favourable conditions for plant growth have been created following the implementation of habitat management measures in 2014. The mean length of individuals in Lake Rūžas in 2015 was virtually the same (t = 1.6; $p = 0.107$) as that measured in the naturalised population of *A. vesiculosa* in Florida [17]. The mean length of individuals in the populations studied in Lithuania was significantly larger (t = 28.9; $p < 0.001$) than in the native population in Poland [89] and significantly larger than in the naturalised populations in Germany (t = 19.5; $p < 0.001$), but almost the same (t = 1.3; $p = 0.190$) as in the naturalised populations in the Czech Republic [90]. This suggests that the habitats in Lithuania are favourable for the growth of *A. vesiculosa* and that individuals develop similarly to those in much warmer climates. The significantly lower mean length of the individuals in the populations studied in Poland could also have been caused by meteorological conditions, as those populations were studied in the 1980s, before any signs of climate warming appeared.

The results of the assessment of the apices of *A. vesiculosa* individuals showed that in Lakes Apvardai and Dysnai, they were significantly more branched, and thus, larger in number. We assumed that habitat conditions have the strongest effect on the branching of individuals. In the shallow and wave-protected areas of Lakes Apvardai and Dysnai, individuals were more branched than in the deeper habitats of Lake Rūžas, where wave action can be quite strong. Waves can fragment individuals of *A. vesiculosa*, resulting in smaller numbers of branches. In addition, the water in shallow areas more strongly and quickly warms up than in the deeper parts of the lake; higher temperatures may encourage branching [89]. Individuals of *A. vesiculosa* are reported to have two–four branches; individuals with eight branches are rare [14]. We assessed the total number of apices, which directly reflects the number of turions formed by an individual. The number of apices in the populations we studied did not differ from the highest values found in naturalised populations in Florida [17]. According to Kamiński [89], individuals of *A. vesiculosa* in populations studied in Poland in the 1980s were less branched than those in all the populations studied in Lithuania. However, because of differences in study methods, we cannot reliably compare Kamiński's [89] data with ours.

Summarising the results, we conclude that individuals of *A. vesiculosa* in Lithuania are well-developed, branch abundantly, and produce many turions, which are essential for the recovery of the population after the cold season and for its long-term stability.

4.2.3. Density and Cover of *Aldrovanda vesiculosa* Individuals

The density of individuals per unit area and the overall size of a population are important indicators of its status. The results of the study revealed that the density of *A. vesiculosa* individuals in Lake Rūžas changed little over the entire study period, and that these changes are probably the result of natural population fluctuations. We found significantly lower densities of individuals in Lakes Apvardai and Dysnai. Whether the lower densities were caused by less favourable habitat conditions or for other reasons that could not be identified after a single survey remains unclear. We found a weak but significant relationship between water depth and density, as well as cover of individuals. We also found that individual *A. vesiculosa* densities were positively affected by the cover of floating-leaved plants, but negatively affected by the cover of submerged and emergent plants (Figure 5, Table 5). We think that floating-leaved plants growing abundantly at greater depths create the micro-conditions required for *A. vesiculosa* individuals to establish. However, in shallow habitats without or with only a few floating-leaved plants, *A. vesiculosa* individuals have nothing on which to anchor and are easily disturbed by waves, so their density is lower. In addition, in shallow habitats, more turions are likely to be lost during the cold season and are easily reached and consumed by plant-eating water birds.

The largest population of *A. vesiculosa* in Lithuania, and possibly in Europe, is in Lake Rūžas. The results of this study indicated that the habitat, which covers approximately 1.1 ha, contained between one and three million individuals in 2022 (Table 2). In Lakes Dysnai and Apvardai, where the suitable habitat and the density of individuals are much smaller than in Lake Rūžas, the total population in 2020 was between 160,000 and 450,000 and between 140,000 and 301,000 individuals, respectively. Thus, the total population of *A. vesiculosa* in Lithuania comprised between 1.30 and 3.75 million individuals. The results of this study are markedly different from the population size estimated by experts in 2012, when the population was assessed to be around 110,000 individuals [51]. The population size estimates differ because we accurately estimated the area covered by the population and the density of individuals, whereas the expert assessment was based on preliminary and small-sample data.

4.3. Habitats of Najas flexilis

We found *N. flexilis* growing in relatively large, stratified, mesotrophic lakes, with bottom sediments mainly consisting of mud with an admixture of fine gravel and stones, at depths ranging from 0.5 m to 3.5 m. *N. flexilis* usually occurred in sparse swards of charophytes and bryophytes. This species has been found under similar conditions in Estonia, where it occurs mostly in mesotrophic lakes in association with *Potamogeton rutilus* [46]. In adjacent Latvia, this species occurs in lakes of varying sizes and depths, even in shallow lakes [38]. In Switzerland, *Najas flexilis* has been recorded mainly in the transition zone between sparse *Potamogeton* stands and shallow vegetation [42]; in Norway, it has been found in mesotrophic water bodies, at depths of 2.9 m to 3.8 m [39]. Further detailed studies are required to assess the distribution of *N. flexilis* and the statuses of its habitats and populations in Lithuania.

4.4. Conservation Status and Proposals

4.4.1. Conservation of *Aldrovanda vesiculosa*

Approximately 90% of the historically known localities of *A. vesiculosa* worldwide were estimated to have become extinct within a century and a half [15]. The main causes of the extinction of many *A. vesiculosa* populations include direct anthropogenic impacts (habitat destruction, drainage, intensive agriculture, eutrophication, municipal pollution, etc.), as well as natural habitat succession, mainly associated with habitat degradation and climate change [14,48,50,91].

In Lithuania, only two special conservation areas have been established for *A. vesiculosa*. A management plan for the habitat of this species in Lake Rūžas was prepared and implemented in 2013–2014. The results of the subsequent studies showed that the positive effects of the management measures have been maintained to date. However, during surveys in September 2022, in some parts of the Lake Rūžas inlet, some *Nuphar lutea* rhizomes had started to be uprooted by methane gas released from soft-bottom sediments or by aquatic animals (*Castor fiber* and *Lutra lutra*) again. The removal of these islands was one of the management measures implemented to reduce the overgrowth of the inlet and ensure increased water movement in the outflowing streamlet.

A concern exists that mats of filamentous algae are re-establishing in the northern part of the inlet, which are particularly abundant in stands of *Myriophyllum verticillatum*. Mats of filamentous algae are an indicator of eutrophication and create unfavourable conditions for the growth of *Aldrovanda vesiculosa* [14]. We found the species to be only sporadic or absent at sites where these algae were abundant. The growth of *Cladophora* and other filamentous algae may be influenced by the slowing of the flow of the watercourse due to the presence of re-established obstacles in the stream bed. In Lakes Apvardai and Dysnai, we also observed the negative effects of filamentous algae on *A. vesiculosa*. We therefore consider that habitat management measures should be implemented every 5–10 years, depending on the results of the periodic assessment of the habitat.

The conservation of *A. vesiculosa* in Lake Alksnas is challenging because the true status of the population remains unknown due to its inaccessibility. A survey of *A. vesiculosa* in the accessible part of the site at the mouth of the stream flowing out of Lake Liūneliai revealed that it was almost extinct. The drastic decline in the abundance of *A. vesiculosa* was caused by the overgrowth of the shallow stream bed and its transformation into a swamp. The survival of the species in this area is limited, as all the surrounding habitats are protected under the Habitats Directive, and drastic measures of management (e.g., restoration of the stream), which are incompatible with the requirements of the other habitats, cannot be applied.

To conserve *A. vesiculosa* populations, periodic assessments must be conducted of the status of their populations. Regular monitoring of the status and abundance of the species is required, as are systematic surveys of the hydrochemical parameters of the water in the habitat of the species. The hydrochemical parameters established from water samples obtained from the deepest part of the lake according to the requirements of the Water Directive do not reflect the conditions in enclosed shallow inlets. When adverse changes in the habitat and population are detected, a comprehensive management plan must be developed, and the measures outlined in the plan must be rigorously implemented.

4.4.2. Conservation of *Caldesia parnassifolia*

C. parnassifolia was recently assessed as a species of least concern globally [23], as it occurs in many countries and has a wide range. Nevertheless, the species is declining and is protected in many European countries under the Habitats Directive [24]. The need for detailed surveys to identify population trends and threats was also highlighted [23]. In Lithuania, the species has been recently rediscovered and is now considered as critically endangered (CE) on the Red List [71]. To date, no special areas of conservation have been established for the conservation of *C. parnassifolia* in the country.

We found *C. parnassifolia* in Lake Rūžas in a special conservation area designated for *Aldrovanda vesiculosa*. The population of *Caldesia parnassifolia* was in favourable condition in 2015 and 2019; however, in 2022, almost all plants were in a vegetative state and much smaller than in previous years, and many uprooted individuals were floating on the surface (Figure 6). We think that *C. parnassifolia*, as well as other aquatic plants, have been uprooted by *Lutra lutra*, which searches the lake bottom for bivalves for food. Because the individuals of *C. parnassifolia* had formed few turions, the population density will likely continue to widely fluctuate in the future. Furthermore, how fluctuations in water level affect the condition of the plants is unclear.

We recorded a small population of *C. parnassifolia* in Lake Apvardai, which has also been proposed as a special conservation area for *Aldrovanda vesiculosa*. Detailed surveys of the species have not been carried out in this lake. In addition to the designation of special conservation areas in existing localities of *Caldesia parnassifolia*, detailed studies on its biology are needed, and the search for new localities in the southern part of Lithuania needs to be extended.

4.4.3. Conservation of *Najas flexilis*

Despite the results of the latest studies, knowledge is still lacking about the environmental preferences of *N. flexilis*, as they vary from region to region and generally depend on a combination of factors. Regular monitoring is required to obtain additional data for future predictions of the distribution of the species and to implement appropriate conservation measures. Approaches to *N. flexilis* monitoring vary between countries, but, in most cases, surveys should be performed annually for a fixed period, with a break of several years thereafter [79]. Because the plant is small and the accuracy of boat-based surveys is low, monitoring should be performed by diving or scuba diving, but this requires special training and additional resources [79]. The probability of finding rare species in aquatic environments is low, which additionally hinders the searches [92]. All these factors complicate the conservation of annual aquatic plant species. Monitoring of the hydrochemical parameters of the lakes in which *N. flexilis* grows and strict protection from anthropogenic pollution are essential prerequisites for its conservation.

5. Conclusions

Three aquatic plant species protected under the European Habitat Directive, *Aldrovanda vesiculosa*, *Caldesia parnassifolia*, and *Najas flexilis*, currently occur in Lithuania. All their localities are concentrated in the northeastern part of the country.

The results of targeted surveys in various lakes between 2019 and 2021 confirmed the presence of *Aldrovanda vesiculosa* in four lakes (Lakes Rūžas, Apvardai, Dysnai, and Alksnas) where it was previously recorded; we detected no new populations. We recently found *Caldesia parnassifolia* in two lakes (Lakes Apvardai and Kampiniškiai) in addition to the previously recorded occurrence in Lake Rūžas. We discovered three new localities of *Najas flexilis* (Lakes Avilys, Ažvintis, and Dūkštas), in addition to the previously recorded locality in Lake Sągardas.

The habitats of *Aldrovanda vesiculosa* and *Caldesia parnassifolia* include shallow eutrophic and dystrophic lakes that are completely or partially surrounded by mires. We mainly found both species in communities of floating-leaved (ass. *Nymphaeo albae-Nupharetum luteae* and ass. *Potametum natantis*) and free-floating (ass. *Stratiotetum aloidis*) plants, which were formed at depths of 1 m or more. The identified habitats of *Najas flexilis* were relatively deep, dimictic, stratified, mesotrophic lakes with limited anthropogenic pressure.

The habitat of *Aldrovanda vesiculosa* in Lake Rūžas covered an area of approximately 3 ha, and we estimated the total population to be one–three million individuals. Over the eight-year study period, the population fluctuated little in terms of the density of individuals. The population of *Caldesia parnassifolia* in Lake Rūžas was abundant; however, the number of generative individuals significantly varied between years. All populations of *Najas flexilis* were small, although the potential habitats in the study lakes covered relatively large areas.

The population of *Aldrovanda vesiculosa* has a favourable conservation status in Lake Rūžas. This lake was already designated as an SAC, and a habitat management plan has been implemented. The populations of *Najas flexilis* are now considered to have an unfavourable conservation status, and all lakes in which the species occurs need to be designated as SACs.

Supplementary Materials: The following supporting information can be downloaded at: https://www.mdpi.com/article/10.3390/d15020185/s1, Table S1: Species diversity of plant communities with *Aldrovanda vesiculosa* and *Caldesia parnassifolia* in Lithuania.

Author Contributions: Conceptualisation, Z.S. and Z.G.; methodology, Z.S. and Z.G.; formal analysis, L.P. and Z.G.; investigation, Z.S., L.K.-B., L.P. and Z.G.; data curation, Z.S.; writing—original draft preparation, Z.S. and Z.G.; writing—review and editing, Z.S., L.K.-B., L.P. and Z.G.; visualisation, Z.G.; supervision, Z.S. All authors have read and agreed to the published version of the manuscript.

Funding: This study was partly funded by the State Service for Protected Areas of the Ministry of Environment of Lithuania under the LIFE-IP PAF-NATURALIT project LIFE16 IPE/LT/016 (contract no. F4-2019-022, 28 February 2019).

Institutional Review Board Statement: Not applicable.

Informed Consent Statement: Not applicable.

Data Availability Statement: The data that support the findings of this study are available on request from the corresponding author.

Acknowledgments: We are sincerely grateful to volunteer Jovita Papučkaitė for her help during the field work in 2015.

Conflicts of Interest: The authors declare no conflict of interest.

Appendix A

Table A1. Diversity and frequency of species recorded in communities with *Aldrovanda vesiculosa* and *Caldesia parnassifolia* in studied lakes and in different years of the study.

Species	Rūžas			Apvardai	Dysnai	Frequency (%)
Study year	2015	2019	2022	2020	2020	
Number of relevés	20	20	20	10	15	
Aldrovanda vesiculosa	100	100	100	100	100	100
Caldesia parnassifolia	40	40	30	40		80
Hydrocharis morsus-ranae	5	10	10	30	53	100
Myriophyllum cf. *verticillatum*	70	50	50	20	93	100
Nymphaea candida	40	10	25	50	20	100
Nuphar lutea	90	80	100	50	53	100
Potamogeton natans	85	85	80	100	87	100
Potamogeton × *bambergensis*	45	60	60	60	47	100
Stratiotes aloides	65	75	55	70	73	100
Utricularia vulgaris s.l.	30	25	5	20	7	100
Ceratophyllum demersum	10	5		10	53	80
Elodea canadensis	5	10	25	20		80
Hottonia palustris	80	70	90	50		80
Myriophyllum cf. *spicatum*	10	55	20	10		80
Sagittaria sagittifolia	20	20	25	20		80
Sparganium emersum	55	45	25	10		80
Nitellopsis obtusa	10	5	10			60
Phragmites australis	5	5		20		60
Potamogeton lucens	5		5		7	60
Ranunculus circinatus	30	5	25			60
Sparganium natans	20	40	20			60
Typha angustifolia		5	10		47	60
Carex lasiocarpa			5	20		40
Carex rostrata			5	10		40
Comarum palustre			5	10		40
Lemna minor				10	20	40
Lemna trisulca			5		20	40
Spirodela polyrhiza			10	10	13	40
Thelypteris palustris			5	30		40
Utricularia minor				10	13	40
Carex pseudocyperus					7	20
Cicuta virosa					13	20
Potamogeton crispus		5				20
Potamogeton perfoliatus					7	20
Potamogeton praelongus		5				20
Schoenoplectus lacustris				10		20
Typha latifolia				10		20
Filamentous algae				40	2	20

References

1. Sala, O.E.; Stuart Chapin, F.I.I.I.; Armesto, J.J.; Berlow, E.; Bloomfield, J.; Dirzo, R.; Wall, D.H. Global biodiversity scenarios for the year 2100. *Science* **2000**, *287*, 1770–1774. [CrossRef] [PubMed]
2. Díaz, S.M.; Settele, J.; Brondízio, E.; Ngo, H.; Guèze, M.; Agard, J.; Arneth, A.; Balvanera, P.; Brauman, K.; Butchart, S. *The Global Assessment Report on Biodiversity and Ecosystem Services: Summary for Policy Makers. 3947851138*; Intergovernmental Science-Policy Platform on Biodiversity and Ecosystem Services: Bonn, Germany, 2019.
3. Grzybowski, M.; Glińska-Lewczuk, K. Principal threats to the conservation of freshwater habitats in the continental biogeographical region of Central Europe. *Biodivers. Conserv.* **2019**, *28*, 4065–4097. [CrossRef]
4. Janse, J.H.; Van Dam, A.A.; Hes, E.M.; de Klein, J.J.; Finlayson, C.M.; Janssen, A.; vanWijk, D.; Mooij, W.M.; Verhoeven, J.T.A. Towards a global model for wetlands ecosystem services. *Curr. Opin. Environ. Sustain.* **2019**, *36*, 11–19. [CrossRef]
5. Mitsch, W.J.; Gosselink, J.G. The value of wetlands: Importance of scale and landscape setting. *Ecol. Econ.* **2000**, *35*, 25–33. [CrossRef]
6. Preston, C.D.; Sheail, J.; Armitage, P.; Davy-Bowker, J. The long-term impact of urbanisation on aquatic plants: Cambridge and the River Cam. *Sci. Total Environ.* **2003**, *314*, 67–87. [CrossRef]
7. Brauns, M.; Gücker, B.; Wagner, C.; Garcia, X.F.; Walz, N.; Pusch, M.T. Human lakeshore development alters the structure and trophic basis of littoral food webs. *J. Appl. Ecol.* **2011**, *48*, 916–925. [CrossRef]
8. Heino, J.; Virkkala, R.; Toivonen, H. Climate change and freshwater biodiversity: Detected patterns, future trends and adaptations in northern regions. *Biol. Rev.* **2009**, *84*, 39–54. [CrossRef]
9. Antonelli, A.; Fry, C.; Smith, R.J.; Simmonds, M.S.J.; Kersey, P.J.; Pritchard, H.W.; Abbo, M.S.; Acedo, C.; Adams, J.; Ainsworth, A.M.; et al. *State of the World's Plants and Fungi 2020*; Royal Botanic Gardens, Kew: Richmond, UK, 2020; 100p. [CrossRef]
10. Bolpagni, R.; Laini, A.; Stanzani, C.; Chiarucci, A. Aquatic plant diversity in Italy: Distribution, drivers and strategic conservation actions. *Front. Plant Sci.* **2018**, *9*, 116. [CrossRef]
11. Cwener, A.; Krawczyk, R.; Michalczuk, W. Nowe stanowisko *Caldesia parnassifolia* (Alismataceae) w Polsce. *Fragm. Florist. Geobot. Polon.* **2016**, *23*, 165–169. (In Polish)
12. Wassen, M.J.; Venterink, H.O.; Lapshina, E.D.; Tanneberger, F. Endangered plants persist under phosphorus limitation. *Nature* **2005**, *437*, 547–550. [CrossRef]
13. Röschel, L.; Noebel, R.; Stein, U.; Naumann, S.; Romão, C.; Tryfon, E.; Gaudillat, Z.; Roscher, S.; Moser, D.; Ellmauer, T.; et al. *State of Nature in the EU-Methodological Paper. Methodologies under the Nature Directives Reporting 2013–2018 and Analysis for the State of Nature 2000. ETC/BD Report to the EEA.*; European Environment Agency: Copenhagen, Denmark, 2020.
14. Adamec, L. Biological flora of Central Europe: *Aldrovanda vesiculosa* L. *Perspect. Plant Ecol. Evol.* **2018**, *35*, 8–21. [CrossRef]
15. Cross, A.; Adamec, L. *Aldrovanda vesiculosa*. The IUCN Red List of Threatened Species 2020: E.T162346A83998419. Available online: https://www.iucnredlist.org/species/162346/83998419 (accessed on 24 October 2022).
16. Lamont, E.E.; Sivertsen, R.; Doyle, C.; Adamec, L. Extant populations of *Aldrovanda vesiculosa* (Droseraceae) in the New World. *J. Torrey Bot. Soc.* **2013**, *140*, 517–522. [CrossRef]
17. Cross, A.T.; Skates, L.M.; Adamec, L.; Hammond, C.M.; Sheridan, P.M.; Dixon, K.W. Population ecology of the endangered aquatic carnivorous macrophyte *Aldrovanda vesiculosa* at a naturalised site in North America. *Freshw. Biol.* **2015**, *60*, 1772–1783. [CrossRef]
18. Ngangbam, R.D.; Devi, N.P.; Devi, M.H.; Singh, P.K. Rediscovery of *Aldrovanda vesiculosa* L. (Droseraceae), an endangered plant, from Manipur in India after six decades, with studies on micromorphology and physico-chemistry of water. *Reinwardtia* **2019**, *18*, 71–80. [CrossRef]
19. Shiga, T.; Khaliunaa, K.; Baasanmunkh, S.; Oyuntsetseg, B.; Midorikawa, S.; Choi, H.J. New Mongolian records of two genera, seven species, and two hybrid nothospecies from Khar-Us Lake and its associated wetlands. *J. Asia Pac. Biodivers.* **2020**, *13*, 443–453. [CrossRef]
20. Lukash, O.V.; Popruha, V.M.; Kupchyk, O.Y.; Strilets, S.I. Phytocenotic and hydrochemical conditions of the new localities of *Aldrovanda vesiculosa* (Droseraceae) in Chernihiv Polissya. *Ukr. Bot. J.* **2020**, *77*, 466–471. [CrossRef]
21. Cameron, K.M.; Wurdack, K.J.; Jobson, R.W. Molecular evidence for the common origin of snap-traps among carnivorous plants. *Am. J. Bot.* **2002**, *89*, 1503–1509. [CrossRef]
22. Fleischmann, A. *Aldrovanda vesiculosa* L. neu in der Oberpfalz, und eine Übersicht zur natürlichen und neophytischen Verbreitung der Art in Deutschland. *Ber. Bayer. Bot. Ges.* **2021**, *91*, 267–285.
23. Gupta, A.K.; Beentje, H.J.; Lansdown, R.V. *Caldesia parnassifolia* (L.) Parl. The IUCN Red List of Threatened Species 2018. e.T162381A120123874. Available online: https://www.iucnredlist.org/species/162381/120123874 (accessed on 15 October 2022).
24. Kamiński, R. Caldesia parnassifolia (L.) Parl. (Kaldezja dziewięciornikowata). *Aldrovanda vesiculosa* L. (Aldrowanda pęcherzykowata). In *Polska Czerwona Księga Roślin: Paprotniki i Rośliny Kwiatowe*, 3rd ed.; Kaźmierczakowa, R., Zarzycki, K., Mirek, Z., Eds.; PAN: Kraków, Poland, 2014; pp. 237–239, 562–564. (In Polish)
25. Mirek, Z.; Piękoś-Mirkowa, H.; Zając, A.; Zając, M. *Vascular Plants of Poland an Annotated Checklist*; Polish Academy of Sciences: Krakow, Poland, 2020; pp. 1–526.
26. Mesterházy, A. *Caldesia parnassifolia* (L.) Parl. új előfordulása Magyarországon. *Kitaibelia* **2014**, *19*, 50–55. (In Hungarian)
27. Sinkevičienė, Z. Širdžialapė kaldezija (*Caldesia parnassifolia* (L.) Parl.). In *Lietuvos Raudonoji Knyga*; Rašomavičius, V., Ed.; Lututė: Vilnius, Lithuania, 2007; pp. 404–405. (In Lithuanian)

28. Fenu, G.; Siniscalco, C.; Bacchetta, G.; Cogoni, D.; Pinna, M.S.; Sarigu, M.; Abeli, T.; Barni, E.; Bartolucci, F.; Bouvet, D.; et al. Conservation status of the Italian flora under the 92/43/EEC 'Habitats' Directive. *Plant Biosyst.* **2021**, *155*, 1168–1173. [CrossRef]

29. Skuratovič, A.N. Alismataceae Vent. In *Flora Belarusi*; Parfionov, V.I., Ed.; Belaruskaja Nauka: Minsk, Belarus, 2013; Volume 2, pp. 82–93.

30. Hultén, E.; Fries, M. *Atlas of North European Vascular Plants North of the Tropic Cancer*; Koeltz Scientific Books: Königstein, Germany, 1986; p. 486.

31. Maiz-Tome, L. *Najas flexilis*. The IUCN Red List of Threatened Species 2016: E.T162310A78457107. Available online: https://www.iucnredlist.org/species/162310/78457107 (accessed on 15 November 2022).

32. Bishop, I.J.; Bennion, H.; Sayer, C.D.; Patmore, I.R.; Yang, H. Filling the "data gap": Using paleoecology to investigate the decline of *Najas flexilis* (a rare aquatic plant). *Geo Geogr. Environ.* **2019**, *6*, e00081. [CrossRef]

33. Lansdown, R.V. *Najas flexilis* (Europe assessment). The IUCN Red List of Threatened Species 2011: E.T162310A5572090. Available online: https://www.iucnredlist.org/species/162310/5572090 (accessed on 15 November 2022).

34. Casper, S.J.; Krausch, H.D. *Pteridophyta u. Anthophyta, 1. Teil, Süßwasserflora von Mitteleuropa*; Ettl, H., Gerlof, J., Heyning, H., Eds.; Gustav Fischer Verlag: Stuttgart, Germany; New York, NY, USA, 1980; Volume 23, 145p.

35. Philippi, G. Najadaceae—Nixenkrautgewächse. In *Die Farn- und Blütenpflanzen Baden-Württembergs Band 7*; Sebald, O., Seybold, S., Philippi, G., Wörz, A., Eds.; Ulmer: Stuttgart, Germany, 1998; pp. 47–52.

36. Zalewska, J. The genus *Najas* (Najadaceae) in Poland: Remarks on taxonomy, ecology, distribution and conservation. *Fragm. Flor. Geobot.* **1999**, *44*, 401–422.

37. Roden, C.; Murphy, P.; Ryan, J.B. A study of lakes with Slender Naiad (*Najas flexilis*). *Ir. Wildl. Man.* **2021**, *132*, 1–90.

38. Suško, U.; Čakare, M.; Jēkabsone, J.; Vītola, I.; Grīnberga, L.; Zviedre, E.; Līcīte, V.; Skrinda, I.; Evarts-Bunders, P. New Records of Najas flexilis, Najas tenuissima and other Najas species in Latvia. In *11th International Conference on Biodiversity Research, Daugavpils, Latvia, 20–22 October 2022*; Daugavpils University Academic Press "Saule": Daugavpils, Latvia, 2022; pp. 130–131.

39. Rørslett, B. Mykt havfrugras, *Najas flexilis* i Norge. *Blyttia* **1981**, *39*, 1–6. (In Norwegian)

40. Sinkevičienė, Z. *Najas flexilis* and *Najas minor* in Lithuania. *Bot. Lith.* **2001**, *7*, 203–208. (In Lithuanian)

41. Olsson, K.-A. *Åtgärdsprogram för Bevarande av Sjönajas (Najas flexilis)*; Naturvårdsverke: Bromma, Sweden, 2006; p. 31. (In Swedish)

42. Pall, K. *Najas flexilis* (Najadaceae or Hydrocharitaceae) a Natura 2000 species—New for Austria. *Neilreichia* **2011**, *6*, 11–26.

43. Issakainen, J.; Kemppainen, E.; Mäkelä, K.; Hakalisto, S.; Koistinen, M. Hentonäkinruoho (*Najas tenuissima*) ja notkeanäkinruoho (*Najas flexilis*) Suomen uhanalaisia lajeja. *Suom. Ympäristö* **2011**, *13*, 1–223. (In Finnish)

44. Dubovik, D.V. Najadaceae Juss. In *Flora Belarusi*; Parfionov, V.I., Ed.; Belaruskaj navuka: Minsk, Belarus, 2013; Volume 2, pp. 72–77. (In Russian)

45. Dubovik, D.V. Caulinia flexilis. In *Red Data Book of Belarus Plants*, 4th ed.; Kachanovskii, I.M., Nikiforov, V.I., Parfenov, V.I., Eds.; Belaruskia Encyklapedya imia Petrusia Brouki: Minsk, Belarus, 2015; pp. 64–65. (In Russian)

46. Mäemets, H. Commented list of rare and protected vascular plants of inland water bodies of Estonia. *Nat. Conserv. Res.* **2016**, *1*, 85–89. [CrossRef]

47. Renzi, J.J.; He, Q.; Silliman, B.R. Harnessing positive species interactions to enhance coastal wetland restoration. *Front. Ecol. Evol.* **2019**, *7*, 131. [CrossRef]

48. Gudžinskas, Z.; Taura, L. *Scirpus radicans* (Cyperaceae), a newly-discovered native species in Lithuania: Population, habitats and threats. *Biodivers. Data J.* **2021**, *9*, e65674. [CrossRef]

49. Sinkevičienė, Z.; Gudžinskas, Z. Revision of the Characeae (Charales, Charophyceae) species and their distribution in Lithuania. *Botanica* **2021**, *27*, 95–124. [CrossRef]

50. Taura, L.; Kamaitytė-Bukelskienė, L.; Sinkevičienė, Z.; Gudžinskas, Z. Study on the Rare Semiaquatic Plant *Elatine hydropiper* (Elatinaceae) in Lithuania: Population Density, Seed Bank and Conservation Challenges. *Front. Biosci.* **2022**, *27*, 162. [CrossRef]

51. Sinkevičienė, Z.; Gudžinskas, Z. Conservation plan of *Aldrovanda vesiculosa* L. *Raudoni Lapai* **2012**, *14*, 75–85. (In Lithuanian)

52. Otto-Bruc, C.; Haury, J.; Lefeuvre, J.C.; Dumeige, B.; Pinet, F. Variations temporelles des populations de *Caldesia parnassifolia* (L.) Parl, dans lcs étangs de la Brenne (Indre, France). *Acta Bot. Gall.* **2000**, *147*, 375–397. [CrossRef]

53. Robert, G.W.; Qing-Feng, W.; Yong, W.; You-Hao, G. Reproductive biology and prospects for conservation of *Caldesia parnassifolia* (Alismataceae)—A threatened monocot in China. *Wuhan Univ. J. Nat. Sci. A* **2003**, *8*, 117–124. [CrossRef]

54. Wang, Q.; Haynes, R.; Hellquist, C.B. Caldesia. In *Flora of China*; Wu, Z., Raven, P.H., Eds.; Missouri Botanical Garden Press: St. Louis, MO, USA, 2010; Volume 23, p. 87.

55. Grīnberga, L.; Cepurīte, B.; Opmanis, A.; Suško, U. Sirdslapu kaldēzija *Caldesia parnassifolia* (Bassi ex L.) Parl.—Jauna suga Latvijas florā. *Latv. Veģetācija* **2022**, *32*, 42–48. (In Latvian)

56. Kaźmierczakowa, R.; Bloch-Orłowska, J.; Celka, Z.; Cwener, A.; Dajdok, Z.; Michalska-Hejduk, D.; Pawlikowski, P.; Szczęśniak, E.; Ziarnek, K. *Polska Czerwona Lista Paprotników i Roślin Kwiatowych. [Polish Red List of Pteridophytes and Flowering Plants]*; Ss. 44; Instytut Ochrony Przyrody PAN: Kraków, Poland, 2016; p. 48.

57. Česnulevičius, A. Aukštaičių aukštuma [Aukštaičiai Upland]. In *Visuotinė Lietuvių Enciklopedija, T. II (Arktis-Beketas)*; Mokslo ir enciklopedijų leidybos institutas: Vilnius, Lithuania, 2002; p. 229. (In Lithuanian)

58. Galvonaitė, A.; Valiukas, D.; Kilpys, J.; Kitrienė, Z.; Misiūnienė, M. *Climate Atlas of Lithuania*; Hydrometeorological Service of Lithuania: Vilnius, Lithuania, 2013; pp. 1–176.

59. Kilkus, K.; Stonevičius, E. *Lietuvos Vandenų Geografija*; Apyaušris: Vilnius, Lithuania, 1998; p. 248.

60. Environmental Protection Agency. *Monitoring Results of Lakes and Ponds*; Environmental Protection Agency: Vilnius, Lithuania. Available online: https://aaa.lrv.lt/lt/veiklos-sritys/vanduo/upes-ezerai-ir-tvenkiniai/valstybinis-upiu-ezeru-ir-tvenkiniu-monitoringas/ezeru-ir-tvenkiniu-monitoringo-rezultatai (accessed on 25 September 2022). (In Lithuanian)
61. Braun-Blanquet, J. *Pflanzensociologie: Grundzuge der Vegetation-Skunde*; 3te aufl.; Springer: Viena, Austria, 1964; pp. 1–865.
62. Šumberová, K. Vegetace volně plovoucích vodních rostlin (Lemnetea). Vegetace vodních rostlin zakořeněných ve dně (Potametea). [Vegetation of free floating aquatic plants. Vegetation of aquatic plants rooted in the bottom]. In *Vegetace České Republiky. 3. Vodní a Mokřadní Vegetace [Vegetation of the Czech Republic 3. Aquatic and wetland vegetation]*; Chytrý, M., Ed.; Academia: Praha, Czech Republic, 2011; pp. 43–99, 100–245. (In Czech)
63. Euro+Med, 2022. The Euro+Med PlantBase. Available online: https://europlusmed.org/node/129 (accessed on 20 November 2022).
64. Hammer, Ø.; Harper, D.A.T.; Ryan, P.D. PAST: Paleontological statistics software package for education and data analysis. *Palaeontol. Electron.* **2001**, *4*, 9.
65. Šarkinienė, I. Rytų ir pietų Lietuvos TSR ežerų makrofitų floristinė ir morfologinė-ekologinė analizė. *Lietuvos TSR Aukštųjų Mokyklų Mokslo Darbai. Biologija* **1961**, *1*, 159–194. (In Lithuanian)
66. Diels, L. Droseraceae. In *Das Pflanzenreich Regni Vegetabilis Conspectus*; Engler, A., Ed.; Verlag von Wilhelm Engelmann: Leipzig, Germany, 1906; p. 61.
67. Kamiński, R. *Restytucja Aldrowandy Pęcherzykowatej (Aldrovanda vesiculosa L.) w Polsce Irozpoznanie Czynników Decydujacych o jej Przetrwaniu w Klimacie Umiarkowanym [Restitution of Waterwheel plant (Aldrovanda vesiculosa L.) in Poland and Recognition of Factors Deciding on Its Survival in Temperate Climate]*; Adiunkt w Ogrodzie Botanicznym—Uniwersytet Wrocławski: Wrocław, Poland, 2006. (In Polish)
68. Gorski, S.B. Botanische Bemerkungen (Wyliczenie roślin naczyniowych Litewskich postrzeganych od r. 1820–1829). In *Naturhistorische Skizze von Lithauen, Volhynien und Podolien in Geognostisch-Mineralogischer Botanischer und Zoologischer Hinsicht*; Eichvald, E.K., Ed.; Zawadzki: Wilna, Litauen, 1830; pp. 105–184.
69. Wolfgang, J.F. Wiadomość o świeżo odkrytych lub rzadszych roślinach przybylych do Flory Litewskiej w. r. 1821. *Pamiętnik Farm. Wileński* **1822**, *2*, 649–653. (In Polish)
70. Obelevičius, S. *Aldrovanda vesiculosa* in Lithuania. *Bot. Lith.* **2001**, *7*, 385–386. (In Lithuanian)
71. Sinkevičienė, Z. *Aldrovanda vesiculosa* L. In *Red Data Book of Lithuania. Animals, Plants, Fungi*; Rašomavičius, V., Ed.; Lututė: Vilnius, Lithuania, 2021; p. 479. (In Lithuanian)
72. Wolfgang, J.F. Wiadomość o nowo przybylych roślinach do Flory Litewskiej w r. 1822 i 1823. *Dziennik Medycyny Chirurgii i Farmacji* **1824**, *2*, 647–656. (In Polish)
73. Pabrieža, A. *Botanika Arba Taislius Auguminis*; V.J.Stagaro ir Co: Shenandoah, VA, USA, 1900; p. 166.
74. Pabrėža, J.A. *Taislius Augyminis*; Lietuvių kalbos institutas: Vilnius, Lithuania, 2014; p. 84.
75. Natkevičaitė, M. Daugų ežero vakarų kranto ir Banduragio įlankos aukštesniosios augalijos fitocenologiniai tyrimai. *Vilniaus Universiteto Mokslo Darbai Gamtos-Matematikos Mokslų Serija* **1954**, *2*, 141–156. (In Lithuanian)
76. Sinkevičienė, Z. Aquatic plant species of European importance in Lithuania. In Proceedings of the 11th EWRS International Symposium of Aquatic Weeds, Moliets et Maâ, France, 2–6 September 2002; pp. 315–317.
77. Gavrilova, G.; Kuusk, V.; Sinkevičienė, Z. Alismataceae Vent. In *Flora of the Baltic Countries 3, Kuusk, V., Tabaka, L., Jankevičienė, R., Eds.*; Estonian Agricultural University Institute of Zoology and Botany: Tartu, Estonia, 2003; pp. 190–193.
78. Sinkevičienė, Z. *Caldesia parnassifolia*—Not extinct in Lithuania. *Bot. Lith.* **2016**, *22*, 49–52. [CrossRef]
79. Wingfield, R.; Murphy, K.J.; Gaywood, M. Assessing and predicting the success of *Najas flexilis* (Willd.) Rostk. & Schmidt, a rare European aquatic macrophyte, in relation to lake environmental conditions. *Hydrobiologia* **2006**, *570*, 79–86. [CrossRef]
80. Kilkus, K. *Ežerotyra*; VU leidykla: Vilnius, Lithuania, 2005; p. 20. (In Lithuanian)
81. Povilaitis, A.; Taminskas, J.; Gulbinas, Z.; Linkevičienė, R.; Pileckas, M. *Lietuvos Šlapynės ir jų Vandensauginė Reikšmė*; Apyaušris: Vilnius, Lithuania, 2011; p. 69. (In Lithuanian)
82. Tretyakov, D.I. Aldrovanda vesiculosa. In *Red Data Book of Belarus Plants*, 4th ed.; Kachanovskii, I.M., Nikiforov, V.I., Parfenov, V.I., Eds.; Belaruskia Encyklapedya imia Petrusia Brouki: Minsk, Belarus, 2015; pp. 108–109. (In Russian)
83. Fishman, J.R.; Orth, R.J. Effects of predation on *Zostera marina* L. seed abundance. *J. Exp. Mar. Biol. Ecol.* **1994**, *198*, 11–26. [CrossRef]
84. Liu, X.; Wang, J.-Y.; Wang, Q.-F. Current Status and conservation strategies for *Isoetes* in China a case study for the conservation of threatened aquatic plants. *Oryx* **2005**, *39*, 335–338. [CrossRef]
85. Vejříková, I.; Vejřík, L.; Leps, J.; Kočvara, L.; Sajdlova, Z.; Čtvrtlíková, M.; Peterka, J. Impact of herbivory and competition on lake ecosystem structure: Underwater experimental manipulation. *Sci. Rep.* **2018**, *8*, 12130. [CrossRef]
86. Bukantis, A.; Rimkus, E. Climate variability and change in Lithuania. *Acta Zool. Litu.* **2005**, *15*, 100–104. [CrossRef]
87. Dubyna, D.V. *Higher Aquatic Vegetation*; Phytosociocentre: Kyiv, Ukraine, 2006; pp. 1–412.
88. Sculthorpe, C.D. *The Biology of Aquatic Vascular Plants*; Edward Arnold: London, UK, 1967; p. 610.
89. Kamiński, R. Studies on the ecology of *Aldrovanda vesiculosa* L. I. Ecological differentiation of *A. vesiculosa* population under the influence of chemical factors in the habitat. *Ekol. Pol.* **1987**, *35*, 559–590.

90. Horstmann, M.; Heier, L.; Kruppert, S.; Weiss, L.C.; Tollrian, R.; Adamec, L.; Westermeier, T.; Speck, T.; Poppinga, S. Comparative prey spectra analyses on the endangered aquatic carnivorous waterwheel plant (*Aldrovanda vesiculosa*, Droseraceae) at several naturalized microsites in the Czech Republic and Germany. *Integr. Org. Biol.* **2019**, *1*, oby012.
91. Gudžinskas, Z.; Taura, L. Confirmed occurrence of the native plant species *Eleocharis ovata* (Cyperaceae) in Lithuania. *Botanica* **2021**, *27*, 44–52. [CrossRef]
92. Pollock, K.H.; Marsh, H.; Bailey, L.L.; Farnsworth, G.L.; Simons, T.R.; Alldredge, M.W. Separating components of detection probability in abundance estimation: An overview with diverse examples. In *Sampling Rare or Elusive Species: Concepts, Designs, and Techniques for Estimating Population Parameters*; Thompson, W.L., Ed.; Island Press: Washington, DC, USA, 2004; pp. 43–58.

 *diversity*

Article

Waterbodies in the Floodplain of the Drava River Host Species-Rich Macrophyte Communities despite *Elodea* Invasions

Igor Zelnik *, Mateja Germ, Urška Kuhar and Alenka Gaberščik

Department of Biology, Biotechnical Faculty, University of Ljubljana, Jamnikarjeva 101, 1000 Ljubljana, Slovenia
* Correspondence: igor.zelnik@bf.uni-lj.si

Abstract: The contribution discusses macrophyte communities in natural and man-made waterbodies located on the active floodplain along the Drava river (Slovenia). We presumed that these different types of wetlands host a great number of macrophyte species, but this diversity may be affected by the presence of alien invasive species *Elodea canadensis* and *E. nuttallii*. Presence, relative abundance, and growth forms of plant species along with selected environmental parameters were monitored. Correlation analyses and direct gradient analyses were performed to reveal the possible relations between the structure of macrophyte community and environmental parameters. Number of macrophytes in surveyed water bodies varied from 1 to 23. Besides numerous native species we also recorded *Elodea canadensis* and *E. nuttallii*, which were present in 19 out of 32 sample sites, with *E. nuttallii* prevailing. The less invasive *E. canadensis* was absent from ponds and oxbow lakes but relatively abundant in side-channels, while *E. nuttallii* was present in all types but dominant in ponds. The most abundant native species were *Myriophyllum spicatum* and *M. verticillatum*, *Ceratophyllum demersum* and *Potamogeton natans*. Correlation analyses showed no negative effect of the invasive alien *Elodea* species to the species richness and diversity of native flora. Positive correlation between the abundance of *E. nuttallii* and temperature of the water was obtained.

Keywords: macrophytes; small water bodies; wetlands; alien invasive species; *Elodea*; flood plain

Citation: Zelnik, I.; Germ, M.; Kuhar, U.; Gaberščik, A. Waterbodies in the Floodplain of the Drava River Host Species-Rich Macrophyte Communities despite *Elodea* Invasions. *Diversity* **2022**, *14*, 870. https://doi.org/10.3390/d14100870

Academic Editor: Jesús M. Castillo

Received: 28 September 2022
Accepted: 11 October 2022
Published: 14 October 2022

1. Introduction

Rivers are complex ecosystems that change in time and space due to ecological and hydro-morphological processes [1]. River flow determines processes that affect the shape and distribution of habitats, and thus associated biotic communities [2]. The diversity of river channel and floodplain wetlands support the diversity of biotic communities. These habitats enable the establishment and dispersal of organisms, which ultimately affects biodiversity patterns [3], especially in macrophytes that have relatively low dispersal ability [4].

The global increase in energy and water demand of the human population resulted in alterations of river channels that affected the function of rivers and adjacent floodplains, as well as wetlands along these rivers such as oxbow lakes, side channels, backwaters, ponds etc. [5]. Williams et al. [6] emphasized that such small waterbodies can contribute significantly to regional biodiversity, including macrophyte communities [7], and are important for the conservation and management of the local biodiversity [8–10]. Oxbow lakes may be especially rich since they represent a transition between lotic and lentic ecosystems [11]. Sustainable catchment management should be based on the knowledge of the biodiversity in different water bodies within these catchments [12] and its vulnerability. Such water bodies have a great potential for the conservation of biological diversity and are recognised for their importance for ecosystem services [13], even though they have received relatively little attention. This situation is different along the Danube river, where these water bodies were studied by many researchers (e.g., Otahelová et al. [14]; Schmidt-Mumm and Janauer [15]; Vukov et al. [16]; Gyosheva et al. [17]). The increase in the proportion of urban and agricultural land-use within the catchment areas of mentioned habitats results in a decrease in species richness, thus, for efficient conservation of their biodiversity, actions at a local and regional spatial scale are required [18].

Macrophytes are an important element of the aquatic ecosystem since they are the basis for energy flow and nutrient cycling, and they affect sedimentation processes [19,20]. Macrophyte stands are habitats, refugia and a source of organic material for a range of other organisms [21,22]. Brysiewicz et al. [23] also discovered that species occurrences and abundances of fish fauna in small waterbodies were associated with the amount of macrophytes growing in them. High macrophyte abundance may significantly alter the chemical and biological structure of the ecosystem [24], while degradation of macrophyte communities may cause a reduction in diversity of organisms dwelling in these macrophyte stands [25]. By nutrient uptake from water and sediment, macrophytes ameliorate water quality and affect the quality water and sediment [20,21]. The affinity for specific water and sediment properties in different species make them valuable indicators of water and sediment quality [26–29].

Small sized water bodies are often subjected to extreme water level fluctuations, which are more pronounced in hydrologically isolated systems, where accelerated succession often occurs [30]. These water level fluctuations may also cause the dieback of some plant species, and consequently a release of nutrients [31], and a decrease in biodiversity. They effect aquatic vegetation, the trophic state of the ecosystem, and consequently affect the diversity and abundance of macroinvertebrates within the macrophyte stands [32]. Another threat to local biodiversity, especially diversity of freshwater biota, is the spread of alien invasive species that is often reported as one of the major factors for its decline [33]. Vukov et al. [16] report that *Elodea canadensis* and *E. nuttallii* have been rapidly spreading along the whole Danube, which was documented in [34]. However, the negative influence of these species is not always significant and there may also be some positive effects on target ecosystems [35,36], like new habitat formation for aquatic fauna or cyanobacterial blooms prevention, especially in lentic waterbodies. The extent of the effect of a certain invasive species for the functioning of target ecosystems largely depends on its abundance [37].

In this paper, we studied macrophyte communities in natural and man-made shallow waterbodies located on the active floodplain along the Drava river in Slovenia, within the section with generally preserved morphological conditions but with modified hydrology, to estimate their potential for the conservation value for macrophyte biodiversity. Since these habitats represent different types of wetlands, we hypothesised that they harbour a great number of macrophyte species and so mitigate their loss in other sections of the river, which are affected by numerous hydropower-plants and their impoundments. We also hypothesised that species diversity may be affected by the presence of invasive alien species of the genus *Elodea*. Deeper understanding of such water bodies will facilitate effective conservation and management of floodplains and support their ecosystem services.

2. Materials and Methods

2.1. Study Area

Studied wetlands are found within the active floodplain of the Drava river in northeast Slovenia, between the town Maribor and the state border with Croatia (Figure 1). The river Drava is among the biggest tributaries of the Danube river and gathers water from Italy, Austria, Slovenia, Croatia and Hungary. The hydrological regime of the surveyed section of the river has been modified, since a great proportion of the water from the river Drava is diverted into artificial channels that supply the water for Hydropower-plants. The advantage of this fact is that there have been no major changes in the morphological conditions of the old river channel and adjacent floodplain, where the studied wetlands occur. As the reference habitats, four reaches in the main channel of the Drava river were also surveyed. Beyond the edges of the floodplain, the land use is characterized by intensive agriculture with cultivated and uncultivated land mosaic.

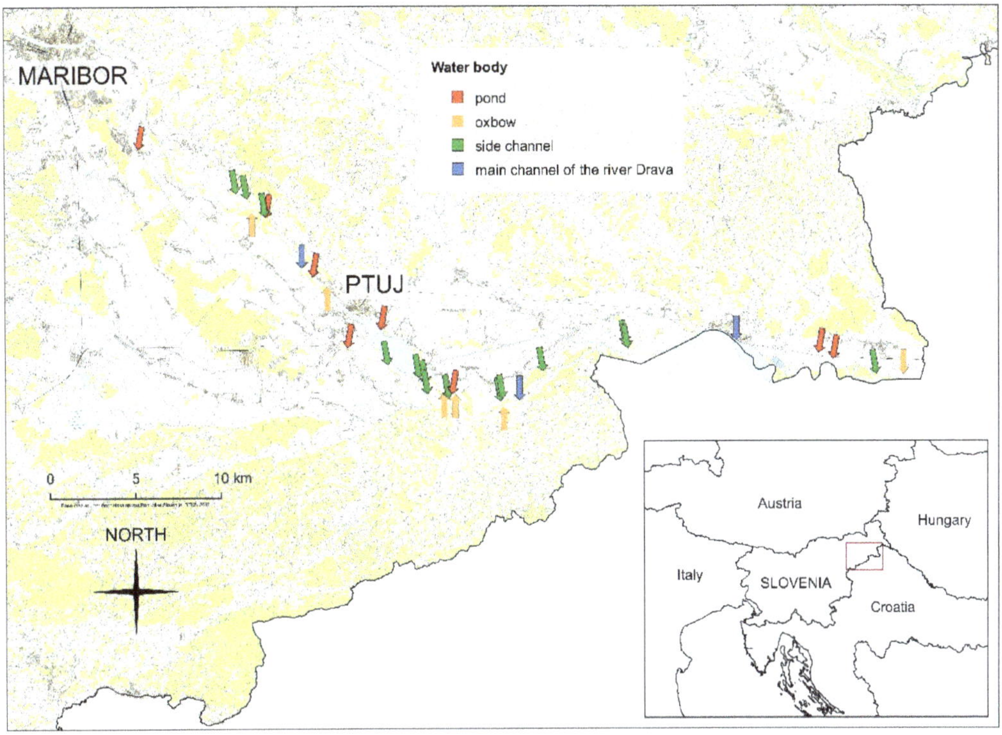

Figure 1. Distribution of the studied waterbodies in the floodplain of the Drava river (Base map source: Surveying and Mapping Authority of the republic of Slovenia (2016).

2.2. Macrophyte Data Set

Surveys were carried out in the years 2015–2016. Since we surveyed different types of the waterbodies the approaches were combined [22]—in ponds, the entire length (<100 m) was examined, whereas in oxbows, side-channels and the river we examined at least 100 m long sections. We recorded emergent, floating-leaved and submerged vascular plants, bryophytes and charophytes. The presence and abundance of macrophytes were evaluated from the boat or from the bank and collected with a stick with hooks. Macrophyte species abundance was estimated as a relative plant biomass using a five-degree scale, namely 1—very rare, 2—rare, 3—commonly present, 4—frequent, and 5—predominant, as proposed by Kohler and Janauer [38]. These values were transformed by the function x^3, as suggested by Schneider and Melzer [27]. The plants that were sampled in the phenological phase, which prevented identification to the species level, were only recorded on the genus level (e.g., *Carex*, *Callitriche*). Species names followed the nomenclature of Euro+Med Plantbase [39].

We classified the macrophytes into the following growth forms: natant (leaves or whole plants floating at the water surface); submerged (assimilation areas submerged in water column); amphiphytes (having the ability to produce terrestrial and aquatic growth forms, or aquatic and aerial leaves); and helophytes (anchored in the water-saturated sediment, with plant assimilation areas permanently in the air). For the purpose of correlation analyses and comparisons of average abundances, the ordinal values of the Kohler-scale were transformed into quantitative values ("quantities") [40]. We equalized the transformed values as percentage cover-abundance values according to [41].

2.3. Environmental Parameters

The assessment of environmental conditions was performed in the sites as the survey of macrophytes. We also assessed parameters like land-use type beyond the riparian zone, characteristics of the riparian zone (width, completeness, and vegetation type), and morphology (bank structure) [42]. Each parameter includes four categories comprising quality gradient, coded numerically from 1 to 4:1 presented good, close to natural condition, while quality gradient values from 2 to 4 indicate worsening of environmental conditions. Results of the assessment of the land-use and the width of riparian zone are presented in Table 1. Apart from the mentioned environmental parameters, we also recorded the basic physical and chemical parameters (temperature of the water, pH, conductivity, concentration of dissolved O_2, saturation with O_2) with the multimeter (Eutech PCD-650, Singapore).

Table 1. Characteristics of the catchment area of specific waterbodies. Width of the riparian zone, where woody or herbaceous wetland vegetation is thriving and prevailing land-use behind the riparian zone is presented.

Location	Width of Riparian Zone	Land-Use behind the Riparian Zone	Type
1	1–5 m	arable land, grassland, houses	river
2	1–5 m	arable land, grassland, houses	pond
3	1–5 m	arable land, grassland, houses	channel
4	1–5 m	arable land, grassland, houses	channel
5	1–5 m	arable land, grassland, houses	oxbow
6	<1 m	mainly arable land or urban area	pond
7	1–5 m	mainly arable land or urban area	channel
8	<1 m	arable land, grassland, houses	river
9	<1 m	mainly arable land or urban area	pond
10	1–5 m	mainly arable land or urban area	oxbow
11	<1 m	mainly arable land or urban area	pond
12	5–30 m	grassland, forest and/or wetland, some arable land	pond
13	5–30 m	grassland, forest and/or wetland, some arable land	channel
14	1–5 m	mainly arable land or urban area	channel
15	<1 m	arable land, grassland, houses	channel
16	1–5 m	arable land, grassland, houses	channel
17	<1 m	arable land, grassland, houses	oxbow
18	<1 m	mainly arable land or urban area	channel
19	5–30 m	arable land, grassland, houses	pond
20	<1 m	mainly arable land or urban area	oxbow
21	1–5 m	arable land, grassland, houses	channel
22	1–5 m	arable land, grassland, houses	channel
23	<1 m	arable land, grassland, houses	oxbow
24	1–5 m	arable land, grassland, houses	river
25	<1 m	arable land, grassland, houses	channel
26	<1 m	mainly arable land or urban area	channel
27	<1 m	mainly arable land or urban area	channel
28	5–30 m	arable land, grassland, houses	river
29	1–5 m	arable land, grassland, houses	pond
30	1–5 m	arable land, grassland, houses	pond
31	1–5 m	grassland, forest and/or wetland, some arable land	channel
32	1–5 m	mainly arable land or urban area	oxbow

2.4. Statistical Analyses

Correlation analyses between species and parameters was calculated with PAST, version 2.17c [43]. Kendall *tau* correlation coefficients were calculated.

Detrended correspondence analysis (DCA) was performed in the first step of gradient analyses. This analysis also informed us whether the gradients in the matrix of plant species are linear or unimodal, and which direct gradient analysis to use in further analyses. When we performed DCA with the matrix of functional types/growth forms, the eigenvalue for the first axis was lower than 0.4 (0.08) and we selected Redundancy analysis (RDA), as suggested by ter Braak and Verdonschot [44]. These results provided the information about the relationships between environmental factors and the structure of macrophyte community and their growth forms, respectively.

We used forward selection of the variables (499 permutations were performed) to rank the relative importance of explanatory variables. Only the variables with significance $p < 0.05$ were considered in further analyses. All analyses were performed with CANOCO for Windows 4.5 program package [45].

3. Results

The entire list of macrophytes comprised of 73 plant taxa, while the number of macrophytes in specific waterbodies varied from 1 to 23 (Figure 2). Beside numerous native species, waterbodies also host two invasive alien species of *Elodea*, namely *Elodea canadensis* and *E. nuttallii*, present in 19 out of 32 sample sites (Figure 3), with *E. nuttallii* prevailing. The most abundant native species were *Myriophyllum spicatum* and *M. verticillatum*, *Ceratophyllum demersum* and *Potamogeton natans*, which were also present in various locations (Figure 4). Natant species *Nuphar luteum* was present at one site only.

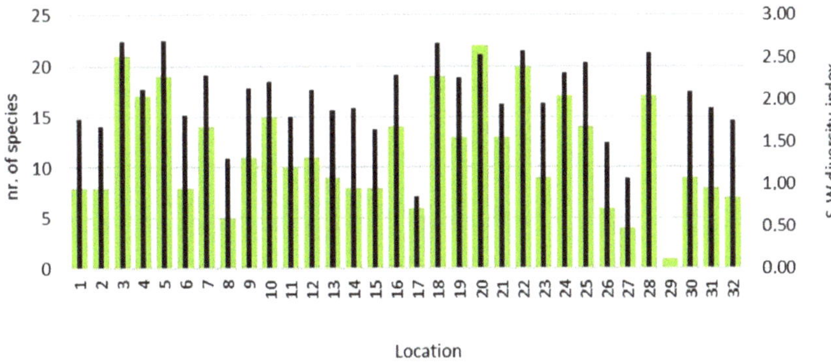

Figure 2. Total species number (green bars) and Shannon–Wiener (S–W) diversity index (black lines) in surveyed waterbodies.

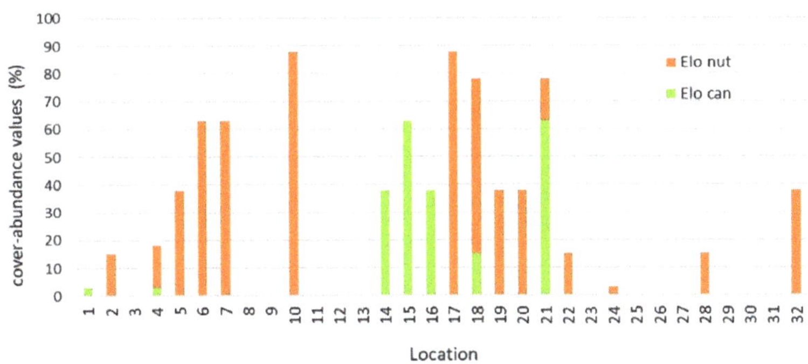

Figure 3. Relative abundance of invasive alien species *Elodea canadensis* and *E. nuttallii* in surveyed waterbodies.

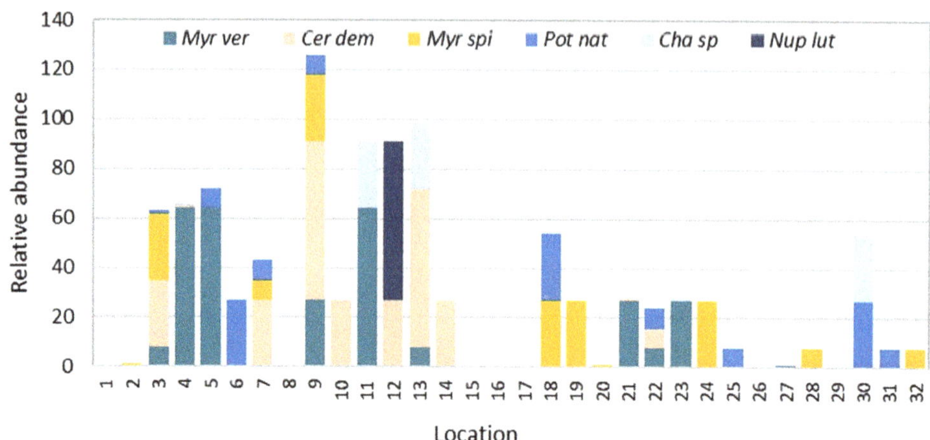

Figure 4. Relative abundance of native hydrophyte species with total abundance more than 2% present in surveyed waterbodies: *Myr ver—Myriophyllum verticillatum, Cer dem—Ceratophyllum demersum, Myr spi—Myriophyllum spicatum, Pot nat—Potamogeton natans, Cha* sp—*Chara* species, *Nup lut—Nuphar luteum*.

E. nuttallii was the most abundant among all hydrophyte species, with more than 25% of the total abundance, followed by *M. verticillatum, C. demersum, E. canadensis* and *M. spicatum* (Figure 5).

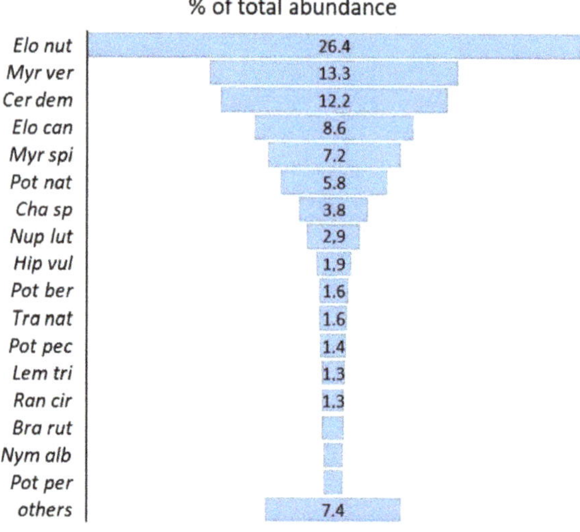

Figure 5. Ratio of total relative abundance of hydrophyte species that presented more than 1% in surveyed waterbodies: *Elo nut—Elodea nuttallii, Myr ver—Myriophyllum verticillatum, Cer dem—Ceratophyllum demersum, Elo can—Elodea canadensis, Myr spi—Myriophyllum spicatum, Pot nat—Potamogeton natans, Cha* sp—*Chara* species, *Nup lut—Nuphar luteum, Hip vul—Hippuris vulgaris, Pot ber—Potamogeton bertholdii, Tra nat—Trapa natans, Pot pec—Potamogeton pectinatus, Lem tri—Lemna trisulca, Ran cir—Ranunculus circinatus, Bra rut—Brachythecium rutabulum, Nym alb—Nymphaea alba, Pot per—Potamogeton perfoliatus, others—*the sum of abundances of species with less than 1%.

The DCA analysis shows the similarity of surveyed sites in terms of the structure of macrophyte communities in the peak vegetation period. The closer sites are on the ordination plot, the more similar are the macrophyte communities. It is evident that the type of aquatic ecosystem does not dictate the macrophyte community structure (Figure 6).

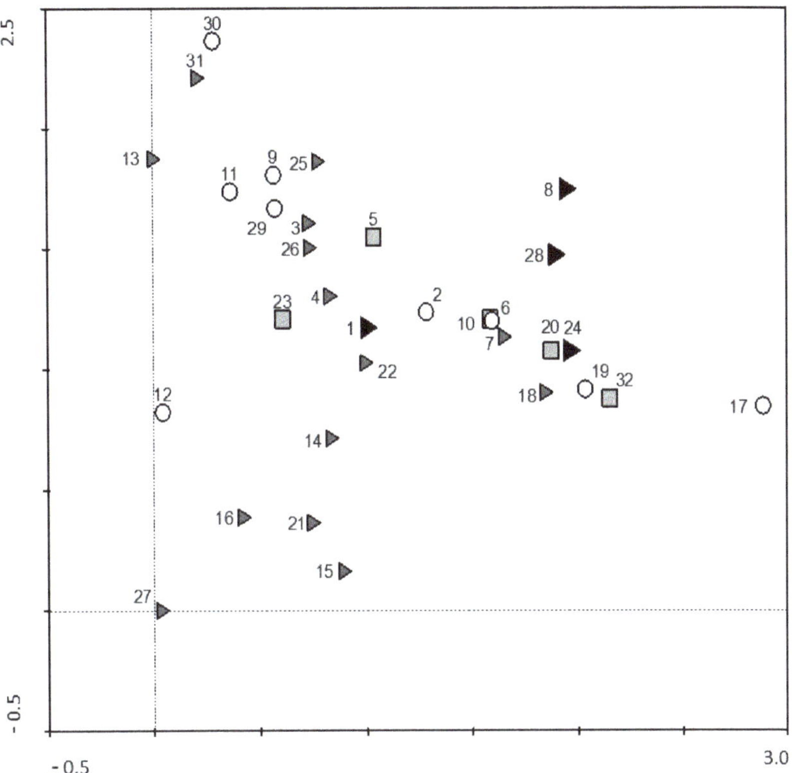

Figure 6. Detrended correspondence analysis ordination diagram showing the similarity of the macrophyte stands in different sites. Numbers from 1 to 32 indicate the location of the site with respect to the river Drava flow direction. Different symbols indicate different water bodies (white circles—ponds, light grey squares—oxbows, small dark grey right triangles—side channels, black right triangle—sample sites in the main channel of the river Drava.

Redundancy analysis revealed that the presence and abundance of *Elodea* affected the presence of native groups of macrophytes (Figure 7). When testing both species of *Elodea* separately, *E. nuttallii* explained 9% and *E. canadensis* 3%. However, when we tested sum of abundances of both species, this parameter explained 14% of macrophytes species parameters variability. Vectors representing the number and abundance of plant groups are in the opposite direction to *Elodea* vector. The distribution of the locations along this vector shows that the locations 10 and 17 are the most abundant with *Elodea*, as is also evident from Figure 3.

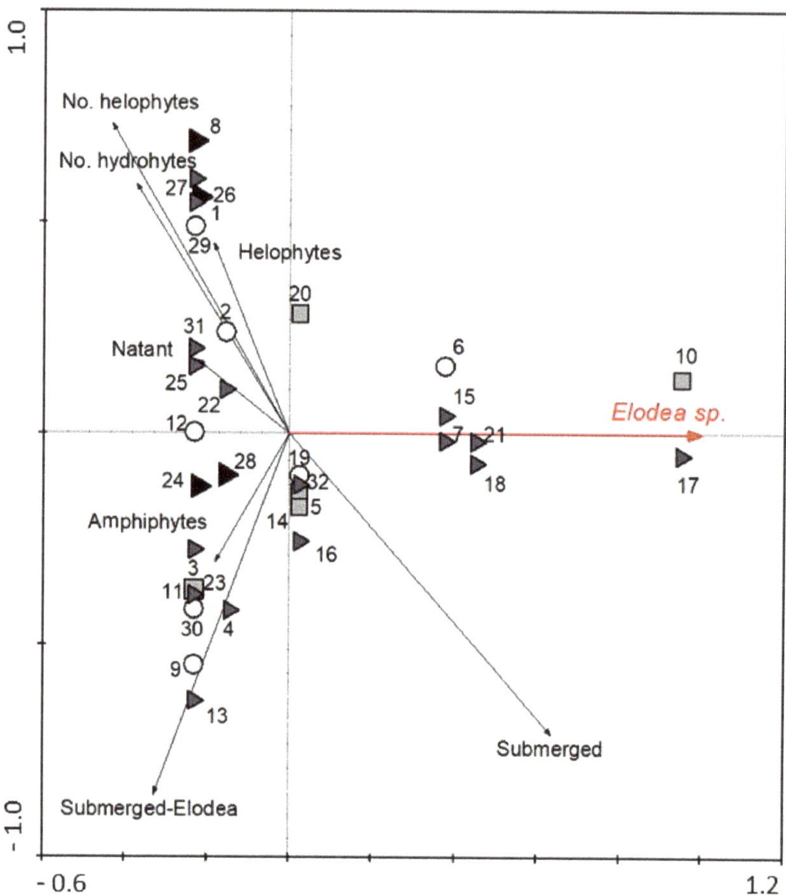

Figure 7. Redundancy analysis plot showing the relationship between the number of different ecological groups of macrophytes and relative abundance of their growth forms and *Elodea* species abundance. Abundance of submerged plants is represented by two parameters, including and excluding *Elodea* (-Elodea). Different symbols indicate different water bodies (white circles—ponds, light grey squares—oxbows, small dark grey right triangles—side channels, black right triangle—sites in the main channel of the river Drava.

Correlation analyses revealed no significant negative effect of the alien *Elodea* species to the native flora of the studied water bodies (Table 2). We calculated positive correlation between the abundance of *E. nuttallii* and temperature of the water, and the share of arable land in the catchment areas of the studied wetlands.

The abundance of *E. nuttallii* was negatively correlated with the sum of abundances of floating-leaved macrophytes (*Nymphaea alba* and *Spirodela polyrhiza*).

Average values for specific types of the studied waterbodies are listed in Table 3. *E. canadensis* was absent in lentic ecosystems but relatively abundant in side-channels, while *E. nuttallii* was present in all types but was dominant in ponds, where its average cover-abundance value was 48%. Despite this fact the species richness and diversity of native flora was not lower, but even higher in ponds than in larger oxbow lakes.

Table 2. Correlation coefficients (Kendall *tau*) between the abundance of *Elodea canadensis, E. nuttallii* and the sum of both species with diversity indices of native flora as well as with selected environmental parameters. Only significant ($p < 0.05$) correlations are shown. (* $p = 0.05$).

Variable	*E. canadensis*	*E. nuttallii*	*E. canadensis* **and** *nuttallii*
Number of native taxa	n.s.	n.s.	n.s.
Total abundance of plants	n.s.	0.2679	0.2669
Shannon–Wiener diversity index	n.s.	n.s.	n.s.
Concentration of O_2 [mg/L]	n.s.	0.2679	n.s.
Temperature of the water [°C]	n.s.	0.2614 *	n.s.
Cover of floating-leaved macrophytes	n.s.	−0.2782	n.s.
Land-use in the catchment	n.s.	0.2988	0.2805
Abundance of *Nymphaea alba*	n.s.	−0.2617	−0.3079
Abundance of *Spirodela polyrhiza*	n.s.	−0.2886	n.s.

Table 3. Average abundances (in %) of IAS *Elodea canadensis* and *E. nuttallii* in four types of waterbodies and average values of species-richness and diversity of native flora.

	Oxbows	Ponds	Side-Channels	River
E. canadensis	0	0	15.7	0.75
E. nuttallii	14.5	48	12.2	4.5
Nr. of native taxa	8	10.5	11.4	9.5
S–W diversity index	1.7	1.9	2	1.9

4. Discussion

Small waterbodies are often intact and unpolluted, and as such they present a refuge for species which have disappeared from larger, more disturbed, water bodies [13]. In the case of the Drava river, the sections upstream the studied area are degraded and converted into the chain of reservoirs for HPPs. The surveyed ecosystems occur within the active floodplain, which remained relatively intact in terms of morphological alterations. However, in different river-fed wetlands, the flood regimes affect macrophytes community traits and thus the structure and function of the wetlands [46]. The entire set of the studied waterbodies hosted 73 macrophyte species, which is a rather high number in comparison to river habitats, where in over 1000 reaches of 33 Slovenian rivers 87 species were recorded [47]. In similar studies within the Danube river corridor, Schmidt-Mumm and Janauer [15] recorded 78 species in 49 transects sampled in oxbows and side-channels in Austria, while Gyosheva et al. [17] recorded 112 species within a much larger set of 144 samples from Bulgaria.

The study of macrophytes in subtropical ponds revealed that pond size was positively related to richness of emergent and floating species, and the isolation of the pond negatively affected the richness of amphibious species, which was a consequence of their dispersal strategies [48], but our results do not confirm such relations. Emergent species in the studied waterbodies presented a great share of species, namely 43. This high helophyte diversity may be a consequence of relative naturalness of the adjacent parts of their catchment areas within the floodplain and supported by rich seed banks as shown in a case of small ponds [7].

Diverse and abundant stands of helophytes provide a protection for hydrophytes in the water since they act as their buffer zones. Despite this protection, native species could be endangered by alien *Elodea*, as both species of *Elodea* found in surveyed water bodies usually exhibit high growth rates with a high tolerance to a wide range of environmental conditions, low vulnerability to grazing and other stress factors, high distribution and reproduction potential [49]. Our results revealed no influence from both *Elodea* species on the floristic composition of aquatic vegetation (Figure 2), which is stronger in the case of *E. nuttalli* since its abundance is responsible for 9% of variability of macrophyte stands. Besides, *E. nuttallii* is also documented to replace *E. canadensis* from several waterbodies where it has established before the invasion of *E. nuttallii* [50,51]. The ecophysiological differences between both species explain the invasion success of *E. nuttallii* over *E. canadensis* [52]. Szabó et al. [53] found out that under more eutrophic conditions, *E. nuttallii* grows quicker and reaches the water surface sooner in comparison to *E. canadensis*. In addition, intensive branching outcompetes all other plants, including *E. canadensis*.

In our case, no evident impact on the native plant diversity was confirmed, neither in case of single *Elodea* species, nor when both species were present (Table 2). One of the reasons is that the hydromorphological characteristics of the majority of studied waterbodies has not been modified, except the ponds that are of anthropogene origin. This also explains the highest abundance of *E. nuttallii* in the ponds (Table 3). Otahelová et al. [14] report that new man-modified aquatic habitats have been successfully invaded by *E. nuttallii*. Mazej-Grudnik and Germ [54] report that *E. nuttallii* can cause severe problems in water bodies that are heavily modified due to human activity. The reason for lower competitive ability of *E. nuttallii* over *E. canadensis*, as well as other submerged species, is connectivity of these waterbodies with the main course of the Drava river [55] that floods the entire floodplain during the extreme events. Vukov et al. [16] report that both species are characteristic for aquatic habitats with lower levels of connectivity with main channels. There is also a difference between these two species in their preference to the reaction of the water, according to Ellenberg indicator values (EIV) for reaction [56]. *E. canadensis* has a relatively high EIV = 7 (out of 9) for water pH, while *E. nuttallii* is indifferent to pH (x). Since the catchment area above the studied section of the Drava river is mostly in the Central Alps and built mainly by non-carboniferous rocks, the sediments forming the floodplain of the Drava river and the river itself contain relatively low contents of basic cations and provide habitats with relatively low pH. This might explain lower abundance values of *E. canadensis* in comparison to *E. nuttallii* in these ecosystems and its absence from lentic habitats (Table 3).

We obtained positive correlation between the abundance of *E. nuttallii* and temperature of the water, thus further increasing in temperatures of these small waterbodies may favour the spread of *E. nuttalli*. Mazej Grudnik et al. [57] report that one can expect a more invasive character of *E. nuttallii* in the years with higher temperatures in January and March.

On the other hand, the abundance of *E. nuttallii* was negatively correlated with the sum of abundances of floating macrophytes (e.g., *Nymphaea alba* and *Spirodela polyrhiza*). However, it is not a case for all floating species, since a study with *Lemna* revealed the opposite result, especially under low nutrient concentrations [58]. *Elodea* is effective in using nutrients as phosphorus and nitrogen, which results in nutrient deficiency for other primary producers [59]. In our case it seems that nutrients were not a problem, since the conductivity was relatively high; therefore, better position regarding light conditions in floating-leaved macrophytes in comparison to submerged *Elodea* deprived this species. We presume that the spreading of floating species may help to suppress the spread of these two IAS in the waterbodies. Netten et al. [60] found out that the free-floating *Salvinia natans* in mesocosms benefited from increased temperature and increased nutrient concentrations and lowered the potential of submerged *E. nuttallii* to take advantage of such conditions. Their results also indicate that with global warming, invasive free-floating plants might become more successful and cause decline of submerged plants. The environment below floating plants is poorer with light [61]. Deliberate introduction of the pleustophyte *Spirodela polyrhiza*, which is a native species and distributed in waterbodies in the studied region, would be easy to imply, but it may also affect other submerged species so the use of such a measure for mitigation of the spread of these invasive species should be studied in advance. The study in Slovenian watercourses revealed that the abundance of *E. canadensis* is negatively related to abundance of *M. spicatum* [62], which is also highly invasive in USA [63]. This effect was only partly confirmed in the present study. According to this knowledge, we strongly discourage the removal of the stands of *M. spicatum* as well as any other native macrophyte.

5. Conclusions

Surveyed water bodies along the river Drava harboured a high number of species. Against expectations, the presence of alien *E. nuttallii* and *E. canadensis* exerted no effect on presence and abundance of hydrophytes, possibly due to water level fluctuations in these water bodies. *E. nuttallii* reached the highest abundance in ponds, which are the only group of anthropogenic ecosystems. We can conclude that maintenance of good ecological status of waterbodies, including their morphology, is among the most important measures to prevent the spreading of invasive species.

Different hydrological dynamics, as one of the consequences of the climate changes, has led to increased frequency of the floods as well as droughts. Small water bodies and wetlands, respectively, found within the floodplains can mitigate both types of events since they enable substantially longer water retention than do regulated rivers. The maintenance and hydrological connectivity of these waterbodies in sufficient extent could not only contribute to higher species diversity but could also reduce the flood-waves and supply the water for the baseflow at lower water levels due to their retention capacity.

Author Contributions: Conceptualization, A.G. and M.G.; methodology, A.G. and M.G.; validation, M.G., A.G. and U.K.; formal analysis, A.G. and I.Z.; investigation, M.G., A.G. and U.K.; data curation, A.G., M.G. and I.Z.; writing—original draft preparation, A.G. and I.Z.; writing—review and editing, M.G., A.G. and I.Z.; visualization, A.G. and I.Z.; supervision, A.G.; funding acquisition, A.G. and I.Z. All authors have read and agreed to the published version of the manuscript.

Funding: This research was funded by the Slovenian Research Agency, within the core research funding Nr. P1-0212, 'Biology of Plants'.

Institutional Review Board Statement: Not applicable.

Informed Consent Statement: Not applicable.

Data Availability Statement: Data can be provided upon reasonable request.

Acknowledgments: Authors thank Matej Holcar for the creation of Figure 1.

Conflicts of Interest: The authors declare no conflict of interest.

References

1. Hynes, H.B.N. *The Ecology of Running Waters*; University of Toronto Press: Toronto, ON, Canada, 1970.
2. Zeiringer, B.; Seliger, C.; Greimel, F.; Schmutz, S. River Hydrology, Flow Alteration, and Environmental Flow. In *Riverine Ecosystem Management*; Springer: Cham, Switzerland, 2018; pp. 67–89.
3. Tonkin, J.D.; Altermatt, F.; Finn, D.S.; Heino, J.; Olden, J.D.; Pauls, S.U.; Lytle, D.A. The Role of Dispersal in River Network Metacommunities: Patterns, Processes, and Pathways. *Freshw. Biol.* **2018**, *63*, 141–163. [CrossRef]
4. Padial, A.A.; Ceschin, F.; Declerck, S.A.J.; de Meester, L.; Bonecker, C.C.; Lansac-Tôha, F.A.; Rodrigues, L.; Rodrigues, L.C.; Train, S.; Velho, L.F.M.; et al. Dispersal Ability Determines the Role of Environmental, Spatial and Temporal Drivers of Metacommunity Structure. *PLoS ONE* **2014**, *9*, e111227. [CrossRef] [PubMed]
5. Molnár, Z. Types and Characteristics of the Oxbow-Lakes in Lower-Tisza-Valley—Classification from Landscape Planning Perspective. *Landscape Environ.* **2013**, *7*, 19–25.
6. Williams, P.; Whitfield, M.; Biggs, J.; Bray, S.; Fox, G.; Nicolet, P.; Sear, D. Comparative Biodiversity of Rivers, Streams, Ditches and Ponds in an Agricultural Landscape in Southern England. *Biol. Conserv.* **2004**, *115*, 329–341. [CrossRef]
7. Shutoh, K.; Yamanouchi, T.; Kato, S.; Yamagishi, H.; Ueno, Y.; Hiramatsu, S.; Nishihiro, J.; Shiga, T. The Aquatic Macrophyte Flora of a Small Pond Revealing High Species Richness in the Aomori Prefecture, Japan. *J. Asia Pac. Biodivers.* **2019**, *12*, 448–458. [CrossRef]
8. Kubiak, A.P.; Krawczyk, R. Diversity of Macrophytes in Riverine Aquatic Habitats: Comparing Active River Channel and Its Cut-Offs. *Ann. UMCS Biol.* **2014**, *69*, 49–57. [CrossRef]
9. Waldon-Rudzionek, B. Is the Flora of Oxbow Lakes Different from That of Fishponds? A Comparison of Two Types of Water Reservoirs in the Noteć River Valley and Bydgoszcz Canal Valley (NW Poland). *Ecol. Quest.* **2017**, *25*, 27. [CrossRef]
10. Panzeca, P.; Troia, A.; Madonia, P. Aquatic Macrophytes Occurrence in Mediterranean Farm Ponds: Preliminary Investigations in North-Western Sicily (Italy). *Plants* **2021**, *10*, 1292. [CrossRef] [PubMed]
11. Obolewski, K. Biodiversity of Macroinvertebrates in Oxbow-Lakes of Early Glacial River Basins in Northern Poland. In *Ecosystems Biodiversity*; InTech: Rijeka, Croatia, 2011.

12. Schneiders, A.; Verheyen, R. A Concept of Integrated Water Management Illustrated for Flanders (Belgium). *Ecosyst. Health* **1998**, *4*, 256–263. [CrossRef]

13. Biggs, J.; von Fumetti, S.; Kelly-Quinn, M. The Importance of Small Waterbodies for Biodiversity and Ecosystem Services: Implications for Policy Makers. *Hydrobiologia* **2017**, *793*, 3–39. [CrossRef]

14. Oťaheľová, H.; Valachovič, M.; Hrivnák, R. The Impact of Environmental Factors on the Distribution Pattern of Aquatic Plants along the Danube River Corridor (Slovakia). *Limnologica* **2007**, *37*, 290–302. [CrossRef]

15. Schmidt-Mumm, U.; Janauer, G.A. Macrophyte Assemblages in the Aquatic-Terrestrial Transitional Zone of Oxbow Lakes in the Danube Floodplain (Austria). *Folia Geobot.* **2016**, *51*, 251–266. [CrossRef]

16. Vukov, D.; Ilić, M.; Ćuk, M.; Igić, R.; Janauer, G. The Relationship between Habitat Factors and Aquatic Macrophyte Assemblages in the Danube River in Serbia. *Arch. Biol. Sci.* **2017**, *69*, 427–437. [CrossRef]

17. Gyosheva, B.; Kalchev, R.; Beshkova, M.; Valchev, V. Relationships between Macrophyte Species, Their Life Forms and Environmental Factors in Floodplain Water Bodies from the Bulgarian Danube River Basin. *Ecohydrol. Hydrobiol.* **2020**, *20*, 123–133. [CrossRef]

18. Akasaka, M.; Higuchi, S.; Takamura, N. Landscape- and Local-Scale Actions Are Essential to Conserve Regional Macrophyte Biodiversity. *Front. Plant Sci.* **2018**, *9*, 599. [CrossRef] [PubMed]

19. Bornette, G.; Puijalon, S. Macrophytes: Ecology of Aquatic Plants. In *Encyclopedia of Life Sciences (ELS)*; Wiley: Chichester, UK, 2009. [CrossRef]

20. Preiner, S.; Dai, Y.; Pucher, M.; Reitsema, R.E.; Schoelynck, J.; Meire, P.; Hein, T. Effects of Macrophytes on Ecosystem Metabolism and Net Nutrient Uptake in a Groundwater Fed Lowland River. *Sci. Total Environ.* **2020**, *721*, 137620. [CrossRef] [PubMed]

21. Glińska-Lewczuk, K.; Burandt, P. Effect of River Straightening on the Hydrochemical Properties of Floodplain Lakes: Observations from the Łyna and Drwęca Rivers, N Poland. *Ecol. Eng.* **2011**, *37*, 786–795. [CrossRef]

22. Ambrožič, Š.; Gaberščik, A.; Vrezec, A.; Germ, M. Hydrophyte Community Structure Affects the Presence and Abundance of the Water Beetle Family Dytiscidae in Water Bodies along the Drava River. *Ecol. Eng.* **2018**, *120*, 397–404. [CrossRef]

23. Brysiewicz, A.; Czerniejewski, P.; Bonisławska, M. Effect of Diverse Abiotic Conditions on the Structure and Biodiversity of Ichthyofauna in Small, Natural Water Bodies Located on Agricultural Lands. *Water* **2020**, *12*, 2674. [CrossRef]

24. Baattrup-Pedersen, A.; Riis, T. Macrophyte Diversity and Composition in Relation to Substratum Characteristics in Regulated and Unregulated Danish Streams. *Freshw. Biol.* **1999**, *42*, 375–385. [CrossRef]

25. Hansen, J.P.; Wikström, S.A.; Axemar, H.; Kautsky, L. Distribution Differences and Active Habitat Choices of Invertebrates between Macrophytes of Different Morphological Complexity. *Aquat. Ecol.* **2011**, *45*, 11–22. [CrossRef]

26. Haslam, S.M. *River Plants of Western Europe: The Macrophytic Vegetation of Watercourses of the European Economic Community*; Cambridge University Press: Cambridge, UK, 1987; ISBN 0-521-26427-8.

27. Schneider, S.; Melzer, A. The Trophic Index of Macrophytes (TIM)—A New Tool for Indicating the Trophic State of Running Waters. *Int. Rev. Hydrobiol.* **2003**, *88*, 49–67. [CrossRef]

28. Szpakowska, B.; Świerk, D.; Pajchrowska, M.; Gołdyn, R. Verifying the Usefulness of Macrophytes as an Indicator of the Status of Small Waterbodies. *Sci. Total Environ.* **2021**, *798*, 149279. [CrossRef]

29. Kuhar, U.; Germ, M.; Gaberščik, A.; Urbanič, G. Development of a River Macrophyte Index (RMI) for Assessing River Ecological Status. *Limnologica* **2011**, *41*, 235–243. [CrossRef]

30. van Geest, G.J.; Coops, H.; Roijackers, R.M.M.; Buijse, A.D.; Scheffer, M. Succession of Aquatic Vegetation Driven by Reduced Water-Level Fluctuations in Floodplain Lakes. *J. Appl. Ecol.* **2005**, *42*, 251–260. [CrossRef]

31. Lu, J.; Bunn, S.E.; Burford, M.A. Nutrient Release and Uptake by Littoral Macrophytes during Water Level Fluctuations. *Sci. Total Environ.* **2018**, *622–623*, 29–40. [CrossRef] [PubMed]

32. Bogut, I.; Vidaković, J.; Palijan, G.; Čerba, D. Benthic Macroinvertebrates Associated with Four Species of Macrophytes. *Biologia* **2007**, *62*, 600–606. [CrossRef]

33. Pyšek, P.; Richardson, D.M. Invasive Species, Environmental Change and Management, and Health. *Annu. Rev. Environ. Resour.* **2010**, *35*, 25–55. [CrossRef]

34. Janauer, G.; Exler, N.; Anačkov, G.; Barta, V.; Berczik, Á.; Boža, P.; Dinka, M.; Georgiev, V.; Germ, M.; Holcar, M.; et al. Distribution of the Macrophyte Communities in the Danube Reflects River Serial Discontinuity. *Water* **2021**, *13*, 918. [CrossRef]

35. Ricciardi, A.; Hoopes, M.F.; Marchetti, M.P.; Lockwood, J.L. Progress toward Understanding the Ecological Impacts of Nonnative Species. *Ecol. Monogr.* **2013**, *83*, 263–282. [CrossRef]

36. Katsanevakis, S.; Wallentinus, I.; Zenetos, A.; Leppäkoski, E.; Çinar, M.E.; Oztürk, B.; Grabowski, M.; Golani, D.; Cardoso, A.C. Impacts of Invasive Alien Marine Species on Ecosystem Services and Biodiversity: A Pan-European Review. *Aquat. Invasions* **2014**, *9*, 391–423. [CrossRef]

37. Vimercati, G.; Probert, A.F.; Volery, L.; Bernardo-Madrid, R.; Bertolino, S.; Céspedes, V.; Essl, F.; Evans, T.; Gallardo, B.; Gallien, L.; et al. The EICAT+ Framework Enables Classification of Positive Impacts of Alien Taxa on Native Biodiversity. *PLoS Biol.* **2022**, *20*, e3001729. [CrossRef] [PubMed]

38. Kohler, A.; Janauer, G.A. Zur Methodik Der Untersuchung von Aquatischen Makrophyten in Flie_gewassern. In *Handbuch Angewandte Limnologie*; Steinberg, C.H., Bernhardt, H., Klapper, H., Eds.; Ecomed: Landsberg am Lech, Germany, 1995; pp. 3–22.

39. Euro+Med PlantBase—The Information Resource for Euro-Mediterranean Plant Diversity. Available online: http://ww2.bgbm. org/EuroPlusMed/ (accessed on 1 September 2022).

40. Schaumburg, J.; Schranz, C.; Foerster, J.; Gutowski, A.; Hofmann, G.; Meilinger, P.; Schneider, S.; Schmedtje, U. Ecological Classification of Macrophytes and Phytobenthos for Rivers in Germany According to the Water Framework Directive. *Limnologica* **2004**, *34*, 283–301. [CrossRef]
41. Braun-Blanquet, J. *Pflanzensoziologie, Grundzüge Der Vegetationskunde*, 3rd ed.; Springer: Berlin, Germany, 1964.
42. Germ, M.; Janež, V.; Gaberščik, A.; Zelnik, I. Diversity of Macrophytes and Environmental Assessment of the Ljubljanica River (Slovenia). *Diversity* **2021**, *13*, 278. [CrossRef]
43. Hammer, Ø.; Harper, D.A.T.; Ryan, P.D. PAST: Paleontological Statistics Software Package for Education and Data Analysis. *Palaeontol. Electron.* **2001**, *4*, 1–9.
44. ter Braak, C.J.F.; Verdonschot, P.F.M. Canonical Correspondence Analysis and Related Multivariate Methods in Aquatic Ecology. *Aquat. Sci.* **1995**, *57*, 255–289. [CrossRef]
45. ter Braak, C.J.F.; Smilauer, P. *CANOCO Reference Manual and CanoDraw for Windows User's Guide: Software for Canonical Community Ordination (Version 4.5)*; Canoco: Ithaca, NY, USA, 2002.
46. Huang, Y.; Chen, X.-S.; Li, F.; Hou, Z.-Y.; Li, X.; Zeng, J.; Deng, Z.-M.; Zou, Y.-A.; Xie, Y.-H. Community Trait Responses of Three Dominant Macrophytes to Variations in Flooding During 2011–2019 in a Yangtze River-Connected Floodplain Wetland (Dongting Lake, China). *Front. Plant Sci.* **2021**, *12*, 604677. [CrossRef] [PubMed]
47. Zelnik, I.; Kuhar, U.; Holcar, M.; Germ, M.; Gaberščik, A. Distribution of Vascular Plant Communities in Slovenian Watercourses. *Water* **2021**, *13*, 1071. [CrossRef]
48. Pereira, K.M.; Hefler, S.M.; Trentin, G.; Rolon, A.S. Influences of Landscape and Climatic Factors on Aquatic Macrophyte Richness and Composition in Ponds. *Flora* **2021**, *279*, 151811. [CrossRef]
49. Zehnsdorf, A.; Hussner, A.; Eismann, F.; Rönicke, H.; Melzer, A. Management Options of Invasive Elodea Nuttallii and Elodea Canadensis. *Limnologica* **2015**, *51*, 110–117. [CrossRef]
50. Barrat-Segretain, M.-H. Invasive Species in the Rhône River Floodplain (France): Replacement of Elodea Canadensis Michaux by E. Nuttallii St. John in Two Former River Channels. *Arch. Hydrobiol.* **2001**, *152*, 237–251. [CrossRef]
51. Barrat-Segretain, M.; Elger, A. Experiments on Growth Interactions between Two Invasive Macrophyte Species. *J. Veg. Sci.* **2004**, *15*, 109–114. [CrossRef]
52. Szabó, S.; Peeters, E.T.H.M.; Borics, G.; Veres, S.; Nagy, P.T.; Lukács, B.A. The Ecophysiological Response of Two Invasive Submerged Plants to Light and Nitrogen. *Front. Plant Sci.* **2020**, *10*, 01747. [CrossRef]
53. Szabó, S.; Peeters, E.T.H.M.; Várbíró, G.; Borics, G.; Lukács, B.A. Phenotypic Plasticity as a Clue for Invasion Success of the Submerged Aquatic Plant Elodea Nuttallii. *Plant Biol.* **2019**, *21*, 54–63. [CrossRef]
54. Mazej Grudnik, Z.; Germ, M. Spatial Pattern of Native Species Myriophyllum Spicatum and Invasive Alien Species Elodea Nuttallii after Introduction of the Latter One into the Drava River (Slovenia). *Biologia* **2013**, *68*, 202–209. [CrossRef]
55. Barrat-Segretain, M.-H.; Elger, A.; Sagnes, P.; Puijalon, S. Comparison of Three Life-History Traits of Invasive Elodea Canadensis Michx. and Elodea Nuttallii (Planch.) H. St. John. *Aquat. Bot.* **2002**, *74*, 299–313. [CrossRef]
56. Ellenberg, H.; Weber, H.; Düll, R.; Wirth, V.; Werner, W.; Paulißen, D. *Zeigerwertevon Pflanzen in Mitteleuropa*; Verlag Erich Goltze: Göttingen, Germany, 1992.
57. Grudnik, Z.M.; Jelenko, I.; Germ, M. Influence of Abiotic Factors on Invasive Behaviour of Alien Species Elodea Nuttallii in the Drava River (Slovenia). *Ann. Limnol. Int. J. Limnol.* **2014**, *50*, 1–8. [CrossRef]
58. Szabo, S.; Scheffer, M.; Roijackers, R.; Waluto, B.; Braun, M.; Nagy, P.T.; Borics, G.; Zambrano, L. Strong Growth Limitation of a Floating Plant (Lemna Gibba) by the Submerged Macrophyte (Elodea Nuttallii) under Laboratory Conditions. *Freshw. Biol.* **2010**, *55*, 681–690. [CrossRef]
59. van Donk, E.; Gulati, R.D.; Iedema, A.; Meulemans, J.T. Macrophyte-Related Shifts in the Nitrogen and Phosphorus Contents of the Different Trophic Levels in a Biomanipulated Shallow Lake. In *Nutrient Dynamics and Retention in Land/Water Ecotones of Lowland, Temperate Lakes and Rivers*; Springer: Dordrecht, The Netherland, 1993; pp. 19–26. [CrossRef]
60. Netten, J.J.C.A.; Nes, E.H.; Scheffer, M.; Roijackers, R.M.M. Effect of Temperature and Nutrients on the Competition between Free-Floating Salvinia Natans and Submerged Elodea Nuttallii in Mesocosms. *Fundament. Appl. Limnol.* **2010**, *177*, 125–132. [CrossRef]
61. Klančnik, K.; Iskra, I.; Gradinjan, D.; Gaberščik, A. The Quality and Quantity of Light in the Water Column Are Altered by the Optical Properties of Natant Plant Species. *Hydrobiologia* **2018**, *812*, 203–212. [CrossRef]
62. Kuhar, U.; Germ, M.; Gaberščik, A. Habitat Characteristics of an Alien Species Elodea Canadensis in Slovenian Watercourses. *Hydrobiologia* **2010**, *656*, 205–212. [CrossRef]
63. Hoff, H.K.; Thum, R.A. Hybridization and Invasiveness in Eurasian Watermilfoil (*Myriophyllum spicatum*): Is Prioritizing Hybrids in Management Justified? *Invasive Plant Sci. Manag.* **2022**, *15*, 3–8. [CrossRef]

Article

Effects of Sediment Types on the Distribution and Diversity of Plant Communities in the Poyang Lake Wetlands

Jie Li [1,†], Yizhen Liu [1,2,†], Ying Liu [3], Huicai Guo [4], Gang Chen [1], Zhuoting Fu [5], Yvying Fu [1] and Gang Ge [1,2,*]

[1] School of Life Science, Nanchang University, Nanchang 330031, China; leecan66@163.com (J.L.); liuyizhen@ncu.edu.cn (Y.L.); 41193266@163.com (G.C.); fyyncu@163.com (Y.F.)
[2] Key Laboratory of Poyang Lake Environment and Resource Utilization, Ministry of Education, Nanchang University, Nanchang 330031, China
[3] Institute of Life Science, Nanchang University, Nanchang 330031, China; yingliu@ncu.edu.cn
[4] Jiangxi Poyang Lake Nanji Wetland National Natural Reserve Authority, Nanchang 330100, China; guohuicai2008@126.com
[5] Queen Mary School, Nanchang University, Nanchang 330031, China; fu17922022@163.com
* Correspondence: econcu@163.com
† These authors contributed equally to this work.

Abstract: At small scales, sedimentary deposition types mediate hydrological changes to drive wetland vegetation distribution patterns and species diversity. To examine the effects of sediment types on the distribution and diversity of plant communities in a wetland region, 150 quadrats were investigated (elevation range of 10.5–12.5 m) in the lake basin areas of Poyang Lake. We divided the surface soil into three sediment types (lacustrine sediments, fluvio-lacustrine sediments, and fluvial sediments), and then compared and analyzed the distribution and species diversity of the wetland plants among them. The results revealed the following findings: (i) within this elevation range, *Carex cinerascens*, *Carex cinerascens–Polygonum criopolitanum*, *Polygonum criopolitanum*, and *Phalaris arundinacea* communities exist; (ii) from lacustrine sediments to fluvial sediments, the distribution of plant communities showed a transition trend—with the *Carex cinerascens* and *Phalaris arundinacea* communities shifting into the *Polygonum criopolitanum* community; (iii) detrended correspondence analysis and redundancy analysis demonstrated that the soil particle composition and flood duration in 2017 generated a differential wetland plant distribution under the conditions of three sediment types along the littoral zones of Poyang Lake; and (iv) the plant communities on the lacustrine sediments had a higher species diversity than those established on the fluvio-lacustrine sediments and fluvial sediments.

Keywords: sediment types; wetland plant; distribution; diversity; Poyang Lake

Citation: Li, J.; Liu, Y.; Liu, Y.; Guo, H.; Chen, G.; Fu, Z.; Fu, Y.; Ge, G. Effects of Sediment Types on the Distribution and Diversity of Plant Communities in the Poyang Lake Wetlands. *Diversity* **2022**, *14*, 491. https://doi.org/10.3390/d14060491

Academic Editor: Mario A. Pagnotta

Received: 10 May 2022
Accepted: 14 June 2022
Published: 16 June 2022

Publisher's Note: MDPI stays neutral with regard to jurisdictional claims in published maps and institutional affiliations.

1. Introduction

In a wetland ecosystem, examples of which include the littoral zones of lakes or the riparian wetlands along rivers, sediment properties play a key role in the growth of plants and the distribution of vegetation [1–3]. Sedimentation has multiple effects on the germination, distribution, and diversity of plant communities; this is mainly due to the physical and chemical characteristics of sediments [4,5]. Surface sediments accumulate nutrient and soil particles that have been transported via water flow from upland areas [6], and determine the soil conditions where plants thrive [7]. Currently, most of the research is focused on the nutrient retention effects of topsoil on plants. The nitrogen [8,9], phosphorus [10], and potassium [11] content in sediments are assumed to affect the distribution and diversity of wetland plant communities by governing plant growth and reproduction [12,13]. Furthermore, the differing textures, structures, and performance of soil types are closely related to the different soil grain-size compositions among sediment types [14]. These physical properties of soil have significant impacts on the moisture retention capacity of sediments [15]. The soil water content is strongly considered to be a vital element in the distribution patterns and specific assemblage of wetland plant communities [16]. However, there appears to be a lack of knowledge about the effects of sediment types with differing grain-size compositions on plant communities and species composition.

Poyang Lake is the largest freshwater lake in China and is also a wetland ecosystem of international conservation significance [17,18]. Studies of Poyang Lake are significant to the management of the lake wetlands in the middle and lower reaches of the Yangtze River, and even to those of freshwater lake wetlands around the world. As a river-connected lake, one of the characteristics of Poyang Lake is its large seasonal water level fluctuation, which has vital impacts on the transport and distribution of sediment. When the lake enters a period of low flow (November–March of the next year), the sediment settles and provides a place for wetland plants to grow [19]. The regular alternation of water and land phases in Poyang Lake creates a specific hydrological environment for wetland plants, forming zonally-distributed plant communities [20]. However, some researchers have indicated that the wetland plant communities here are distributed in patches of varying sizes instead of in regular zonations at small scales [21,22]. In a particular elevation range with similar hydrological characteristics, sediment types with differing soil textures determine the growth of vegetation. The role played by the sediment types in the distribution and diversity of wetland plant communities is crucial [23]. However, there is little research on this topic [13,24], and the existing research lacks in-depth data mining and analysis.

Accordingly, using the Poyang Lake wetlands as our research area, in this paper, we investigated and analyzed the impact of sediment types on the wetland plant community distribution and diversity along the littoral wetlands of Poyang Lake. The elevation range was 10.5–12.5 m, the elevation at which the lake and land conservation in pace with the transition of high and low water periods was the highest. We posited and tested three hypotheses: (i) under the joint influence of hydrological and topographic factors, diverse sedimentary deposition types might exist in the shoaly lands of Poyang Lake; (ii) wetland plant communities have various reactions to sedimentary deposition types; and (iii) within sedimentary deposition types, divergent diversity characteristics might exist among plant communities. This study not only broadens our perspective on the formation mechanisms determining the distribution patterns and biodiversity in wetland ecosystems, but also has theoretical value and is of practical significance for improving wetland management and promoting its sustainable development.

2. Materials and Methods

2.1. Study Area

Poyang Lake is located on the south bank of the middle and lower reaches of the Yangtze River. Its geographical location is 115°9′–116°46′ E and 28°11′–29°51′ N. It is a typical subtropical monsoon climate: the mean annual temperature is 17.6 °C and the average annual precipitation is 1528 mm [25]. The entire lake is 173 km long from north to south, spanning 74 km at its widest areas and just 2.8 km at its narrowest part [26]. Under the dual influence of five rivers—Ganjiang River, Fu River, Xin River, Rao River, and Xiu River—around Poyang Lake and the backwater effect of the Yangtze River, the annual and interannual water levels of Poyang Lake experience significant fluctuations [27]. These unique environmental conditions help to develop the species-rich wetland floras found in the Poyang Lake wetlands.

2.2. Field Investigation

The field investigation was conducted in the winter (December 2020 to January 2021), when Poyang Lake was in a low water-level period. A combination of transect and quadrat-based methods was used for sampling. A total of 12 transects were established along the littoral zones (elevation range of 10.5–12.5 m), where the variation in the plant community is obvious (Table S1). Along each transect, 10–15 quadrats were taken at about 100 m intervals depending on the length of each transect and vegetation zonation (Figure 1). A total of 150 quadrats that were 1 × 1 m in size were established. All of the species in each quadrat were recorded, and the coverage abundance and aboveground biomass for each species were measured [28]. The plant species coverage was estimated using a visual method. The abundance of each plant species was assessed using the Drude scale [28]. The aboveground biomass of each species was harvested and weighed in the field. Moreover, the locations of the quadrats were recorded using the global positioning system (GPS), and animal activities were recorded as well. The species were identified, and the nomenclature of the plant communities was standardized according to the Flora of China [29]. After the quadrats had been investigated, a soil sample (approximately 500 g) was collected from each quadrat at a depth of 0~10 cm from the topsoil using a five-spot-sampling method and mixed into one sample. The selected soil depth represented the amount deposited in the past 50 years, with an average sedimentation rate of 2.2 mm a^{-1} estimated using ^{137}Cs and ^{210}Pb methods [30]. All soil samples were put into polyethylene-sealed bags, and the compositions of their soil particles determined in our laboratory.

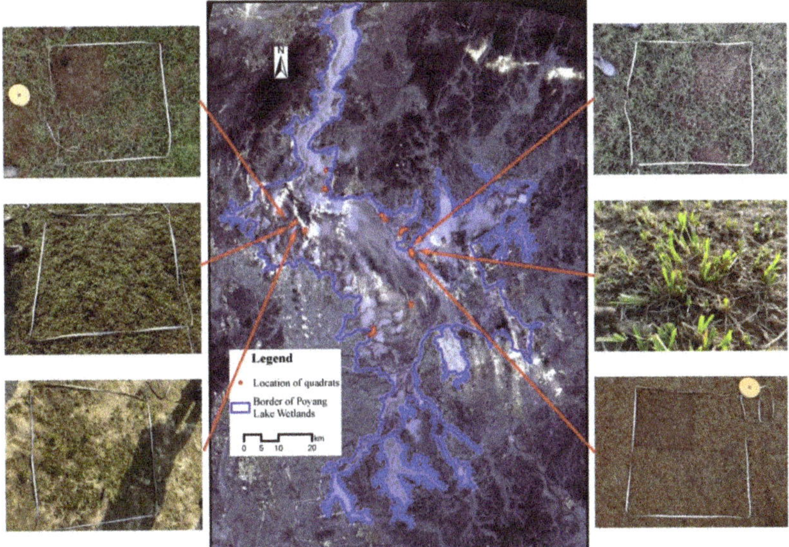

Figure 1. Locations of quadrats in our survey.

2.3. Environmental Data Collection

The soil samples were preprocessed before the particle composition was determined. Preprocessing included removing any impurities and foreign matter from the sample and air-drying in a cool place. Then, the processed samples were pretreated with the addition of 10% H_2O_2 and HCl to remove any organic matter or inorganic carbon matter. The soil particle composition for each soil sample was analyzed and determined using a laser diffraction particle size analyzer (LS 13320, Beckman, Brea, CA, USA), with a measuring range of 0–2000 μm. The proportions of sand (>50 μm), silt (2–50 μm), and clay (<2 μm) were divided and calculated according to the American soil texture classification system (USAD) [31].

Furthermore, we calculated eleven ecohydrological parameters for every quadrat based on a digital elevation model (DEM) of Poyang Lake. The spatial resolution of the DEM was 5 m, and the contour interval was 0.2 m. A difference value of DEM > 0 indicated that the quadrat was flooded, and the opposite result indicated that it was not flooded. We counted the annual flood duration from 2013 to 2020 based on the days with a value > 0. Additionally, the mean and longest annual flood durations over ten years and the mean annual flood durations of the last five years could be counted. The elevation of each quadrat was also extracted from the DEM.

2.4. Data Analysis

We categorized the life forms of plant species based on the Raunkiaer system [32] and determined the sediment-tolerant types for plants according to the ability of the plants to adapt to sedimentation [33]. To classify and identify the types of wetland plant communities in the littoral zones of Poyang Lake, we applied the agglomerative hierarchical cluster using Ward's linkage method. The Jaccard index was then used to indicate the dissimilarity between quadrats for cluster analysis. The validity of the resulting clusters was improved by approximating natural clustering using the silhouette algorithm [34]. We assessed how the dissimilarity changed within and between the assigned classification clusters using a detrended correspondence analysis (DCA) [35]. The DCA parameters were the default settings: 4 ordination axes and 26 segments. The multi-response permutation procedure (MRPP) and the analysis of similarities (ANOSIM) with Bray–Curtis distance were implemented to test the distinguishability of the accepted classification results. The chi-square test was used to analyze the association between sediment types and plant community clusters.

Next, we used a cluster and principal component analysis of the grain-size compositions and average particle sizes to categorize the sediments into three types: lacustrine sediments (44 quadrats), fluvio-lacustrine sediments (81 quadrats), and fluvial sediments (25 quadrats) (Figure 2). Additionally, we applied the chi-square test to analyze the association between the sediment types and plant community clusters.

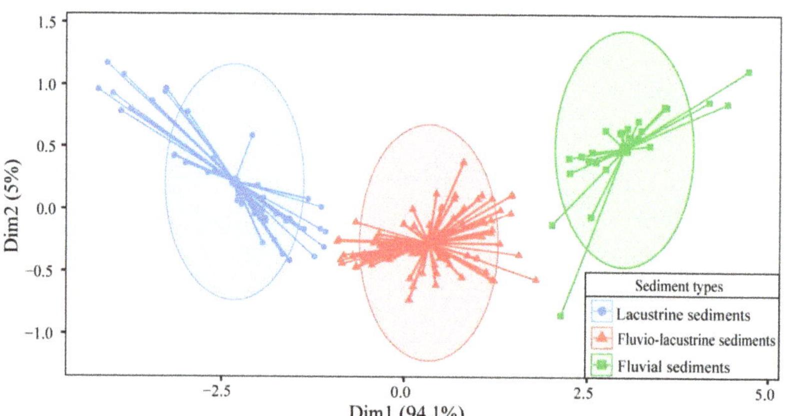

Figure 2. Sediment types were assigned using cluster and principal component analysis.

To explore the major environmental gradients determining the distribution of the plant communities, we used constrained ordination. The DCA results revealed that the gradient length of the first axis (2.31) was less than 3 SD (standard deviation units), indicating that redundancy analysis (RDA) could be used. The environmental factors included ecohydrological parameters, elevation, and the proportions of sand, silt, and clay. Forward selection was conducted to reduce the collinearity of the environmental variables.

We analyzed the effects of the sedimentary deposition types on the species diversity of plant communities. The species diversity metrics included species richness (SR) and the Shannon index (H) [36,37]. H is one of the most widely used measures of diversity and is based on information theory. H can be calculated as follows:

$$H = -\sum p_i \ln p_i \tag{1}$$

where SR is the number of species, and p_i is the relative abundance of a species in a quadrat, namely $p_i = \frac{n_i}{N}$ for which n_i is the abundance of a species in a quadrat and N is the summed abundance of all species in that quadrat.

Next, we used the non-parametric Kruskal–Wallis test to compare and analyze the differences in species diversity among the different plant communities and sediment types.

We conducted all of the data analyses and constructed the graphs using the R v4.1.0 platform [38]. The following R packages were used: vegan v2.5-7 [39], adiv v 2.1.1 [40], pgirmess v1.7.0 [41], and ggplot2 [42]. Vegan v2.6-2 was used to conduct the plant community classification and ordination; adiv v 2.1.1 was used to calculate the diversity indexes; pgirmess v1.7.0 was used for the multiple comparison tests; and ggplot2 was used to visualize the results.

3. Results

3.1. Species Compositions of Plant Communities in the Wetlands

According to the field investigation, sixteen plant species were recorded in the wetland plant communities and were found to belong to seven families (Table 1). Most of those species were from Polygonaceae, with the least species coming from Ranunculaceae and Rosaceae. Among those plants, both *Carex cinerascens* and *Polygonum criopolitanum* were the most common and widely distributed and thus the most dominant. In terms of life forms, there were six therophyte species, six cryptophyte species, one hemicryptophyte species, and three chamaephyte species. Moreover, most of the species showed a positive response and were able to adapt to the sedimentation.

Table 1. Sample survey of wetland plant species.

S/N	Species	Family	Life Form	Sediment-Tolerant Types
1	*Cardamine lyrata*	Brassicaceae	Cryptophyte	Sediment-dependence
2	*Cardamine impatiens*	Brassicaceae	Therophyte	Sediment-dependence
3	*Heleocharis valleculosa*	Cyperaceae	Cryptophyte	Sediment-dependence
4	*Carex cinerascens*	Cyperaceae	Cryptophyte	Sediment-tolerance
5	*Phalaris arundinacea*	Gramineae	Cryptophyte	Sediment-tolerance
6	*Potentilla limprichtii*	Rosaceae	Chamaephyte	Sediment-tolerance
7	*Polygonum pubescens*	Polygonacae	Therophyte	Sediment-tolerance
8	*Polygonum criopolitanum*	Polygonacae	Therophyte	Sediment-tolerance
9	*Rumex acetosa*	Polygonacae	Chamaephyte	Sediment-tolerance
10	*Rumex acetosella*	Polygonacae	Chamaephyte	Sediment-tolerance
11	*Artemisia selengensis*	Asteraceae	Cryptophyte	Sediment-dependence
12	*Lapsana apogonoides*	Asteraceae	Therophyte	Sediment-sensitivity
13	*Gnaphalium affine*	Asteraceae	Therophyte	Sediment-sensitivity
14	*Hemarthria altissima*	Gramineae	Hemicryptophyta	Sediment-sensitivity
15	*Kalimeris indica*	Asteraceae	Cryptophyte	Sediment-dependence
16	*Ranunculus polii*	Ranunculaceae	Therophyte	Sediment-dependence

3.2. Quantitative Classification of Plant Communities

The phytosociological classification results showed that the average silhouette width was 0.73, and the 150 community quadrats were divided into four clusters (Figures S1 and S2). The first cluster was the *Carex cinerascens* community, with 74 quadrats and an average silhouette width of 0.60. The second cluster was the *Carex cinerascens–Polygonum criopolitanum* community, with 14 quadrats and an average silhouette width of 0.80; it was often located in the ecotone between the *Polygonum criopolitanum* communities and the *Carex cinerascens* communities. The third cluster was the *Polygonum criopolitanum* community, with 56 quadrats and an average silhouette width of 0.92. The fourth cluster was the *Phalaris arundinacea* community, with 6 quadrats and an average silhouette width of 0.56.

The analysis results of the quantitative classification were then ranked by DCA (Figure 3). The proportion of the variation in the first two axes was 64.53%, and the variation among the different community groups exceeded that within groups. The four community types were well distinguished by different ranking spaces (ANOSIM: R = 0.885, *p* < 0.0001; MRPP: A = 0.566, *p* < 0.0001). According to Figure 3, the quadrats of the *Polygonum criopolitanum* community were distributed on the left of the ordination plot, while those of the *Phalaris arundinacea* community were distributed on its upper right corner. These two communities had the highest average dissimilarity. The *Carex cinerascens–Polygonum criopolitanum* community was distributed between the *Polygonum criopolitanum* and *Carex cinerascens* communities on the ordination plot, indicating that it was a transitional group between those two communities.

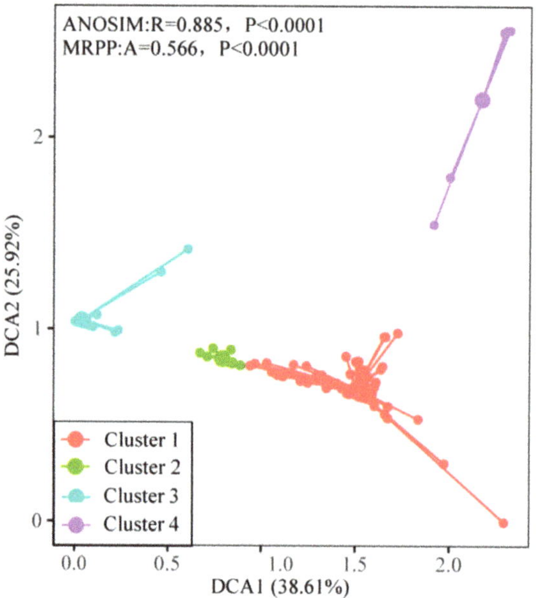

Figure 3. DCA of the plant communities on the quadrats. Cluster 1: *Carex cinerascens* community; Cluster 2: *Carex cinerascens–Polygonum criopolitanum* community; Cluster 3: *Polygonum criopolitanum* community; Cluster 4: *Phalaris arundinacea* community.

3.3. Sediment Types and Plant Community Distribution

The chi-square test was used to analyze the correlation between the sediment types and plant community types. These results revealed a significant correlation between sediment types and plant types in the quadrats (χ^2 = 77.586, *p* < 0.0001).

In the fluvial sediments, the *Polygonum criopolitanum* community predominated, and a small portion of the *Carex cinerascens* community was present; apart from the *Phalaris arundinacea* community, the other communities were distributed in the fluvio-lacustrine sediments. Mostly *Carex cinerascens* and *Phalaris arundinacea* communities were distributed in the lacustrine sediments, with the latter being restricted to that sediment type in our surveyed plant community quadrats.

DCA was performed on the plant community quadrats according to the three sediment types (Figure 4). The results showed that the inter-group variation in the plant communities among the different sedimentary depositions was greater than their intra-group variation, leading to their pronounced separation in the DCA sorting space. On the ordination map, the plant communities of the lacustrine sediments were on the right side, while those of the fluvial sediments were on the left side. Accordingly, those of the fluvio-lacustrine sediments were located between the lacustrine sediments and fluvial sediments on the ordination map. The differences in the plant communities among the three sediment types were analyzed using ANOSIM and MRPP. The results of both tests (ANOSIM: R = 0.267, $p < 0.0001$; MRRP: A = 0.19, $p < 0.0001$) indicated that the plant communities changed significantly across the sediment types.

Before this ranking analysis, the environmental variables that had significant effects were selected by forward selection and included flood duration in 2017 (x17fd) ($p < 0.05$), silt content ($p < 0.05$), and clay content ($p < 0.01$). The total variance of the forward selection was 2300.1, and the constrained variance was 611.3. The three significant factors explained 26.58% of the variance. The total inertia of this RDA was 2300.1, its restricted inertia was 519.6, its unrestricted inertia was 1780.5, and the explanatory power of the environmental factors was 22.59% according to this ranking. Altogether, the first three RDA axes explained 100% of the variance in the data, but the first two axes accounted for 99.91%; hence, these first two axes in the sorting space could explain the vast majority of the variance present. As seen in the RDA two-dimensional ordination map (Figure 5) and the biplot scores for the constraining variables (Table 2), a strong correlation existed between the plant communities and the soil grain-size composition.

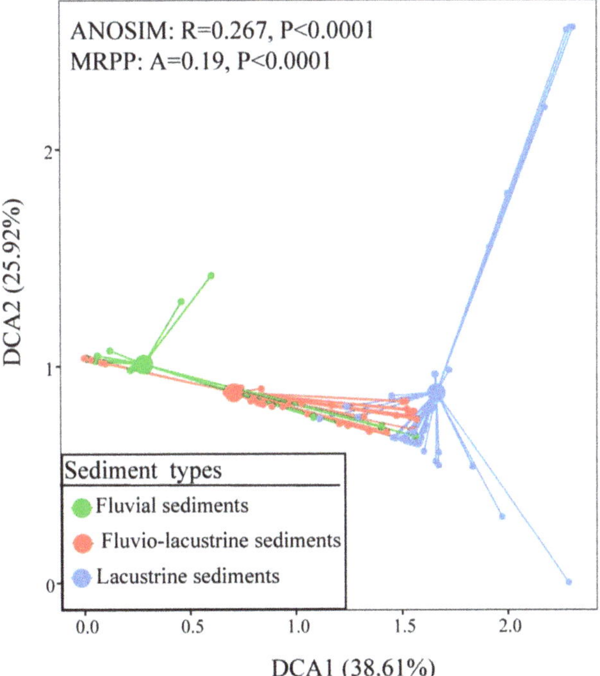

Figure 4. DCA of plant communities in three sediment types.

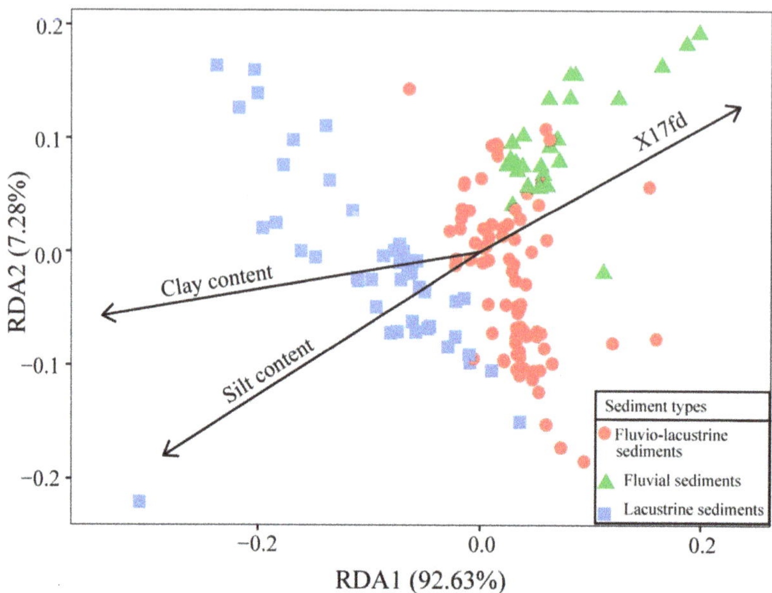

Figure 5. RDA of plant communities on three sediment types. X17fd indicates flood duration in 2017.

Table 2. The biplot scores for the constraining variables in RDA.

RDA Axes	RDA1	RDA2	RDA3
Silt content	−0.71	−0.45	0.53
Clay content	−0.85	−0.14	0.50
X17fd	0.58	0.32	0.74

3.4. Comparison of Plant Species Diversity among Three Sediment Types

Kruskal–Wallis tests were used to compare the differences in the diversity of the plant communities among the three sediment types. Statistically significant differences among the SR and H of the three sediment types were observed (SR: $p < 0.0001$; H: $p < 0.05$). The two indexes were much greater for the lacustrine sediments than they were for either the fluvio-lacustrine sediments or the fluvial sediments (Figure 6a,b). However, no significant difference was observed in either of the two indexes between the fluvio-lacustrine and fluvial sediments (Figure 6a,b). In general, there might be a discernible gradient change in plant species diversity across the three sediment types. The species diversity level of the lacustrine sediments was significantly higher than that of the fluvio-lacustrine or fluvial sediments.

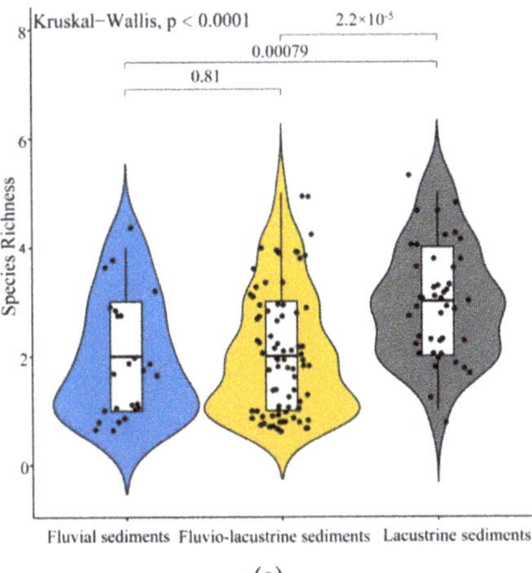

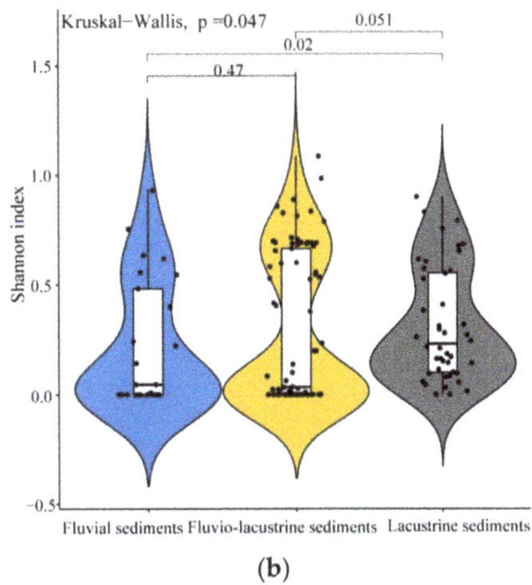

(a)   (b)

Figure 6. Alpha diversity (two metrics) of the plant communities was compared among the three soil deposition types: (**a**) species richness (SR); (**b**) Shannon index (H).

4. Discussion

4.1. The Distribution of Wetland Plants Differs among Sedimentary Deposition Types along the Beaches of Poyang Lake

In this study, 150 quadrats from the littoral zones of Poyang Lake were classified into three sediment types, namely lacustrine sediments, fluvio-lacustrine sediments, and fluvial sediments. Among these, the fluvio-lacustrine sediments accounted for the vast majority, which is consistent with the distribution of Poyang Lake's sedimentary environment reported by Gan et al. (2019) [43]. The distribution of plants among the sediment types was not the same: from the fluvial sediments to the lacustrine sediments, wetland plant communities transition from those dominated by *Carex cinerascens* and *Phalaris arundinacea* to a *Polygonum criopolitanum* community. There is actually an inseparable correlation between the formation of different sediment types and the complex hydrological characteristics of the Poyang Lake area [19,44]. During the dry season, when the water falls into the trough of Poyang Lake, the slope between the upper and lower reaches increases [17]. This intensifies the hydrodynamic force in the lake area. Then, a host of small-size particles such as clay and silt are suspended and transported by water flow, whereas the larger and heavier sand becomes settled [45]. The sand is exposed when the water level drops, forming the fluvial sediments; however, in the areas where the hydrodynamic environment is mild, clay and silt can also be deposited such that lacustrine sediments are formed. Concerning the formation of fluvio-lacustrine sediments, these represent a transitional state between the fluvial and lacustrine sediments. This discrepancy in particle composition leads to a differentiation in the soil texture among sediment types, which in turn triggers changes in other physical properties of the substrate and forms a distinctive soil microenvironment [46,47]. These can profoundly impact the survival, growth, distribution, and evolution of wetland plants [48]. The clay and silt soil in the lacustrine sediments sustains a high nutrient content and moisture content [48]. *Carex cinerascens* and *Phalaris arundinacea*—with robust rhizomes—are better suited to the lacustrine sediments [49]. *Polygonum criopolitanum*, a therophyte species, characterizes low nutrient requirements [50]. The plant can adapt well to the fluvio-lacustrine and fluvial sediments. Additionally, most wetland vegetation is accustomed to distribution in predominantly wet soils [51] and *Polygonum criopolitanum* achieves growth dominance easily in the sandy soil of fluvial sediments. It is worth emphasizing that the formation of sediments requires time [52]. Flooding in the past can have a significant impact on the current physical and chemical characteristics of sediments [53].

Then, the sediments mediate hydrological changes to affect the wetland vegetation distribution patterns and species diversity. Our RDA results showed that the composition of sediment particles and flood duration in 2017 are indeed closely related to the differences in the distribution of plants among sediment types. However, in our RDA, the differentiation in the soil moisture contents and groundwater levels across sediment types that could impose irreplaceable effects on plant communities [54] were not considered, resulting in the slightly low explanatory power of environmental factors (22.59%). In addition, in this research, we did not conduct an in-depth analysis of the relationship between changes in the soil fertility across sedimentary types and the distribution of wetland plants, either. The responses of the wetland vegetation distribution to soil fertility and water conditions influenced by grain-size compositions across soil-deposition types merit discussion and analysis in future studies. Additionally, they could help to further explain the influence of sediment types on the distribution of wetland plant communities.

4.2. Species Diversity of Wetland Plant Communities Differs among Sediment Types in the Poyang Lake Wetlands

Because aspects of community diversity include the synthesis of the quantitative characteristics of different communities, they convey the differences among communities well [55]. The α, β, and γ indexes are commonly used to evaluate the species diversity of plant communities, and the α index can best reflect community-level changes in a local habitat [56]. Comparing the α diversity indexes of the community quadrats across the three sedimentary deposition types, the species diversity of the lacustrine sediments significantly surpassed those of the fluvial sediments and fluvio-lacustrine sediments. This shows that a predictable gradient might span the three types of soil deposition in different plant communities. Many of the *Polygonum criopolitanum* community quadrats were distributed in the fluvial sediments. *Polygonum criopolitanum* is a typical "c-strategy" plant that is able to grow rapidly and can make the most of the limited resources in the environment [50,57]. Over the course of its life, it relies on a strong asexual reproduction ability to form a large number of seedlings [29], resulting in a population growth advantage. Then, it will effectively occupy the living space of other wetland plants and replace them, which results in lower species diversity in fluvial sediments. Further, the compositions of soil particles also affect the diversity of plant communities [58]. Tiny particulate matter, such as clay and silt, is conducive to the formation of agglomerate structures in soil [59], thereby promoting the establishment and growth of plants [60,61]. From the lacustrine sediments to the fluvial sediments, the clay and silt contents decrease while the sand content increases. This enables lacustrine sediments to sustain a higher level of species diversity, whereas the fluvial sediments have a relatively low level of diversity. In addition, within the same elevation range, the sediment types are subject to varying degrees of hydrological disturbance. The greater the disturbance intensity, the lower the level of plant community diversity [62]. Fluvial sediments are generally more disturbed by hydrology than either lacustrine sediments or fluvio-lacustrine sediments. Therefore, in moving from lacustrine sediments to fluvial sediments, the diversity of wetland plant communities tends to decline.

5. Conclusions

The complex hydrological situation of Poyang Lake has led to the formation of the sediment types on its littoral zones which, coupled with its hydrological factors, have shaped the distribution patterns and diversity of the wetland plant communities there. Our results revealed that the three sediment types present in the Poyang Lake wetlands differ in their soil texture which, when combined with flood duration, influences the distribution of vegetation in the littoral zone. Additionally, the distribution of plant communities shifts from the *Carex cinerascens* and *Phalaris arundinacea* communities to the *Polygonum criopolitanum* community from the lacustrine sediments to the fluvial sediments. Furthermore, there is higher species diversity in the lacustrine sediments. Future research should investigate the relationship between wetland vegetation and soil deposition types. An analysis of the combined effects of sedimentary factors and hydrological factors can provide a conducive reference for the management and planning of Poyang Lake's wetlands, as well as other freshwater wetlands, from a more scientific and comprehensive perspective.

Supplementary Materials: The following supporting information can be downloaded at: https://www.mdpi.com/article/10.3390/d14060491/s1, Figure S1: Silhouette width of plant groups; Figure S2: Cluster results of plant communities; Table S1: Information about the 12 sampled field transects.

Author Contributions: Conceptualization, G.G.; methodology, J.L. and Y.L. (Yizhen Liu); investigation, J.L., Y.L. (Yizhen Liu), Y.L. (Ying Liu), H.G., G.C. and Y.F.; data curation, J.L. and Y.L. (Yizhen Liu); writing—original draft preparation, J.L.; writing—review and editing, J.L., Y.L. (Yizhen Liu), Y.L. (Ying Liu), Z.F. and G.G.; visualization, J.L. and Y.L. (Yizhen Liu). All authors have read and agreed to the published version of the manuscript.

Funding: This research was funded by the National Natural Science Foundation of China (31860125) and the Youth Science Fund Project in Jiangxi Province (2018BA214004).

Institutional Review Board Statement: Not applicable.

Data Availability Statement: All data are available in this article.

Acknowledgments: We acknowledge technical assistance by Xiaoguo Xiang, Mingfei Wang, Kaifeng Xing, Qingshan Yang, and Yue Su. In addition, we would like to thank the anonymous reviewers for their helpful remarks.

Conflicts of Interest: The authors declare no conflict of interest.

References

1. De Wilde, M.; Puijalon, S.; Bornette, G. Sediment type rules the response of aquatic plant communities to dewatering in wetlands. *J. Veg. Sci.* **2017**, *28*, 172–183. [CrossRef]
2. Huai, W.X.; Li, S.L.; Katul, G.G.; Liu, M.Y.; Yang, Z.H. Flow dynamics and sediment transport in vegetated rivers: A review. *J. Hydrodyn.* **2021**, *33*, 400–420. [CrossRef]
3. Yu, H.; Wang, L.; Liu, C.; Yu, D.; Qu, J. Effects of a spatially heterogeneous nutrient distribution on the growth of clonal wetland plants. *BMC Ecol.* **2020**, *20*, 59. [CrossRef] [PubMed]
4. Wang, C.; Zheng, S.S.; Wang, P.F.; Hou, J. Interactions between vegetation, water flow and sediment transport: A review. *J. Hydrodyn.* **2015**, *27*, 24–37. [CrossRef]
5. Zheng, P.R.; Li, C.H.; Ye, C.; Wang, H.; Wei, W.W.; Zheng, Y.; Zheng, X.Y. Characteristic and affecting factors of wetland herbs' distribution in the radiant belt toward land of lake-terrestrial ecotone in Tibet, China. *Environ. Sci. Eur.* **2022**, *34*, 14. [CrossRef]
6. Lo, E.; McGlue, M.; Silva, A.; Bergier, I.; Yeager, K.; Macedo, H.; Swallom, M.; Assine, M. Fluvio-lacustrine sedimentary processes and landforms on the distal Paraguay fluvial megafan (Brazil). *Geomorphology* **2019**, *342*, 163–175. [CrossRef]
7. Fan, H.X.; Xu, L.G.; Wang, X.L.; Jiang, J.H.; Feng, W.J.; You, H.L. Relationship between vegetation community distribution patterns and environmental factors in typical wetlands of Poyang Lake, China. *Wetlands* **2019**, *39*, S75–S87. [CrossRef]
8. Zheng, Z.; Ma, P.F. Changes in above and belowground traits of a rhizome clonal plant explain its predominance under nitrogen addition. *Plant Soil* **2018**, *432*, 415–424. [CrossRef]
9. Guan, B.; Xie, B.H.; Yang, S.S.; Hou, A.X.; Chen, M.; Han, G.X. Effects of five years' nitrogen deposition on soil properties and plant growth in a salinized reed wetland of the Yellow River Delta. *Ecol. Eng.* **2019**, *136*, 160–166. [CrossRef]
10. Di Luca, G.A.; Maine, M.A.; Mufarrege, M.M.; Hadad, H.R.; Pedro, M.C.; Sanchez, G.C.; Caffaratti, S.E. Phosphorus distribution pattern in sediments of natural and constructed wetlands. *Ecol. Eng.* **2017**, *108*, 227–233. [CrossRef]
11. Liu, Y.; Bachofen, C.; Lou, Y.J.; Ding, Z.; Jiang, M.; Lu, X.G.; Buchmann, N. The effect of temperature changes and K supply on the reproduction and growth of Bolboschoenus planiculmis. *J. Plant Ecol.* **2021**, *14*, 337–347. [CrossRef]
12. Chen, X.S.; Li, X.; Xie, Y.H.; Li, F.; Hou, Z.Y.; Zeng, J. Combined influence of hydrological gradient and edaphic factors on the distribution of macrophyte communities in Dongting Lake wetlands, China. *Wetl. Ecol. Manag.* **2015**, *23*, 481–490. [CrossRef]
13. Ren, G.H.; Deng, B.; Shang, Z.H.; Hou, Y.; Long, R.J. Plant communities and soil variations along a successional gradient in an alpine wetland on the Qinghai-Tibetan Plateau. *Ecol. Eng.* **2013**, *61*, 110–116. [CrossRef]
14. Reza, S.K.; Nayak, D.C.; Chattopadhyay, T.; Mukhopadhyay, S.; Singh, S.K.; Srinivasan, R. Spatial distribution of soil physical properties of alluvial soils: A geostatistical approach. *Arch. Agron. Soil Sci.* **2016**, *62*, 972–981. [CrossRef]
15. Chen, Y.P.; Xia, J.B.; Zhao, X.M.; Zhuge, Y.P. Soil moisture ecological characteristics of typical shrub and grass vegetation on Shell Island in the Yellow River Delta, China. *Geoderma* **2019**, *348*, 45–53. [CrossRef]
16. Weiher, E.; Keddy, P.A. The Assembly of Experimental Wetland Plant Communities. *Oikos* **1995**, *73*, 323–335. [CrossRef]

17. Gao, J.H.; Jia, J.; Kettner, A.J.; Xing, F.; Wang, Y.P.; Xu, X.N.; Yang, Y.; Zou, X.Q.; Gao, S.; Qi, S.; et al. Changes in water and sediment exchange between the Changjiang River and Poyang Lake under natural and anthropogenic conditions, China. *Sci. Total Environ.* **2014**, *481*, 542–553. [CrossRef]

18. Ye, X.; Li, Y.; Li, X.; Zhang, Q. Factors influencing water level changes in China's largest freshwater lake, Poyang Lake, in the past 50 years. *Water Int.* **2014**, *39*, 983–999. [CrossRef]

19. Dai, Z.J.; Mei, X.F.; Darby, S.E.; Lou, Y.Y.; Li, W.H. Fluvial sediment transfer in the Changjiang (Yangtze) river-estuary depositional system. *J. Hydrol.* **2018**, *566*, 719–734. [CrossRef]

20. Dai, X.; Wan, R.R.; Yang, G.S.; Wang, X.L.; Xu, L.G.; Li, Y.Y.; Li, B. Impact of seasonal water-level fluctuations on autumn vegetation in Poyang Lake wetland, China. *Front. Earth Sci.* **2019**, *13*, 398–409. [CrossRef]

21. Hu, Z.; Ge, G.; Liu, C.; Chen, F.; Li, S. Structure of Poyang Lake wetland plants ecosystem and influence of lake water level for the structure. *Resour. Environ. Yangtze Basin* **2010**, *19*, 597–605.

22. Fu, H.; Zhong, J.Y.; Yuan, G.X.; Guo, C.J.; Ding, H.J.; Feng, Q.; Fu, Q. A functional-trait approach reveals community diversity and assembly processes responses to flood disturbance in a subtropical wetland. *Ecol. Res.* **2015**, *30*, 57–66. [CrossRef]

23. Qin, T.; Guan, Y.-T.; Zhang, M.-X.; Li, H.-L.; Yu, F.-H. Sediment type and nitrogen deposition affect the relationship between Alternanthera philoxeroides and experimental wetland plant communities. *Mar. Freshw. Res.* **2018**, *69*, 811. [CrossRef]

24. Zhang, Y.; Chen, L.-B.; Qu, X.-D.; Kong, W.-J.; Meng, W. Environmental Factors and Community Characteristics of Aquatic Macrophytes in Taizi River Tributaries of Liaoning Province. *Plant Sci. J.* **2011**, *29*, 552–560. [CrossRef]

25. Feng, W.J.; Mariotte, P.; Xu, L.G.; Buttler, A.; Bragazza, L.; Jiang, J.H.; Santonja, M. Seasonal variability of groundwater level effects on the growth of Carex cinerascens in lake wetlands. *Ecol. Evol.* **2020**, *10*, 517–526. [CrossRef]

26. Zhang, Q.; Ye, X.C.; Werner, A.D.; Li, Y.L.; Yao, J.; Li, X.H.; Xu, C.Y. An investigation of enhanced recessions in Poyang Lake: Comparison of Yangtze River and local catchment impacts. *J. Hydrol.* **2014**, *517*, 425–434. [CrossRef]

27. Dai, X.; Wan, R.R.; Yang, G.S. Non-stationary water-level fluctuation in China's Poyang Lake and its interactions with Yangtze River. *J. Geogr. Sci.* **2015**, *25*, 274–288. [CrossRef]

28. Fang, J.; Wang, X.; Shen, Z.; Tang, Z.; He, J.; Yu, D.; Jiang, Y.; Wang, Z.; Zheng, C.; Zhu, J.; et al. Methods and protocols for plant community inventory. *Biodivers. Sci.* **2009**, *17*, 533–548. [CrossRef]

29. Flora of China. Available online: http://www.iplant.cn/foc/ (accessed on 5 March 2021).

30. Ye, C. Contrasting Investigation by ^{137}Cs Method and ^{210}Pb Method for the Present Sedimentation Rate of Poyang Lake, Jiangxi. *Acta Sedimentol. Sin.* **1991**, *9*, 106.

31. Barman, U.; Choudhury, R.D. Soil texture classification using multi class support vector machine. *Inf. Process. Agric.* **2020**, *7*, 318–332. [CrossRef]

32. Di Biase, L.; Pace, L.; Mantoni, C.; Fattorini, S. Variations in Plant Richness, Biogeographical Composition, and Life Forms along an Elevational Gradient in a Mediterranean Mountain. *Plants* **2021**, *10*, 2090. [CrossRef] [PubMed]

33. Pan, Y.; Xie, Y.-H.; Chen, X.-S.; Li, F. Adaptation of wetland plants to sedimentation stress: A review. *Shengtaixue Zazhi* **2011**, *30*, 155–161.

34. Murtagh, F.; Legendre, P. Ward's Hierarchical Agglomerative Clustering Method: Which Algorithms Implement Ward's Criterion? *J. Classif.* **2014**, *31*, 274–295. [CrossRef]

35. Hill, M.O.; Gauch, H.G. Detrended Correspondence Analysis: An Improved Ordination Technique. *Vegetatio* **1980**, *42*, 47–58. [CrossRef]

36. Shaltout, K.H.; Eid, E.M.; Al-Sodany, Y.M.; Heneidy, S.Z.; Shaltout, S.K.; El-Masry, S.A. Effect of Protection of Mountainous Vegetation against Over-Grazing and Over-Cutting in South Sinai, Egypt. *Diversity* **2021**, *13*, 113. [CrossRef]

37. Han, W.J.; Cao, J.Y.; Liu, J.L.; Jiang, J.; Ni, J. Impacts of nitrogen deposition on terrestrial plant diversity: A meta-analysis in China. *J. Plant Ecol.* **2019**, *12*, 1025–1033. [CrossRef]

38. R Core Team. *R: A Language and Environment for Statistical Computing*; R Foundation for Statistical Computing: Vienna, Austria, 2012. Available online: https://www.R-project.org/ (accessed on 8 April 2021).

39. Oksanen, J. Vegan Community Ecology Package Version 2.6-2. 2021. Available online: https://CRAN.R-project.org/package=vegan (accessed on 8 April 2021).

40. Pavoine, S. Adiv: Analysis of Diversity. R Package Version 2.1.1. 2021. Available online: https://CRAN.R-project.org/package=adiv (accessed on 20 April 2021).

41. Giraudoux, P.; Antonietti, J.P.; Beale, C.; Lancelot, R.; Pleydell, D.; Treglia, M. Pgirmess: Spatial Analysis and Data Mining for Field Ecologists. R Package Version 1.7.0. 2021. Available online: https://CRAN.R-project.org-/package=pgirmess (accessed on 22 April 2021).

42. Wickham, H. *Ggplot2: Elegant Graphics for Data Analysis*, 2nd ed.; Springer International Publishing AG: Houston, TX, USA, 2016.

43. Gan, J.; Wan, S.; Li, J.; Tang, C.; Yang, T.; Luozang, Q. Classification of lacustrine soft soil based on sedimentary environment and engineering characteristics—A case study of four typical soft soil in Poyang Lake basin. *Geol. Rev.* **2019**, *65*, 983–992. [CrossRef]

44. Bracken, L.J.; Turnbull, L.; Wainwright, J.; Bogaart, P. Sediment connectivity: A framework for understanding sediment transfer at multiple scales. *Earth Surf. Processes Landf.* **2015**, *40*, 177–188. [CrossRef]

45. Selim, T.; Hesham, M.; Elkiki, M. Effect of sediment transport on flow characteristics in non-prismatic compound channels. *Ain Shams Eng. J.* **2022**, *13*, 101771. [CrossRef]

46. Juma, N.G. Interrelationships between soil structure/texture, soil biota/soil organic matter and crop production. *Geoderma* **1993**, *57*, 3–30. [CrossRef]
47. Smaill, S.J.; Clinton, P.W.; Allen, R.B.; Leckie, A.C.; Davis, M.R. Coarse soil can enhance the availability of nutrients from fine soil. *J. Plant Nutr. Soil Sci.* **2014**, *177*, 848–850. [CrossRef]
48. Mobilian, C.; Craft, C.B. Wetland Soils: Physical and Chemical Properties and Biogeochemical Processes. *Ref. Modul. Earth Syst. Environ. Sci.* **2021**, *3*, 157–168. [CrossRef]
49. Ren, B. Analysis on the Typical Phytocoenoses and Its Principle Factors of Distribution Paterns in East Dongting Lake Wetland. Ph.D. Thesis, Hunan Agricultural University, Changsha, China, 2012.
50. Zhou, D.M.; Luan, Z.Q.; Guo, X.Y.; Lou, Y.J. Spatial distribution patterns of wetland plants in relation to environmental gradient in the Honghe National Nature Reserve, Northeast China. *J. Geogr. Sci.* **2012**, *22*, 57–70. [CrossRef]
51. Ballantine, K.; Schneider, R.; Groffman, P.; Lehmann, J. Soil Properties and Vegetative Development in Four Restored Freshwater Depressional Wetlands. *Soil Sci. Soc. Am. J.* **2012**, *76*, 1482. [CrossRef]
52. Nyarko, B.K. Wetland river flow interaction in a sedimentary formation of the white Volta basin, Ghana. *Earth Sci. Res.* **2019**, *9*, 15–30. [CrossRef]
53. Feng, W.J.; Santonja, M.; Bragazza, L.; Buttler, A. Shift in plant-soil interactions along a lakeshore hydrological gradient. *Sci. Total Environ.* **2020**, *742*, 140254. [CrossRef]
54. Castelli, R.M.; Chambers, J.C.; Tausch, R.J. Soil-plant relations along a soil-water gradient in great basin riparian meadows. *Wetlands* **2000**, *20*, 251–266. [CrossRef]
55. Mougi, A.; Kondoh, M. Diversity of Interaction Types and Ecological Community Stability. *Science* **2012**, *337*, 349–351. [CrossRef]
56. Ram, J.; Singh, J.S.; Singh, S.P. Plant biomass, species diversity and net primary production in a central Himalayan high altitude grassland. *J. Ecol.* **1989**, *77*, 456–468. [CrossRef]
57. Grime, J.P. Evidence for the Existence of Three Primary Strategies in Plants and Its Relevance to Ecological and Evolutionary Theory. *Am. Nat.* **1977**, *111*, 1169–1194. [CrossRef]
58. Rodrigues, P.M.S.; Schaefer, C.E.G.R.; de Oliveira Silva, J.; Ferreira Júnior, W.G.; dos Santos, R.M.; Neri, A.V. The influence of soil on vegetation structure and plant diversity in different tropical savannic and forest habitats. *J. Plant Ecol.* **2018**, *11*, 226–236. [CrossRef]
59. Papadopoulos, A. Soil Aggregates, Structure, and Stability. In *Encyclopedia of Agrophysics*; Gliński, J., Horabik, J., Lipiec, J., Eds.; Springer: Dordrecht, the Netherlands, 2011; pp. 736–740.
60. Stirzaker, R.J.; Passioura, J.B.; Wilms, Y. Soil structure and plant growth: Impact of bulk density and biopores. *Plant Soil* **1996**, *185*, 151–162. [CrossRef]
61. Li, Y.; Wang, D.; Xin, Z. Spatial distribution of vegetation and soil in aquatic-terrestrial ecotone, Li River. *Trans. Chin. Soc. Agric. Eng.* **2013**, *29*, 121–128. [CrossRef]
62. Limberger, R.; Wickham, S.A. Disturbance and diversity at two spatial scales. *Oecologia* **2012**, *168*, 785–795. [CrossRef] [PubMed]

Article

Effects of Environment and Human Activities on Plant Diversity in Wetlands along the Yellow River in Henan Province, China

Zhiliang Yuan [1,†], Man Xiao [1,†], Xiao Su [1], He Zhao [2], Yushan Li [1], Huiping Zhang [1], Ziyu Zhou [1], Rui Qi [3], Yun Chen [1,*] and Wei Wang [4,*]

[1] College of Life Sciences, Henan Agricultural University, No.63 Agricultural Road, Zhengzhou 450002, China; yzlsci@163.com (Z.Y.); xxmmmandy@163.com (M.X.); suxiao198127@163.com (X.S.); liyushan123456@163.com (Y.L.); zhp20208090@163.com (H.Z.); zzy246296@163.com (Z.Z.)

[2] Henan Institute of Geological Survey, No.81 Science Avenue, Zhengzhou 450053, China; zh2015ecology@sina.com

[3] College of Forestry, Henan Agricultural University, No.63 Agricultural Road, Zhengzhou 450002, China; qicailianyi@yeah.net

[4] Zhengzhou Yellow River Wetland Nature Reserve Management Center, Donggang Middle Road, Huayuankou Town, Zhengzhou 450040, China

* Correspondence: cyecology@163.com (Y.C.); hhwetlands@163.com (W.W.); Tel.: +151-3856-8663 (W.W.)

† These authors contributed equally to this work.

Citation: Yuan, Z.; Xiao, M.; Su, X.; Zhao, H.; Li, Y.; Zhang, H.; Zhou, Z.; Qi, R.; Chen, Y.; Wang, W. Effects of Environment and Human Activities on Plant Diversity in Wetlands along the Yellow River in Henan Province, China. *Diversity* **2022**, *14*, 470. https://doi.org/10.3390/d14060470

Academic Editors: Mateja Germ, Igor Zelnik, Matthew Simpson and Michael Wink

Received: 21 April 2022
Accepted: 9 June 2022
Published: 12 June 2022

Publisher's Note: MDPI stays neutral with regard to jurisdictional claims in published maps and institutional affiliations.

Abstract: *Background and Objectives:* The Yellow River is the sixth longest river in the world, and it is considered the mother river of China. Biodiversity conservation in the middle and lower reaches of the Yellow River is an urgent concern due to the impact of topography, sediment deposition, and human activities. Therefore, in this study, we aimed to investigate the diversity of plant communities in wetlands along the middle and lower reaches of the Yellow River from the perspectives of the natural environment and human disturbance. *Materials and Methods:* In this study, 830 plots were set up in seven nature reserves in the middle and lower reaches of the Yellow River to investigate wetland plant diversity. The distribution characteristics of plant diversity and the effects of environmental and human activities on plant diversity were analyzed. *Results:* (1) A total of 184 plant species belonging to 52 families and 135 genera were found in the seven nature reserves. Network analysis showed that the connectance index was 0.3018. (2) Betadisper analysis followed by ANOVA revealed differences in the community composition of the wetland plants ($F = 21.123$, $p < 0.001$) in the different nature reserves. (3) Analysis of variation partitioning indicated that the effects of pure environmental factors (elevation, precipitation, evaporation, and temperature) on the beta diversity of the wetland plants in the nature reserves was the strongest (15.45% and 17.08%, respectively), followed by the effects of pure human disturbance factors (population density, industrial output value, and agricultural output value) (15.13% and 16.71%, respectively). *Conclusions:* Variations occurred in the assemblage characteristics of the wetland plants in the different Yellow River wetland nature reserves. The wetland species exhibited strong associations with the reserves in the Yellow River wetland in Henan Province. Elevation, longitude, precipitation, and evaporation were important factors that affected the diversity of wetland plants in the middle and lower reaches of the Yellow River in China. The findings provide insights into plant biodiversity conservation in riverine wetlands.

Keywords: riverine wetland; species diversity; space distribution; habitat heterogeneity; species protection

1. Introduction

Wetland plant diversity plays a vital role in maintaining wetland ecological functions and ecosystem stability [1,2]. The wetland ecosystem in some areas has been seriously degraded due to the misuse and over-exploitation of wetland resources, leading to the

---

destruction of wetland resources [3,4]. Research on wetland plant diversity has always been a popular issue among ecologists.

The Yellow River basin (32–42° N, 96–119° E) covers an area of 800×10^3 km^2 and spans many geographically distinct regions [5]. Given that the topography, climate, hydrology, and human activities in the various sections of the Yellow River basin vary greatly [6–8], evident differences in the presence of plant species exist among wetlands. Existing studies have shown that the vegetation distribution of the upper reaches of the Yellow River wetland is mainly affected by climate change [9], and the plant diversity of the wetland in the lower Yellow River is significantly related to water depth and human activities [10–12]. Therefore, examining the composition of wetland plants and the determinants of plant diversity is crucial for the protection of river ecosystems.

Henan Province has a large population and is among the main food production areas in China. Over-cultivation of land during agriculture development has affected the ecological environment of the riparian wetlands in the middle and lower reaches of the Yellow River [13]. At the same time, the Yellow River basin is continuously affected by human activities and industrial agglomeration. These factors lead to unprecedented challenges in the wetland plant diversity of the basin [14,15]. Current research on wetlands along the Yellow River in Henan Province mainly involves changes in the patterns of the Yellow River wetlands [16], the ecological environment of the Yellow River wetlands [17,18], countermeasures for forestry development in the lower reaches of the Yellow River [19], and local wetland vegetation diversity [20]. However, the impact of human disturbance and natural environment on the wetland plant diversity of the Yellow River in Henan Province is unclear.

Ecological protection and high-quality development of the Yellow River basin have become national strategies. This study is expected to provide a reference for ecological protection of the Yellow River basin. It investigates the diversity of plant communities in the middle and lower reaches of the Yellow River from the perspectives of the natural environment and human disturbance. This study aims to (1) determine the spatial distribution characteristics of wetland plant communities and (2) explore the main driving factors that affect wetland plant diversity.

2. Materials and Methods

2.1. Study Site and Sampling

The Yellow River is the sixth longest river in the world. The Chinese call the Yellow River their "mother river." The Yellow River basin is designated as the "the cradle of Chinese civilization" and has played an important role in China's social, cultural, economic, and political development [21]. The upper reaches of the Yellow River is the birthplace of the river, and it has high mountains, valleys with steep slopes, a large drop, and abundant hydraulic resources [22]. The middle reaches of the Yellow River have a large water flow and a high content of suspended solids, coming from the fine particulate sediments, thus making the Yellow River a renowned sandy river. The lower reaches of the Yellow River, with low terrain and slow water flow, represent the estuary of the Yellow River [23]. The middle and lower reaches of the Yellow River make up the transition zone from high mountains to plains; the zone has a large elevation drop, large water flow, and high accumulation of sediments, and it forms the world-famous "Hanging River" [24].

The study area was the Yellow River wetland in Henan Province located in the middle and lower reaches of the Yellow River (34°34′–36°08′ N, 110°22′–116°07′ E). The Yellow River wetland in Henan Province spans 711 km, and the overall elevation gradually decreases from west to east, with an average elevation of 124 m [25]. The climate of the studied area in Henan province is warm temperate and semi-humid monsoon. The annual temperature is 12–16 °C, the annual precipitation is 500–900 mm, the annual evaporation is 1300–2100 mm, and the annual sunshine is 2083–2246 h [26,27].

To protect the ecosystem and biological groups in the Yellow River wetlands, the Chinese government has established different nature reserves along the banks of the Yellow

River. National nature reserves are a combination of one or more ecosystems approved and established by the Chinese government. Local nature reserves at all levels are nature reserves approved and established by the local government in China. They are important biodiversity-rich areas, vital habitats for species or other protected objects, and have areas of protection value.

To protect the species diversity of the wetlands in the middle and lower reaches of the Yellow River, the Chinese government has established seven nature reserves along the banks of the Yellow River in Henan Province. Henan Yellow River wetland national nature reserve (HHG) is located in the middle and lower reaches of the Yellow River in the northwest of Henan Province, starting from the junction of Shaanxi and Henan in the west (34°34′35.10″–35°58′12″ N, 110°22′49″–112°47′56.03″ E). Sanmenxia reservoir area wetland (SMX) is located at the junction of Henan, Shaanxi, and Shanxi provinces (34°34′17.2″–34°48′23.1″ N, 110°22′48″–111°15′25.9″ E). Henan Zhengzhou Yellow River wetland provincial nature reserve (ZZS) is located in the north of Zhengzhou City (34°50′04″–34°57′59″ N, 112°54′49″–113°54′59″ E). Henan Zhengzhou Yellow River national wetland park (ZZG) is located in the north of Zhengzhou City (34°54′25″–34°55′30″ N, 113°29′23″–113°39′24″E). Henan Kaifeng Liuyuankou wetland provincial nature reserve (KFS) is located in the east of Henan Province and north of Kaifeng City (34°28′24″–34°59′49″ N, 114°15′57″–114°49′55″ E). Henan Xinxiang Yellow River wetland birds national nature reserve (XXG) is located in the east of Xinxiang City, Henan Province (34°55′46.4″–35°56′13.9″ N, 114°25′30.1″–115°00′23.6″ E). Meanwhile, Henan Puyang Yellow River wetland provincial nature reserve (PYS) belongs to the upper reaches of the lower part of the Yellow River and is located in the south of Puyang City, Henan Province (34°56′41.4″–35°25′38.5″ N, 114°42′27.8″–115°10′04.4″ E).

Through a comprehensive investigation, 830 plots were set up in the Yellow River wetland. Our quadrats were randomly selected based on the vegetation types of the nature reserves, and the detailed sampling site distribution is shown in Figure 1. The herb plot size was 1 m × 1 m, the shrub plot size was 5 m × 5 m, and the tree plot size was 20 m × 20 m. The collected data included species name of all plants, number of individual plants, average height, and average coverage. A detailed description of the seven nature reserves is given in Supplementary Table S1. A list of plant species names and plant stand data is shown in Supplementary Table S2.

2.2. Spatial, Environmental and Antropogenic Variables

The spatial factors, such as the longitude, latitude, and elevation of the plots were recorded by GPS. The environmental factors, such as precipitation, evaporation, and temperature data were obtained from the meteorological stations in the various nature reserves of the Yellow River in Henan Province, China "The China Meteorological Data Service Center. Available online: https://data.cma.cn/ (accessed on 8 June 2022)". Population density (PD), industrial output value (IOV), and agricultural output value (AOV) to some extent reflect the intensity of human activities. Population density (10,000 people/Km^2) is the number of people living on a unit area of land. The industrial output value refers to the quality of the final industrial products and the total price of the industrial labor activities in monetary form produced by industrial enterprises. The calculation method of total agricultural output value is usually obtained by adding the output value of agricultural products, forestry products, animal husbandry products, and fishery products [28]. Data on population density, industrial output value, and agricultural output value come from local government statistical reports "Henan Province Bureau of Statistics. Available online: http://tjj.henan.gov.cn/ (accessed on 8 June 2022).

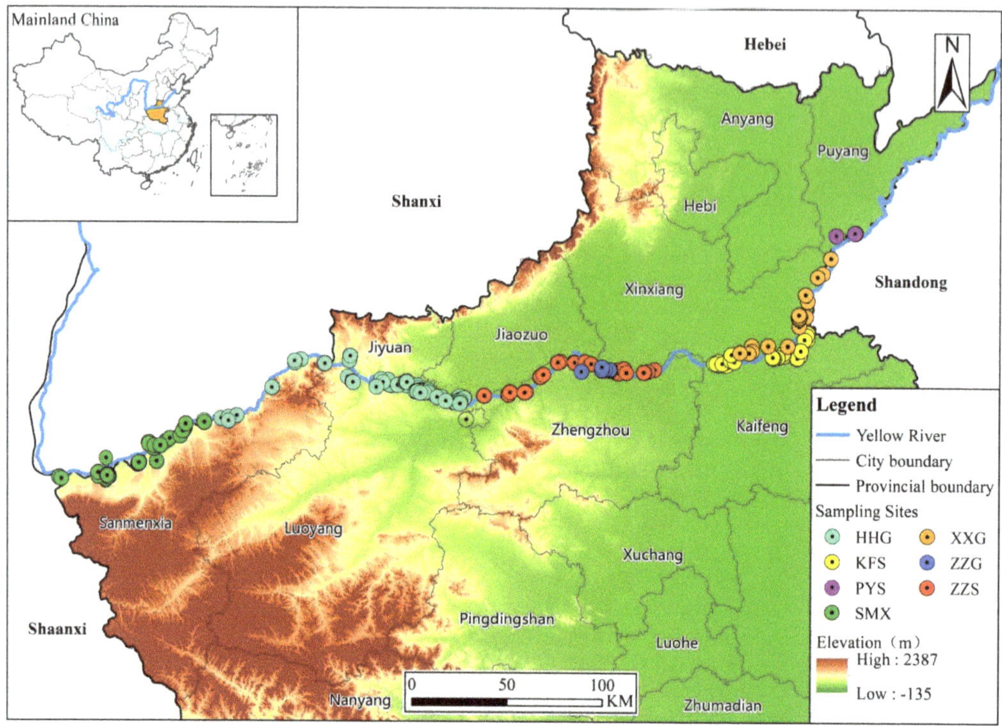

Figure 1. Geographical location distribution of the group of the 830 quadrats set in this study. HHG, Henan Yellow River wetland national nature reserve. SMX, Sanmenxia reservoir area wetland. ZZS, Henan Zhengzhou Yellow River wetland provincial nature reserve. ZZG, Henan Zhengzhou Yellow River national wetland park. KFS, Henan Kaifeng Liuyuankou wetland provincial nature reserve. XXN, Henan Xinxiang Yellow River wetland birds national nature reserve. PYS, Henan Puyang Yellow River wetland provincial nature reserve.

2.3. Data Analysis

In this study, the Shannon–Wiener index (H') and the Pielou evenness index (E) were used to compare the species diversity of plant communities in the seven nature reserves [29,30]. Jaccard's index (βj) [31] and Sorenson's index (βs) [32] were adopted to measure the beta diversity of plant communities in the seven nature reserves [33]. The diversity index was calculated using the Vegan package in R 3.5.2 [34].

A Venn diagram was created to show the number of wetland plants in the different nature reserves [35]. A correlation network was used to visualize the relationships between wetland plants and nature reserves. We evaluated the structure of the network by using the H2' metric of specialization and the connectance index [36]. The architecture of the wetland plants and habitat network was visualized with Gephi 0.9.2 software [37].

Indicator species analysis was conducted using the "indicspecies" package of R to delineate the indicator species of wetland plants in the seven nature reserves [38]. The dependent variable in the indicator species analysis was the species abundance matrix of plants.

The Kruskal–Wallis test was employed to explore the alpha diversity differences among the wetland plants of the seven nature reserves ($p < 0.05$ level of significance). We assessed the impact of the nature reserves on the beta diversity of wetland plants by running the betadisper function. ANOVA was conducted to test the significant differences in beta diversity among the wetland plants in the seven nature reserves. A betadisper test was conducted using the betadisper command in the Vegan package [39,40].

Moreover, a Mantel test [41] was performed to examine the linkage between environmental factors (longitude, latitude, elevation, precipitation, evaporation, temperature, PD, IOV, and AOV) and the beta diversity of the wetland plant community. Pearson correlation analysis was performed to estimate the autocorrelation among environmental factors. A correlation diagram was plotted using the "ggcor" package in R 3.4.0.

We used variance partitioning [42] to evaluate the relative importance of spatial distance (longitude and latitude), environmental factors (elevation, precipitation, evaporation, and temperature), and human disturbance (population density, industrial output value, and agricultural output value) in the beta diversity of the wetland plant community. We utilized beta diversity (βj and βs) as the response variable and three explanatory variables, namely, spatial distance, environmental factors, and human disturbance. A correlation diagram was plotted using the Vegan package in R [43].

For the analysis of invasive alien species, we used Venn diagrams to show the number of invasive alien species in the different nature reserves [35] and a box plot to visualize the species richness of the invasive alien species in the seven nature reserves [44].

3. Results

3.1. Species Composition in Nature Reserves

A total of 184 plant species belonging to 52 families and 135 genera were found in the seven nature reserves. HHG had the most plant species, with 117 wetland plants. PYS had the fewest plant species, with 17 wetland plants. Forty-seven and two species of wetland plants were found in only one community in HHG and ZZS, respectively. Furthermore, six plant species, namely, *Phragmites australis*, *Cynodon dactylon*, *Typha latifolia*, *Polygonum hydropiper*, *Erigeron canadensis*, and *Glycine soja* were shared by the seven nature reserves in Henan Province, accounting for 3.26% of the total number of species (Figure 2).

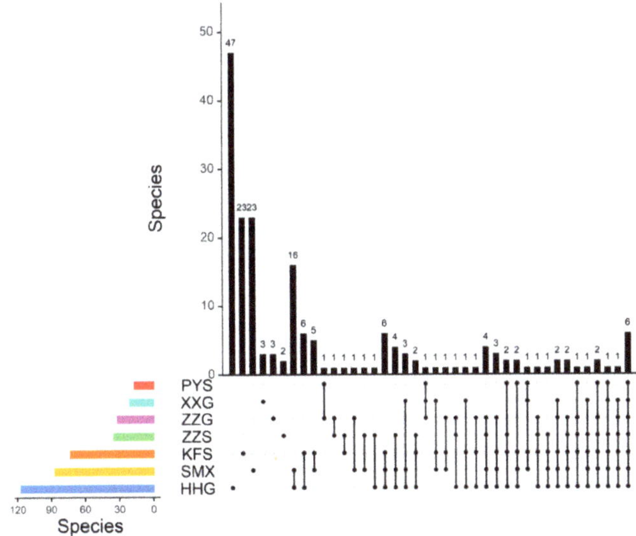

Figure 2. Venn diagram of wetland plants in the seven nature reserves of the Yellow River in Henan Province, China. HHG, Henan Yellow River wetland national nature reserve. SMX, Sanmenxia reservoir area wetland. ZZS, Henan Zhengzhou Yellow River wetland provincial nature reserve. ZZG, Henan Zhengzhou Yellow River national wetland park. KFS, Henan Kaifeng Liuyuankou wetland provincial nature reserve. XXN, Henan Xinxiang Yellow River wetland birds national nature reserve. PYS, Henan Puyang Yellow River wetland provincial nature reserve.

According to the network analysis, 30.18% of the interactions occurred in the observed interactions between wetland plants and nature reserves on the basis of the connectance index (Figure 3).

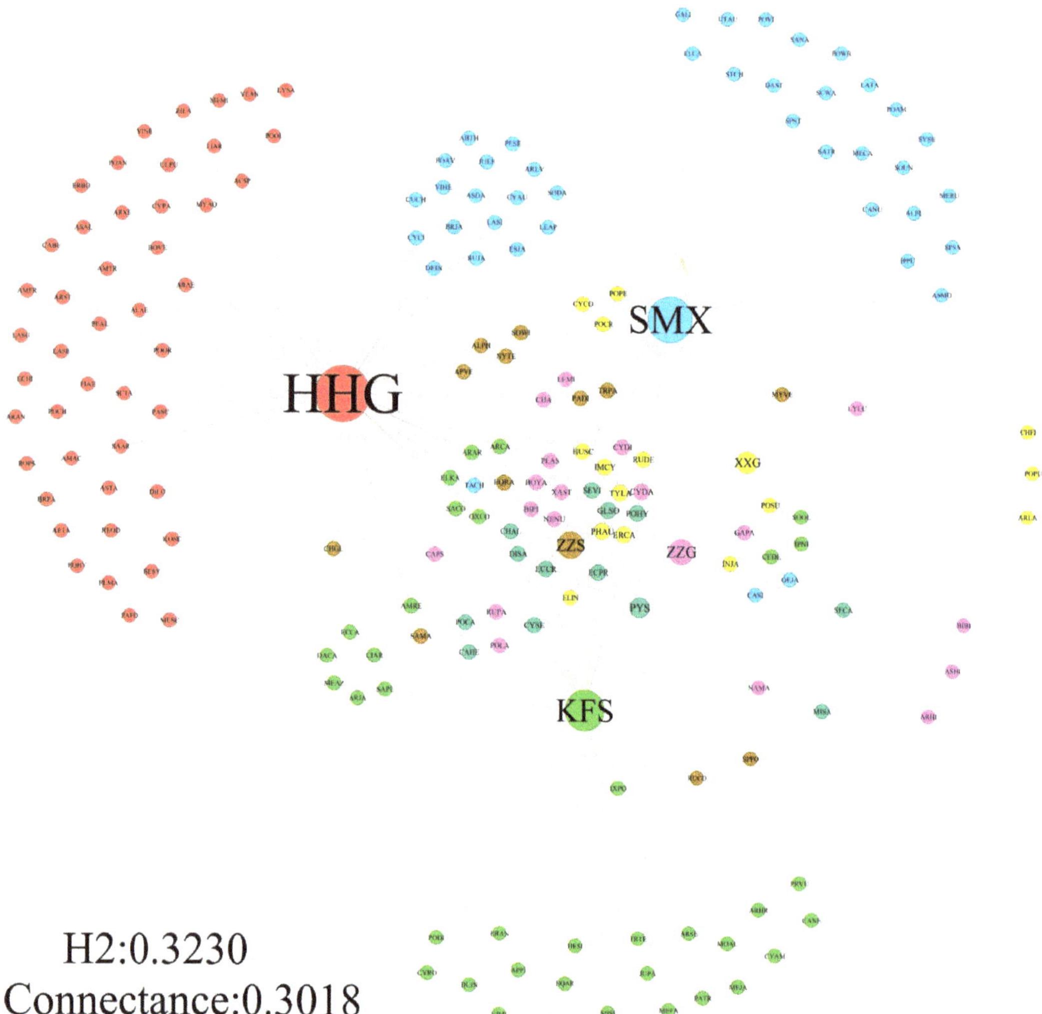

Figure 3. Network analysis of wetland plant composition in the seven nature reserves of the Yellow River in Henan Province, China. The size of the dot indicates the abundance of species. Small letters in small circles indicate plant species. Different colors represent different nature reserves. HHG, Henan Yellow River wetland national nature reserve. SMX, Sanmenxia reservoir area wetland. ZZS, Henan Zhengzhou Yellow River wetland provincial nature reserve. ZZG, Henan Zhengzhou Yellow River national wetland park. KFS, Henan Kaifeng Liuyuankou wetland provincial nature reserve. XXN, Henan Xinxiang Yellow River wetland birds national nature reserve. PYS, Henan Puyang Yellow River wetland provincial nature reserve.

The indicator species analysis showed that the indicator species of the wetland plant community varied among the seven nature reserves. For example, the indicator species of HHG were primarily composed of *Aster altaicus, Setaria viridis,* and *Erigeron canadensis.* The indicator species of SMX primarily consisted of *Cynanchum auriculatum, Apocynum venetum,* and *Datura stramonium.* The indicator species of ZZS were primarily composed of *Calamagrostis pseudophragmites, Paspalum distichum,* and *Cyperus serotinus.* Meanwhile, the indicator species of ZZG primarily consisted of *Bidens biternata, Morus alba,* and *Echinochloa crus-galli.* The indicator species of KFS were primarily composed of *Miscanthus sinensis, Erigeron canadensis,* and *Calystegia hederacea.* The indicator species of XXG consisted of *Chenopodium ficifolium, Cyperus difformis,* and *Potamogeton crispus.* The indicator species of PYS were *Miscanthus sacchariflorus* and *Eclipta prostrata* (Table 1).

3.2. Species Differences among Nature Reserves

The Kruskal–Wallis test revealed differences in the Shannon–Wiener and Pielou evenness indices of plants among the seven nature reserves ($p < 0.01$). Moreover, the two indices of wetland plant alpha diversity in ZZG were higher than those in HHG, SMX, ZZS, KFS, and XXG. Meanwhile, the two indices of wetland plant alpha diversity in KFS were lower than those in XXG and PYS (Figure 4). The betadisper analysis followed by ANOVA showed differences in the community composition of the wetland plants (F = 21.123, $p < 0.001$) among the different nature reserves. Among the seven nature reserves, HHG and SMX had a similar species composition (Figure 5).

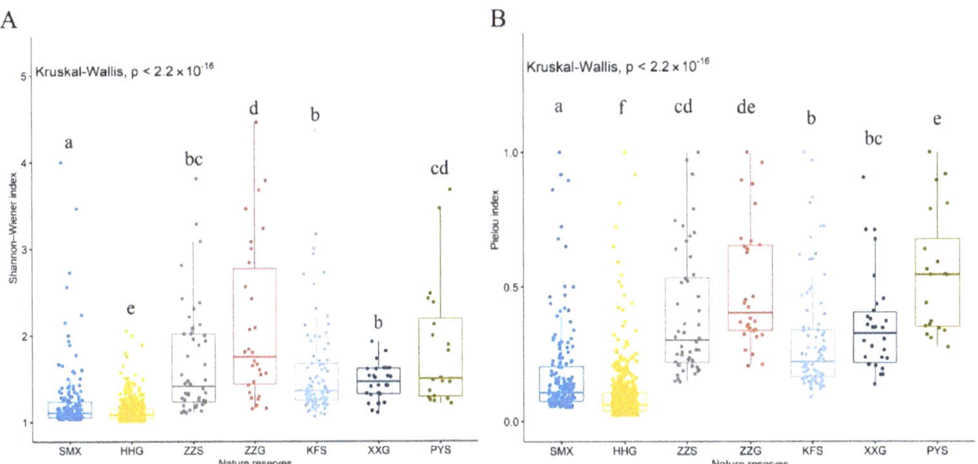

Figure 4. Differences in wetland plant diversity in the seven nature reserves of the Yellow River in Henan Province, China. The number on the black line represents the significance value. Letters are used to distinguish whether there is a significant difference between the protected areas, and different letters indicate that there is a display difference between the protected areas ($p < 0.05$). (**A**) Differences in the Shannon–Wiener index of wetland plants in the seven nature reserves; (**B**) Differences in the Pielou index of wetland plants in the seven nature reserves. HHG, Henan Yellow River wetland national nature reserve. SMX, Sanmenxia reservoir area wetland. ZZS, Henan Zhengzhou Yellow River wetland provincial nature reserve. ZZG, Henan Zhengzhou Yellow River national wetland park. KFS, Henan Kaifeng Liuyuankou wetland provincial nature reserve. XXN, Henan Xinxiang Yellow River wetland birds national nature reserve. PYS, Henan Puyang Yellow River wetland provincial nature reserve.

3.3. Correlation between Plant Diversity and Environmental Factors

The results of the Mantel test indicated that longitude, elevation, and evaporation had the strongest correlation with wetland plant species (Figure 6). Longitude, elevation, evaporation, and population density had positive effects on the beta diversity of the wetland plants in the nature reserves. The beta diversity of the wetland plants in the nature reserves was negatively affected by precipitation and industrial output value.

Table 1. Indicator species analysis in seven nature reserves of the Yellow River in Henan Province, China.

Name of the Seven Nature Reserves	Species	*p*. Value
Henan Yellow River wetland national nature reserve (HHG)	*Aster altaicus*	0.007
	Setaria viridis	0.006
	Erigeron canadensis (IAS)	0.004
	Oxalis corniculata	0.018
	Rumex dentatus	0.017
	Plantago asiatica	0.016
	Bidens pilosa (IAS)	0.020
	Polygonum lapathifolium	0.022
	Chenopodium glaucum (IAS)	0.046
	Myosoton aquaticum (IAS)	0.025
	Artemisia anethoides	0.036
Sanmenxia reservoir area wetland (SMX)	*Cynanchum auriculatum*	0.018
	Lycopus lucidus	0.031
	Apocynum venetum	0.009
	Datura stramonium (IAS)	0.009
	Polygonum persicaria	0.038
	Polygonum aviculare	0.032
	Eschenbachia japonica	0.030
	Equisetum ramosissimum	0.046
Henan Zhengzhou Yellow River wetland provincial nature reserve (ZZS)	*Calamagrostis pseudophragmites*	0.012
	Paspalum distichum (IAS)	0.012
	Cyperus serotinus	0.009
	Tripolium pannonicum	0.008
	Imperata cylindrica	0.012
	Glycine soja	0.046
	Phragmites australis	0.045
Henan Zhengzhou Yellow River national wetland park (ZZG)	*Bidens biternata*	0.011
	Morus alba	0.049
	Echinochloa crus-galli	0.013
Henan Kaifeng Liuyuankou wetland provincial nature reserve (KFS)	*Miscanthus sinensis*	0.012
	Erigeron canadensis (IAS)	0.012
	Calystegia hederacea	0.009
	Parthenocissus tricuspidata	0.009
	Avena fatua (IAS)	0.014
	Digitaria sanguinalis	0.009
	Cirsium arvense var. *integrifolium*	0.011
	Duchesnea indica	0.022
	Erigeron annuus (IAS)	0.013
	Cyperus rotundus	0.024
	Gaura parviflora (IAS)	0.024
	Cynodon dactylon	0.008
	Carex neurocarpa	0.047
	Amaranthus retroflexus (IAS)	0.023
	Artemisia argyi	0.033
	Elymus kamoji	0.042
Henan Xinxiang Yellow River wetland birds national nature reserve (XXG)	*Chenopodium ficifolium (IAS)*	0.043
	Cyperus difformis	0.046
	Potamogeton crispus	0.028
Henan Puyang Yellow River wetland provincial nature reserve (PYS)	*Miscanthus sacchariflorus*	0.012
	Eclipta prostrata (IAS)	0.023

p ≤ 0.05 level of significance. IAS—invasive alien species.

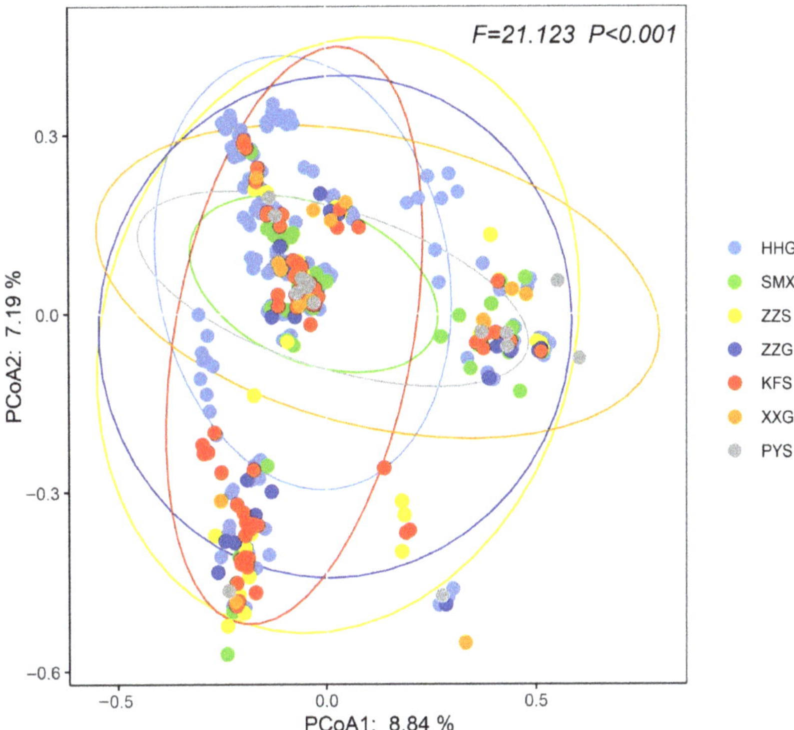

Figure 5. Effect of different nature reserves on the beta diversity of wetland plants determined with the betadisper function. ANOVA is applied to test how these distances differed among different nature reserves. Different colors represent different nature reserves. HHG, Henan Yellow River wetland national nature reserve. SMX, Sanmenxia reservoir area wetland. ZZS, Henan Zhengzhou Yellow River wetland provincial nature reserve. ZZG, Henan Zhengzhou Yellow River national wetland park. KFS, Henan Kaifeng Liuyuankou wetland provincial nature reserve. XXN, Henan Xinxiang Yellow River wetland birds national nature reserve. PYS, Henan Puyang Yellow River wetland provincial nature reserve.

The analysis of variation partitioning indicated that the effects of pure environmental factors (elevation, precipitation, evaporation, and temperature) on the beta diversity of the wetland plants in the nature reserves was the strongest (15.45% and 17.08%, respectively), followed by the effects of pure human disturbance factors (population density, industrial output value, and agricultural output value) (15.13% and 16.71%, respectively). The pure spatial factors, namely, longitude and latitude, had the lowest explanation rate for the beta diversity (βj and βs) of the wetland plants in the nature reserves; the values were 13.05% and 13.71%, respectively (Figure 7A,B).

3.4. Invasive Alien Species

A total of 28 invasive alien species belonging to 12 families and 21 genera were found in the seven nature reserves (Supplementary Table S3) and accounted for 14.97% of the total species. HHG had the most invasive alien species, with 18 invasive alien species. XXG had the fewest invasive alien species, with 2 invasive alien species. Eight invasive alien species were found in only one community in HHG. Furthermore, one invasive alien species, namely, *Eclipta prostrata*, was shared by the seven nature reserves and accounted for 3.57% of the total invasive alien species (Figure 8A). The box plot in Figure 8B shows that the differences in species richness among the seven nature reserves are not significant.

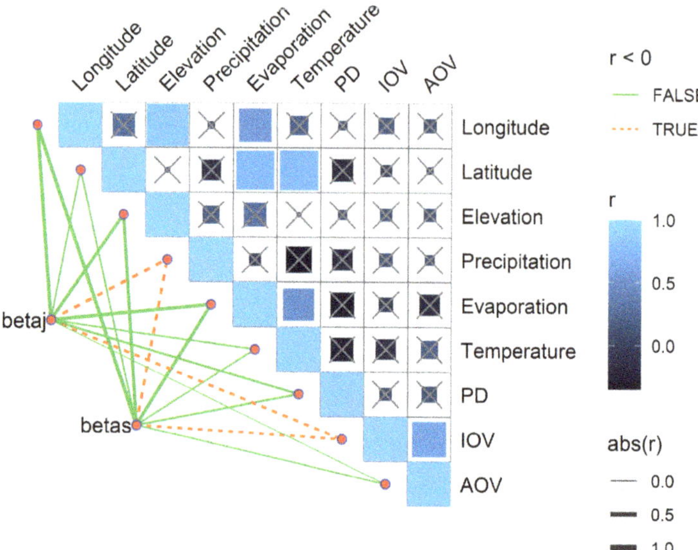

Figure 6. Relationship between natural and human environment variables and beta diversity of wetland plants. The color gradient represents the correlation coefficient. The thickness of the line indicates the correlation. The thicker the line, the stronger the correlation; the thinner the line, the weaker the correlation. PD, population density. IOV, industrial output value. AOV, agricultural output value.

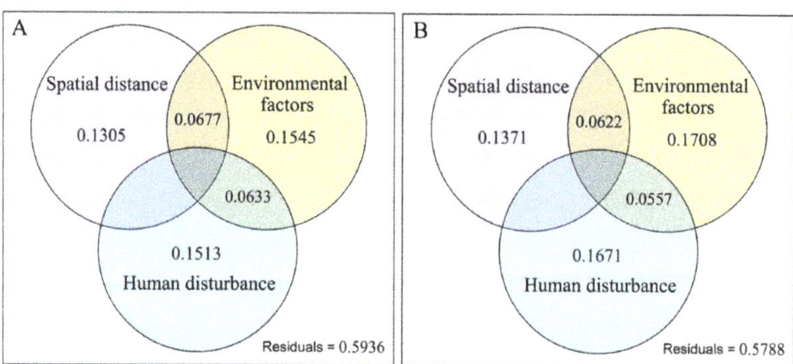

Figure 7. Main drivers of plant diversity in the seven nature reserves of the Yellow River in Henan Province, China. (**A**,**B**) Variance partitioning for the effects of spatial distance, environmental factors, and human disturbance on the beta diversity (βj and βs) of wetland plants in the nature reserves. Values less than zero are not shown. Spatial distance: longitude, latitude. Environmental factors: elevation, precipitation, evaporation, temperature. Human disturbance: population density, industrial output value, agricultural output value.

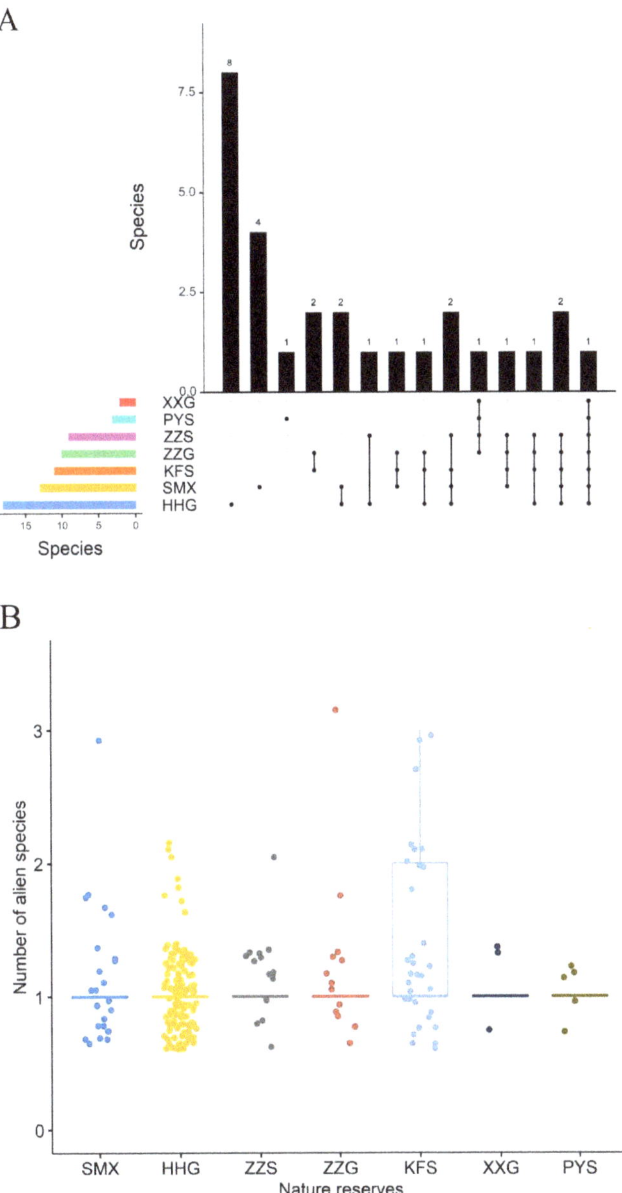

Figure 8. Composition of invasive alien species in the seven nature reserves of the Yellow River in Henan Province, China. (**A**) Venn diagram of invasive alien species in the seven nature reserves; (**B**) Number of invasive alien species in the seven nature reserves. HHG, Henan Yellow River wetland national nature reserve. SMX, Sanmenxia reservoir area wetland. ZZS, Henan Zhengzhou Yellow River wetland provincial nature reserve. ZZG, Henan Zhengzhou Yellow River national wetland park. KFS, Henan Kaifeng Liuyuankou wetland provincial nature reserve. XXN, Henan Xinxiang Yellow River wetland birds national nature reserve. PYS, Henan Puyang Yellow River wetland provincial nature reserve.

4. Discussion

4.1. Spatial Distribution of Plant Diversity

Our results revealed variations in the assemblage characteristics of the wetland plants in the different Yellow River wetland nature reserves ($F = 21.123$, $p < 0.001$). Plant diversity gradually decreases from west to east in the Yellow River wetland in Henan Province [45]. This phenomenon is mainly due to the fact that the middle and lower reaches of the Yellow River have the landform characteristics of western mountainous areas, central hills, and eastern plains [46]. The mountains have a large elevation difference and cover diverse topographic and climatic gradients, whereas the habitats in the plain areas are relatively simple, which may be an important reason for the gradually decreasing pattern of plant diversity from west to east [47,48]. HHG located in the western mountainous area is a national-level nature reserve with a large area and diverse habitats. It is a synthesis of multiple ecosystems, which may also be one of the reasons for the large diversity in the western mountainous area.

Our study found that ZZG had the highest plant diversity. This may be because ZZG located at the boundary of the middle and lower reaches of the Yellow River. Its unique geographical location makes the development of agriculture, forestry, and animal husbandry here faster, the annual average industrial output value and agricultural output value were 395.23 million yuan and 171.83 million yuan, respectively. ZZG is also the only characteristic area of National wetland park, which is a wetland landscape with special ecological and biological diversity values. Therefore, the plant diversity of ZZG was the highest in the studied wetland reserve.

Our network analysis suggests that a strong connection exists between the wetland plants and reserves (connectance index: 0.3018). The habitats of the different reserves vary greatly. For example, HHG reserve has a mountainous terrain, ZZG reserve is hilly, and KFS reserve is a plain. In addition, some differences exist in the precipitation and average temperature of the different reserves [49]. The indicator species of the different nature reserves are also different. For example, *Aster altaicus* and *Setaria viridis* are the main indicator species for HHG, whereas *Calamagrostis pseudophragmites* and *Paspalum distichum* are the main indicator species for ZZS. These findings are consistent with the discussion of Chen (2018), who reported that plants choose specific areas on the basis of their habitat traits. Therefore, wetland species have strong associations with reserves in the Yellow River wetland in Henan Province.

4.2. Determinants of Plant Diversity

In this study, elevation, precipitation, evaporation, and longitude were determined to be the main factors driving wetland species distribution in the Yellow River wetland in Henan Province. Elevation gradually decreases from the mountains in the west to the plains in the east in Henan Province [25]. Precipitation is an important factor that affects the spatial distribution and diversity of plants [50]. Influenced by the southeast monsoon and seaway location, the precipitation of the Yellow River wetland gradually decreases from east to west in Henan Province [51]. Studies have shown that 81.6% of the annual precipitation in the Yellow River basin evaporates [52]. Therefore, elevation, longitude, precipitation, and evaporation are important factors that affect the diversity of wetland plants in the middle and lower reaches of the Yellow River in China.

4.3. Effects of Human Disturbance on Plant Diversity

Aside from topography and climate playing a major role in plant diversity in the Yellow River in Henan Province, this study also found that human disturbance is an important factor that influences plant species diversity in the wetland. The explanation rates of pure human disturbance factors for the wetland plant beta diversity in the nature reserves were 15.13% and 16.71%, respectively. The wetland in the lower reaches of the Yellow River in Henan is a plain grain-producing area that is densely populated [53]. For historical reasons, the Loess Plateau in the middle reaches of the Yellow River and the

North China Plain in the lower reaches have a wide distribution of cultivated land [54]. According to national census data released in 2020, the permanent population of Henan Province is about 99.366 million, showing an increase of 5.68% compared with the figure for 2010. Additionally, agricultural irrigation influences wetland plant diversity. The vast river floodplain in this section has been developed into grazing sites and cultivated land for planting corn, wheat, and other field crops [55,56]. Previous results have indicated that frequent human activities, to some extent, adversely affect the wetland plant diversity of the middle and lower reaches of the Yellow River in China [57–59].

In our study, we also found that population density had a positive effect on the beta diversity of the wetland plants. A high population density indicates the possibility of strong human interference. Human disturbance results in landscape fragmentation and diverse habitats in the studied wetlands. The plant composition of different habitats exhibits great differences [60]. Therefore, population density has an important impact on plant diversity in wetlands along the Yellow River.

4.4. Invasive Alien Species Diversity

High population densities mean great human disturbance, which results in high beta diversity [60]. Another possible reason may be the inclusion of alien species, which are often positively correlated with high population densities or large settlements [61]. The invasive alien species of the Yellow River wetland in Henan Province invaded the local area early and have already adapted to the local habitat. They have a wide distribution in the local area. For example, *Eclipta prostrata* are distributed in all seven nature reserves, and according to historical data, *Eclipta prostrata* may have been introduced to China from abroad during the Tang dynasty (618–907). Therefore, the diversity of invasive alien species among the seven nature reserves in Henan province does not differ considerably. According to our field investigation and relevant literature, invasive plants do not affect the Yellow River ecosystem on a large scale in Henan Province. However, future research should pay careful attention to the impact of invasive alien species on the local ecological environment. The risks posed by invasive alien species should be evaluated thoroughly, monitored, and managed. In addition, preventive and protective measures must be implemented in advance.

5. Conclusions and Implications

From the perspectives of the natural environment and human disturbance, this study analyzed the characteristics of the plant community distribution in the middle and lower reaches of the Yellow River in China. It revealed the distribution of plant species in the middle and lower reaches of the Yellow River wetland in China and its driving factors.

Our results showed variations in the assemblage characteristics of wetland plants in the different Yellow River wetland nature reserves. The wetland species exhibited strong associations with the reserves in the Yellow River wetland in Henan Province. Elevation, longitude, precipitation, and evaporation were important factors that affected the diversity of wetland plants in the middle and lower reaches of the Yellow River in China. Human disturbance was another important factor that influenced plant species diversity in the wetland.

Wetland plants were strongly associated with the nature reserves in our study. Therefore, different conservation strategies should be adopted for different reserves and plants. Some protected-area endemic species require more attention than relatively common species during biodiversity conservation. Common wetland plants are widely distributed and have strong adaptability, so they can be used for wetland restoration. Reducing the farmland area and human activities is also an important measure for wetland restoration. This work is expected to provide a theoretical basis and possible data support for the high-quality development of ecological protection in the Yellow River basin.

Supplementary Materials: The following supporting information can be downloaded at: https://www.mdpi.com/article/10.3390/d14060470/s1. Table S1: The detailed description of the seven nature reserves; Table S2: The list of plant species names and plant stand data; Table S3: List of Invasive Alien Species Information.

Author Contributions: Data curation, Z.Y. and Y.C.; Investigation, M.X., Z.Z. and W.W.; Methodology, Y.C.; Project administration, Z.Y.; Software, M.X. and R.Q.; Supervision, Z.Y. and W.W.; Visualization, M.X., X.S., Y.L. and H.Z. (Huiping Zhang); Writing—original draft, M.X.; Writing—review & editing, Z.Y., H.Z. (He Zhao) and Y.C. All authors have read and agreed to the published version of the manuscript.

Funding: Project funded by China Postdoctoral Science Foundation (2021M693400); Youth Foundation of Natural Science Foundation of Henan Province (212300410153); Young Talents project funded by Henan Agricultural University (111/30500744); Natural Science Foundation of Henan Province (No. 212300410166).

Institutional Review Board Statement: Not applicable.

Informed Consent Statement: Not applicable.

Data Availability Statement: The data that support the findings of this study are available from the corresponding author upon reasonable request.

Conflicts of Interest: The authors declare no conflict of interest.

References

1. Engelhardt, K.A.; Ritchie, M.E. Effects of macrophyte species richness on wetland ecosystem functioning and services. *Nature* **2001**, *411*, 687–689. [CrossRef] [PubMed]
2. Tilman, D.; Lehman, C.L.; Thomson, K.T. Plant diversity and ecosystem productivity: Theoretical considerations. *Proc. Natl. Acad. Sci. USA* **1997**, *94*, 1857–1861. [CrossRef] [PubMed]
3. Meng, W.; He, M.; Hu, B.; Mo, X.; Li, H.; Liu, B.; Wang, Z. Status of wetlands in China: A review of extent, degradation, issues and recommendations for improvement. *Ocean. Coast. Manag.* **2017**, *146*, 50–59. [CrossRef]
4. Yu, X.; Ding, S.; Zou, Y.; Xue, Z.; Lyu, X.; Wang, G. Review of rapid transformation of floodplain wetlands in northeast China: Roles of human development and global environmental change. *Chin. Geogr. Sci.* **2018**, *28*, 654–664. [CrossRef]
5. Fang, L.; Wang, L.; Chen, W.; Sun, J.; Cao, Q.; Wang, S.; Wang, L. Identifying the impacts of natural and human factors on ecosystem service in the Yangtze and Yellow River Basins. *J. Clean. Prod.* **2021**, *314*, 127995. [CrossRef]
6. Jiang, W.; Yuan, L.; Wang, W.; Cao, R.; Zhang, Y.; Shen, W. Spatio-temporal analysis of vegetation variation in the Yellow River Basin. *Ecol. Indic.* **2015**, *51*, 117–126. [CrossRef]
7. Zhang, W.; Wang, L.; Xiang, F.; Qin, W.; Jiang, W. Vegetation dynamics and the relations with climate change at multiple time scales in the Yangtze River and Yellow River Basin, China. *Ecol. Indic.* **2020**, *110*, 105892. [CrossRef]
8. Shi, F.; Liu, S.; Sun, Y.; An, Y.; Zhao, S.; Liu, Y.; Li, M. Ecological network construction of the heterogeneous agro-pastoral areas in the upper Yellow River basin. *Agric. Ecosyst. Environ.* **2020**, *302*, 107069. [CrossRef]
9. Li, H.; Liu, G.; Fu, B. Response of vegetation to climate change and human activity based on NDVI in the Three-River Headwaters region. *Acta Ecol. Sin.* **2011**, *31*, 5495–5504.
10. Li, F.; Xie, Y.; Chen, X.; Deng, Z.; Li, X. Composition of aquatic plants and their niche characteristics in wetlands of the Yellow River Delta. *Acta Ecol. Sin.* **2009**, *29*, 6257–6265.
11. Liu, Y.; Han, M.; Pan, B.; Liu, Y. The Spatial Distribution of Vegetation Biomass and Soil Salinity in New-born Wetlands of the Yellow River Delta. *Wetl. Sci.* **2017**, *15*, 364–368. [CrossRef]
12. Wang, Y.; Zhang, J.; Lv, Q.; Lv, Z.; Wang, Y.; Gao, Y.; Ren, Z. Effects of water and sediment regulation scheme on phytoplankton community and abundance in the Yellow River Estuary. *Mar. Environ. Sci.* **2021**, *40*, 369–378. [CrossRef]
13. Guofu, L.; Shengyan, D.; Zhiheng, L. Regional agricultural landscape pattern changes along the Yellow River in Henan Province from 1987 to 2002. *J. Geogr. Sci.* **2005**, *15*, 415–422. [CrossRef]
14. Guofu, L.; Shengyan, D. Impacts of human activity and natural change on the wetland landscape pattern along the Yellow River in Henan Province. *J. Geogr. Sci.* **2004**, *14*, 339–348. [CrossRef]
15. Liu, L.; Wei, J.; Ma, Y.; Luo, Q.; Wang, Z. Temporal and spacial changes for ecological environment of Yellow River in Henan Province. *Environ. Sci. Manag.* **2021**, *46*, 169–173.
16. Liang, S.; Li, C.; Li, C.; Ren, J.; Song, L.; Qiao, X. Distribution and Change of Wetlands in Henan Province in 1990, 2000 and 2007. *Wetl. Sci.* **2011**, *9*, 94–96. [CrossRef]
17. Yu, J.; Chen, X.; Sun, Z.; Xie, W.; Mao, P.; Wu, C.; Dong, H.; Mou, X.; Li, Y.; Guan, B.; et al. The spatial distribution characteristics of soil nutrients in new-born coastal wetland in the Yellow River delta. *Acta Sci. Circumstantiae* **2010**, *30*, 855–861. [CrossRef]

18. Lin, M.; Liu, R.; Wang, Y.; Luo, X.; Zhou, F. Spatial Heterogeneity of Soil Nutrients and The Relationship between Soil Nutrients and Plant Community in *Tamarix chinensis* Forest Farm Wetland of Yellow River Delta Wetland. *Wetl. Sci.* **2010**, *8*, 92–97. [CrossRef]

19. Zhao, Y. Potentialities and Countermeasures of Forestry Development in the Lower Reaches of the Yellow River in Henan. *Areal Res. Dev.* **2004**, *23*, 61–65.

20. He, X.; Zheng, D.; Guo, H.; Ma, G. The Vegetation Species Diversity and Its Response to the Human Activities in Zhengzhou Yellow River Wetland Reserve. *Wetl. Sci.* **2014**, *12*, 459–463. [CrossRef]

21. Wohlfart, C.; Kuenzer, C.; Chen, C.; Liu, G. Social–ecological challenges in the Yellow River basin (China): A review. *Environ. Earth Sci.* **2016**, *75*, 1066. [CrossRef]

22. Hao, F.; Zhang, X.; Ouyang, W.; Skidmore, A.K.; Toxopeus, A.G. Vegetation NDVI linked to temperature and precipitation in the upper catchments of Yellow River. *Environ. Modeling Assess.* **2012**, *17*, 389–398. [CrossRef]

23. Chen, Y.; Syvitski, J.P.; Gao, S.; Overeem, I.; Kettner, A.J. Socio-economic impacts on flooding: A 4000-year history of the Yellow River, China. *Ambio* **2012**, *41*, 682–698. [CrossRef] [PubMed]

24. Ge, J.; Hu, Y. The Yellow River and the Yellow River Civilization. In *A Historical Survey of the Yellow River and the River Civilizations*; Springer: Singapore, 2021; pp. 1–48.

25. Zhang, J.; Zhu, W.; Zhu, L.; Cui, Y.; He, S.; Ren, H. Topographical relief characteristics and its impact on population and economy: A case study of the mountainous area in western Henan, China. *J. Geogr. Sci.* **2019**, *29*, 598–612. [CrossRef]

26. Liao, B. Research on Plant Diversity and It's Dynamics along the Environmental Gradient in Henan Province Section of the Yellow River Basin. Ph.D. Thesis, Henan University, Kaifeng, China, 2013.

27. Guo, J. Research on the Species Diversity of Compositae along the Different Gradient to Henan Section of the Yellow River Watershed. Master's Thesis, Henan University, Kaifeng, China, 2011.

28. Zuo, Q.; Wu, Z. Risk-Based Evaluation Index of Water Resource Renewability in the Yellow River Basi. Ph.D. Thesis, Zhengzhou University, Zhengzhou, China, 2003.

29. Benayas, R.; José, M.; Scheiner, S.M. Plant diversity, biogeography and environment in Iberia: Patterns and possible causal factors. *J. Veg. Sci.* **2002**, *13*, 245–258. [CrossRef]

30. Chesson, P. Mechanisms of maintenance of species diversity. *Annu. Rev. Ecol. Syst.* **2000**, *31*, 343–366. [CrossRef]

31. Jaccard, P. The distribution of the flora in the alpine zone. 1. *New Phytol.* **1912**, *11*, 37–50. [CrossRef]

32. Sorensen, T.A. A method of establishing groups of equal amplitude in plant sociology based on similarity of species content and its application to analyses of the vegetation on Danish commons. *Biol. Skar.* **1948**, *5*, 1–34.

33. Jost, L. Partitioning diversity into independent alpha and beta components. *Ecology* **2007**, *88*, 2427–2439. [CrossRef]

34. Dixon, P. VEGAN, a package of R functions for community ecology. *Journal of Vegetation Science* **2003**, *14*, 927–930. [CrossRef]

35. Conway, J.R.; Lex, A.; Gehlenborg, N. UpSetR: An R package for the visualization of intersecting sets and their properties. *Bioinformatics* **2017**, *33*, 2938–2940. [CrossRef] [PubMed]

36. Blüthgen, N.; Menzel, F.; Hovestadt, T.; Fiala, B.; Blüthgen, N. Specialization, constraints, and conflicting interests in mutualistic networks. *Curr. Biol.* **2007**, *17*, 341–346. [CrossRef] [PubMed]

37. Bastian, M.; Heymann, S.; Jacomy, M. Gephi: An open source software for exploring and manipulating networks. In Proceedings of the Third International Conference on Weblogs and Social Media, San Jose, CA, USA, 17–20 May 2009; Volume 3, pp. 361–362.

38. De Caceres, M.; Jansen, F.; De Caceres, M.M. Package 'Indicspecies'. *Indicators*. 2016, 8. Available online: https://emf-creaf.github.io/indicspecies/ (accessed on 8 June 2022).

39. Chen, Y.; Yuan, Z.; Bi, S.; Wang, X.; Ye, Y.; Svenning, J.C. Macrofungal species distributions depend on habitat partitioning of topography, light, and vegetation in a temperate mountain forest. *Sci. Rep.* **2018**, *8*, 13589. [CrossRef]

40. Oksanen, J. Vegan: Community Ecology Package. R Package Version 1.17-9. 2011. Available online: http://cran.r-project.org/package=vegan (accessed on 8 June 2022).

41. Hollister, E.B.; Engledow, A.S.; Hammett, A.J.M.; Provin, T.L.; Wilkinson, H.H.; Gentry, T.J. Shifts in microbial community structure along an ecological gradient of hypersaline soils and sediments. *ISME J.* **2010**, *4*, 829–838. [CrossRef]

42. Legendre, P.; Mi, X.; Ren, H.; Ma, K.; Yu, M.; Sun, I.F.; He, F. Partitioning beta diversity in a subtropical broad-leaved forest of China. *Ecology* **2009**, *90*, 663–674. [CrossRef]

43. Oksanen, J.; Kindt, R.; Legendre, P.; O'Hara, B.; Stevens, M.H.H.; Oksanen, M.J.; Suggests, M.A.S.S. The vegan package. *Community ecology package.* **2007**, *10*, 719.

44. Kassambara, A.; Kassambara, M.A. Package 'Ggpubr'. In R Package Version 0.1. 2020, Volume 6. Available online: https://www.sthda.com/english/rpkgs/ggpubr (accessed on 8 June 2022).

45. Xu, H.; Cao, M.; Wu, Y.; Cai, L.; Cao, Y.; Wu, J.; Lei, J.; Le, Z.; Ding, H.; Cui, P. Disentangling the determinants of species richness of vascular plants and mammals from national to regional scales. *Sci. Rep.* **2016**, *6*, 21988. [CrossRef]

46. Eryong, Z.; Yingchun, S.; Cunrong, G.; Xinwei, H.; Zhantao, H.; Hongmei, Z.; Qianqing, D.; Xingchun, L.; Baogui, L.; Runlian, Z.; et al. Regional geology and hydrogeology of the Yellow River basin. *Bull. Geol. Surv. Jpn.* **2009**, *60*, 19–32. [CrossRef]

47. Becker, A.; Körner, C.; Brun, J.J.; Guisan, A.; Tappeiner, U. Ecological and land use studies along elevational gradients. *Mt. Res. Dev.* **2007**, *27*, 58–65. [CrossRef]

48. Moeslund, J.E.; Arge, L.; Bøcher, P.K.; Dalgaard, T.; Svenning, J.C. Topography as a driver of local terrestrial vascular plant diversity patterns. *Nord. J. Bot.* **2013**, *31*, 129–144. [CrossRef]

49. Zhao, Q.; Ma, L.; Liu, Q.; Ding, S.; Tang, Q.; Lu, X. Plant species diversity and its response to environmental factors in typical river riparian zone in the middle and lower reaches of Yellow River. *Chin. J. Ecol.* **2015**, *34*, 1325–1331. [CrossRef]
50. Stella, J.C.; Rodríguez-González, P.M.; Dufour, S.; Bendix, J. Riparian vegetation research in Mediterranean-climate regions: Common patterns, ecological processes, and considerations for management. *Hydrobiologia* **2013**, *719*, 291–315. [CrossRef]
51. Shi, B.; Zhu, X.; Hu, Y.; Yang, Y. Drought characteristics of Henan province in 1961-2013 based on Standardized Precipitation Evapotranspiration Index. *J. Geogr. Sci.* **2017**, *27*, 311–325. [CrossRef]
52. Liu, C.; Wang, S.; Liang, Y. Analysis of pan evaporation change and the influence factors in the Yellow River Basin in 1961–2010. *Progress. Inquisitiones DE Mutat. Clim.* **2013**, *9*, 327. [CrossRef]
53. Li, H.; Wang, J.; Zhang, J.; Qin, F.; Hu, J.; Zhou, Z. Analysis of characteristics and driving factors of wetland landscape pattern change in Henan Province from 1980 to 2015. *Land* **2021**, *10*, 564. [CrossRef]
54. Sun, Y.; Hu, W.; Yao, S.; Sun, Y.; Deng, J. Geographic patterns and environmental determinants of angiosperm and terrestrial vertebrate species richness in the Yellow River basin. *Biodivers. Sci.* **2020**, *28*, 1523. [CrossRef]
55. Liang, S.; Luo, X.; Meng, X.; Sun, R. Characteristics of herbaceous plant community diversity along the Yellow River wetland of Henan. *Pearl River* **2018**, *39*, 9–13. [CrossRef]
56. Zhao, X.; Zhao, Y. Current Situation, Protection and Management of Wetlands along the Yellow River in Henan. *Wetl. Sci. Manag.* **2016**, *12*, 27–29.
57. Cousins, S.A.; Eriksson, O. The influence of management history and habitat on plant species richness in a rural hemiboreal landscape, Sweden. *Landsc. Ecol.* **2002**, *17*, 517–529. [CrossRef]
58. Chen, Y.; Xu, N.; Yu, Q.; Guo, L. Ecosystem service response to human disturbance in the Yangtze River economic belt: A case of Western Hunan, China. *Sustainability* **2020**, *12*, 465. [CrossRef]
59. Arias, M.E.; Cochrane, T.A.; Norton, D.; Killeen, T.J.; Khon, P. The flood pulse as the underlying driver of vegetation in the largest wetland and fishery of the Mekong Basin. *Ambio* **2013**, *42*, 864–876. [CrossRef] [PubMed]
60. Chen, J.; Guo, Y.; Lu, X.; Ding, S.; Su, S.; Guo, J.; Li, Q. Species diversity of herbaceous communities in the Yiluo River Basin. *Acta Ecol. Sin.* **2012**, *32*, 3021–3030. [CrossRef]
61. Wu, X.; Luo, J.; Chen, J.; Li, B. Spatial patterns of invasive alien plants in China and its relationship with environmental and anthropological factors. *J. Plant Ecol.* **2006**, *30*, 576–584.

Article

Barium, Lithium and Titanium Content in Herbs of Mid-Field Wet Depressions in East-Central Poland

Elżbieta Malinowska [1,*] and Jan Novak [2]

1   Faculty of Agrobioengineering and Animal Husbandry, Siedlce University of Natural Sciences and Humanities (UPH), Prusa 14 Street, 08-110 Siedlce, Poland

2   Faculty of Agrobiology and Food Resources, Slovak University of Agriculture in Nitra, Tr. A. Hlinku 2, 949 76 Nitra, Slovakia; jan.novak460@gmail.com

*   Correspondence: elzbieta.malinowska@uph.edu.pl

Abstract: This paper presents the results of research on the Ba, Li and Ti content in six species of herbs sampled from mid-field wet depressions and from the soil. These temporary flooded depressions were surrounded by arable crops, permanent grassland and shrubby vegetation. The research area was located in the eastern part of the Mazovian Voivodeship, east-central Poland. The following plants were used in the experiment: corn mint (*Mentha arvensis* L.), purple marshlocks (*Comarum palustre* L.), silverweed (*Potentilla anserina* L.), yarrow (*Achillea millefolium* L.), yellow loosestrife (*Lysimachia vulgaris* L.) and gypsy-wort (*Lycopus europaeus* L.). The Li, Ba and Ti content of plants, bottom sediment and soil was determined by the ICP-AES method after previous dry mineralization. Of the six herb species, *Mentha arvensis* L. was with the greatest accumulation potential of the chemical elements. However, no excessive Ba, Li and Ti content was found in herbs growing at different distances from arable fields, permanent grassland and shrubby vegetation. The highest Ba content was found in periodically flooded soil (zone II), while the highest amounts of Li and Ti were recorded in non-flooded soil (zone III).

Keywords: trace elements; herbal plants; soil; natural aquatic ecosystems

Citation: Malinowska, E.; Novak, J. Barium, Lithium and Titanium Content in Herbs of Mid-Field Wet Depressions in East-Central Poland. *Diversity* **2022**, *14*, 189. https://doi.org/10.3390/d14030189

Academic Editors: Mateja Germ, Igor Zelnik, Matthew Simpson and Michael Wink

Received: 27 January 2022
Accepted: 4 March 2022
Published: 5 March 2022

Publisher's Note: MDPI stays neutral with regard to jurisdictional claims in published maps and institutional affiliations.

1. Introduction

With the socio-economic development of rural areas and as a consequence of structural changes in agriculture, environmental degradation is observed, eliminating natural biocenoses and having a direct impact on the deterioration of water and air quality [1–3]. The share of such habitats in the landscape is difficult to determine so their ecological analysis and impact on the functioning of neighboring terrestrial ecosystems is important. Vegetation in small aquatic ecosystems and adjacent areas is usually exposed to excessive amounts of nutrients and heavy metals, resulting from improper farming methods. According to many authors [4,5], chemical composition of plants forming communities in such areas is affected by soil properties. environmental pollution and climatic conditions. Determination of the chemical composition of plants growing in small aquatic ecosystems is extremely important due to the possibility of obtaining herbal raw materials in their natural state [6–9]. The collection of medicinal plants in their natural state in agricultural, wasteland and other areas is supervised in Poland by the Regional Directorate for Environmental Protection. According to literature reports [10], herbal plants have a number of properties exerting a beneficial effect on human physical and mental health; they have antioxidant, antibacterial and anti-inflammatory properties, regulating digestion and preserving food. The aim of the paper is to assess the content of Ba, Li and Ti in bottom sediment, soil and in six herbal species growing in mid-field depressions of the Siedlce Plateau.

The choice of the chemical elements was determined by the fact that there were few scientific reports on Ba, Li and Ti in herbs from natural and agricultural ecosystems. The literature was mainly concerned with the content of heavy metals. The present research proves that even in unpolluted areas, the bioaccumulation of these elements is very diverse and requires constant monitoring.

2. Materials and Methods

The material was sampled between mid-June and the end of July, most often at the beginning of the flowering and full flowering stages. Only the aboveground parts of plants were used. Six species of herbs growing in the mid-field depressions of Siedlce Plateau were selected: silverweed (*Potentilla anserina* L.), corn mint (*Mentha arvensis* L.), yarrow (*Achillea millefolium* L.), purple marshlocks (*Comarum palustre* L.), yellow loosestrife (*Lysimachia vulgaris* L.) and gypsy-wort (*Lycopus europaeus* L.). Natural aquatic ecosystems were located at different places, with three mid-field ponds (A and B) in the commune of Jabłonna Lacka, in the village of Bujały Mikosze, and one area (C) in the commune of Sabnie, the village of Grodzisk (Figure 1). Each mid-field aquatic depression was surrounded by a different type of land cover: A—arable fields, B—permanent grassland and C—wild shrubs. Plants and soil were sampled from three different soil moisture zones: flooded (I), periodically flooded (II) and non-flooded (III). A mid-field location of a pond, diversity of growing vegetation and vicinity of agricultural crops were taken into account while selecting a study location. The pool area ranged from 15,000 m^2 (Grodzisk—C) to 850 m^2 (Bujały Mikosze—A). In administrative terms, the ponds were located in the eastern part of the Mazovian voivodeship, east-central Poland. The area constitutes an ecoregion called Green Lungs. According to the National Ecological Network 'Econet-Polska' [11], it is situated in the central part of the postglacial zone. Soils in that part of the country mainly developed from glacial tills during the central Polish glaciation [12]. They are dominated by luvisols, gleysols and brown soils, made from loamy sand and dusty loams. There are also rusty soils in the area [13].

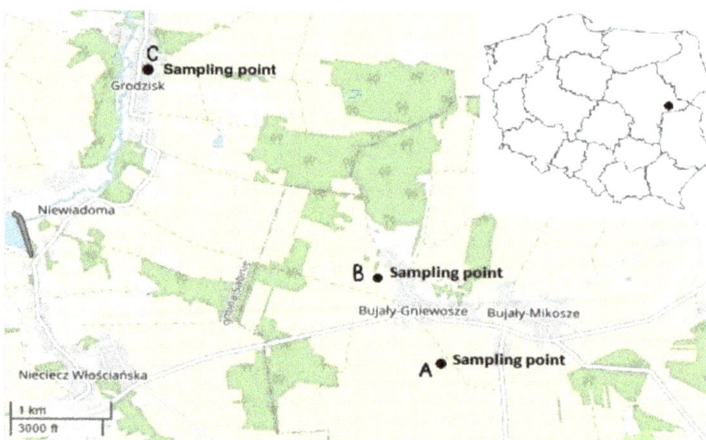

Figure 1. Locations of the sampling points (https://www.openstreetmap.org/#map=12/52.1036/22.1357 (accessed on 4 March 2022)) [14].

From each moisture zone, 85 plant samples representing each species were collected. They were not rinsed with water before chemical analyses. Chemical composition of the above-ground part of the plant was determined. Samples of soil and bottom sediment from each mid-field depression were also collected. The outer layer of bottom sediment was sampled with a corer. After draining its water, the sediment was homogenized and dried at first at room temperature and then at 105 °C to constant weight. The plant material was dried in the same way. The plant material was ground to a diameter of 0.25 mm and 1 g was put into a porcelain crucible, after which the organic substance was dry-oxidized in a muffle furnace, at 450 °C for about 15 h. When oxidation was completed, what remained in the ash was minerals in the form of carbonates and oxides, and partly in the form of phosphates. Then, 10 mL of diluted HCl (1:1) was added to the crucible and the contents were evaporated on a sand bath to decompose carbonates and separate silica. After the addition of 5 mL of 10% HCl, crucible contents were passed through a hard filter to a 100 mL volume flask and supplemented with distilled water up to the mark. In the presence of hydrochloric acid, silicic acid was precipitated from silicates in the form of white gelatinous mass with high water content ($H_2SiO_3 \cdot nH_2O$), which at a temperature of 100–200 °C converted into sparingly soluble SiO_2 [15]. The content of selected chemical elements was determined using Inductively Coupled Plasma-Atomic Emission Spectrometry (ICP–AES), while calibration was performed using standard Merck solutions. An internal quality control procedure was used to verify the accuracy of the methods. Two measurements were taken for each series of samples with the recovery being within the 85–115% range. The limit of detection for Ba, Li and Ti was 0.01 mg kg^{-1}.

Soil samples were dried at 105 °C, then ground and sifted through a sieve with a mesh diameter of 1mm [15]. Samples of soil, bottom sediment and plant material were mineralized in the same way. In the soil and in the outer layer of sediment, pH in 1 mol KCl dm^{-3} was determined by the potentiometric method.

Accumulation coefficient (AC) of selected chemical elements in herbs were calculated on the basis of their content in plant biomass and soil. The following formula was used [16]:

$$AC = Cp/Cs$$

(Cp—content of the metal in the plant, Cs—content of the metal in the soil):
AC < 0.01—no accumulation;
AC < 0.1—slight accumulation;
AC = 1—medium accumulation;
AC > 1—high accumulation.

The results were statistically processed, with the significance of the results assessed using variance analysis. Tukey's test at 5% significance level ($p < 0.05$) was used to determine significant differences between means. Statistica version 10.0 StatSoft was used for the calculations [17]. The values of linear correlation coefficients between the Ba, Li and Ti content of the soil from the individual moisture zones and their content in the aboveground parts of selected herbs were calculated. The standard deviation was calculated to determine the variability of Ba, Li and Ti content in plants and the soil.

3. Results and Discussion

The Ba content of the six herbs ranged between 6.7 and 82.1 mg kg^{-1} (Table 1). It significantly varied depending on the plant species and its habitat (moisture zone). The greatest variation of Ba content was recorded in *Potentilla anserina* L., *Mentha arvensis* L. and *Comarum palustre* L. in all sampling locations. The SD values for those plants were higher than for other herbs. As the average of all moisture zones, the highest content of Ba was recorded in the biomass of *Potentilla anserina* L. and *Mentha arvensis* L. The average content of Ba in *Potentilla anserina* L. was the largest in depression A (73.5 mg kg^{-1}), surrounded by arable fields, while in the *Mentha arvensis* L., it was the largest in depression B (43.1 mg kg^{-1}), surrounded by permanent grassland. The highest SD value for this chemical element was recorded for plants sampled from the depression surrounded by arable fields (A). Among the moisture zones, the highest standard deviation of Ba content, average from all locations, was recorded in herbs collected in the periodically flooded zone (II). The Ba content of *Lycopus europaeus* L., *Lysimachia vulgaris* L. and *Achillea millefolium* L. was several times lower than the average, not exceeding 14 mg kg^{-1}. Kuziemska et al. [18] recorded higher Ba content in *Dactylis glomerata* L. growing on non-limed soil than in plants on limed soil, with 17.57 and 20.6 mg kg^{-1}, respectively. According to Kabata-Pendias and Pendias [19], Ba concentration in plants is most often between 10 and 150 mg kg^{-1}. In the present experiment, the average content of this element was within the same range.

Table 1. Ba concentration ($mg\ kg^{-1}$) in the biomass of some herbs.

Species	A					B					C					SD		
	I	II	III	Mean	SD	I	II	III	Mean	SD	I	II	III	Mean	SD	I	II	III
								Ba										
Potentilla anserina L.	60.3	82.1	78.0	73.5 F	6.8	51.3	63.2	50.9	55.1 E	5.4	56.0	55.4	60.7	57.4 D	4.5	4.6	10.4	11.2
Mentha arvensis L.	32.1	35.2	43.1	36.8 E	3.7	41.3	49.9	38.1	43.1 D	4.3	45.7	41.9	40.1	42.6 C	3.2	4.2	8.5	2.1
Achillea millefolium L.	18.8	15.2	16.9	16.9 C	1.3	7.9	6.7	8.0	7.6 A	1.0	10.9	12.3	11.8	11.7 A	1.5	5.4	5.4	6.4
Comarum palustre L.	30.2	39.4	30.5	33.4 D	3.7	29.9	30.1	28.9	29.7 C	1.6	31.5	32.1	30.9	31.5 B	1.4	1.8	4.2	1.2
Lysimachia vulgaris L.	10.2	12.5	11.2	11.3 A	1.1	13.6	13.0	13.4	13.3D	0.80	14.2	13.1	12.9	13.4 A	1.0	2.3	1.0	1.0
Lycopus europaeus L.	11.2	13.9	14.1	13.1 B	2.4	12.4	12.9	13.1	12.8 B	1.4	12.0	12.9	12.9	12.6 A	0.45	1.2	0.61	1.4
mean	27.1 a	33.0 c	32.3 b	30.8	3.8	26.0 b	29.3 c	25.4 a	26.9	2.4	28.4 a	27.9 a	28.2 a	28.2	2.0	3.3	7.8	3.8

A—area with the cultivated field; B—area with permanent grasslands; C—area with bushes; I—flooded zone; II—periodically flooded zone; III—non-flooded zone; D—standard deviation; different lowercase letters in the same row or different uppercase letters on the same column indicate significant differences between treatments.

The physiological role of Ba in plants has not yet been clarified, but the toxic effects of its high doses on the human body (gastrointestinal disorders, muscular hypoplasia and difficulty breathing) have been recorded [19,20]. For humans, the lethal dose of barium chloride (LD_{50}) is approx. $14\ mg\ kg^{-1}$. The mechanism of the toxicity of this metal consists in the displacement of potassium and the binding of sulphate anions [19]. When studying the effect of sampling location, it was found that the highest average concentration of Ba was in herbs from the depression surrounded by arable fields (A) and the lowest from that surrounded by permanent grassland (B). This might be due to the fact that the accumulation of this element, due to its chemical properties, was to some extent limited by the presence of Ca and Mg, acting antagonistically to Ba [21,22].

Varied Li content in the herb plants was recorded, depending on the species and moisture zone (Table 2). The greatest variation of Li content, based on the calculated SD values, was found for *Achillea millefolium* L. and *Mentha arvensis* L. in all sampling locations. In the case of moisture zones, the highest standard deviation of Li content was in herbs of the periodically flooded (II) and not-flooded (III) zones. As the average of all moisture zones, the highest content of Li was recorded in the biomass of *Achillea millefolium* L., *Mentha arvensis* L. and *Lycopus europaeus* L. This content varied depending on the surroundings of the mid-field ponds. In the area surrounded by arable fields (A), the most Li was found in the biomass of *Achillea millefolium* L., *Mentha arvensis* L. and *Lycopus europaeus* L., in the area surrounded by permanent grassland (B) in the biomass of *Mentha arvensis* L., *Lycopus europaeus* L. and *Achillea millefolium* L. and in the area surrounded by shrubs (C) in the biomass of *Achillea millefolium* L., *Lycopus europaeus* L. and *Mentha arvensis* L. Kabata-Pendias and Pendias [19] report that the accumulation of Li in plants varies between species, often proportional to its concentration in the soil. It is assumed that a content higher than $35\ mg\ kg^{-1}$ Li in dry matter is toxic to plants [19]. In their research, Kabata-Pendias and Pendias [19] recorded the accumulation of this element in halophilic plants, reaching up to $1000\ mg\ kg^{-1}$. This group included *Solanacea* and *Rosacea* families. In *Datura inoxia* (*Solanacea*) and *Potentilla palustris* (*Rosacea*), both of which are medicinal plants, Li concentrations did not exceed $2.4\ mg\ kg^{-1}$ and $2.0\ mg\ kg^{-1}$, respectively [23]. According to other data, the Li content of cereals and vegetables was $0.5–3.4\ mg\ kg^{-1}$ [24]. The studies of Szilagyi et al. [25] confirm that Li is toxic. Kasperczyk and Wiśniowska-Kielian [26] reported that Li content in dicotyledonous plants was higher than in monocotyledonous ones. This chemical element regulates growth and the activity of certain enzymes, hormones, vitamins and translocation factors [27]. When analyzing moisture zones, it was found that herbs growing in the periodically flooded zone accumulated Li more intensively compared to the other zones of the aquatic depressions, especially in locations A (surrounded by arable fields) and B (grassland). At the location overgrown with shrubs (C), the highest average concentration of Li was recorded in the non-flooded zone.

Table 2. Li concentration (mg kg^{-1}) in the biomass of some herbs.

Species	A					B					C					SD		
	I	II	III	Mean	SD	I	II	III	Mean	SD	I	II	III	Mean	SD	I	II	III
								Li										
Potentilla anserina L.	0.25	0.24	0.24	0.24 B	0.001	0.17	0.24	0.30	0.24 B	0.03	0.20	0.27	0.39	0.89 B	0.03	0.05	0.01	0.10
Mentha arvensis L.	2.9	3.2	3.0	3.1 E	1.0	5.4	9.1	6.1	6.9 E	2.4	3.9	2.5	3.1	3.2 D	1.0	1.3	3.4	1.9
Achillea millefolium L.	3.0	4.2	2.2	3.1 E	1.9	3.8	1.5	3.0	2.8 D	1.3	5.9	10.1	10.1	8.9 E	3.0	1.7	5.0	5.8
Comarum palustre L.	1.0	1.2	1.4	1.2 C	0.18	1.6	1.6	1.8	1.7 C	0.50	1.2	1.02	1.5	1.2 C	0.3	0.31	0.30	0.20
Lysimachia vulgaris L.	0.12	0.15	0.13	0.13 A	0.002	0.20	0.19	0.18	0.19 A	0.001	0.18	0.19	0.17	0.18 A	0.001	0.03	0.02	0.02
Lycopus europaeus L.	2.6	3.0	3.3	2.9 B	1.0	3.5	3.9	3.6	3.6 F	0.80	2.6	3.9	3.9	3.5 D	0.95	1.1	0.61	0.35
mean	1.7 a	2.0 b	1.7 ab	1.8	0.68	2.4 a	2.8 b	2.5 ab	2.6	0.84	2.7 a	3.1 b	3.2b	3.0	0.885	0.75	1.6	1.4

A—area with the cultivated field; B—area with permanent grasslands; C—area with bushes; I—flooded zone; II—periodically flooded zone; III—non-flooded zone; SD—standard deviation; different lowercase letters in the same row or different uppercase letters on the same column indicate significant differences between treatments.

The accumulation of Ti in the plants from different habitats varied significantly under the influence of experimental factors (Table 3). On average, the most Ti was found in *Mentha arvensis* L., almost 4–5 times higher than in the other herbs. The highest SD of Ti was recorded for *Mentha arvensis* L. and *Lycopus europaeus* L. sampled from the depression surrounded by arable fields (A). In subsequent sampling locations (B) and (C), the greatest variation in Ti content was found for *Mentha arvensis* L., *Achillea millefolium* L. and *Comarum palustre* L. The highest standard deviation of Ti content in plants was recorded in the non-flooded zone (III). Depending on the plant species and their locations, the concentration of Ti ranged from 1.9 to 22.9 mg kg^{-1}. Malinowska and Kalembasa [28] confirmed its varied accumulation depending on the plant species. According to the authors, *Lolium multiflorum* accumulated much more of this element than corn and sunflower. On average, the largest Ti accumulation was recorded in herbs sampled from the depression surrounded by permanent grassland (B) (7.3 mg kg^{-1}) and the smallest from that covered with shrubs (C) (5.9 mg kg^{-1}). Analyzing moisture zones around the natural aquatic ecosystems, it was found that the highest concentration of Ti was in plants of the periodically flooded zone in area A (arable fields) and B (permanent grassland), and in the plants of the flooded zone in area C (wild shrubs). The accumulation of Ti in various crops is significantly influenced by pH of the soil and its content of organic substance [28]. The authors reported that in forage plants from control, Ti content was higher than in plants treated with municipal sewage sludge. There are many factors that influence its uptake. With a long retention of water on the soil (more than 48 h), anaerobic processes begin to occur. Periodic flooding of the soil leads to many changes in its environment, which affects nutrient uptake by plants. According to the literature, Ti content of plants varies greatly between 0.2 and 80 mg kg^{-1} DM, while some species, such as horsetail, moss and nettle, may contain more than 300 mg Ti kg^{-1} DM [19,29].

Table 3. Ti concentration (mg kg^{-1}) in the biomass of some herbs.

Species	A					B					C					SD		
	I	II	III	Mean	SD	I	II	III	Mean	SD	I	II	III	Mean	SD	I	II	III
								Ti										
Potentilla anserina L.	3.1	3.4	4.9	3.8 A	1.3	1.9	4.1	3.6	3.2 B	1.0	4.6	2.7	4.0	3.8 A	0.41	1.3	1.1	2.0
Mentha arvensis L.	18.5	20.5	13.0	17.3 D	3.8	20.8	22.9	19.5	21.0 F	2.1	17.5	15.5	14.7	15.9 D	2.0	1.9	2.0	4.8
Achillea millefolium L.	3.8	4.0	4.6	4.1 B	1.0	5.5	9.5	7.5	7.5 E	2.9	4.6	3.8	3.0	3.8 A	1.1	1.7	3.2	3.1
Comarum palustre L.	4.5	2.5	3.1	3.4 A	1.2	5.2	4.9	5.7	5.3 D	1.1	3.6	4.2	4.7	4.2 C	1.3	2.0	1.4	1.2
Lysimachia vulgaris L.	3.0	2.6	4.0	3.2 A	1.3	2.8	2.3	2.1	2.4 A	0.32	4.1	3.9	3.5	3.9 B	0.70	1.6	1.0	1.0
Lycopus europaeus L.	3.7	8.3	7.4	6.5 C	3.0	4.1	4.5	4.9	4.6 C	0.58	3.7	3.9	3.7	3.8 A	0.42	1.2	2.4	2.0
mean	6.1 a	6.9 b	6.2 ab	6.4	1.9	6.7 a	8.1 b	7.2c	7.3	1.3	6.3b	5.7 ab	5.6a	5.9	0.83	1.6	1.9	2.4

A—area with the cultivated field; B—area with permanent grasslands; C—area with bushes; I—flooded zone; II—periodically flooded zone; III—non-flooded zone; SD—standard deviation; different lowercase letters in the same row or different uppercase letters on the same column indicate significant differences between treatments.

The content of Ba, Li and Ti in the bottom sediment of the three mid-field ponds located in the Siedlce Plateau varied (Table 4). Most of the Li was recorded in the sediment from the pond surrounded by permanent grassland (B), with the greatest amounts of Ba and Ti in sediment of the depression surrounded by arable fields (A). The top layer of the bottom sediment from the three midfield ponds was characterized by similar pH_{KCl} values, ranging from 6.4 to 6.5.

Table 4. Ba, Li and Ti concentration in bottom sediment (mg kg^{-1} DM).

Metal	A	B	C	Mean
Ba	47.1	22.2	6.5	25.3
Li	2.2	3.6	0.66	2.2
Ti	52.3	13.8	36.8	34.3
pH_{KCl}	6.5	6.4	6.5	-

A—area with the cultivated field; B—area with permanent grasslands; C—area with bushes.

The Ba, Li and Ti content of the soil sampled from three mid-field depressions varied significantly (Table 5). Most Li and Ti was recorded in the soil of the non-flooded zone (III), on average 2.6 and 37.1 mg kg^{-1}, respectively, and Ba in transitional zone soil (II), with 31.4 mg kg^{-1}. Soil Ba content ranged from 10.1 to 51.1 mg kg^{-1}, with its greatest accumulation in herbs collected in the non-flooded zone (Table 1). On average across sampling locations, Ba content was 28.2 mg kg^{-1}, 2.3 mg kg^{-1} for Li and 29.2 mg kg^{-1} for Ti. According to Kabata–Pendias and Pendias [19], most Polish soils contain less than 50 mg Ba kg^{-1}. The content of Ba ions in the soil is generally low as this element easily enters the soil sorption complex and is retained there [22]. Kabata–Pendias and Pendias [19] report that Ba is strongly bound by clay minerals, iron-manganese and phosphate concretions and sulfur compounds. Increased soil acidity and sulfur content can affect the uptake of this chemical element by plants. Barium easily migrates in soils together with circulating water, and is leached deep into the soil profile, or is subject to concentration in the top surface layer [19]. In non-contaminated soils, the amount of Ti usually ranges between 1.5 and 60 mg kg^{-1}, and the amount of Li between 0.01 and 40 mg kg^{-1}. However, some soils may contain even very significant amounts of Ti [19]. The lowest pH was recorded in the soil of flooded zones (I) of all depressions, due to the predominance of anaerobic conditions and biological sorption processes (Table 5). In the periodically flooded and non-flooded zones, soil pH was slightly higher, but it did not exceed 6.2. On average, the highest SD value for all chemical elements was recorded in soil from the depression surrounded by permanent grassland (B).

Table 5. Ba, Li and Ti concentration in soil (mg kg^{-1} DM).

Area	Ba					Li					Ti					pH_{KCl}		
	I	II	III	Mean	SD	I	II	III	Mean	SD	I	II	III	Mean	SD	I	II	III
A	35.4	50.1	17.8	34.4 AB	15.1	0.95	2.5	1.1	1.5 A	1.0	28.9	36.1	12.8	25.9 A	8.5	5.6	5.8	6.2
B	29.4	34.1	51.1	38.2 B	20.0	3.1	3.2	4.9	3.7 B	1.2	13.9	14.7	49.4	25.9 AB	18.3	5.9	6.0	6.0
C	11.2	10.1	14.9	12.1 A	3.4	1.6	1.5	1.9	1.7 C	0.50	30.2	27.1	49.2	35.6 B	15.0	5.4	5.9	6.1
mean	25.3 a	31.4 b	27.9 ab	28.2	12.8	1.9 a	2.4 ab	2.6 b	2.3	0.91	24.4 a	26.1 ab	37.1 b	29.2	13.9	-	-	-

A—area with the cultivated field; B—area with permanent grasslands; C—area with bushes I—flooded zone; II—periodically flooded zone; III—non-flooded zone; SD—standard deviation; different lowercase letters in the same row or different uppercase letters on the same column indicate significant differences between treatments.

The accumulation coefficient of Ba, Li and Ti was significantly dependent on the plant species and moisture zones (Tables 6–8). Thus, the accumulation coefficient of Ba and Li in herbs was estimated as average, while for Ti it was low [16]. The coefficient values for plants indicated varied accumulation of metals from the soil, which was also demonstrated in many other studies [19,30–34]. The highest Ba, Li and Ti accumulation coefficient was recorded in *Mentha arvensis* L., in addition to Ba in *Potentilla anserina* L. (Table 6) and Li in

Achillea millefolium L. (Table 7). Coefficient values for some elements can be high, ranging from 1 to 10 [35]. However, for the elements studied in the present experiment those values were rather low. The highest accumulation coefficient of Ba and Li in the soil-plant system was recorded in plants from the area adjacent to the pond overgrown with shrubs (C), with 2.4 and 1.8, respectively (Tables 6 and 7). For Ti, the highest value (0.39) was recorded in plants harvested from area B, with permanent grassland (Table 8). To date, little has been known about the content of Ba, Li and Ti in food and their amounts entering the human body, as the data available in the literature are incomplete.

Table 6. Ba accumulation coefficient in the plants.

Species	A				B				C			
	I	II	III	Mean	I	II	III	Mean	I	II	III	Mean
						Ba						
Potentilla anserina L.	1.7	1.6	4.4	2.6 C	1.7	1.9	0.99	1.5 F	5.0	5.5	4.1	4.9 D
Mentha arvensis L.	0.91	0.70	2.4	1.3 B C	1.4	1.5	0.75	1.2 E	4.1	4.2	2.7	3.6 C
Achillea millefolium L.	0.53	0.30	0.95	0.59 A	0.27	0.19	0.16	0.21 A	0.98	1.2	0.79	0.99 A
Comarum palustre L.	0.85	0.79	1.7	1.1 B	1.0	0.88	0.57	0.82 D	2.8	3.2	2.1	2.7 B
Lysimachia vulgaris L.	0.29	0.25	0.63	0.39 A	0.46	0.38	0.26	0.37 C	1.3	1.3	0.86	1.1 A
Lycopus europaeus L.	0.32	0.28	0.72	0.46 A	0.42	0.38	0.26	0.35 AB	1.1	1.3	0.87	1.1 A
mean	0.77 ab	0.66 a	1.8 b	1.1	0.89 b	0.86 ab	0.49 a	0.75	2.5 a	2.8 b	1.9 c	2.4

A—area with the cultivated field; B-area with permanent grasslands; C—area with bushes; I—flooded zone; II—periodically flooded zone; III—non-flooded zone, different lowercase letters in the same row or different uppercase letters on the same column indicate significant differences between treatments.

Table 7. Li accumulation coefficient in the plants.

Species	A				B				C			
	I	II	III	Mean	I	II	III	Mean	I	II	III	Mean
						Li						
Potentilla anserina L.	0.26	0.09	0.22	0.19 B	0.06	0.07	0.06	0.06 A	0.13	0.19	0.20	0.17 AB
Mentha arvensis L.	3.0	1.3	2.8	2.4 D	1.8	2.8	1.3	1.9 F	2.6	1.7	1.6	1.9 D
Achillea millefolium L.	3.2	1.7	1.9	2.3 D	1.2	0.45	0.70	0.79 D	3.8	7.5	5.2	5.5 E
Comarum palustre L.	1.1	0.49	1.3	0.96 C	0.52	0.51	0.37	0.47 C	0.79	0.69	0.76	0.75 C
Lysimachia vulgaris L.	0.13	0.06	0.12	0.10 A	0.07	0.06	0.04	0.06 AB	0.11	0.13	0.09	0.11 A
Lycopus europaeus L.	2.7	1.2	3.0	2.3 D	1.1	1.2	0.73	1.0 E	1.7	2.7	2.0	2.1 DE
mean	1.7 b	0.82 a	1.6 ab	1.4	0.80 ab	0.86 a	0.53 a	0.73	1.5 a	2.1 b	1.6 ab	1.8

A—area with the cultivated field; B—area with permanent grasslands; C—area with bushes; I—flooded zone; II—periodically flooded zone; III—non-flooded zone; different lowercase letters in the same row or different uppercase letters on the same column indicate significant differences between treatments.

Table 8. Ti accumulation coefficient in the plants.

Species	A				B				C			
	I	II	III	Mean	I	II	III	Mean	I	II	III	Mean
						Ti						
Potentilla anserina L.	0.11	0.09	0.38	0.13 A	0.14	0.28	0.07	0.164 A	0.15	0.09	0.08	0.11 A
Mentha arvensis L.	0.64	0.57	1.0	0.74 D	1.5	1.6	0.39	0.145 A	0.58	0.57	0.29	0.48 B
Achillea millefolium L.	0.13	0.11	0.36	0.20 B	0.39	0.64	0.15	0.397 C	0.15	0.14	0.06	0.12 A
Comarum palustre L.	0.16	0.07	0.25	0.16 A	0.38	0.33	0.12	0.275 B	0.12	0.17	0.09	0.13 A
Lysimachia vulgaris L.	0.10	0.07	0.31	0.16 A	0.21	0.16	0.04	0.135 A	0.14	0.15	0.07	0.12 A
Lycopus europaeus L.	0.13	0.23	0.58	0.31 C	0.29	0.32	0.09	0.239 B	0.12	0.15	0.08	0.12 A
mean	0.21 b	0.19 a	0.48 c	0.29	0.48 b	0.55 c	0.15 a	0.392	0.21 b	0.21 ab	0.11 a	0.18

A—area with the cultivated field; B—area with permanent grasslands; C—area with bushes; I—flooded zone; II—periodically flooded zone; III—non-flooded zone; different lowercase letters in the same row or different uppercase letters on the same column indicate significant differences between treatments.

The calculated values of the correlation coefficients between the Ba, Li and Ti content of the soil and in herbs showed no significant relationship between Ba and Ti (Table 9). A lack of significant correlations between the content of the elements in plants and in the soil may be related to the soil pH, ranging from 5.4 to 6.2 (Table 5), which could have affected the soluble forms of these elements. On the other hand, a positive significant correlation was found between the Li content of the soil and its amounts in the biomass of *Mentha arvensis* L. and *Comarum palustre* L. The values of the correlation coefficient (r) were, respectively, r = 0.711 and r = 0.815.

Table 9. Linear correlation coefficient between the Ba, Li and Ti content of the soil and of herb biomass.

Species	Soil		
	Ba	**Li**	**Ti**
Potentilla anserina L.	0.144	0.069	0.073
Mentha arvensis L.	−0.459	0.711 *	−0.130
Achillea millefolium L.	−0.080	−0.314	−0.253
Comarum palustre L.	0.263	0.815 *	0.173
Lysimachia vulgaris L.	−0.145	0.592	−0.206
Lycopus europaeus L.	−0.139	0.391	−0.143

$p \leq 0.05$, $n = 9$, * significance.

4. Conclusions

The content of those chemical elements varied significantly depending on the plant species and the place of growing. As the average of all moisture zone, the highest content of Ba was found in the biomass of *Potentilla anserina* L. and *Mentha arvensis* L. and Li in the biomass of *Achillea millefolium* L., *Mentha arvensis* L. and *Lycopus europaeus* L.

The content of Ti was several times higher in the biomass of *Mentha arvensis* L. than in other herbs. There were no excessive levels of Ba, Li and Ti in the biomass of the herbs growing at various distances from arable fields, permanent grassland and habitats overgrown with wild shrubs.

The content of Ba, Li and Ti in the soil of different moisture zones surrounding the three mid-field depressions of the Siedlce Plateau varied significantly. The highest Ba content was recorded in periodically flooded soil (zone—II), while the highest Li and Ti contents were recorded in the non-flooded soil (zone—III). In the soil of the flooded zone (I), the content of these elements was the lowest.

The results indicated that various systems of management of agricultural areas in the central-eastern region of Poland did not cause excessive bioaccumulation of the studied chemical elements in herbal plants and in the soil.

Author Contributions: E.M. and J.N. have equally contributed to this article. All authors have read and agreed to the published version of the manuscript.

Funding: The results of the research carried out under the research theme No. 134/21/B were financed from the science grant, granted by the Ministry of Science and Higher Education.

Institutional Review Board Statement: Not applicable.

Data Availability Statement: The datasets supporting the conclusions of the article are included within the article.

Conflicts of Interest: The authors declare no conflict of interest.

References

1. Nicolet, P.; Ruggiero, A.; Biggs, J. Second European pond workshop: Conservation of pond biodiversity in a changing European landscape. *Ann. Limnol. Int. J. Limnol.* **2007**, *43*, 77–80. [CrossRef]
2. Boix, D.; Biggsb, J.; Cereghino, R.; Hull, A.; Kalettka, T.; Oertli, B. Pond research and management in Europe: Small is beautiful. *Hydrobiologia* **2012**, *689*, 1–9. [CrossRef]

3. Gruchelski, M.; Niemczyk, J. Requirements and activities in the area of rationalization of polish agricultural policy (in a socio-economic and environmental protection aspects). *Post. Techn. Przetw. Spożyw.* **2018**, *2*, 103–107. (In Polish)
4. Chan, K. Some aspects of toxic contaminants in herbal medicines. *Chemosphere* **2003**, *52*, 1361–1371. [CrossRef]
5. Zheljazkova, V.D.; Jeliazkova, E.A.; Kovacheva, N.; Dzuhurmanski, A. Metal uptake by medicinal plant species grown in soils contaminated by a smelter. *Environ. Exp. Bot.* **2008**, *64*, 207–216. [CrossRef]
6. Regulations of the Ministry of the Environment (Dz. U. z 2004 r. Nr 168, poz. 1764). 2004. Available online: http://isap.sejm.gov.pl/isap.nsf/DocDetails.xsp?id=WDU20041681764 (accessed on 1 April 2021).
7. Deng, H.; Ye, Z.H.; Wong, M.H. Accumulation of lead, zinc, copper and cadmium by 12 wetland plants species thriving in metal-contaminated sites in China. *Environ. Pollut.* **2004**, *132*, 29–40. [CrossRef]
8. Garcia-Rico, L.; Leyva-Perez, J.; Jara-Marini, M.E. Content and daily intake of copper, zinc, lead, cadmium and mercury from dietary supplements in Mexico. *Food Chem. Toxicol.* **2007**, *45*, 1599–1605. [CrossRef]
9. Arslan, H.; Güleryüz, G.; Leblebici, Z.; Kırmızı, S.; Aksoy, A. *Verbascum bombyciferum* Boiss. (Scrophulariaceae) as a possible bio-indicator for the assessment of heavy metals in the environment of Bursa, Turkey. *Environ. Monit. Assess.* **2010**, *1631*, 105–113. [CrossRef]
10. Bielicka-Giełdoń, A.; Ryłko, E. Estimation of metallic elements in herbs and species available on the Polish market. *Pol. J. Environ. Stud.* **2013**, *22*, 1251–1256.
11. Liro, A. *The Concept of the National Ecological Network ECONET—Poland*; IUCN Poland Foundation: Warsaw, Poland, 1995.
12. Kuźnicki, F.; Białousz, S.; Składowski, P.; Szafranek, A.; Kamińska, K.; Ziemińska, A. Physicochemical properties of soils of southeastern part of the Masovian Lowland as a criterion of their typology. *Soil Sci. Annu. Wars.* **1979**, *30*, 3–20. (In Polish)
13. IUSS Working Group WRB. *World Reference Base for Soil Resources 2014, Update 2015. International Soil Classification System for Naming Soils and Creating Legends for Soil Maps*; World Soil Resources Reports No. 106; FAO: Rome, Italy, 2015.
14. Location of the Analysed Samples. Available online: http://www.openstreetmap.org/#map=12/52.1036/22.1357 (accessed on 16 September 2021).
15. Kalembasa, S.; Kalembasa, D.; Żądełek, J. *Soil Science and Agricultural Chemistry*; Practice Script; WSRP: Siedlce, Poland, 1989; p. 434. (In Polish)
16. Wesołowski, M.; Radecka, I. Medicinal plants. Elemental composition, source of minerals for plants, indicators of environmental contamination by heavy metals. *Pol. Pharm.* **2003**, *59*, 911–919. (In Polish)
17. StatSoft, Inc. 2011 STATISTICA (Data Analysis Software System), Version 10. Available online: www.statsoft.com (accessed on 2 April 2021).
18. Kuziemska, B.; Jaremko, D.; Bik, B.; Jakubicka, M. Content of lead and barium in biomass of cocksfoot (*Dactylis glomerata* L.) cultivated on soil contaminated with nickel under different liming and organic fertilization. *Zesz. Nauk. UP Wrocław Roln. CV* **2013**, *594*, 27–36. (In Polish)
19. Kabata-Pendias, A.; Pendias, H. *Biogeochemistry of Trace Elements*; PWN: Warsaw, Poland, 1999; p. 398. (In Polish)
20. Migaszewski, Z.M.; Gałuszka, A. *Fundamentals of Environmental Geochemistry*; WNT: Warsaw, Poland, 2007. (In Polish)
21. Kalembasa, D.; Jaremko, D. Total content of lithium, barium and strontium in soils of South Podlasie Lowland. *Pol. J. Environ. Stud.* **2006**, *15*, 320–325.
22. Jaremko, D.; Kalembasa, D. A comparison of methods for the determination of cation exchange capacity of soils. *Ecol. Chem. Eng. S* **2014**, *21*, 487–498.
23. Lovkova, M.Y.; Sokolova, S.M.; Buzuk, G.N. Lithium-concentrating plant species and their pharmaceutical usage. *Dokl. Biol. Sci.* **2007**, *412*, 64–66. [CrossRef]
24. Armendáriz, C.R.; Gonzalez-Weller, D.; Gutierrez Fernandez, A.J.; Hardisson De La Torre, A. Levels of lithium in the six most taken groups of food among the canarian population. *Abstr. Toxicol. Lett.* **2010**, *196S*, 298. [CrossRef]
25. Szilagyi, M.; Balogh, I.; Anke, A.; Szentmihalyi, S. Biochemical aspects of lithium toxicity. *Acta Med. Leg. Soc.* **1985**, *35*, 167–180.
26. Kasperczyk, M.; Wiśniowska-Kielian, B. The influence of fertilization and undersowing on cadmium, nickel and lithium contents in meadow sward. *Zesz. Probl. Post. Nauk Rol.* **1997**, *448*, 207–212. (In Polish)
27. Rogóż, A.; Wiśniowska-Kielian, B. *Lithium Content and Circulation in the Natural Environment*; AGH: Kraków, Poland, 2019; p. 216. (In Polish)
28. Malinowska, E.; Kalembasa, S. The influence of sewage sludge doses and liming on the content of Li, Ti, Ba, Sr and As in the test plants biomass. *J. Ecol. Eng.* **2011**, *27*, 110–119. (In Polish)
29. Kabata-Pendias, A.; Szteke, B. *Trace Elements in the Geo- and Biosphere*; Wyd. IUNG-PIB: Puławy, Poland, 2012; pp. 229–233. (In Polish)
30. Królak, E. Heavy Metal Content in Filling Dusts, Soil and Dandelion (*Taraxacum officinale* Webb.) in Suthern Podlasie Lowland. *EJPAU.* 2001. Environ Dev 4,1. Available online: www.ejpau.pl/series/volume4/environmental/art-01.html (accessed on 10 April 2021).
31. Lock, K.; Jansen, C.R. Influence of Aging on Metal availability in Soil. *Rev. Environ. Contam. Toxicol.* **2003**, *178*, 1–21.
32. Rosselli, W.; Keller, C.; Bosch, K. Phytoextraction capacity of trees growing a metal contaminated soil. *Plant Soil* **2003**, *256*, 265–272. [CrossRef]
33. Malinowska, E.; Jankowski, K. Impact of agricultural chemicals on selected heavy metals accumulation in herb plants. *Appl. Ecol. Environ. Res.* **2016**, *14*, 479–487. [CrossRef]

34. Malinowska, E.; Jankowski, K. Copper and zinc concentrations of medicinal herbs and soil surrounding ponds on agricultural land. *Landsc. Ecol. Eng.* **2017**, *13*, 83–188. [CrossRef]
35. Kloke, A.; Sauerbeck, D.R.; Vetter, H. The contamination of plants and soils with heavy metals and the transport of metals in the terrestrial food chains. In *Changing Metal Cycles and Human Health*; Nriagu, J.O., Ed.; Springer: Berlin/Heidelberg, Germany; New York, NY, USA; Tokyo, Japan, 1984; pp. 113–141.

Article

Aquatic Insects in Habitat-Forming Sponges: The Case of the Lower Mekong and Conservation Perspectives in a Global Context

Nisit Ruengsawang [1], Narumon Sangpradub [2] and Renata Manconi [3,*]

[1] Division of Biology, Faculty of Science and Technology, Rajamangala University of Technology Krungthep, Bangkok 10120, Thailand
[2] Applied Taxonomic Research Center, Department of Biology, Faculty of Science, Khon Kaen University, Khon Kaen 40002, Thailand
[3] Department of Veterinary Medicine, University of Sassari, 07100 Sassari, Italy
* Correspondence: rmanconi@uniss.it or renata.manconi@uniss.it

Citation: Ruengsawang, N.; Sangpradub, N.; Manconi, R. Aquatic Insects in Habitat-Forming Sponges: The Case of the Lower Mekong and Conservation Perspectives in a Global Context. *Diversity* **2022**, *14*, 911. https://doi.org/10.3390/d14110911

Academic Editors: Mateja Germ, Igor Zelnik, Matthew Simpson and Simon Blanchet

Received: 15 July 2022
Accepted: 11 October 2022
Published: 27 October 2022

Publisher's Note: MDPI stays neutral with regard to jurisdictional claims in published maps and institutional affiliations.

Abstract: Shallow water sponges settled on a raft along the Pong River (Lower Mekong Basin, Thailand) were investigated to highlight the taxonomic richness, composition, relative abundance and lifestyle of sponge-dwelling aquatic Insecta. The three-dimensional biogenic structures of the model sponges hosted 4 orders of Insecta, belonging to 10 families and 19 genera/species, able to strictly coexist at the level of the sponges in aquiferous canals and/or at the body surface, and/or dwelling in the extracellular matrix. On the basis of the identified 379 larvae and pupae, Trichoptera and Diptera were found to be the dominant inhabitants of *Corvospongilla siamensis* (Demospongiae: Spongillida), endemic to Southeast Asia. In the focused lotic ecosystem, dominated by soft bottoms, sponges play a functional role. Insecta use sponges as a substratum, nursery ground, food source, and shelter microhabitat, protecting them from predation and environmental aggression. Moreover, their feeding behavior indicates the insects' adaptive traits to recycle sponge siliceous spicules as a source of exogenous material to strengthen the larval–pupal cases and the digestive system. The results of the Thai sponge model contribute to the inventory of global engineering species richness, ecosystem types, and biogeographic diversity, thus raising awareness for freshwater biodiversity conservation. In this regard, the present data, along with the worldwide inventory, focus on sponges as (a) key habitat-forming species for aquatic insect assemblages, (b) ecosystem engineers in river/lake/wetland ecosystems, providing water purification, the processing of organic matter, recycling of nutrients, and freshwater–terrestrial coupling, and (c) promising candidates in restoration projects of tropical freshwater ecosystems by bioremediation.

Keywords: entomofauna diversity; tropical Insecta-Porifera association; global inventory; larval microhabitat and behavior; restoration and bioremediation

1. Introduction

Spongillida (Demospongiae) inhabit and settle on a wide array of natural and artificial substrata in lotic and lentic water and wetlands in all continents, except Antarctica [1–3]. In the food web of freshwater ecosystems, sponges play a functional role as filter feeders, consuming the pelagic resources linking the pelagic and benthic trophic webs [4]. A sponge can filter up to 35 mL min^{-1}(cm sponge)$^{-3}$ [5]. This water volume is related to the sponge volume, oscule diameter, and water pumping rates, and could be affected by the temperature, food concentration, suspended sediment concentration, water flow, and viscosity [6–8]. This behavior was highlighted also in stressed conditions of sponge microcosms living in water rich with organic and bacterial loads in polluted sites during in situ experimental sponge aquaculture [9].

During their long-lasting lifecycle, continental sponges' growth forms range from encrusting to massive and arborescent in the sessile adult phase, with a body size from a few millimeters to more than 1 m in diameter [1,2,10].

It is well known that the Spongillida (families Lubomirskiidae, Metaniidae, Potamolepidae, and Spongillidae) are associated with a diversified entomofauna worldwide, i.e., Ephemeroptera, Plecoptera, Odonata, Hemiptera, Megaloptera, Neuroptera, Trichoptera, Lepidoptera, Coleoptera, and Diptera, plus several other phyla ranging from bacteria to fishes and amphibians [1,2,11–13]. However, little is known about the diversity of sponge-dwelling aquatic insects and their relationship in Southeast Asia [14].

Investigations were carried out in the Lower Mekong Basin (Thailand, Oriental–Himalayan Region) to study (a) the composition, relative abundance, and taxonomic richness of sponge-dwelling aquatic insects, (b) their behavioral and ecological relationship, and (c) their biogeographic diversity. The data of insect–sponge associations are discussed also in comparison with the global dataset to highlight the sponges' ecosystem services and to raise awareness of the biodiversity conservation of these peculiar living freshwater habitats.

2. Materials and Methods

2.1. Study Area

This study was carried out in the Pong River (16°46′19.9″ N, 102°38′3.25″ E) which is a tributary of the Lower Mekong Basin (Figure 1) by the largest hydrographic basin of Northeastern Thailand. This regulated river provides an important water resource for agriculture, domestic uses, aquaculture, recreation, electricity generation, and industrial purposes, especially in Khon Kaen Province. Water is released from the outlet of the Ubolratana Dam to the lower part of the Pong River for water management, such as supplying water for agriculture and preventing water degradation by the operation of the Electricity Generating Authority Thailand, both in the wet (June–November) and dry (December–May) seasons. This eutrophic river, dominated by silty bottoms and having the most altered water quality of this region, since 1993, has been polluted from multiple sources, e.g., agricultural run-off, a pulp and paper mill, and a fertilizer factory [15,16]. Nevertheless, the water quality was in a generally fair condition, according to the surface water quality standard and water quality index [17].

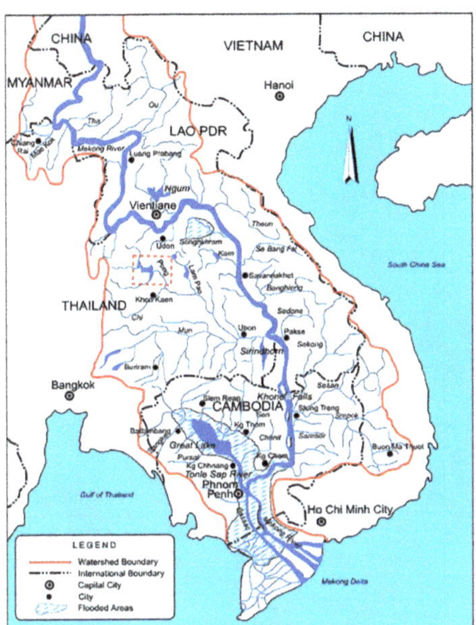

Figure 1. Map of the Lower Mekong River hydrographic basin with the study site along Pong River (indicated by a red dashed-line square) in Northeastern Thailand, Southeast Asia (modified from van Zalinge et al. [18]).

The study site, Ban Huai Sai, is located at 165 m asl (16°46′19.9″ N, 102°38′3.25″ E) ~1.5 km downstream from the outlet of the Ubolratana Dam (Figure 2). The riverbed is characterized by a maximum depth of 8 m and width of 70 m. The submerged substrata range from silty bottoms, timbers, and aquatic plants to artificial manufacture along the banks, e.g., fish cage nets, submerged structures of rafts, and nylon lines. The riparian vegetation mainly consists of the rain tree, *Samanea saman* (Jacq.) Merr. The aquatic vegetation mainly consists of *Colocasia* sp., *Hydrilla verticillata* (L. f.) Royle, *Cyperus* sp., and *Nephrolepis* sp. Sponges of the order Spongillida (Demospongiae) were found on the twigs and the bottom of the riverbed [19]. Flourishing and abundant sponge populations were also settled on the submerged artificial substrata, floating raft construction, buoys, and recreational fishing nylon lines trapped in the raft buoys at the study site (Figure 2a).

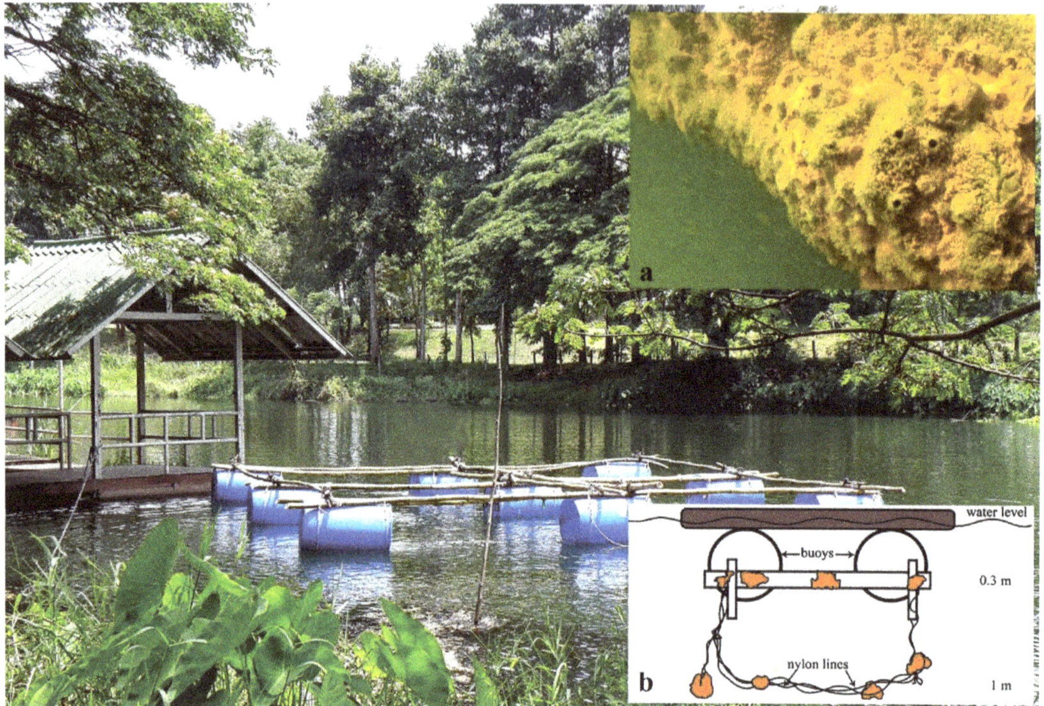

Figure 2. View of the study site. (**a**) *Corvospongilla siamensis* on artificial substrata. (**b**) Schematic diagram of a raft, submerged structures, and nylon lines colonized by sponges (orange) at 0.3–1 m water depth in the study site.

2.2. Sampling Methods

Fourteen specimens of sponges were collected in February 2008 (dry season, northeast monsoon) at two water depths, i.e., 0.3 m from the submerged raft construction (seven sponges), and 1 m on the nylon lines (seven sponges) (Figure 2b). These specimens were ascribed to *Corvospongilla siamensis*, Manconi & Ruengsawang, 2012. The identification of the sponges was based on the growth form, skeletal architecture, and microtraits. Detailed analyses of the gemmular theca and its gemmuloscleres, together with the skeletal microscleres and megascleres, were carried out [20–22]. All the sponges considered in the study were covered with a hand net of 450 µm mesh size, carefully detached from the substrate, and then placed into plastic bags. To allow most of the aquatic insects to leave their hosts, each sample with its river water was placed in a plastic tray for half an hour (ambient temperature), and then it was preserved in 70% ethanol. Each sponge was photographed in the field. In the laboratory, the samples were washed through a 500 µm mesh-size sieve, and the associated insects remaining on the surface of the sponges and sieve were collected by forceps. Each sponge was also carefully dissected to collect the insects that remained within its body. All the aquatic insects were sorted, identified to the lowest possible taxonomic level, and counted by screening at the microscope. The taxonomic identification of the insects from Ban Huai Sai was based on Morse et al. [23], Epler [24], Sangpradub and Boonsoong [25], and Cranston [26]. The similarities in the aquatic insect composition between the screened sponge specimens were analyzed by cluster analysis, using Jaccard's similarity index, by PC-ORD software version 5.10.

The gut contents of the dominant insect taxa were investigated qualitatively by light microscopy for their primary dietary components. Thirty specimens of each taxon were dissected under a stereomicroscope; their digestive tracts were mounted on slides and examined under a compound microscope. The cases of the trichopterans ($n = 15$) and tubes of the chironomids ($n = 15$) were examined and photographed by a scanning electron microscope (SEM, LEO model 1450VP) to determine whether these insects use the sponge spicules to construct their cases or tubes.

The water quality parameters were evaluated at two shallow water depths, (0.3 m and 1 m), during the dry season–northeast monsoon. The water quality parameters, such as dissolved oxygen (DO), electrical conductivity, total dissolved solids (TDS), water temperature, pH and flow velocity, were measured in the field, along with the collections of sponges at each depth, by portable digital instruments (a dissolved oxygen meter YSI model 57 (Yellow Spring Instrument Co. Inc., Yellow Spring, MP, USA); conductivity meter Fisher Scientific model 09-326-2 (Fisher Scientific International Inc., Pittsburgh, PA, USA); pH meter HACH sension[TM]1 (HACH Company, Loveland, CO, USA); Genuine Gurley[®] current meter model D625 digital pygmy meter with Model 1100 Flow Velocity Indicator (Gurley Precision Instruments, New York, NY, USA)). The suspended solids (SS), turbidity, nitrate, and phosphate were measured in the laboratory by the HACH DR/2010 Spectrophotometer (HACH Company, Loveland, CO, USA). The biological oxygen demand (5 days, 20 °C) (BOD_5) was also evaluated. The difference in the environmental parameters between the two water depths was determined by the Mann–Whitney U test with SPSS for Windows software, version 16.

3. Results

3.1. Physicochemical Parameters

The mean values of most of the physicochemical parameters were not significantly different between 0.3 m and 1 m, except flow velocity and DO, which were significantly higher at 1.0 m ($p < 0.05$, Supplementary Materials Table S1). Due to unavailable data from the deep zone, the comparative analyses on the suspended solids, turbidity, nitrate, and phosphate at the different water depths were not carried out.

3.2. Composition and Abundance of Associated Aquatic Insects

A total of 379 larvae and pupae of aquatic insects belonging to 4 orders, 10 families, and 19 taxa were recorded, by microscopical screening, from 14 sponge samples at two water depths at Ban Huai Sai (Table 1; Figures 3–6). The dominant larvae colonizing *C. siamensis* were Trichoptera (48.81%) and Diptera (47.75%) (Figure 3b–d), whereas Neuroptera and Ephemeroptera were less represented. The relative abundance of associated insect taxa varied from 3 to 94 individuals per sponge specimen and ranged from 140 specimens (14 taxa) at 0.3 m depth to 239 specimens (16 taxa) at 1 m depth. The egg jelly envelops of *Povilla heardi* Hubbard, 1984 (Ephemeroptera, Polymitarcyidae) were also found on the sponge surface (Figure 6a,b).

The most dominant taxa in the Pong River were *Ecnomus* spp., followed by *Amphipsyche meridiana* Ulmer, 1909 (Trichoptera), *Polypedilum* sp., and *Xenochironomus* sp. (Diptera), making up 77.8% of the specimens. The larval stages of first three genera are clingers, but *Xenochironomus* is a burrower in sponges. Eleven taxa were shared by both water depths, i.e., *Polypedilum* sp. was present in all the samples, followed by *Ecnomus* and *Xenochironomus*, respectively. The chironomid larvae *Ablabesmyia* and *Dicrotendipes* were found from six out of the seven sponges at the 0.3 m water depth. The larvae of *Sisyra* sp. were only 2.3% ($n = 9$), the pupae of chironomids 2.3% ($n = 9$), and the caddisflies 1% ($n = 4$) were also recorded (Table 1). The cluster analysis, based on Jaccard's similarity index, relying on presence/absence data, showed differences in the aquatic insect composition among the fourteen sponge samples and separated them into two main clusters (Figure 7). One cluster comprised the shallow-water samples (0.3 m) and the other comprised all but one of those from the deep water (1 m).

Table 1. Checklist of aquatic insects associated with *Corvospongilla siamensis* in the Pong River (Lower Mekong, Thailand). Total number of larvae and/or pupae in each sponge sample at two water depths. Presence of eggs is indicated by asterisk (*).

Taxa	Number of Individuals														Total
	Depth 0.3 m							Depth 1.0 m							
	1	2	3	4	5	6	7	1	2	3	4	5	6	7	
Ephemeroptera															
Baetidae gen. sp.	-	-	-	-	-	-	-	1	-	-	-	-	-	-	1
Caenidae															
Caenodes sp.	-	2	-	-	-	1	-	-	-	-	-	-	-	-	3
Polymitarcyidae															
Povilla heardi	-	-	-	-	-	-	-	*	-	-	-	-	-	-	-
Neuroptera															
Sisyridae															
Sisyra sp.	-	3	1	-	1	-	-	2	-	1	1	-	-	-	9
Trichoptera															
Ecnomidae															
Ecnomus spp.	-	8	3	1	4	1	4	21	27	12	7	1	2	1	92
Hydropsychidae															
Amphipsyche meridiana	-	-	-	-	-	-	-	62	20	-	-	-	-	2	84
Hydroptilidae gen. sp.	-	1	-	-	-	-	-	4	1	1	-	-	-	-	7
Trichoptera fam. gen.	-	-	-	-	-	-	1	-	-	1	-	-	-	-	2
Diptera															
Chironomidae															
Ablabesmyia sp.	1	3	3	1	1	1	-	-	-	-	-	-	-	-	10
Demicryptochironomus sp.	-	2	-	-	-	-	1	-	-	-	-	-	-	-	3
Dicrotendipes sp.	-	2	6	1	2	1	1	-	-	-	-	-	1	-	14
Nanocladius sp.	-	-	-	-	-	1	-	1	-	-	-	-	1	-	3
Paramerina sp.	-	3	2	-	-	-	1	-	-	-	1	-	1	-	8
Polypedilum sp.	2	12	9	5	5	1	7	3	4	8	6	8	3	3	76
Rheotanytarsus sp.	-	-	-	-	-	-	-	-	5	1	2	1	-	-	9
Tanytarsus sp.	-	1	-	2	-	-	2	-	-	1	-	-	-	-	6
Xenochironomus sp.	-	9	10	1	3	-	2	-	7	3	3	2	1	2	43
Chironomidae gen. sp.	-	1	4	-	-	-	-	-	1	-	1	1	-	-	8
Ceratopogonidae															
Bezzia sp.	-	-	-	-	-	-	-	-	-	1	-	-	-	-	1
Total	3	47	38	11	16	6	19	94	65	29	21	13	9	8	379

3.3. Gut Content Analysis

The gut content analysis of the four dominant taxa, i.e., *Ecnomus* spp., *A. meridiana*, *Polypedilum* sp., and *Xenochironomus* sp., respectively, showed a predominance of detritus, followed by diatoms, fine detritus material, rotifers, and sponge siliceous spicules (Figure 8c,d). Generally, the entire strongyle megascleres characterized by rounded tips were commonly found in the digestive tract, whereas fragments of these spicules rarely occurred. The percentage of spicules occurring in the gut contents of the representative dominant genera showed a wide range, i.e., *Ecnomus* spp. (70%; Trichoptera), *A. meridiana* (56.6%; Trichoptera), *Xenochironomus* sp. (6.6%; Chironomidae), and *Polypedilum* sp. (6.6%; Chironomidae). In addition, megascleres and microscleres of sponges were also found in the chironomids *Ablabesmyia* sp. and *Paramerina* sp.

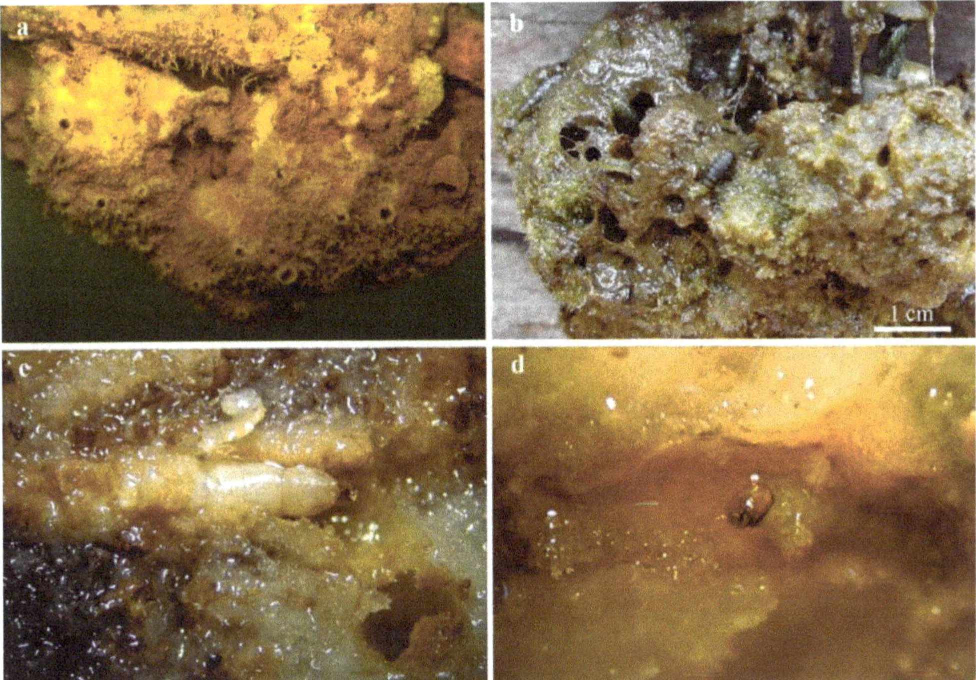

Figure 3. *Corvospongilla siamensis* (Porifera: Spongillida) in vivo settled on the raft artificial substrata of Pong River (Lower Mekong). (**a**) Outline of sponge conulose body surface with a variety of microhabitats at level of irregularly scattered crevices and ridges, and tube-like exhalant aquiferous openings (oscules). (**b**) Caddisfly larvae (*Amphipsyche meridiana*, Hydropsychidae, Trichoptera) crawling on a cavity-rich surface of a greenish sponge. (**c,d**) Chironomid larvae (Chironomidae, Diptera) burrowing in the inner sponge body.

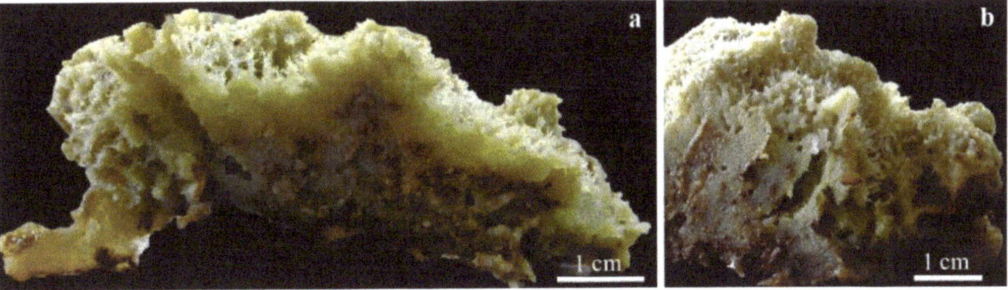

Figure 4. *Corvospongilla siamensis* morphology (Pong River, Lower Mekong). (**a,b**) Heterogeneous microhabitat at the sponge irregular surface with ridges and large tube-like exhalant aquiferous aperture openings.

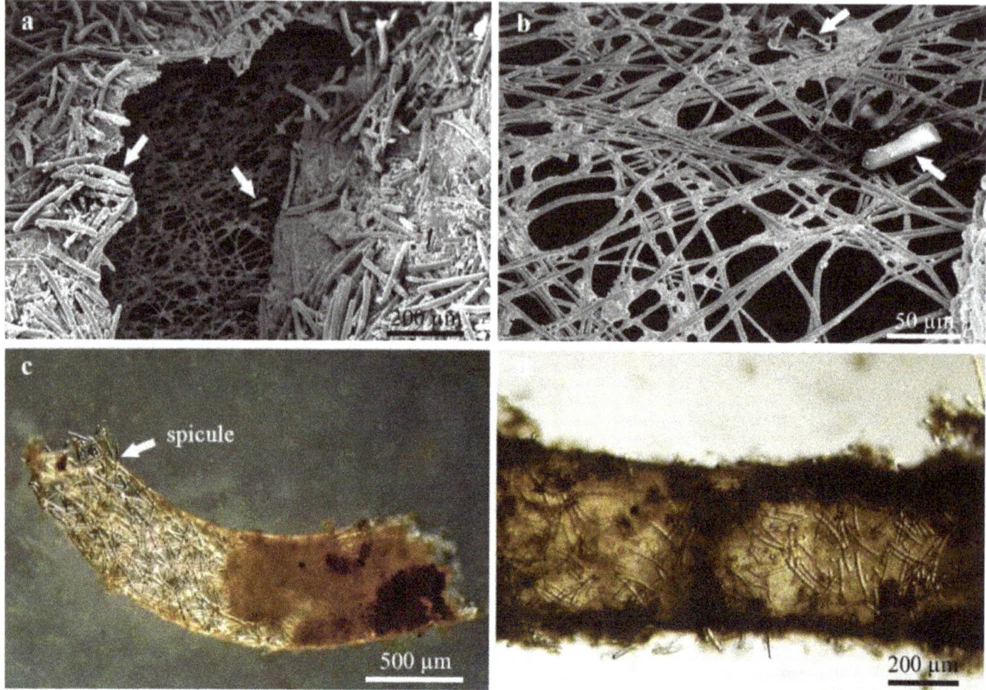

Figure 5. Sponge-dwelling caddisfly and non-biting midge larvae recycling siliceous spicules of *Corvospongilla siamensis* (Lower Mekong, Southeast Asia). (**a**,**b**) Pupae case of spongivorous *Amphipsyche meridiana* (Trichoptera, Hydropsychidae) reinforced by megascleres and microscleres (arrows, SEM) adhering to the silk network of the inner and outer cases. (**c**,**d**) Case of a chironomid larva (Chironomidae, Diptera) with a dense network of spicules (arrow, LM).

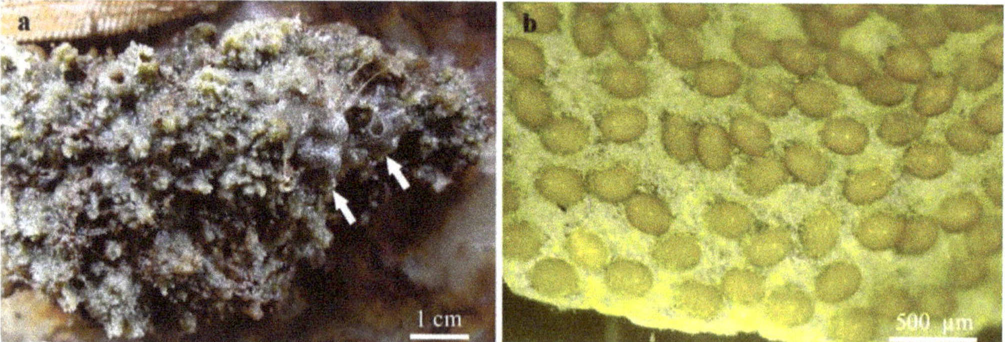

Figure 6. Eggs of *Povilla heardi* (Ephemeroptera, Polymitarcyidae) on *Corvospongilla siamensis* (Lower Mekong, Southeast Asia). (**a**) Egg jelly envelopes (arrows) on the irregular surface of a live sponge. (**b**) Eggs viewed by light microscopy.

The texture of the pupal case of the *A. meridiana* studied by an SEM showed that silk thread is a major component of the case, strengthened by two types of sponge spicules (microscleres, gemmuloscleres, and megascleres) on both the inner and outer layers of the cases (Figure 5a,b). The observations on the tubes of the chironomid larvae and pupae stages highlighted that both are made, in part, by sponge spicules (Figure 5c,d).

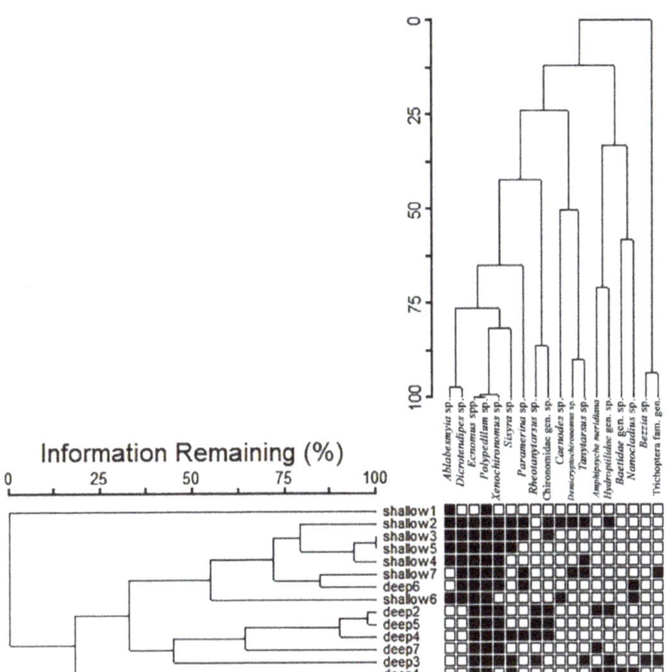

Figure 7. Occurrence and composition of sponge-dwelling larvae of aquatic insects associated with 14 samples of *Corvospongilla siamensis* (Pong River, Lower Mekong). One cluster comprised the shallow-water samples (0.3 m) and the other all but one of those from deep water (1 m).

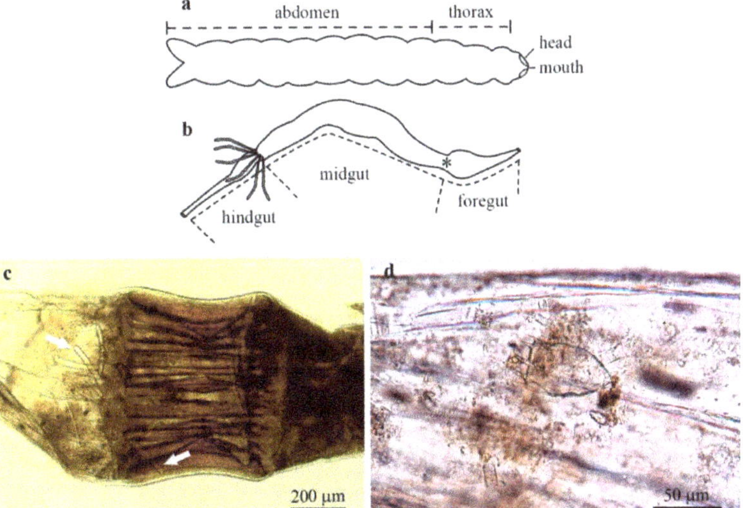

Figure 8. *Amphipsyche meridiana* (Trichoptera, Hydropsychidae) associated with *Corvospongilla siamensis* (Lower Mekong, Southeast Asia). (**a**,**b**) Scheme of body and digestive tract of the sponge-eating caddisfly larva (adapted from Sangpradub and Giller [27]). The position of the proventriculus (*) is indicated in (**b**). (**c**) Siliceous spicules with blunt tips (arrows) in the proventriculus. (**d**) Food debris in the midgut, i.e., diatoms and a rotifer.

4. Discussion

4.1. Freshwater Sponges as Microhabitat for Associated Insects

The data highlight that many insects can inhabit sponges in the Mekong River Basin (Table 1), matching the previous studies on sponge-dwelling entomofauna, reporting 10 orders of Insecta inhabiting almost 20 Spongillida species from all the biogeographic regions, except Antarctica (Table 2). The flourishing population of *Corvospongilla* (Figure 9) is a temporary microhabitat for the long-lasting larval and pupal stages of many insects, either at the body surface and canals, or within the extracellular matrix, suggesting that, also, other Spongillida species play the same functional role as the habitat-forming species in the same site, e.g., *Oncosclera asiatica* [28].

Table 2. Worldwide inventory of aquatic Insecta inhabiting Spongillida (Porifera) so far reported. Sponge species belong to families Lubomirskiidae, Metaniidae, Potamolepidae, and Spongillidae. Taxa and countries listed in alphabetical order. References listed in chronological order.

Insect Taxa	Sponge Taxa	References	Biogeographic Region/Country
Ephemeroptera (Baetidae, Caenidae, Polymitarcyidae) Neuroptera (Sisyridae) Trichoptera (Ecnomidae, Hydroptilidae, Hydropsychidae) Diptera (Ceratopogonidae, Chironomidae)	*Corvospongilla siamensis* *Eunapius carteri* *Radiospongilla crateriformis* *Spongilla alba* Spongillida fam. gen. sp.	**Present paper**, Annandale [29], Schröder [30]	**Oriental-Himalayan** Thailand; Borneo; India; Java; Philippines
—	*Spongilla alba* Spongillida fam. gen. sp.	Schröder [30], Mansell [31]	**Afrotropical** Madagascar; South Africa
Trichoptera (Hydroptilidae)	Spongillida fam. gen. sp.	Wells [32], Wells & Johanson [33], Forteath & Osborne [34], Forteath et al. [35]	**Australasian** Australia; New Zealand; New Caledonia; Tasmania
Diptera (Chironomidae) Ephemeroptera (Baetidae, Caenidae) Neuroptera (Sisyridae) Megaloptera (Sialidae) Trichoptera (Apataniidae, Brachycentridae, Ecnomidae, Hydropsychidae, Hydroptilidae, Leptoceridae, Limnephilidae, Polycentropodidae, Psychomyiidae)	*Ephydatia fluviatilis* *Ephydatia muelleri* *Eunapius fragilis* *Heteromeyenia* sp. *Lubomirskia baikalensis* *Spongilla lacustris* Spongillida fam. gen. sp.	Pavesi [36], Arndt [37], Rezvoi [38], Berg [39], Gaumont [40], Moretti & Corallini-Sorcetti [41], Konopacka & Socinski [42], Kamaltynov et al. [43], Weissmair & Mildner [44], Gugel [45], Corallini & Gaino [46,47], Gaino et al. [48], Wallace et al. [49], Weinberg et al. [50], Loru et al. [51], Schiffels [52], Sokolova & Palatov [53], Palatov & Sokolova [54], Zvereva et al. [55], Ostrovsky [56]	**Palaearctic** Algeria; Austria; China; East Siberia; Denmark; England; France; Germany; Italy; Japan; Mongolia; Poland; Russia; Scandinavia; Serbia; Spain
Diptera (Chironomidae, Simuliidae) Coleoptera (Dytiscidae) Ephemeroptera (Polymitarcyidae, Tricorythidae) Hemiptera (Belostomatidae) Lepidoptera (Pyralidae) Neuroptera (Sisyridae) Odonata (Coenagrionidae, Libellulidae) Trichoptera (Hydropsychidae, Polycentropodidae)	*Acalle recurvata* *Drulia uruguayensis* *Corvoheteromeyenia australis* *Corvospongilla seckti* *Heteromeyenia cristalina* *Metania spinata* *Metania reticulata* *Oncosclera navicella* *Oncosclera spinifera* *Radiospongilla inesi* *Trochospongilla paulula*	Melão & Rocha [57], Roque et al. [58], Fusari et al. [59–62], Clavier et al. [63], Hamada et al. [64], Bowles [65], Ardila-Camacho & Martins [66], Assmar & Salles [67], Fisher et al. [68], da Silva Fernandes [69]	**Neotropical** Argentina; Belize; Bolivia; Brazil; Chile; Cuba; French Guiana; Guatemala; Guyana; Honduras; Mexico; Panama; Paraguay; Peru; Suriname; Uruguay; Venezuela
Neuroptera (Sisyridae) Megaloptera Trichoptera (Leptoceridae)	*Eunapius fragilis* *Ephydatia muelleri* *Ephydatia fluviatilis* *Eunapius igloviformis* *Dosilia radiospiculata* *Heteromeyenia argirosperma* *Heteromeyenia baileyi* *Heteromeyenia repens* (as *H. baileyi*) *Racekiela ryderi* (as *Heteromeyenia ryderi*) *Radiospongilla crateriformis* *Spongilla lacustris* *Ephydatia robusta* (as *Meyenia subdivisa*) *Trochospongilla horrida* *Trochospongilla leidii* (as *T. leidyi*) *Trochospongilla muelleri* *Trochospongilla pennsylvanica* (as *Tubella pennsylvanica*)	Brown [70], Parfin & Gurney [71], Roback [72], Heiman & Knight [73], Poirrier [74], Poirrier & Arceneaux [75], Resh [76], Pennak [77], Matteson & Jacobi [78], Stoakes et al. [79], Clark [80], Pupedis [81], Whitlock & Morse [82], Bowles [65,83], Cover & Resh [84], Rothfuss & Heilveil [85], Fisher et al. [68], Bowles & Courtney [86]	**Nearctic** United States of America Canada

Figure 9. Dense assemblage of freshwater sponges with massive, irregular growth form as engineering species on farm nets in the Pong River (Lower Mekong, Southeast Asia). (**a**) The red area outlining the body of *Corvospongilla siamensis* is ~800 cm^2. (**b**) Living *Corvospongilla* sp. with associated algae. (**c**) Dry freshwater sponges on net.

4.2. Associated Insects and Environmental Conditions

Even though the dissolved oxygen (DO) was significantly different between the two depths, it did not seem to affect the sponge-dwelling insect assemblage structure, i.e., the relative abundance and taxonomic richness.

The flow-velocity data also support the fact that the DO is not a limiting factor to the sponges' inhabitants in the study site, which is located near the outlet of the dam. This could be due to both the oxygen contribution from the sponge algal symbiont (endocellular) and the hydrodynamic condition of the sponge microenvironment. Indeed, the sponges' pumping activity, involving conspicuous amounts of water, creates a microcirculation, both within the sponge body (inner canals and cavities) and at its irregular surface (inhalant openings with incurrent and oscules with outcurrent water), matching the previous data [87,88].

Although the composition of the associated aquatic insects between the shallow and deep zones is similar in sharing 11 taxa, it seems that the flow-velocity values could be a factor that affects the number of *Amphipsyche meridiana*, inhabiting only two samples in the deeper zone, while the higher values may favor these net-spinning larvae. In addition, the division between the "shallow and deeper parts" of the cluster differs in the greater abundance of Trichoptera, from the genus *Ecnomus* and *Amphipsyche*, requiring higher oxygen concentrations than the chironomids, which is also confirmed by the environmental data and could be the consequence of the higher flow velocity.

The colonization of *Polypedilum* sp. in all the samples indicates that *Corvospongilla* is a favorable habitat matching the known wide habitat range of this chironomid genus [24,69]. The present data, reporting Chironomidae larvae fairly commonly in the sponges from both the two shallow zones, match those of the Chironomidae [48,61,69], which is well represented in the Pong River's soft bottom, ranging from pristine to impacted water quality conditions [89]. Further in-depth analyses of the taxonomy and diversity of the chironomid larvae in the Pong River will be essential to clarify the sponge–midge interactions.

The present data also corroborate those of Gugel [45], who reported that the high larvae number of *Ecnomus tenellus* (Rambur, 1842) and *Hydropsyche* sp. occurring on European sponges was higher than their occurrence independently from sponges. The results confirm that sponges represent a favorable habitat for these larvae, as reported also by Annandale [29,90] for other hydrographic basins of the Asian tropics.

4.3. Aquatic Insect Abundance, Sponge Size and Body Architecture

The high number of sponge-dwelling aquatic insects in this study occurred in a silty riverbed with a dense population of long-living, massive sponges, characterized by a 3D body architecture, rich in large inner canals/cavities and exhalant apertures (Figures 3a,4a,b and 9), matching the investigations on marine sponge canals diameter and body morphology inhabited by invertebrates [91,92]. The sponge body architecture, rich in cavities, seems to provide a suitable space for some aquatic insects to segregate and avoid predators, as suggested also by the high density of freshwater African *A. meridiana* in vesicular rocks (i.e., volcanic rock pitted with many cavities/vesicles at its surface and inside), which was about 10 times higher than those of non-vesicular rocks [93].

Only the biomass and numbers of three species of amphipods inhabiting sponges are known to be correlated with the weight of a Baikalian sponge species, according to Kamaltynov et al. [43], but the other fauna including insects showed no correlation.

4.4. Interspecific Relationships between Aquatic Insects and Sponges

Several genera of Insecta have been recognized as obligatory associated with sponges, i.e., *Ceraclea* (Trichoptera), *Demeijerea*, *Oukuriella*, *Xenochironomus* (Chironomidae), *Climacea*, and *Sisyra* (Neuroptera) [46,47,53,54,60,69,72,76,82,94] (see Table 2). Spongivory is, however, rare, probably due to the production of sponge toxins and repellents [13,95]. Sisyridae larvae are well known as obligate predators (spongivory) of freshwater sponges [71,76,84,96].

In the present study, the type of sponge–insect relationship ranges from incidental to obligate. The larvae of *Sisyra* and *Xenochironomus* we discovered in sponges had never been found in the Pong River's macroinvertebrate community [89,97,98]. Weissmair and Warninger [99] stated that the *Sisyra* adults need rich riparian habitats, which occur rarely along the Pong River. This datum reinforces the opinion that the larvae of these taxa are strictly associated with freshwater sponges.

Little information has been published in Asia on the *Sisyra* biology, ecology, and phylogeny [100,101]. Only two species, *Sisyra indica* Needham, 1909, and *Sisyra maculata* Monserrat, 1981, are known from Thailand [83,84,102,103], suggesting further studies on neglected neuropterans and *Xenochironomus* species and their life cycles in tropical sponges of Asia would be worthwhile.

4.5. Sponges and Gut Content of Their Associated Aquatic Insects

Roback [72] reported that the digestive tracts of caddisfly larvae belonging to the genus *Athripsodes* were filled with debris and sponge spicules. These larvae ingest the sponge tissues and do not merely use them as an occasional substrate or case-building material. In this study, the gut contents of the caddisfly *A. meridiana* larvae (Figure 8a–d), showing high percentages of diatoms, may indicate that these net-spinning caddisfly larvae are filtering collectors. Our results agree with Boon [104], who concluded that the well-developed gastric mill plates of *A. meridiana* are ideal for crushing diatoms. Although it is known that sponges are not edible for most animals, due to their chemical deterrence, with hard skeletal components and bioactive compounds, the occurrence of many spicules in the digestive tracts may also suggest that *A. meridiana* feeds on the tissues of the sponge when moving on its surface, which matches the previous reports [47,72]. Entire siliceous strongyle spicules in the digestive tract could improve the grinding or filtering of foods, as suggested by not breaking down—even passing—the proventriculus, which consists of large tooth plates ranging from 533 to 656 µm in length (Figure 8c). Moreover, it seems that the spicules of *C. siamensis*, with typical blunt tips, cannot create a risk of damage to the gut wall of spongivor insects, contrasting with the condition of caddisfly *Ceraclea fulva* (Rambur, 1842), displaying an abnormally thick peritrophic membrane shielding its soft internal tissues from ingested spicules, with acute tips, of the sponge *Ephydatia fluviatilis* [46,47].

As for the pupal case of *A. meridiana*, an SEM analysis showed that sponge spicules were interwoven in the case wall, which is similar to those observed in *C. fulva*. So, these spicules seem to be useful material to reinforce the cases of the caddisfly larvae (Figure 5a,b). The chironomid larvae and pupae also utilized the sponge spicules for constructing their tubes and cases in the Pong River (Figure 5c,d), thus corroborating previous data on the occurrence of sponge spicules in the gut [94].

4.6. Biodiversity and Conservation Perspectives

The results of this study expand our knowledge of freshwater sponges in Southeast Asia, which is considered a biodiversity hotspot, highlighted by the 3007 new species of vascular plants, fishes, amphibians, reptiles, birds, and mammals discovered in the Greater Mekong region between 2007 and 2020 [105,106]. The description of three new Spongillida species and a new record in Thailand and Vietnam [20,28,107,108], and ongoing discoveries in Thailand (Ruengsawang unpublished data), indicate that the sponges inhabiting rivers, lakes, and wetlands in this region remain unexplored and are underestimated in their biodiversity. Environmental impacts on freshwater sponges have been reported, i.e., temperature, heavy metal contamination [109], and habitat change, as one of biodiversity loss drivers in Southeast Asia [110]. Therefore, a diversity assessment of this neglected fauna is urgently needed.

In the case of the Pong River, the biofouling by sponges in fish culture, resulting in clogging of the nets and decreased water circulation, was frequently found in the previous and current studies [19]. Fish farmers lift their cage nets from the water to dry and remove sponges for maintenance after fish harvesting (Figures 9 and 10). Skin irritation caused by sponge spicules was also reported by the farmers. The advantages and disadvantages of sponges biofouling in the fish culture in the Pong River need further research to fully understand to demonstrate the ecological value of these animals.

Figure 10. Fish cage maintenance after harvesting in the Pong River. (**a**) Fisherman with dried nets. (**b,c**) Freshwater sponges encrusting on farm nets.

From the conservation aspect, it is clear that freshwater sponges play an important role in aquatic ecosystems as biofilters, through their filter-feeding and water-pumping activities and are vulnerable habitats for aquatic insects. The opportunistic insect from this study, A. *meridiana*, adopts a variety of strategies to colonize and to use sponges as a microhabitat, which agrees with Boon [93]. Despite freshwater sponges not being threatened globally [1,10,109,111], the conservation and monitoring of freshwater sponges in tropical Asia are needed to understand their status.

Synthesis

The data indicate that, worldwide, freshwater sponges perform functional roles as ecosystem engineers, sensu Jones et al. [112]. They supply to insects: (a) persistent heterogeneous substrata and microclimatic conditions related to the sponges' massive to arborescent growth form and their physiological performance, (b) ecological refugia and breeding/nursery grounds as a sheltering microhabitat for settling and the development of eggs, (c) food source for detritivores by the continuous water-flow, trapping particulate organic matter, as a foraging area, for eating on sponge-associated algae, and as prey for predatory species on other hosted insects and a plethora of other invertebrates, (d) the morpho-functional role of the siliceous spicules to strengthen the larval insect's digestive system and the cases of larvae and pupae, (e) protection against predators, (f) shelter for withstanding floods, high siltation, and desiccation (dry season) during water level fluctuations. Despite the sponges' functional role as biofilters, on occasion, local people consider these animals responsible for economic damage, e.g., on the health of fish in farming cages along the Lower Mekong River. In agreement with the restoration science proposed by Ormerod [113], we suggest that sponge farming in inland water could be an effective nature-based solution to reduce biodiversity loss, as a management tool, by enhancing the water quality. This is particularly relevant in tropical freshwater to support bioremediation projects in restoring the environmental value, also in giving economic benefits to the local people. Another key point could be a targeted action on the people's environmental education in order to raise awareness of freshwater bioresources conservation.

Supplementary Materials: The following supporting information can be downloaded at: https://www.mdpi.com/article/10.3390/d14110911/s1; Table S1: Mean values (±SD) for physicochemical parameters during the dry season (northeast monsoon) at two shallow water depths inhabited by a flourishing population of *Corvospongilla siamensis* and its associated insects in the Pong River (Lower Mekong, Thailand).

Author Contributions: All authors contributed equally to the conceptualization, methodology, validation, formal analysis, investigation, resources, data curation, writing—original draft preparation, writing—review and editing, visualization, supervision, project administration, and funding acquisition. All authors have read and agreed to the published version of the manuscript.

Funding: This work was supported by the Thailand Research Fund (TRF), in collaboration with the Commission on Higher Education, under grant number MRG6080200, the Office of the Higher Education Commission, Thailand, under the program "Strategic Scholarships for Frontier Research Network" for the Joint Ph.D. Program Thai Doctoral degree. Funding to R. Manconi was provided by the Università di Sassari [grant Fondo di Ateneo per la Ricerca 2019-2021], Fondazione di Sardegna [grant FdS/RAS-2016/CUP J86C1800082005], Regione Autonoma della Sardegna [grant RAS2012-LR7/2007-CRP-60215], and Parco Nazionale dell'Asinara [grant PNA-2016].

Institutional Review Board Statement: Not applicable.

Informed Consent Statement: Not applicable.

Data Availability Statement: Not applicable.

Acknowledgments: We would like to thank Roberto Pronzato and Martin Spies for their valuable comments. We thank also Boonsong Kongsook (Applied Taxonomic Research Center, Division of Biology, Faculty of Science, Khon Kaen University) for the SEM assistance. Special thanks go to Chawalit Wongsanorasaet for the study site data and Paiboon Getwongsa for the field assistance.

Conflicts of Interest: The authors declare no conflict of interest.

References

1. Manconi, R.; Pronzato, R. Global Diversity of Sponges (Porifera: Spongillina) in Freshwater. *Hydrobiologia* **2008**, *595*, 27–33. [CrossRef]
2. Manconi, R.; Pronzato, R. Chapter 8—Phylum Porifera. In *Thorp and Covich's Freshwater Invertebrates*, 4th ed.; Thorp, J.H., Rogers, D.C., Eds.; Academic Press: Boston, MA, USA, 2015; pp. 133–157. ISBN 978-0-12-385026-3.
3. Manconi, R.; Pronzato, R. Chapter 3—Phylum Porifera. In *Thorp and Covich's Freshwater Invertebrates*, 4th ed.; Thorp, J.H., Rogers, D.C., Eds.; Academic Press: Boston, MA, USA, 2016; pp. 39–83. ISBN 978-0-12-385028-7.
4. Skelton, J.; Strand, M. Trophic Ecology of a Freshwater Sponge (*Spongilla Lacustris*) Revealed by Stable Isotope Analysis. *Hydrobiologia* **2013**, *709*, 227–235. [CrossRef]
5. Weisz, J.B.; Lindquist, N.; Martens, C.S. Do Associated Microbial Abundances Impact Marine Demosponge Pumping Rates and Tissue Densities? *Oecologia* **2008**, *155*, 367–376. [CrossRef]
6. Reiswig, H.M. Particle Feeding in Natural Populations of Three Marine Demosponges. *Biol. Bull.* **1971**, *141*, 568–591. [CrossRef]
7. Morganti, T.M.; Ribes, M.; Yahel, G.; Coma, R. Size Is the Major Determinant of Pumping Rates in Marine Sponges. *Front. Physiol.* **2019**, *10*, 1474. [CrossRef] [PubMed]
8. Morganti, T.M.; Ribes, M.; Moskovich, R.; Weisz, J.B.; Yahel, G.; Coma, R. In Situ Pumping Rate of 20 Marine Demosponges Is a Function of Osculum Area. *Front. Mar. Sci.* **2021**, *8*, 583188. [CrossRef]
9. Ledda, F.D.; Pronzato, R.; Manconi, R. Mariculture for Bacterial and Organic Waste Removal: A Field Study of Sponge Filtering Activity in Experimental Farming. *Aquac. Res.* **2014**, *45*, 1389–1401. [CrossRef]
10. Manconi, R.; Pronzato, R. How to Survive and Persist in Temporary Freshwater? Adaptive Traits of Sponges (Porifera: Spongillida): A Review. *Hydrobiologia* **2016**, *782*, 11–22. [CrossRef]
11. Frost, T.M.; Reiswig, H.M.; Ricciardi, A. Porifera. In *Ecology and Classification of North American Freshwater Invertebrates*; Thorp, J.H., Covich, A.P., Eds.; Academic Press: New York, NY, USA, 2001; pp. 97–133.
12. Pronzato, R.; Manconi, R. Atlas of European Freshwater Sponges. *Ann. Mus. Civ. Storia Nat. Ferrara* **2001**, *4*, 3–64.
13. Reiswig, H.M.; Frost, T.M.; Ricciardi, A. Porifera. In *Ecology and Classification of North American Freshwater Invertebrates*, 3rd ed.; Thorp, J.H., Covich, A.P., Eds.; Academic Press: London, UK, 2010; pp. 91–123.
14. Manconi, R.; Ruengsawang, N.; Vannachak, V.; Hanjavanit, C.; Sangpradub, N.; Pronzato, R. Biodiversity in South East Asia: An Overview of Freshwater Sponges (Porifera: Demospongiae: Spongillina). *J. Limnol.* **2013**, *72*, e15. [CrossRef]
15. Mizuno, T.; Mori, S. Preliminary Hydrobiological Survey of Some Southeast Asian Inland Waters. *Biol. J. Linn. Soc.* **1970**, *2*, 77–118. [CrossRef]
16. Sangsurasak, C.; Hsieh, H.-N.; Wongphathanakul, W.; Wirojanagud, W. Water Quality Modeling in the Nam Pong River, Northeast Thailand. *ScienceAsia* **2006**, *32*, 071. [CrossRef]
17. Pollution Control Department. *Thailand State of Pollution Report 2007*; Pollution Control Department: Bangkok, Thailand, 2007.
18. Van Zalinge, N.; Degen, P.; Pongsri, C.; Nuov, S.; Jensen, J.G.; Nguyen, V.H.; Choulamany, X. *The MekongRiverSystem*; RAP publication 16; FAO Regional Office for Asia and the Pacific: Phnom Penh, Cambodia, 2004.
19. Ruengsawang, N. The Distribution, Growth, Reproduction, and Ecology of Freshwater Sponges in the Pong River. Ph.D. Dissertation, Khon Kaen University, Khon Kaen, Thailand, 2013.
20. Ruengsawang, N.; Sangpradub, N.; Hanjavanit, C.; Manconi, R. Biodiversity Assessment of the Lower Mekong Basin: A New Species of *Corvospongilla* (Porifera: Spongillina: Spongillidae) from Thailand. *Zootaxa* **2012**, *3320*, 47–55. [CrossRef]
21. Manconi, R.; Erpenbeck, D.; Fromont, J.; Wörheide, G.; Pronzato, R. Discovery of the Freshwater Sponge Genus *Corvospongilla* Annandale (Porifera: Spongillida) in Australia with the Description of a New Species and Phylogeographic Implications. *Limnology* **2022**, *23*, 73–87. [CrossRef]
22. Manconi, R.; Pronzato, R. The Genus *Corvospongilla* Annandale, 1911 (Porifera: Demospongiae: Spongillida) from Madagascar Freshwater with Description of a New Species: Biogeographic and Evolutionary Aspects. *Zootaxa* **2019**, *4612*, 544–554. [CrossRef] [PubMed]
23. Morse, J.C.; Yang, L.; Tian, L. *Aquatic Insects of China Useful for Monitoring Water Quality*, 1st ed.; Hohai University Press: Nanjing, China, 1994.
24. Epler, J.H. *Identification Manual for the Larval Chironomidae (Diptera) of North and South Carolina*; North Carolina Department of Environment and Natural Resources: Raleigh, NC, USA, 2001.
25. Sangpradub, N.; Boonsoong, B. *Identification of Freshwater Invertebrates of the Mekong River and Its Tributaries*; Mekong River Commission: Vientiane, Laos, 2006; ISBN 978-92-95061-00-2.
26. Cranston, P.S. The Chironomidae Larvae Associated with the Tsunami-impacted Waterbodies of the Coastal Plain of Southwestern Thailand. *Raffles Bull. Zoo.* **2007**, *55*, 231–244.
27. Sangpradub, N.; Giller, P.S. Gut Morphology, Feeding Rate and Gut Clearance in Five Species of Caddis Larvae. *Hydrobiologia* **1994**, *287*, 215–223. [CrossRef]
28. Manconi, R.; Ruengsawang, N.; Ledda, F.D.; Hanjavanit, C.; Sangpradub, N. Biodiversity Assessment in the Lower Mekong Basin: First Record of the Genus *Oncosclera* (Porifera: Spongillina: Potamolepidae) from the Oriental Region. *Zootaxa* **2012**, *3544*, 41–51. [CrossRef]

29. Annandale, N. Notes on the freshwater fauna of India. No. 5. Some animals found with *Spongilla carteri* in Calcutta. *J. Asiat. Soc. Bengal* **1906**, *2*, 187–196.

30. Schröder, K. Spongillidenstudien VII, Susswasserschwamme von Neuseeland, Borneo und Madagaskar. *Zool. Anz.* **1935**, *109*, 97–106.

31. Mansell, M.W. Neuropterology in Southern Africa. In *Recent Research in Neuropterology: Proceedings of the 2nd International Symposium on Neuropterology (Insecta, Megaloptera, Raphidioptera, Planipennia), Hamburg, Germany, 20–26 August 1984*; Gepp, J., Aspock, H., Hölzel, H., Eds.; Gedruckt von Druckhaus Thalerhof: Österreich, Austria, 1986; pp. 45–52.

32. Wells, A. New Caledonian Hydroptilidae (Trichoptera), with New Records, Descriptions of Larvae and a New Species. *Aquat. Insects* **1995**, *17*, 223–239. [CrossRef]

33. Wells, A.; Johanson, K. Review of the New Caledonian Species of Acritoptila Wells, 1982 (Trichoptera, Insecta), with Descriptions of 3 New Species. *ZooKeys* **2014**, *397*, 1–23. [CrossRef]

34. Forteath, G.N.R.; Osborn, A.W. Biology, Ecology and Voltinism of the Australian Spongillafly Sisyra Pedderensis Smithers (Neuroptera: Sisyridae). *Pap. Proc. R. Soc. Tasman.* **2012**, *146*, 25–35. [CrossRef]

35. Forteath, G.N.R.; Purser, J.; Osborn, A.W. A New Species of *Sisyra* Burmeister 1839 (Neuroptera: Sisyridae) from Four Springs Lake and Wadley's Dam, Northern Tasmania. *Austral Entomol.* **2015**, *54*, 217–220. [CrossRef]

36. Pavesi, P. Di una spugna d'acqua dolce nuova per l'Italia. *Rendiconti del Reale Istituto Lombardo di Scienze e Lettere Ser. II* **1881**, *14*, 1–6.

37. Arndt, W. Porifera, Schwämme, Spongien. In *Die Tierwelt Deutschlands und der Angrenzenden Meeresteile nach ihren Merkmalen und ihrer Lebensweise 4 (Porifera-Hydrozoa-Coelenterata-Echinodermata)*; Dahl, F., Ed.; Gustav Fischer Verlag: Leipzig, Germany, 1928; pp. 1–94.

38. Rezvoi, P.D. Presnovodnye gubki. [Freshwater sponges]. In *Fauna SSSR [The Fauna of the USSR]*; Rezvoi, P.D., Ed.; Akad Nauk SSSR: Moscow, Russia, 1936; Volume 2. (In Russian)

39. Berg, K. Biological studies on the river Susaa. Faunistic and biological investigations. Porifera. *Folia Limnol. Scand.* **1948**, *4*, 10–23.

40. Gaumont, J. L'appareil digestif de la larve d'un Planipenne associé aux éponges d'eau douce: *Sisyra fuscata*. *Ann. Soc. Entomol. Fr. (NS)* **1965**, *1*, 335–357.

41. Moretti, G.; Corallini-Sorcetti, C. Al Lago Trasimeno è arrivata Ceraclea fulva Ramb. (Trichoptera–Leptoceratidae). *Riv. Idrobiol.* **1980**, *19*, 1–9.

42. Konopacka, A.; Siciński, J. Macrofauna Inhabiting the Colonies of the Sponge *Spongilla lacustris* (L.) in the River Gać. *SIL Proc. 1922–2010* **1985**, *22*, 2968–2973. [CrossRef]

43. Kamaltynov, R.M.; Chernykh, V.I.; Slugina, Z.V.; Karabanov, E.B. The Consortium of the Sponge Lubomirskia Baicalensis in Lake Baikal, East Siberia. *Hydrobiologia* **1993**, *271*, 179–189. [CrossRef]

44. Weissmair, V.W.; Mildner, P. Zur Kenntnis der Schwammfliegen (Neuroptera: Sisyridae), ihrer Wirte und Wohngewässer in Kärnten. *Carinthia II* **1995**, *105*(2), 535–552.

45. Gugel, J. Life Cycles and Ecological Interactions of Freshwater Sponges (Porifera, Spongillidae) in the River Rhine in Germany. *Limnologica* **2001**, *31*, 185–198. [CrossRef]

46. Corallini, C.; Gaino, E. Peculiar Digestion Patterns of Sponge-Associated Zoochlorellae in the Caddisfly Ceraclea Fulva. *Tissue Cell* **2001**, *33*, 402–407. [CrossRef] [PubMed]

47. Corallini, C.; Gaino, E. The Caddisfly *Ceraclea fulva* and the Freshwater Sponge Ephydatia Fluviatilis: A Successful Relationship. *Tissue Cell* **2003**, *35*, 1–7. [CrossRef]

48. Gaino, E.; Lancioni, T.; La Porta, G.; Todini, B. The Consortium of the Sponge *Ephydatia fluviatilis* (L.) Living on the Common Reed Phragmites Australis in Lake Piediluco (Central Italy). *Hydrobiologia* **2004**, *520*, 165–178. [CrossRef]

49. Wallace, I.D.; Wallace, B.; Philipson, G.N. *Keys to the Case-Bearing Caddis Larvae of Britain and Ireland*; Freshwater Biological Association—Scientific Publication: Ambleside, UK, 2003; Volume 61, p. 259.

50. Weinberg, I.; Glyzina, O.; Weinberg, E.; Kravtsova, L.; Rozhkova, N.; Sheveleva, N.; Natyaganova, A.; Bonse, D.; Janussen, D. Types of interactions in consortia of Baikalian sponges. *Boll. Mus. Ist. Biol. Univ. Genova* **2004**, *68*, 655–663.

51. Loru, L.; Pantaleoni, R.A.; Sassu, A. Overwintering Stages of *Sisyra iridipennis* A. Costa, 1884 (Neuroptera Sisyridae). *Ann. Mus. Cívico Stor. Nat. Ferrara* **2007**, *8*, 153–159.

52. Schiffels, S. Commensal and parasitic Chironomidae. *Lauterbornia* **2009**, *68*, 9–33.

53. Sokolova, A.M.; Palatov, D.M. Macroinvertebrate Associations of Sponges (Demospongiae: Spongillidae) from some Fresh Waters in the Palaearctic (in Russian). *Povolzhskiy J. Ecol.* **2014**, *4*, 618–627.

54. Palatov, D.M.; Sokolova, A.M. Freshwater Sponges and Their Associated Invertebrates in the Great Lakes Basin (Mongolia). *Ukr. J. Ecol.* **2017**, *7*, 635–639. [CrossRef]

55. Zvereva, Y.; Medvezhonkova, O.; Naumova, T.; Sheveleva, N.; Lukhnev, A.; Sorokovikova, E.; Evstigneeva, T.; Timoshkin, O. Variation of Sponge-Inhabiting Infauna with the State of Health of the Sponge Lubomirskia Baikalensis (Pallas, 1776) in Lake Baikal. *Limnology* **2019**, *20*, 267–277. [CrossRef]

56. Ostrovsky, A.M. New Finds and Species of Caddisflies (Trichoptera) in Southeastern Belarus. *Inland Water Biol.* **2021**, *14*, 133–140. [CrossRef]

57. Melão, M.G.G.; Rocha, O. Macrofauna Associada a *Metania spinata* (Cárter, 1881), Porifera, Metaniidae. *Acta Limnol. Bras.* **1996**, *8*, 59–64.

58. De Roque, F.O.; Trivinho-Strixino, S.; Couceiro, S.R.M.; Hamada, N.; Volkmer-Ribeiro, C.; Messias, M.C. Species of Oukuriella Epler (Diptera, Chironomidae) inside Freshwater Sponges in Brazil. *Rev. Bras. Entomol.* **2004**, *48*, 291–292. [CrossRef]

59. Fusari, L.M.; Roque, F.O.; Hamada, N. Sponge-Dwelling Chironomids in the Upper Paraná River (Brazil): Little Known but Potentially Threatened Species. *Neotrop. Entomol.* **2008**, *37*, 522–527. [CrossRef]

60. Fusari, L.M.; Roque, F.O.; Hamada, N. Oukuriella Pesae, a New Species of Sponge-Dwelling Chironomid (Insecta: Diptera) from Amazonia, Brazil. *Zootaxa* **2009**, *2146*, 61–68. [CrossRef]

61. Fusari, L.M.; Oliveira, C.S.N.; Hamada, N.; Roque, F.O. New Species of *Ablabesmyia* Johannsen from the Neotropical Region: First Report of a Sponge-Dwelling Tanypodinae. *Zootaxa* **2012**, *3239*, 43–50. [CrossRef]

62. Fusari, L.M.; Bellodi, C.F.; Lamas, C.J.E. A New Species of Sponge-Dwelling *Oukuriella* (Chironomidae) from Brazil. *Zootaxa* **2014**, *3764*, 418. [CrossRef]

63. Clavier, S.; Guillemet, L.; Rhone, M.; Thomas, A. First Report of the Aquatic Genus *Climacia* McLachlan, 1869 in French Guiana (Neuroptera, Sisyridae). *Ephemera* **2011**, *12*, 31–36.

64. Hamada, N.; Pes, A.M.O.; Fusari, L.M. First Record of Sisyridae (Neuroptera) in Rio De Janeiro State, Brazil, with Bionomic Notes on *Sisyra Panama. Fla. Entomol.* **2014**, 281–284. [CrossRef]

65. Bowles, D. New Distributional Records for Neotropical Spongillaflies (Neuroptera: Sisyridae). *Insecta Mundi* **2015**, *0400*, 1–7.

66. Ardila-Camacho, A.; Martins, C.C. First Record of Spongillaflies (Neuroptera: Sisyridae) from Colombia. *Zootaxa* **2017**, *4276*, 129–133. [CrossRef] [PubMed]

67. Assmar, A.C.; Salles, F.F. Taxonomic and Distributional Notes on Spongilla-Flies (Neuroptera: Sisyridae) from Southeastern Brazil with First Interactive Key to the Species of the Country. *Zootaxa* **2017**, *4273*, 80–92. [CrossRef]

68. Fisher, M.L.; Mower, R.C.; Nelson, C.R. *Climacia Californica* Chandler, 1953 (Neuroptera: Sisyridae) in Utah: Taxonomic Identity, Host Association and Seasonal Occurrence. *Aquat. Insects* **2019**, *40*, 317–327. [CrossRef]

69. Da Fernandes, I.S.; Nicacio, G.; Rodrigues, G.G.; Silva, F.L. da Freshwater Sponge-Dwelling Chironomidae (Insecta, Diptera) in Northeastern Brazil. *Biotemas* **2019**, *32*, 39–48. [CrossRef]

70. Brown, H.P. The Life History of *Climacia areolaris* (Hagen), a Neuropterous "Parasite" of Fresh Water Sponges. *Am. Midl. Nat.* **1952**, *47*, 130–160. [CrossRef]

71. Parfin, S.I.; Gurney, A.B. The Spongilla-Flies, with Special Reference to Those of the Western Hemisphere (Sisyridae, Neuroptera). *Proc. U. S. Natl. Mus.* **1956**, *105*, 421–529. [CrossRef]

72. Roback, S.S. Insects Associated with the Sponge Spongilla fragilis in the Savannah River. *Not. Nat.* **1968**, *412*, 1–10.

73. Heiman, D.R.; Knight, A.W. A North American Trichopteran Larva Which Feeds on Freshwater Sponges (Trichoptera: Leptoceridae; Porifera: Spongillidae). *Am. Midl. Nat.* **1970**, *84*, 278–280. [CrossRef]

74. Poirrier, M.A. Some Fresh-Water Sponge Hosts of Louisiana and Texas Spongilla-Flies, with New Locality Records. *Am. Midl. Nat.* **1969**, *81*, 573–575. [CrossRef]

75. Poirrier, M.A.; Arceneaux, Y.M. Studies on Southern Sisyridae (Spongilla-Flies) with a Key to the Third-Instar Larvae and Additional Sponge-Host Records. *Am. Midl. Nat.* **1972**, *88*, 455–458. [CrossRef]

76. Resh, V.H. Life Cycles of Invertebrate Predators of Freshwater Sponge. In *Aspects of Sponge Biology*; Harrison, F.W., Cowden, R.R., Eds.; Academic Press: New York, NY, USA, 1976; pp. 299–314. ISBN 978-0-12-327950-7.

77. Pennak, R.W. *Fresh-Water Invertebrates of the United States*, 2nd ed.; John Wiley & Sons: Hoboken, NJ, USA; Ronald Press: New York, NY, USA, 1978.

78. Matteson, J.D.; JacobiI, G.Z. Benthic Macroinvertebrates Found on The Freshwater Sponge *Spongilla lacustris. Gt. Lakes Entomol.* **1980**, *3*, 162–172.

79. Stoaks, R.D.; Neel, J.K.; Post, R.L. Observations on North Dakota Sponges (Haplosclerina: Spongillidae) and Sisyrids (Neuroptera: Sisyridae). *Gt. Lakes Entomol.* **1983**, *16*, 171–176.

80. Clark, W.H. First Record of Climacia Californica (Neuroptera: Sisyridae) and Its Host Sponge, *Ephydatia mulleri* (Porifera: Spongillidae), from Idaho with Water Quality Relationships. *Gt. Basin Nat.* **1985**, *45*, 2.

81. Pupedis, R.J. Foraging Behavior and Food of Adult Spongila-Flies (Neuroptera: Sisyridae). *Ann. Entomol. Soc. Am.* **1987**, *80*, 758–760. [CrossRef]

82. Whitlock, H.N.; Morse, J.C. Ceraclea Enodis, a New Species of Sponge-Feeding Caddisfly (Trichoptera: Leptoceridae) Previously Misidentified. *J. North Am. Benthol. Soc.* **1994**, *13*, 580–591. [CrossRef]

83. Bowles, D.E. Spongillaflies (Neuroptera: Sisyridae) of North America with a Key to the Larvae and Adults. *Zootaxa* **2006**, *1357*, 1–19.

84. Cover, M.R.; Resh, V.H. Global Diversity of Dobsonflies, Fishflies, and Alderflies (Megaloptera; Insecta) and Spongillaflies, Nevrorthids, and Osmylids (Neuroptera; Insecta) in Freshwater. In *Freshwater Animal Diversity Assessment*; Developments in Hydrobiology; Balian, E.V., Lévêque, C., Segers, H., Martens, K., Eds.; Springer: Dordrecht, The Netherlands, 2008; pp. 409–417. ISBN 978-1-4020-8259-7.

85. Rothfuss, A.H.; Heilveil, J.S. Distribution of Sisyridae and Freshwater Sponges in the Upper-Susquehanna Watershed, Otsego County, New York with a New Locality for *Climacia areolaris* (Hagen). *Am. Midl. Nat.* **2018**, *180*, 298–305. [CrossRef]

86. Bowles, D.E.; Courtney, G.W. The Aquatic Neuropterida of Iowa. *Proc. Entomol. Soc. Wash.* **2020**, *122*, 556–565. [CrossRef]

87. Vogel, S. Current-Induced Flow through Living Sponges in Nature. *Proc. Natl. Acad. Sci. USA* **1977**, *74*, 2069–2071. [CrossRef]

88. Frost, T.M. Clearance Rate Determinations for the Fresh Water Sponge Spongilla Lacustris Effects of Temperature Particle Type and Concentration and Sponge Size. *Arch. Hydrobiol.* **1980**, *90*, 330–356.

89. Hanjavanit, C.; Tangpirotewong, N. Comparison of Benthic Macroinvertebrate Community Structure in Relation to Different Types of Human Disturbance along the Pong River, Khon Kaen Province (Thai). *Asia-Pac. J. Sci. Technol.* **2007**, *12*, 402–419.

90. Annandale, N. Freshwater Sponges, Hydroids and Polyzoa. In *The Fauna of British India, including Ceylon and Burma*; Shipley, A.E., Ed.; Taylor and Francis: London, UK, 1911; pp. 49–94.

91. Koukouras, A.; Russo, A.; Voultsiadou-Koukoura, E.; Arvanitidis, C.; Stefanidou, D. Macrofauna Associated with Sponge Species of Different Morphology. *Mar. Ecol.* **1996**, *17*, 569–582. [CrossRef]

92. Neves, G.; Omena, E. Influence of Sponge Morphology on the Composition of the Polychaete Associated Fauna from Rocas Atoll, Northeast Brazil. *Coral Reefs* **2003**, *22*, 123–129. [CrossRef]

93. Boon, P.J. Adaptive Strategies of Amphipsyche Larvae (Trichoptera: Hydropsychidae) Downstream of a Tropical Impoundment. In *The Ecology of Regulated Streams*; Springer: Berlin/Heidelberg, Germany, 1979; pp. 237–255.

94. De Oliveira Roque, F.; Trivinho-Strixino, S. *Xenochironomus ceciliae* (Diptera: Chironomidae), a New Chironomid Species Inhabiting Freshwater Sponges in Brazil. *Hydrobiologia* **2005**, *534*, 231–238. [CrossRef]

95. Sata, N.U.; Kaneniwa, M.; Masuda, Y.; Ando, Y.; Iida, H. Fatty Acid Composition of Two Species of Japanese Freshwater Sponges *Heterorotula multidentata* and *Sponilla alba*. *Fish. Sci.* **2002**, *68*, 236–238. [CrossRef]

96. Steffan, A.W. Ectosymbiosis in aquatic insects. In *Symbiosis*; Henry, M.S., Ed.; Academic Press: New York, USA, 1967; pp. 207–289.

97. Sangpradub, N.; Inmuong, Y.; Hanjavanit, C.; Inmuong, U. Biotic Indices for Biological Classification of Water Quality in the Pong Catchment Using Benthic Macroinvertebrates. *KKU Sci. J.* **1998**, *6*, 289–304.

98. Sriariyanuwath, E.; Sangpradub, N.; Hanjavanit, C. Diversity of Chironomid Larvae in Relation to Water Quality in the Phong River, Thailand. *Aquac. Aquar. Conserv. Legis.* **2015**, *8*, 933–945.

99. Weissmair, W.; Waringer, J. Identification of the Larvae and Pupae of Sisyra Fuscata (Fabricius, 1793) and Sisyra Terminalis Curtis, 1854 (Insecta: Planipennia: Sisyridae), Based on Austrian Material. *Aquat. Insects* **1994**, *16*, 147–155. [CrossRef]

100. Navás, L. Névroptères et Insectes Voisins. Chine et Pays Environnants. 4e Série. *Notes d'Entomologie Chinoise* **1933**, *1*, 1–23.

101. Lu, X.; Lin, A.; Wang, D.; Liu, X. The First Mitochondrial Genome of Spongillafly from Asia (Neuroptera: Sisyridae: *Sisyra aurorae* Navás, 1933) and Phylogenetic Implications of Osmyloidea. *Mitochondrial DNA Part B* **2021**, *6*, 2369–2370. [CrossRef] [PubMed]

102. Monserrat, V.J. Sobre los Sisiridos de la Región Oriental (Neuroptera, Planipennia, Sisyridae). *Rev. Esp. Entomol.* **1982**, *57*, 165–186.

103. New, T.R. *The Neuroptera of Malesia*; Brill: Leiden, Netherlands, 2003; 204p.

104. Boon, P.J. Gastric Mill Structure and Its Phylogenetic Significance in Larval Hydropsychidae (Trichoptera). *Zool. J. Linn. Soc.* **1985**, *84*, 301–324. [CrossRef]

105. Myers, N.; Mittermeier, R.A.; Mittermeier, C.G.; Da Fonseca, G.A.; Kent, J. Biodiversity Hotspots for Conservation Priorities. *Nature* **2000**, *403*, 853–858. [CrossRef] [PubMed]

106. New Species Discoveries in The Greater Mekong. 2020. Available online: https://www.wwf.org.la/publications/new_species_discoveries_in_the_greater_mekong_2020/ (accessed on 1 September 2022).

107. Ruengsawang, N.; Sangpradub, N.; Artchawakom, T.; Pronzato, R.; Manconi, R. Rare Freshwater Sponges of Australasia: New Record of Umborotula Bogorensis (Porifera: Spongillida: Spongillidae) from the Sakaerat Biosphere Reserve in Northeast Thailand. *Eur. J. Taxon.* **2017**, *260*, 1–24. [CrossRef]

108. Calcinai, B.; Cerrano, C.; Núñez-Pons, L.; Pansini, M.; Thung, D.C.; Bertolino, M. A New Species of Spongilla (Porifera, Demospongiae) from a Karst Lake in Ha Long Bay (Vietnam). *J. Mar. Sci. Eng.* **2020**, *8*, 1008. [CrossRef]

109. Bell, J.J.; McGrath, E.; Biggerstaff, A.; Bates, T.; Cárdenas, C.A.; Bennett, H. Global Conservation Status of Sponges. *Conserv. Biol.* **2015**, *29*, 42–53. [CrossRef]

110. Tan, Y.-L.; Chen, J.-E.; Yiew, T.-H.; Habibullah, M.S. Habitat Change and Biodiversity Loss in South and Southeast Asian Countries. *Environ. Sci. Pollut. Res.* **2022**, *29*, 63260–63276. [CrossRef]

111. Van Soest, R.W.; Boury-Esnault, N.; Vacelet, J.; Dohrmann, M.; Erpenbeck, D.; de Voogd, N.J.; Santodomingo, N.; Vanhoorne, B.; Kelly, M.; Hooper, J.N. Global Diversity of Sponges (Porifera). *PLoS ONE* **2012**, *7*, e35105. [CrossRef]

112. Jones, C.G.; Lawton, J.H.; Shachak, M. Organisms as Ecosystem Engineers. In *Ecosystem Management*; Springer: Berlin/Heidelberg, Germany, 1994; pp. 130–147.

113. Ormerod, S.J. A Golden Age of River Restoration Science? *Aquat. Conserv. Mar. Freshw. Ecosyst.* **2004**, *14*, 543–549. [CrossRef]

Article

Distribution of Freshwater Alien Animal Species in Morocco: Current Knowledge and Management Issues

Abdelkhaleq Fouzi Taybi [1], Youness Mabrouki [2] and Christophe Piscart [3],*

[1] Applied Biology and Biotechnology Research Team B.P. 300, Multidisciplinary Faculty of Nador, Mohammed Premier University, Selouane 62700, Morocco
[2] Conservation and Valorisation of Natural Resources Laboratory, Faculty of Sciences Dhar El Mehraz, Biotechnology, Sidi Mohamed Ben Abdellah University, Fez. B.P. 1796, Fès-Atlas 30003, Morocco
[3] UMR CNRS 6553 Ecosystems, Biodiversity, Evolution (ECOBIO), University of Rennes, F-35000 Rennes, France
* Correspondence: christophe.piscart@univ-rennes.fr

Abstract: This work presents currently available knowledge on alien species (AS) found in the inland waters of Morocco. The objective is to provide an updated list of alien species and identify the main introduction pathways and possible threats to native biodiversity. The dataset was built from an extensive literature search supplemented by our own research work (published or in progress). The main areas harboring xenodiversity in Moroccan freshwaters correspond to protected areas (e.g., Ramsar Site and SIBE). These areas are currently home to 41 confirmed AS belonging to different taxonomic groups. Fish are the most abundant taxonomic group with 21 species, followed by molluscs (7 species) and arthropods (7 species). The presence of 15 more species was also noticed but considered doubtful. Almost half of these AS were introduced intentionally. They correspond to restocking programs and are likely the most serious threat to native biodiversity through predation, competition, and hybridization. Commercial activities around aquarium and ornamental species appear as the second source favoring colonization by AS. Implementing protective regulations regarding the import of exotic species in Morocco appears very urgent to protect local native diversity. In addition, detecting and monitoring the expansion of AS within the colonized areas and studies improving biological and ecological knowledge seem crucial to mitigate their possible impacts on native communities and preserve Moroccan freshwater ecosystems.

Keywords: biological invasions; checklist; Mediterranean biodiversity; alien species; biodiversity hotspot; North Africa

Citation: Taybi, A.F.; Mabrouki, Y.; Piscart, C. Distribution of Freshwater Alien Animal Species in Morocco: Current Knowledge and Management Issues. *Diversity* **2023**, *15*, 169. https://doi.org/10.3390/d15020169

Academic Editors: Mateja Germ, Igor Zelnik, Matthew Simpson and Luc Legal

Received: 18 November 2022
Revised: 16 January 2023
Accepted: 17 January 2023
Published: 25 January 2023

1. Introduction

Thanks to its geographical position and its different natural barriers, Morocco is one of the most interesting biogeographical regions of the Western Mediterranean region, recognized as a hotspot of biodiversity, with a high rate of endemism in its fauna and flora [1]. However, this high endemicity makes communities vulnerable to the introduction of alien and invasive species without a common evolutionary history [2]. Biological invasions are indeed globally considered one of the most important human impacts on a wide range of ecosystems [3–5]. Introduction of and invasion by alien species are one of the top threats to biodiversity and ecosystem functioning worldwide. Alien species can also drive the degradation of ecosystem functions by altering trophic interactions, nutrient cycling, and habitat structures [6]. Freshwater ecosystems are particularly vulnerable to this phenomenon around the globe, with proportionally more invaders than terrestrial systems [7,8]. Many alien and invasive species have often been implicated in native species displacement within freshwater ecosystems [9,10]. The consequences of biological invasions on freshwater biodiversity are most dramatic in Morocco, where native species are restricted by the Atlas Mountains to the east, the Sahara Desert to the south, and the sea to the north and west, so that no displacement is possible.

More and more exotic and invasive animal species are recorded in Moroccan freshwaters, belonging to a wide range of taxonomic groups, including fish, annelids, molluscs, and arthropods [11–16]. As a result, new communities of species are formed, with unknown ecological and evolutionary consequences. Therefore, monitoring the presence and expansion of alien species and studies improving their biological and ecological knowledge seem to be a crucial concern for managing the environment and indigenous communities.

In this work, we summarize currently available knowledge on the alien species (AS) present in the inland waters of Morocco. The objective is to provide the first and most up-to-date list of alien species and to identify the main introduction pathways and possible threats to native biodiversity.

2. Materials and Methods

The list of the AS reported in Morocco (and their geographical coordinates) was compiled from (i) freely available web databases, including the Global Biodiversity Information Facility GBIF (http://www.gbif.org/ (accessed on 1 April 2022)) and the Global Invasive Species Database GISD (http://www.iucngisd.org/gisd/ (accessed on 1 December 2022)); (ii) an extensive literature search through published articles [17–47]; (iii) our own research, whether published or in progress [11–15,48–55]; and other personal observations.

The introduction pathways of each species were designated by an extensive analysis of literature dealing with biological invasions and summarized in seven categories, following the same method as previously published articles on the same topic of alien animals and aquatic species (e.g. [56]).

After a careful review of the data, we classified the species with a confirmed presence in Morocco into two groups, namely "widely distributed" and "locally distributed". The "widely distributed" species have self-sustaining breeding populations in different hydrosystems and watersheds, and their presence is attested to by specialists through publications in scientific journals. The species whose distribution area is still restricted to certain watersheds, a smaller geographical area, and a small number of localities were considered "locally distributed". The taxa whose presence in Morocco is doubtful were marked as "Unconfirmed". A distribution map of AS hotspots in the inland waters of Morocco was built by processing all the gathered data in ArcGIS GIS software (ArcGIS, hot spot analysis, spatial statistics, version 10.2). Occurrence data are available in the Supplementary Materials (Table S1).

3. Results

Forty-one AS are present in Moroccan inland waters, plus 15 considered doubtful or unconfirmed. Only half of them (21 species) have well-established populations across the country (Table 1). About 560 records of alien animal species occupying the inland waters of Morocco have been gathered with the GPS coordinates of the localities. 374 records come from specialized published literature, of which 104 are our publications (among which some are also present in GBIF), followed by 142 new records (presented in the Supplementary Materials of this work), and completed by 24 records coming from GBIF. The geographical positions of alien animal species present in the inland waters of Morocco, collected from the bibliography, in addition to our data (published and unpublished), are offered in Supplementary Materials (Table S1).

Table 1. List of alien animal species known to be present in the inland waters of Morocco.

Taxon	Detection Date	Mode of Introduction	Origin	Status	Reference
Cnidaria					
Craspedacusta sowerbyii (Lankester, 1880)	2015	Ornamental plant trade	Asia	Locally distributed	[41]
Turbellaria					
Girardia tigrina s.l. (Girard, 1850)	2019	Aquarium/ornamental plant trade	America	Locally distributed	[18,50]
Nematoda					
Anguillicola crassus (Kuwahara, Niimi and Itagaki, 1974)	1994	Aquaculture/international trade	Asia	Locally distributed	[24,42]
Annelida					
Helobdella europaea (Kutschera, 1987)	2014	Aquarium/ornamental plant trade	South America	Widely distributed	[11]
Gastropoda					
Physella acuta (Draparnaud, 1805)	1972	Aquarium/ornamental plant trade	North America	Widely distributed	[51,57]
Ferrissia fragilis (Tryon, 1863)	2022	Aquarium/ornamental plant trade/hitchhiking?	North America	Locally distributed	[50]
Helisoma duryi (Wetherby, 1879)	2022	Aquarium trade	North America	Locally distributed	[55]
Melanoides tuberculata (O.F. Müller, 1774)	1934?	Aquarium/ornamental plant trade	Tropical Africa and Asia	Locally distributed	[29,34]
Potamopyrgus antipodarum (J.E. Gray, 1843)	2021	Aquarium/ornamental plant trade/hitchhiking?	New Zealand	Locally distributed	[54]
Bivalva					
Sinanodonta woodiana (Lea, 1834)	2021	Aquaculture	Asia	Locally distributed	[48]
Corbicula fluminea (O.F. Müller, 1774)	2008	Aquaculture	Asia	Widely distributed	[22,51]
Crustacea					
Potamobius astacus (Linnaeus, 1758)	1914	Intentional (restocking)	Europe	Locally distributed	[39,46]
Orconectes limosus (Rafinesque, 1817)	1937	Intentional (restocking)	North America	Widely distributed	[32,39]
Procambarus clarkii (Girard, 1852)	2008	Probably intentional	North America	Widely distributed	[25,44]
Callinectes sapidus (Rathbun, 1896) *	2017	Ballast waters	North America	Widely distributed	[14]
Lernaea cyprinacea (Linnaeus, 1758)	2013	Aquaculture	Asia	Locally distributed	[23]
Hexapoda					
Stegomyia albopicta (Skuse, 1894)	2015	International trafficking /propagation?	Asia	Locally distributed	[20]
Trichocorixa verticalis verticalis (Fieber, 1851)	2010	Aquaculture/propagation from Europe?	North America	Widely distributed	[35,52,53]
Aves					
Oxyura jamaicensis (Gmelin, 1789)	1990	Propagation from Europe	North America	Widely distributed	[45,58]
Piscis					
Lepomis gibbosus (Linnaeus, 1758)	1955	Intentional (restocking)	North America	Widely distributed	[23,39]
Lepomis macrochinus (Rafinesque, 1819)	1966	Intentional (restocking)	North America	Locally distributed	[27,39]
Lepomis microlophus (Rafinesque, 1859)	1966	Intentional (restocking)	North America	Unconfirmed	[39]

Table 1. *Cont.*

Taxon	Detection Date	Mode of Introduction	Origin	Status	Reference
Micropterus salmoides (Lacépède, 1802)	1934	Intentional (restocking)	North America	Established	[19,39]
Pomoxis annularis (Rafinesque, 1818)	1961	Intentional (restocking)	North America	Unconfirmed	[39]
Pomoxis nigromoculatus (Cuvier, 1829)	1964	Intentional (restocking)	North America	Unconfirmed	[39]
Lates niloticus (Linnaeus, 1758)	1954	Intentional (restocking)	Ethiopia (Afrotropic)	Unconfirmed	[19,39]
Perca fluviatilis (Linnaeus, 1758)	1939	Intentional (restocking)	Eurasia	Locally distributed	[27,39]
Sander lucioperca (Linnaeus, 1758)	1939	Intentional (restocking)	Europe	Locally distributed	[27,39]
Alburnus alburnus (Linnaeus, 1758)	2013	Probably intentional	Europe	Widely distributed	[23]
Pseudorasbora parva (Temminck and Schlegel, 1846)	2013	Probably intentional	Asia	Widely distributed	[23]
Carassius auratus (Linnaeus, 1758)	1950	Aquarium trade	Eurasia	Widely distributed	[23,28]
Phoxinus phoxinus (Linnaeus, 1758)	1934	Intentional (restocking)	Europe	Unconfirmed	[39]
Ctenopharyngodon idella (Valenciennes, 1844)	1981	Intentional (restocking)	Asia	Locally distributed	[19,39]
Cyprinus carpio (Linnaeus, 1758)	1924	Intentional (restocking)	Asia	Widely distributed	[19,39]
Hypophthalmichthys nobilis (Richardson, 1845)	1981	Intentional (restocking)	Asia	Widely distributed	[27,39]
Gobio gobio (Linnaeus, 1758)	1935	Intentional (restocking)	Europe	Locally distributed	[27,39]
Hypophthalmichthys molitrix (Valenciennes, 1844)	1981	Intentional (restocking)	Asia	Widely distributed	[27,39]
Rutilus rutilus (Linnaeus, 1758)	1934	Intentional (restocking)	Eurasia	Locally distributed	[27,39]
Scardinius erythrophthalmus (Linnaeus, 1758)	1934	Intentional (restocking)	Eurasia	Locally distributed	[27,39]
Tinca tinca (Linnaeus, 1758)	1934	Intentional (restocking)	North America	Locally distributed	[27,39]
Gambusia holbrooki (Girard, 1859)	1929	Intentional (restocking)	North America	Widely distributed	[12,15,39]
Xiphophorus hellerii (Heckel, 1848)	2019	Aquarium trade	Central America	Locally distributed	[49]
Fundulus heteroclitus (Linnaeus, 1766)	2019	Aquarium trade	North America	Locally distributed	[15]
Esox lucius (Linnaeus, 1758)	1934	Intentional (restocking)	Europe	Widely distributed	[27,39]
Esox masquinongy (Mitchill, 1824)	1964	Intentional (restocking)	North America	Unconfirmed	[19,39]
Esox niger (Lesueur, 1818)	1966	Intentional (restocking)	North America	Unconfirmed	[19,39]
Oncorhynchus mykiss (Walbaum, 1792)	1925	Intentional (restocking)	North America	Widely distributed	[19,27]
Salmo gairdneri (Richardson, 1836)	1925	Intentional (restocking)	North America	Unconfirmed	[39]
Salmo kamloops (Jordan, 1982)	1955	Intentional (restocking)	North America	Unconfirmed	[39]
Salmo clarkii (Richerdson, 1836)	1955	Intentional (restocking)	North America	Unconfirmed	[39]
Salvelinus namayeush (Walbaum, 1792)	1953	Intentional (restocking)	North America	Unconfirmed	[39]
Salvelinus alpinus (Linnaeus, 1758)	1948	Intentional (restocking)	North America	Unconfirmed	[39]
Salvelinus hucho (Linnaeus, 1758)	1953	Intentional (restocking)	North America	Unconfirmed	[39]
Salvelinus fontinalis (Mitchill, 1915)	1941	Intentional (restocking)	North America	Unconfirmed	[39]
Thymallus thymallus (Linnaeus, 1758)	1948	Intentional (restocking)	North America	Unconfirmed	[39]
Reptilia					
Trachemys scripta elegans (Wied, 1839)	2002	Aquarium trade	North America	Locally distributed	[38]

* Marine species but can be found in brackish and slightly brackish waters.

Within the confirmed alien taxa, chordates are most abundant (56%), followed by molluscs (17%) and arthropods (17%); other groups constitute a minority (Figure 1).

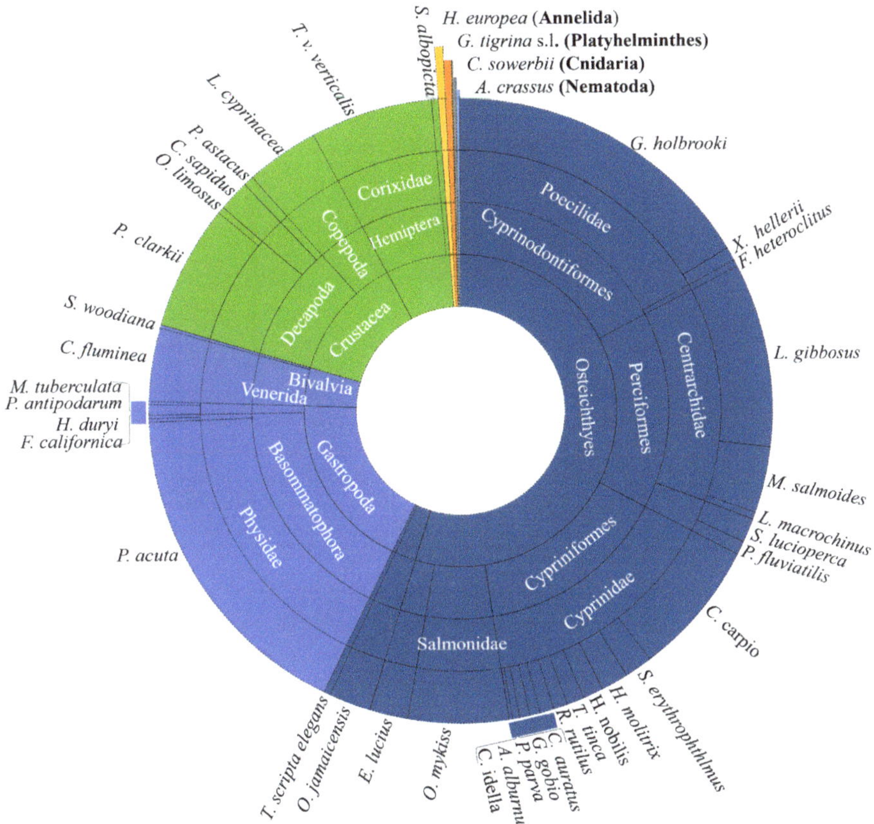

Figure 1. The main faunal groups of AS present in the inland waters of Morocco (according to the number of localities in the list).

The mosquitofish (*Gambusia holbrooki*) is the most reported NI chordate species in the freshwaters of Morocco, followed by sun perch (*Lepomis gibbosus*) and common carp (*Cyprinus carpio*), while the acute bladder snail (*Physella acuta*) is the most widespread NI mollusc species, with the southernmost record (Laayoune-Sakia El Hamra region). Within arthropods, Louisiana crayfish (*Procambarus clarkii*) is the most widely distributed NI crustacean species, and the American boatman (*Trichocorixa verticalis verticalis*) is the most widespread NI aquatic insect. The other branches (Annelida, Platyhelminths, Cnidaria, and Nematodes) are represented by one species each.

Chordates are represented by three classes—bony fish, birds, and reptiles. The last two classes are represented by one species each, i.e., the ruddy duck (*Oxyura jamaicensis*) and the Florida slider (*Trachemys scripta elegans*). Other chordates belonging to the class of mammals, with possible negative impacts on aquatic biodiversity in Morocco, were excluded from this study because of insufficient data, e.g., the Norway rat *Rattus norvegicus* (Berkenhout, 1769) (personal observations).

The vast majority of alien species present in Morocco are alien fish introduced during the 20th century (the first peak was observed between 1940 and 1970), while most invertebrates have been detected recently (the second peak started in 2010) (Figure 2).

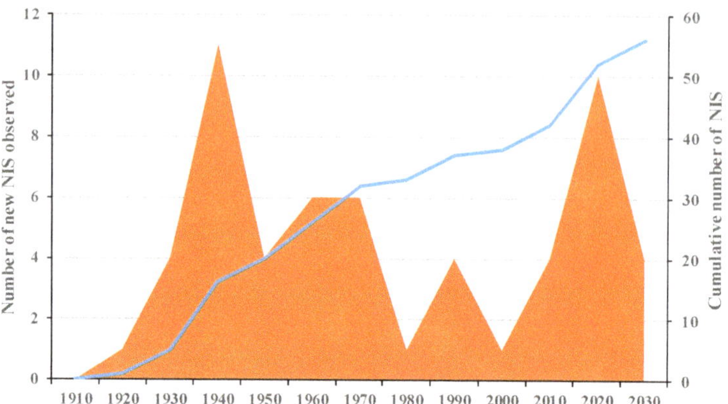

Figure 2. Cumulative number of AS (blue line) and numbers recorded per year (orange curve).

AS are concentrated in the northern part of the country (Figure 3). The highest number of citations was recorded about the Middle Atlas (Fès-Meknes region), followed by the north of the east Regions (Lower Moulouya and Nador lagoon) and the northern part of the Occidental Meseta (the Rabat-Salé-Kenitra region). A file in the Supplementary Materials is also added to specify the names of the invasive species reported for each region (Table S2).

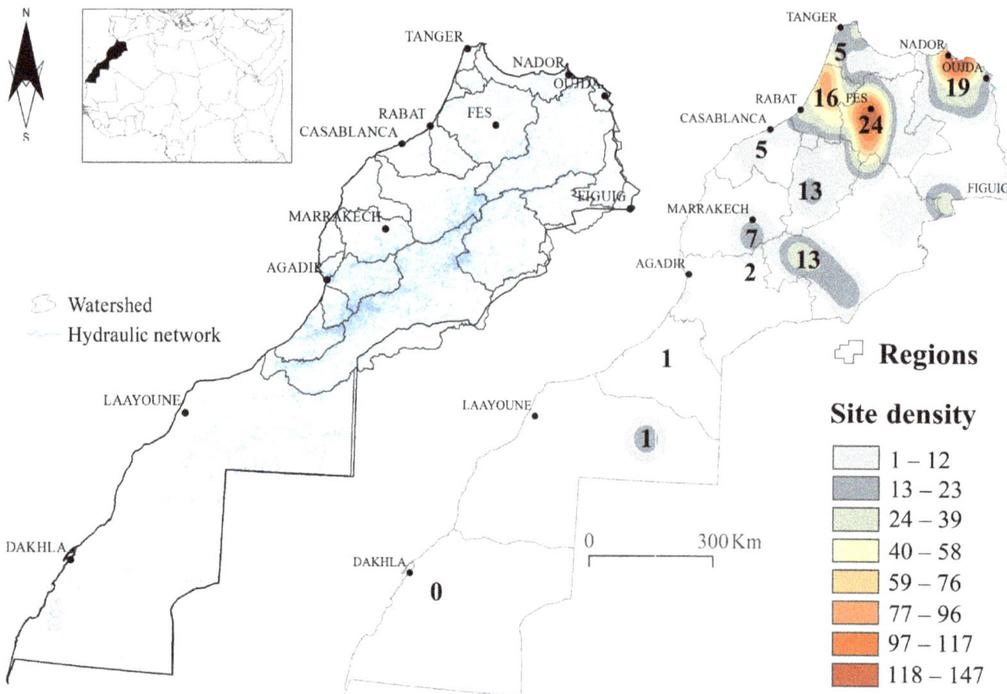

Figure 3. Spatial distribution of alien species in the inland waters of Morocco (**right**). The map was built by interpolating point data using the Spatial Statistics algorithm in ArcGIS software (the numbers from 0 to 24 represent the number of AS per region). In addition to the hydrographic networks of Morocco and the main watersheds (**left**).

Almost half of AS introductions in the freshwater ecosystems of Morocco were intentional through the restocking programs of water bodies with exotic fish (42.85%), followed by aquarium and ornamental animal and plant trades (28.57%), aquaculture (9.52%), and so-called "natural" spread from colonized areas (7.14%) (Figure 4a).

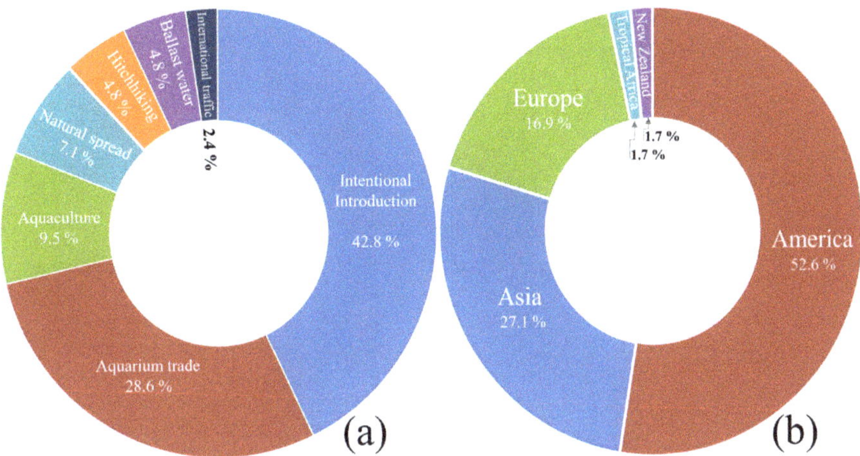

Figure 4. Main routes of introduction of the AS found in the inland waters of Morocco (**a**) and geographical origins (**b**).

More than half of the AS present in Moroccan inland waters are of American origin (52.54%), followed by Asian (27.12%) and European (16.95%) taxa (Figure 4b).

To date, only a few studies have addressed the impact of alien species on native ones or their autoecology in the inland aquatic ecosystems of Morocco. Most of them were published recently or are in progress (Table 2).

Table 2. Data alien species monitoring studies in the inland waters of Morocco.

Taxon	Topic of the Study	References
Anguillicola crassus	Native fish infection.	[42,59]
Physella acuta	Auto-ecology and potential competition with native gastropods.	[16,49,60]
Procambarus clarkii	Carrier of toxins and heavy metals to higher trophic levels. Disturbance of riparian vegetation and rice paddies. Competition with native Decapods.	[44,61] + unpublished data
Callinectes sapidus	Auto-ecology and predation on native aquatic fauna.	[14] + unpublished data
Trichocorixa v. verticalis	Auto-ecology and competition with native Corixidae.	[53,62]
Lepomis gibbosus	Auto-ecology and predation on native aquatic fauna.	[31] + unpublished data
Oxyura jamaicensis	Hybridisation with the white-headed duck (*O. leucocephala*) (Scopoli, 1769).	[21,58]
Gambusia holbrooki	Auto-ecology and predation on native amphibians; competition with native fish.	[12,15]
Fundulus heteroclitus	Auto-ecology and competition with potential *Aphanius* species.	[15]

4. Discussion

Given the lack of a robust national database recording alien species on a regular basis and the huge gaps in research on invasive species in Morocco in particular or in Africa in general, studies of the distribution and detection of invasive species are extremely important and useful for researchers working on the spread, impacts, and management of biological invasions [56]. The present work represents a first step toward the management of AS in Morocco by providing the first annotated list of AS present in the inland waters of the country. This list includes 41 confirmed AS, some of which are ranked in the top hundred worst invasive species worldwide (*Stegomyia albopicta, Micropterus salmoides, Cyprinus carpio, Gambusia holbrooki, Oncorhynchus mykiss, Trachemys scripta elegans*, and others) [63].

Most of these AS are concentrated in the Mediterranean—northern—part of the country (where the human population density is the highest). This could be explained by the relatively high number of surveys carried out in the north of Morocco, but also by the high availability of surface waters compared to the south of the country, which is associated with many commercial activities and the presence of the biggest Moroccan harbors in the area. Surprisingly, most AS originate from the US, not Europe (the closest continent), suggesting that geographical barriers are not a limiting factor [15]. Careful examination of the data showed that AS from America (mainly the US) were introduced recently: only eight American species (25%) were present in Morocco before 1950. On the contrary, all AS from Europe were introduced before 1950, except the common bleak (*Alburnus alburnus*) present only since 2013. Most AS were introduced intentionally for fish restocking or aquaculture between 1914 and 1980. These introductions likely explain why AS are not concentrated along the coast and in big harbors. This observation is also congruent with the fact that the only species introduced from ballast water—the main source of AS worldwide [64]—was the blue crab (*Callinectes sapidus*) in 2017.

Most other AS have more recently come from aquariums or other sources related to the ornamental pet trade, contributing to 72% of the total AS introduced over the last 20 years. These recent occurrences correspond to the second peak of introduction observed since 2000 and explain the strong increase in the proportion of invertebrates among AS. The phenomenon has been described as one of the leading pathways for the introduction of aquatic invasive species around the world [65–67], and the second one in Morocco. For instance, it was the main route of introduction in freshwaters in Morocco of the mummichog (*Fundulus heteroclitus* Linnaeus, 1766), the green swordtail (*Xiphophorus hellerii* Heckel, 1848), and successful freshwater gastropod invaders such as the New Zealand mudsnail (*Potamopyrgus antipodarum*), the American limpet (*Ferrissia californica*), the red-rimmed melania (*Melanoides tuberculata*), the Seminole ramshorn (*Helisoma duryi*), and the acute bladder snail (*Physella acuta*) [15,49–55].

The establishment of AS in Morocco co-occurred in many protected parks, sites of biological and ecological interest (known as SIBE), and Ramsar sites, e.g., Sidi Ali Lake, the lagoon of Nador, the lower Moulouya wetlands, Zerrouka Lake, Sidi Boughaba, and Merja Zerga. These protected areas are the cornerstone of biodiversity conservation. However, they are also environmentally suitable for alien species invasion and establishment [68]. Chordates and particularly fish are potentially the most impacted species. Deliberate stocking by alien fish is still continuous nowadays in the reservoirs, especially by "Asian carps". Some of the reservoirs are also protected areas, e.g., Barrage Mohamed V and Barrage Al Massira, which are also listed as Ramsar sites. These alien and invasive fish not only conceal natural genetic patterns (hence difficulties in discerning evolutionary patterns; [69]); they also potentially represent the most serious threat to native aquatic biodiversity in general and particularly to native fish species and amphibians through predation, competition, and hybridization, sometimes leading to local extinctions [12]. The piscivorous "Nile perch" (*Lates niloticus* Linnaeus, 1758) is a classic example: its introduction in Lake Victoria (East Africa) in the 1950s led to the extinction of over 200 endemic fish species [9].

Our survey highlighted a huge gap between the number of recorded species and the number of studies devoted to their impacts. Only nine species out of all 41 AS have been studied from an autoecological angle or assessed for their impacts on native biodiversity and the aquatic ecosystems of Morocco. Most of these studies were published only recently or are still in progress [12,14,15,31,44,53,61,62] + unpublished data. They all highlight the potential negative role of AS on local environments through bioaccumulation of heavy metals, predation/infection, competition, and hybridization with the native fauna. However, these works remain largely insufficient, and further studies are urgently needed to fully understand the impacts of AS in Morocco. This step is crucial for implementing management strategies that are currently lacking. For instance, there is no eradication program to eliminate or stop the spread of AS in Morocco, except for the ruddy-headed duck (*Oxyura jamaicensis*). Morocco has had a control plan since 2003, which includes four components: (i) a survey of captive and wild birds, detection, and monitoring of favorable sites; (ii) administrative and regulatory procedures to allow access to sites and shooting; (iii) designation of agents dedicated to control actions, purchase of equipment, training and awareness; and (iv) destruction of birds [70]. The same scenario should be applied for the other AS, especially those qualified as invasive or highly invasive.

Developing effective strategies to prevent the ecological and economic impacts of harmful invasive species is considered fundamental to national-scale policies [67–72]. Global warming and salinization of freshwater ecosystems bring about favorable factors for invasive species [73], which may acclimate more easily [15,49,74,75], and end up in a better position against the native fauna [76].

Morocco is in the red zone in terms of climate change predictions, i.e., at risk of water scarcity with decreasing precipitation [77], and Moroccan freshwater ecosystems are likely to become increasingly scarce, especially under a warming climate scenario where higher evapotranspiration rates are likely to intensify saline stress. All of this is exacerbated by anthropogenic disturbances through the withdrawal and diversion of large amounts of water for irrigation, especially during the dry seasons. The majority of rivers suffer from bank alteration caused by agricultural practices and substrate extraction as well as enormous environmental degradation from domestic, industrial, and agricultural wastewater pollution [78–81].

It is more necessary than ever for Morocco to develop adaptive management strategies to identify and minimize the impact of invasive species on the native fauna through the following urgent steps:

1. Intensify research on AS detection, and monitor their expansion within invaded areas, and carry out studies to improve knowledge on their biology and ecology in relation to the local conditions of Moroccan aquatic ecosystems.

2. Question fish stocking programs using AS and consider using native fish species instead to replenish freshwater ecosystems.

3. Enforce strict laws, policies, and procedures about the trade of aquatic species as a preventive measure to preserve the native biodiversity.

4. Take management decisions to eradicate animal AS from the freshwater ecosystems of Morocco to ensure conservation of the native biodiversity.

5. Finally, inform citizens about the importance of the biological endemism that Morocco enjoys, involve them in its conservation and make them aware of the dangers of invasive species on the native fauna and aquatic ecosystems of Morocco.

5. Conclusions

Moroccan freshwaters are colonized by a large number of invasive species, most of which were voluntarily introduced from the USA during the second half of the 20th century. Over the last 20 years, the origin of AS has changed dramatically with the introduction of AS of multiple origins due to the aquarium and ornamental plant trade. To preserve the exceptional diversity of Moroccan freshwaters, it has become very urgent to develop adaptive management strategies to identify and minimize the impact of invasive species on native fauna and to strengthen environmental regulation through new legislation.

Supplementary Materials: The following supporting information can be downloaded at: https://www.mdpi.com/article/10.3390/d15020169/s1, Table S1: GPS points and references of alien aquatic animal species of Morocco. Table S2: Invasive species by region.

Author Contributions: Conceptualization, methodology, software, formal analysis, investigation, resources, data curation, and writing—original draft preparation, Y.M. and A.F.T.; writing—review and editing, visualization, supervision, C.P. All authors have read and agreed to the published version of the manuscript.

Funding: This research received no external funding.

Institutional Review Board Statement: Not applicable.

Data Availability Statement: Not applicable.

Acknowledgments: We are grateful to the editor and the anonymous reviewers for valuable corrections and comments.

Conflicts of Interest: The authors declare no conflict of interest.

References

1. Myers, N.; Mittermeier, R.A.; Mittermeier, C.G.; da Fonseca, G.A.B.; Kent, J. Biodiversity hotspots for conservation priorities. *Nature* **2000**, *403*, 853–858. [CrossRef] [PubMed]
2. Ellender, B.R.; Woodford, D.J.; Weyl, O.L. The invasibility of small headwater streams by an emerging invader, *Clarias Gariepinus*. *Biol. Invasions* **2015**, *17*, 57–61. [CrossRef]
3. Sala, O.E.; Chapin, F.S.; Armesto, J.J.; Berlow, E.; Bloomfield, J.; Dirzo, R.; Huber-Sanwald, E.; Huennek, L.F.; Jackson, R.B.; Kinzig, A.; et al. Global biodiversity scenarios for the year 2100. *Science* **2000**, *287*, 1770–1774. [CrossRef]
4. Taylor, B.W.; Flecker, A.S.; Hall, R.O. Loss of a harvested fish species disrupts carbon flow in a diverse tropical river. *Science* **2006**, *313*, 833–836. [CrossRef] [PubMed]
5. Ricciardi, A. Are modern biological invasions an unprecedented form of global change? *Conserv. Biol.* **2007**, *21*, 329–336. [CrossRef]
6. Strayer, D.L. Eight questions about invasions and ecosystem functioning. *Ecol. Lett.* **2012**, *15*, 1199–1210. [CrossRef]
7. Vitousek, P.M.; D'Antonio, C.M.; Loope, L.; Rejmanek, M.; Westbrooks, R.G. Introduced species, a significant component of human–caused global environmental change. *N. Z. J. Ecol.* **1997**, *21*, 1–16.
8. Gherardi, F.; Gollasch, S.; Minchin, D.; Olenin, S.; Panov, V.E. Alien invertebrates and fish in European inland waters. In *Handbook of Alien Species in Europe DAISIE*; Springer: Dordrecht, The Netherlands, 2009; pp. 81–92.
9. Kitchell, J.F.; Schindler, D.E.; Oguto–Ohwayo, R.; Reinthal, P. The Nile perch in Lake Victoria, interactions betweeen predation and fisheries. *Ecol. Appl.* **1997**, *7*, 653–664. [CrossRef]
10. Strauss, S.Y.; Lau, J.A.; Carroll, S.P. Evolutionary responses of natives to introduced species, what do introductions tell us about natural communities? *Ecol. Lett.* **2006**, *9*, 357–374. [CrossRef]
11. Mabrouki, Y.; Ben Ahmed, R.; Taybi, A.F.; Rueda, J. Annotated checklist of the leech (Annelida, Hirudinida) species of the Moulouya river basin, Morocco with several new distribution records and an historical overview. *Afr. Zool.* **2019**, *54*, 199–214. [CrossRef]
12. Mabrouki, Y.; Taybi, A.F.; Skalli, A.; Sánchez-Vialas, A. Amphibians of the Oriental Region and the Moulouya River Basin of Morocco, distribution and conservation notes. *Basic Appl. Herpetol.* **2019**, *33*, 19–32. [CrossRef]
13. Mabrouki, Y.; Taybi, A.F.; Vila-Farré, M. New to Africa, first record of the globally invasive planarian *Girardia tigrina* (Girard, 1850) in the continent. *Bioinvasion Rec.* **2023**, in press.
14. Taybi, A.F.; Mabrouki, Y. The American blue crab *Callinectes sapidus* Rathbun, 1896 (Crustacea, Decapoda, Portunidae) is rapidly expanding through the Mediterranean coast of Morocco. *Thalassas* **2020**, *36*, 1–5. [CrossRef]
15. Taybi, A.F.; Mabrouki, Y.; Doadrio, I. The occurrence, distribution and biology of invasive fish species in fresh and brackish water bodies of NE Morocco. *Arx. Misc. Zool.* **2020**, *18*, 59–73. [CrossRef]
16. Taybi, A.F.; Mabrouki, Y.; Berrahou, A.; Dakki, A.; Millán, A. Longitudinal distribution of macroinvertebrate in a very wet North African basin, Oued Melloulou (Morocco). *Ann. Limnol.* **2020**, *56*, 17. [CrossRef]
17. Aksissou, M. Dynamique des populations d'Orchestia gammarellus (Pallas, 1766)—Crustacea, Amphipoda, Talitridae—du littoral méditerranéen du Maroc occidentalet impact des aménagements. Ph.D. Thesis, University Abdelmalek Essaâdi, Tetouan, Morocco, 1997.
18. Álvarez, L.B.; El Ouanighi, Y.M.; Bennas, N.; Alami, M.; Riutort, M. The expansion continues, Girardia arrives in Africa. First record of *Girardia sinensis* (Platyhelminthes, Tricladida, Continenticola, Dugesiidae) in Morocco. *Zootaxa* **2022**, *5169*, 497–500. [CrossRef]
19. Azeroual, A. Monographie des Poissons des Eaux inlandes du Maroc, Systématique, Distribution et Ecologie. Ph.D. Thesis, University Mohammed V-Agdal, Rabat, Morocco, 2003.
20. Bennouna, A.; Balenghien, T.; Elrhaffouli, H.; Schaffner, F.; Garros, C.; Gardès, L.; Lhor, Y.; Hammoumi, S.; Chlyeh, G.; Fassifihri, O. First record of *Stegomyia albopicta* (=*Aedes albopictus*) in Morocco, a major threat to public health in North Africa? *Med. Vet. Entomol.* **2016**, *31*, 102–106. [CrossRef]
21. Bergier, P.; Franchimont, J.; Thevenot, M.; MRBC. Rare birds in Morocco—Report of the Moroccan Rare Birds Committee # 2. *Porphyrio* **1997**, *9*, 165–173.

22. Clavero, M.; Araujo, R.; Calzada, J.; Delibes, M.; Fernández, N.; Gutiérrez-Expósito, C.; Revilla, E.; Román, J. The first invasive bivalve in African fresh waters, invasion portrait and management options. *Aquat. Conserv. Mar. Freshw. Ecosyst.* **2012**, *22*, 277–280. [CrossRef]

23. Clavero, M.l.; Esquivias, J.; Qninba, A.; Riesco, M.; Calzada, J.; Ribeiro, F.; Fernández, N.; Delibes, M. Fish invading deserts, alien species in arid Moroccan rivers. *Aquat. Conserv. Mar. Freshw. Ecosyst.* **2015**, *25*, 49–60. [CrossRef]

24. El-Hilali, M.; Yahyaoui, A.; Sadak, A.; Maachi, M.; Taghy, Z. Premières données épidémiologiques sur l'anguillicolose au Maroc. *Bull. Fr. Pêche Piscicult.* **1996**, *340*, 57–60. [CrossRef]

25. El Qoraychy, I.; Fekhaoui, M.; El Abidi, A.; Yahyaoui, A. Biometry and demography of *Procambarus clarkii* in Rharb Region, Morocco. *AACL Bioflux* **2015**, *8*, 751–760.

26. Farid, S.; Ouizgane, A.; Majdoubi, F.Z.; Hasnaoui, M.; Droussi, M. Régime alimentaire de la carpe herbivore (*Ctenopharyngodon idella*, Valenciennes 1844) en stade d'alevinage dans les étangs de la station de pisciculture Deroua, Maroc. *J. Wat. Env. Sci.* **2019**, *3*, 499–511.

27. Ford, M.; Brahimi, A.; Baikeche, L.; Bergner, L.; Clavero, M.; Doadrio, I.; Lopes-Lima, M.; Perea, S.; Yahyaoui, A.; Freyhof, J. Freshwater fish distribution in the Maghreb, a call to contribute. *OSF Preprints* **2020**, Preprint. [CrossRef]

28. Furnestin, J.; Dardignac, J.; Maurin, C.; Vincent, A.; Coupe, R. and Boutiere, H. Données nouvelles sur les poissons du Maroc atlantique. *Rev. Trav. Inst. Pêches Marit.* **1958**, *22*, 379–493.

29. Gauthier, H. Enquête sur la répartition en Algérie des Mollusques susceptibles de véhiculer la bilharziose vésicale. Arch. *Inst. Pasteur Alger.* **1934**, *12*, 305–350.

30. Ghamizi, M.; Idaghdour, M.; Mouahid, A.; Vala, J.C.; El Ouali, M. Chevauchement des habitats entre *Melanopsis praemorsa* Linné (Gastropoda; Melanopsidae) et les autres mollusques des eaux douces des canaux d'irrigation du Haouz ·de Marrakech (Maroc). *Rev. Fac. Sci. Mar.* **1997**, *9*, 21–31.

31. Gkenas, C.; Magalhães, M.F.; Campos-Martin, N.; Ribeiro, F.; Clavero, M. Desert pumpkinseed, diet composition and breadth in a Moroccan river. *Knowl. Manag. Aquat. Ecosyst* **2021**, *422*, 34. [CrossRef]

32. Holdich, D.M.; Haffner, P.; Noël, P. Species files. In *Atlas of Crayfish in Europe*; Muséum national d'Histoire, naturelle; Souty-Grosset, C., Holdich, D.M., Noël, P.Y., Reynolds, J.D., Haffner, P., Eds.; Patrimoines Naturels: Paris, France, 2006.

33. Kharboua, M. Ecologie et Biologie des Mollusques Dulcicoles et Ttat Actuel de la Bilharziose Dans le Maroc Nord Oriental. Ph.D. Thesis, Faculty of Sciences Oujda, Oujda, Marocco, 1994.

34. Laamrani, H.; Khallayoune, K.; Delay, B.; Pointier, J.P. Factors affecting the distribution and abundance of two prosobranch snails in a thermal spring. *J. Freshw. Ecol.* **1997**, *12*, 75–79. [CrossRef]

35. L'mohdi, O.; Bennas, N.; Himmi, O.; Hajji, K.; El Haissoufi, M.; Hernando, C.; Carbonell, J.A.; Millan, A. *Trichocorixa verticalis verticalis* (Fieber, 1851) (Hemiptera, Corixidae, une nouvelle espèce exotique au Maroc. *Boletín SEA* **2010**, *46*, 395–400.

36. Madzivanzira, T.C.; South, J.; Wood, L.E.; Nunes, A.L.; Wey, O.L.F. A Review of Freshwater Crayfish Introductions in Africa. *Rev. Fish. Sci. Aquac.* **2020**, *29*, 218–241. [CrossRef]

37. Maqboul, A.; Aoujdad, R.; Fadli, M.; Fekhaoui, M. Population dynamics of *Physa acuta* (Mollusca, Pulmonata) in the lakes of Rif mountains (Northern Morocco, Ouergha watershed. *J. Entomol. Zool. Stud.* **2014**, *2*, 240–245.

38. Mateo, J.A.; Fahd, S.; Martínez-Medina, F.J.; Pleguezuelos, J.M.; Capítulo, V. *Anfibios y Reptiles en los territorios transfretanos (Ceuta, Melillae islotes en el norte de África*, In *Atlas y libro rojo de los anfibios y reptiles de España*; Pleguezuelos, J.M., Marquez, R., Lizana, M., Eds.; Organismo Autónomo Parques Nacionales: Madrid, Spain, 2002; pp. 382–415.

39. Mouslih, M. L'Introduction de poissons et d'écrevisses au Maroc. *Rev. Hydrobiol. Trop.* **1987**, *20*, 65–72.

40. Nahli, A.; Oubraim, S.; Chlaida, M. Diversité taxonomique et structure du peuplement macrobenthique des eaux de l'Oued Hassar après installation de la station d'épuration de Mediouna (Casablanca, Maroc). *Bull. Inst. Sci. Rabat* **2018**, *40*, 37–56.

41. Oualid, J.A.; Iazza, B.; Tamsouri, N.M.; El Aamri, F.; Moukrim, A.; López–González, P.J. Hidden diversity under morphology–based identifications of widespread invasive species, the case of the 'well–known' hydromedusa *Craspedacusta sowerbii* Lankester 1880. *Anim. Biodivers. Conserv.* **2019**, *42*, 301–316. [CrossRef]

42. Rahhou, L.; Melhaoui, M.; Lecomte-Finiger, R.; Morand, S.; Chergui, H. Abundance and distribution of *Anguillicola crassus* (Nematoda) in eels *Anguilla anguilla* from Moulouya Estuary (Morocco). *Helminthologia* **2001**, *38*, 93–97.

43. Ramdani, M.; Dakki, M.; Kharboua, M.; El Agbani, M.A.; Metge, G. Les Gastéropodes dulcicoles du Maroc, inventaire commenté. *Bull. Inst. Sei. Rabat* **1987**, *11*, 135–140.

44. Saguem, S.; El Alami El Moutaouakil, M. Study on the Spread of *Procambarus clarkii* at Gharb (Morocco) and Its Impact on Rice Growing. *J. Agric. Sci. Technol.* **2019**, *9*, 81–92. [CrossRef]

45. Schollaert, V.; Moumni, T.; Fareh, M.; Gambarotta, C.; Pascon, J.; Franchimont, J. Chronique ornithologique du G.O.MAC. pour 1993. *Porphyrio* **1994**, *6*, 1–108.

46. Benyahia Tabib, M. Etude de la Variabilité Morpho Métrique, du Cycle Biologique, des Glandes Cementaires et de Leur Sécrétion Chez Astacus Astacus (Linné 1758). Ph.D. Thesis, University Sidi Mohamed Benabdellah, Fez, Morocco, 2003.

47. Touabay, M.; Aouad, N.; Mathieu, J. Etude hydrobiologique d'un cours d'eau du Moyen-Atlas, l'oued Tizguit (Maroc). *Ann. Limnol.* **2002**, *38*, 65–80. [CrossRef]

48. Mabrouki, Y.; Taybi, A.F. The first record of the invasive Chinese pond mussel *Sinanodonta woodiana* (Lea, 1834) (Bivalvia, Unionidae) in the African continent. *Nat. Croat.* **2022**, *31*, 393–398. [CrossRef]

49. Mabrouki, Y.; Taybi, A.F.; Bahhou, J.; Doadrio, I. The first record of the green swordtail *Xiphophorus helleri* Heckel, 1848 (Poeciliidae) established in the wild from Morocco. *J. Appl. Ichthyol.* **2020**, *36*, 795–800. [CrossRef]
50. Mabrouki, Y.; Taybi, A.F.; Glöer, P. The first record of the North American freshwater limpet *Ferrissia californica* (Mollusca, Gastropoda) in Morocco. *Nat. Conserv. Res.* **2023**, *8*, 108–112. [CrossRef]
51. Taybi, A.F.; Mabrouki, Y.; Ghamizi, M.; Berrahou, A. The freshwater malacological composition of Moulouya's watershed and Oriental Morocco. *J. Mater. Environ. Sci.* **2017**, *8*, 1401–1416.
52. Taybi, A.F.; Mabrouki, Y.; Chavanon, G.; Millán, A.; Berrahou, A. New data on the distribution of aquatic bugs (Hemiptera) from Morocco with notes on their chorology. *Zootaxa* **2018**, *4459*, 139–163. [CrossRef] [PubMed]
53. Taybi, A.F.; Mabrouki, Y.; Chavanon, G.; Millán, A. The alien boatman *Trichocorixa verticalis verticalis* (Hemiptera, Corixidae is expanding in Morocco. *Limnetica* **2020**, *39*, 49–59. [CrossRef]
54. Taybi, A.F.; Mabrouki, Y.; Glöer, P. First record of the New Zealand Mudsnail *Potamopyrgus antipodarum* (J.E. Gray, 1843) (Tateidae, Mollusca) in Africa. *Graellsia* **2021**, *77*, e140. [CrossRef]
55. Taybi, A.F.; Mabrouki, Y.; Glöer, P. First record of the exotic Seminole rams-horn *Helisoma duryi* (Wetherby, 1879) (Gastropoda, Planorbidae) in Morocco. *Graellsia* **2023**, *79*, 1–6.
56. Marrone, F.; Naselli-Flores, L. A review on the animal xenodiversity in Sicilian inland waters (Italy). *Adv. Oceanogr. Limnol.* **2015**, *6*, 2–12. [CrossRef]
57. GBIF. Global Biodiversity Information Facility. *Physella acuta* (Draparnaud, 1805) in GBIF Secretariat (2022). GBIF Backbone Taxonomy. Available online: GBIF.org. [CrossRef]
58. Cranswick, P.A.; Hall, C. Eradication of the Ruddy Duck *Oxyura jamaicensis* in the Western Palaearctic, a review of progress and a revised Action Plan 2010-2015. In *WWT Report to the Bern Convention*; Wildfowl & Wetlands Trust (WWT): Slimbridge, UK, 2010.
59. El-Hilali, M.; Yahyaoui, A.; Chetto, N. Etude de l'infestation des anguilles (*Anguilla anguilla*) par le parasite (*Anguillicola crassus*) dans l'estuaire du Sebou au nord-ouest du Maroc. *Bull. Inst. Sci. Rabat* **2004**, *26*, 39–42.
60. Mabrouki, Y.; Taybi, A.F.; El Alami, M.; Berrahou, A. Biotypology of stream macroinvertebrates from North African and semi-arid catchment, Oued Za (Morocco). *Knowl. Manag. Aquat. Ecosyst.* **2019**, *420*, 17. [CrossRef]
61. El Qoraychy, I.; Fekhaoui, M.; El Abidi, A.; Yahyaoui, A. Heavy metals in *Procambarus clarkii* of Rharb Region in Morocco and Its Safety for Human consumption. *J. Environ. Sci., Toxicol. Food Technol.* **2015**, *9*, 38–43.
62. Taybi, A.F.; Mabrouki, Y.; Bozdogan, H.; Millan, A. Are aquatic Hemiptera good indicators of environmental river conditions? *Aquat. Ecol.* **2021**, *55*, 791–806. [CrossRef]
63. GISD. Global Invasive Species Database. 100 of the World's Worst Invasive Alien Species". Global Invasive Species Database. Retrieved 5 September 2022. Available online: www.iucngisd.org/gisd/100_worst.php (accessed on 4 March 2016).
64. Lacoursiere-Roussel, A.; Bock, D.G.; Cristescu, M.E.; Guichard, F.; McKindsey, C.W. Effect of shipping traffic on biofouling invasion success at population and community levels. *Biol. Invasions* **2016**, *18*, 3681–3695. [CrossRef]
65. Martin, G.D.; Coetzee, J.A. Pet stores, aquarists and the internet trade as modes of introduction and spread of invasive macrophytes in South Africa. *Water SA WRC* **2011**, *37*, 371–380. [CrossRef]
66. Lockwood, J.L.; Welbourne, D.J.; Romagosa, C.M.; Cassey, P.; Mandrak, N.E.; Strecker, A.; Leung, B.; Stringham, O.C.; Udell, B.; Episcopio-Sturgeon, D.; et al. When pets become pests, the role of the exotic pet trade in producing invasive vertebrate animals. *Front. Ecol. Environ.* **2019**, *17*, 323–330. [CrossRef]
67. Olden, J.D.; Whattam, E.; Wood, S.A. Online auction marketplaces as a global pathway for aquatic invasive species. *Hydrobiologia* **2020**, *848*, 1967–1979. [CrossRef]
68. Liu, X.; Blackburn, T.M.; Song, T.; Wang, X.; Huang, C.; Li, Y. Animal invaders threaten protected areas worldwide. *Nat. Commun.* **2020**, *11*, 2892. [CrossRef]
69. Perea, S.; Al Amouri, M.; Gonzalez, E.G.; Alcaraz, L.; Yahyaoui, A.; Doadrio, I. Influence of historical and human factors on genetic structure and diversity patterns in peripheral populations, implications for the conservation of Moroccan trout. *BioRxiv* 2020. [CrossRef]
70. Mouronva, J.-B.; Maillard, J.-F.; Cugnasse, J.-M. Section A, Bilan des connaissances et des actions mises en œuvre depuis 1996. Plan rédigé par l'Office National de la Chasse et de la Faune Sauvage à la demande du Ministère de l'Ecologie, du Développement Durable et de l'Energie. In *Plan national de lutte contre l'Erismature rousse (Oxyura jamaicensis) 2015–2025*; Office National de la Chasse et de la Faune Sauvage: Paris, France, 2015.
71. Early, R.; Bradley, B.A.; Dukes, J.S.; Lawler, J.J.; Olden, J.D.; Blumenthal, D.M.; Gonzalez, P.; Grosholz, E.D.; Ibanez, I.; Miller, L.P.; et al. Global threats from invasive alien species in the twenty-first century and national response capacities. *Nat. Commun.* **2016**, *7*, 12485. [CrossRef]
72. Turbelin, A.J.; Malamud, B.D.; Francis, R.A. Mapping the global state of invasive alien species, patterns of invasion and policy responses. *Glob. Ecol. Biogeogr* **2017**, *26*, 78–92. [CrossRef]
73. Piscart, C.; Moreteau, J.C.; Beisel, J.N. Biodiversity and structure of macroinvertebrate communities along a small permanent salinity gradient (Meurthe River, France). *Hydrobiologia* **2005**, *551*, 227–236. [CrossRef]
74. Carbonell, J.A.; Millán, A.; Green, A.J.; Céspedes, V.; Coccia, C.; Velasco, J. What traits underpin the successful establishment and spread of the invasive water bug *Trichocorixa verticalis verticalis*? *Hydrobiologia* **2015**, *768*, 273–286. [CrossRef]
75. Krodkiewska, M.; Cieplok, A.; Spyra, A. The Colonization of a Cold Spring Ecosystem by the Invasive Species *Potamopyrgus antipodarum* (Gray, 1843) (Gastropoda: Tateidae) (Southern Poland). *Water* **2021**, *13*, 3209. [CrossRef]

76. Moss, B.; Hering, D.; Green, A.J.; Aidoud, A.; Becares, E.; Beklioglu, M.; Bennion, H.; Boix, D.; Brucet, S.; Carvalho, L.; et al. Climate change and the future of freshwater biodiversity in Europe, A primer for policy-makers. *Freshw Rev.* **2009**, *2*, 103–130. [CrossRef]
77. Sbaa, M.; Vanclooster, M. La gestion des ressources en eau au Maroc face aux changements climatiques, état des lieux et al. ternatives technologiques d'adaptation. *Ann. Sci. Santé.* **2017**, *14*, 24–53.
78. Mabrouki, Y.; Taybi, A.F.; Bensaad, H.; Berrahou, A. Variabilité spatio-temporelle de la qualité des eaux courantes de l'Oued Za (Maroc Oriental). *J. Mater. Environ. Sci.* **2016**, *7*, 231–243.
79. Mabrouki, Y.; Taybi, A.F.; Berrahou, A. L'évolution spatio-temporelle de la qualité des eaux courantes de l'Oued Melloulou (MAROC). *Rev. Des Sci. De L'Eau* **2017**, *30*, 213–225. [CrossRef]
80. Taybi, A.F.; Mabrouki, Y.; Berrahou, A.; Chaabane, K. Évolution spatiotemporelle des paramètres physico-chimiques de la Moulouya. *J. Mater. Environ. Sci.* **2016**, *7*, 272–284.
81. Taybi, A.F.; Mabrouki, Y.; Legssyer, B.; Berrahou, A. Spatio- temporal typology of the physico-chemical parameters of a large North African river: The Moulouya and its main tributaries (Morocco). *Afr. J. Aquat. Sci.* **2020**, *45*, 431–441. [CrossRef]

MDPI
St. Alban-Anlage 66
4052 Basel
Switzerland
www.mdpi.com

Diversity Editorial Office
E-mail: diversity@mdpi.com
www.mdpi.com/journal/diversity